Louis Etienne Lefébure de Fourcy

Elements of trigonometry, plane and spherical

Louis Etienne Lefébure de Fourcy

Elements of trigonometry, plane and spherical

ISBN/EAN: 9783741194917

Manufactured in Europe, USA, Canada, Australia, Japa

Cover: Foto ©Thomas Meinert / pixelio.de

Manufactured and distributed by brebook publishing software (www.brebook.com)

Louis Etienne Lefébure de Fourcy

Elements of trigonometry, plane and spherical

ELEMENTS

OF

TRIGONOMETRY,

PLANE AND SPHERICAL.

BY

LEFEBURE DE FOURCY.

OFFICER OF THE LEGION OF HONOR; PROFESSOR IN THE FACULTY OF SCIENCES IN THE ACADEMY OF PARIS; AND EMERITUS EXAMINER OF CANDIDATES FOR THE POLYTECHNIC SCHOOL.

Translated from the Last French Edition,

BY

FRANCIS H. SMITH, A.M.

SUPERINTENDENT AND PROFESSOR OF MATHEMATICS OF THE VIRGINIA MILITARY INSTITUTE.

WITH

TABLES OF LOGARITHMS OF NUMBERS AND OF SINES AND COSINES FOR EVERY TEN SECONDS OF THE QUADRANT. AND OTHER USEFUL TABLES.

BALTIMORE:
KELLY & PIET, PUBLISHERS,
174 BALTIMORE STREET.
1868.

PRESS OF KELLY & PIET.

TO

GENERAL ROBERT E. LEE.

In submitting to the public a translation of *Lefébure de Fourcy's* admirable treatise on Trigonometry, I have thought it not inappropriate to associate your name with this effort to promote the interests of education.

You have given your influence, by precept and example, to this important cause, and in dedicating this work to you, I only give expression to the general sentiment of those who appreciate the weight of this great influence.

The personal and official relations which it has been my privilege to sustain towards you, during a period of more than thirty years, add a peculiar gratification to me in rendering this tribute of affectionate regard and esteem.

FRANCIS H. SMITH.

Virginia Military Institute,
Lexington, Va., September 19, 1867.

PREFACE.

The elementary treatise on Trigonometry by *Lefébure de Fourcy* has been justly regarded as one of the best ever published; and by its translation, the American student will have access to a work which will prove a valuable auxiliary to the prosecution of the course of the Higher Mathematics.

This work constitutes one of the Mathematical Series of the *Virginia Military Institute.* The following works now comprise the series as published:

Smith's Elementary Arithmetic, for Beginners.
Smith's Arithmetic, for High Schools and Academies.
Smith's Algebra.
Smith's Biot's Analytical Geometry.
Smith's Legendre's Geometry.
Smith's Lefebure de Fourcy's Trigonometry with Tables.

A Descriptive Geometry will immediately follow.

Virginia Military Institute,
Lexington, Va., September 19, 1867.

TABLE OF CONTENTS.

INTRODUCTION.

TRIGONOMETRY.

CHAPTER I.

CHAPTER II.

TRIGONOMETRICAL TABLES.—SOLUTIONS OF TRIANGLES.

Relations between Angles and Sides of Spherical Triangles.

CHAPTER III.

Formulæ used in Higher Mathematics.—Sines and Cosines in Series.—Resolution of Binomial Equations and Equations of Third Degree.

CHAPTER IV.

Promiscuous Examples.

TRIGONOMETRY.

INTRODUCTION.

LOGARITHMS. TABLE OF LOGARITHMS OF NUMBERS. — USE OF TABLE.

1. In the indeterminate equation $a^x = y$, every value, positive and negative, given to x, will lead to a corresponding value of y; and, conversely, for every numerical value assigned to y there will be a corresponding value of x. If, therefore, the quantity a be *arbitrarily* assumed, and be supposed to have a *constant* value, a fixed relation will exist between the numbers represented by x and y, and this relation is expressed by calling x the *logarithm* of y. The constant number a is called the *base* of the system of logarithms.

2. All positive numbers may be produced by the powers of any proposed positive number greater than unity. For, if, in the equation $a^x = y$, a being positive and > 1, x assume successively and continuously all possible values from 0 to ∞, it is clear that y will receive all values from 1 to ∞. Also, if x become negative and equal $-x$, then we have $a^{-x} = y$, or $\frac{1}{a^x} = y$; and now, if x assume all possible values from 0 to ∞, y will receive all values from 1 to 0. Hence, as x changes continuously from $-\infty$ to $+\infty$, a^x changes continuously from 0 to $+\infty$, or produces all positive numbers.

If $a < 1$, writing the equation $a^x = y$ under the form $\left(\frac{1}{a}\right)^x = \frac{1}{y}$, it appears from what has been proved, since $\frac{1}{a} > 1$, that by varying x, $\frac{1}{y}$ and therefore y will assume all positive values.

3. We cannot take unity nor a negative quantity for the base: for all the powers of unity are unity; and the powers

of a negative quantity may be positive, negative, or imaginary, and hence are not subject to the law of continuity and do not reproduce all numbers.

The logarithm of a number to a given base is the exponent of the power to which the base must be raised to become equal to the given number.

4. Since $a^1=a$, $a^0=1$, and $a^{-\infty}=0$, or $a^{+\infty}=0$, according as $a>1$ or <1, we deduce from the definition of logarithms, 1*st*, *that the logarithm of the base itself is unity;* 2*d*, *that the logarithm of unity is* 0; and 3*d*, *that the logarithm of* 0 *is negative infinity or positive infinity according as the base is* >1 *or* <1.

5. When the base is >1, the equation $a^x=y$ shows, that as the number y increases, its logarithm x increases, and that the logarithms of all numbers greater than 1 are positive, and the logarithms of all numbers less than 1 are negative: when the base is <1, the contrary takes place, that is, the logarithms of numbers greater than 1 are negative, and those of numbers less than 1 are positive.

We shall at present assume that, a being any positive quantity, if in the equation $a^x=y$, the value of y be given, the corresponding arithmetical value of x can be found to any degree of approximation we please. (See *Smith's Algebra*, Chap. XII.) If y be assumed to be the exact power of the base a, the corresponding logarithms will be whole numbers, 1, 2, 3, 4, &c. Thus,

$$a^1=y,\ a^2=y^1,\ a^3=y^2,\ a^4=y^3, \text{\&c.}$$

But if y be not equal to an exact power of the base, its logarithm must involve a *decimal* part.

6. If now y assume successive integral values beginning with 1, a remaining the same, and the corresponding values of x be computed from the equation $a^x=y$, and registered in a *Table*, in order, we shall obtain a system of logarithms to the base a. Since a may be any positive quantity, except 1, there may be an infinite number of systems of logarithms. If, however, the logarithms of all numbers in any one system be known, those corresponding to any other given base may be deduced by multiplying the former by a constant factor.

7. Although there is no limit to the number of systems of logarithms which may be formed, there are only two systems which are used, viz., the *Napierian system*, and the *Common system*.

The *Napierian* logarithms, which are always employed in *analytical* investigations, have for a base the incommensurable number $e = 2.7182818 \ldots$

The *Common* logarithms, which are no less constantly employed in *numerical* calculations, have for the base the number 10; and since the number is the base of the common scale of notation in numbers, this system has the advantage that its tables are far more comprehensive than tables of the same size in any other system.

8. If x represents the *common* logarithm of any number y, a being the base, and x' the *Napierian* logarithm of the same number y, e being the base, and if we designate Napierian logarithms always by the dash ($'$); we shall have the relations

$$e^{x'} = y, \text{ and } a^x = y,$$

hence $$e^{x'} = a^x \text{ and } e^{\frac{x'}{x}} = a.$$

Therefore $$\frac{x'}{x} = \text{nap. log } a = \log' a.$$

And we have $$x = x' \times \frac{1}{\log' a}$$

and $$x' = x \div \frac{1}{\log' a}.$$

That is, *knowing the Napierian logarithm x' of any number y, we may obtain the common logarithm x of the same number y*, by *multiplying the Napierian logarithm x' by the multiplier $\frac{1}{\log' a}$*; or, *knowing the common logarithm x of any number y, we may obtain the Napierian logarithm x' by dividing the common logarithm x by the same quantity $\frac{1}{\log' a}$.*

The common factor $\frac{1}{\log' a}$ is called the *modulus* of the system whose base is a, relative to the system whose base is e. Therefore, the *modulus* of the *common system of logarithms is equal to unity divided by the Napierian logarithm of the common base.*

$$\text{When } a = e, \text{ the factor } \frac{1}{\log' a} = \frac{1}{\log' e} = 1.$$

Hence, the *modulus of the Napierian system of logarithms is unity.*

9. Let y, y', y'', &c. represent any numbers whatever, and

x, x', x'', &c. their corresponding logarithms, taken in a system of which a is the base. We shall have

$$a^{x} = y,\ a^{x'} = y',\ a^{x''} = y'', \text{ \&c.} \qquad (1)$$

If we multiply these equations together, member by member, we have

$$a^{x + x' + x'' \text{ \&c.}} = y + y' + y'', \text{ \&c.}$$

But, $x + x' + x''$, &c. $= \log\,(y + y' + y'', \text{ \&c.})$

and $x = \log y,\ x' = \log y',\ x'' = \log y''$, &c.

hence $\log y + \log y' + \log y''$, &c. $= \log\,(y + y' + y'', \text{ \&c.})$ (2)

Therefore, *the logarithm of the product of two or more numbers is equal to the sum of the logarithms of the numbers forming the product.*

10. If we divide two of the equations (1), member by member, we have

$$a^{x-x'} = \frac{y}{y'} \ : \text{ hence } x - x' = \log\left(\frac{y}{y'}\right)$$

or, $$\log y - \log y' = \log\left(\frac{y}{y'}\right)$$

That is, *the logarithm of a quotient is obtained by subtracting the logarithm of the divisor from that of the dividend.*

11. If all the numbers y, y', y'', &c. be equal in equation (2), and there be m of them, this equation becomes

$$m \log y = \log\,(y^{m})$$

which shows, *that the logarithm of the* m[th] *power of a number is equal to* m *times the logarithm of this number.*

12. If we extract the mth root of both members of the equation

$$a^{x} = y,$$

we shall have $$a^{\frac{x}{m}} = \sqrt[m]{y}.$$

But, $$\frac{x}{m} = \log \sqrt[m]{y}, \text{ and } x = \log y.$$

Hence $$\frac{\log y}{m} = \log \sqrt[m]{y}.$$

That is, *the logarithm of the* m[th] *root of a number is equal to the logarithm of the number divided by the index of the root.*

13. Hence, if we have to multiply two numbers together, or to divide one by the other, we may, by taking their logarithms out of the tables and adding or subtracting them,

and then finding in the tables the number whose logarithm is equal to the sum or difference, determine the product or quotient of the numbers, as the case may be.

14. Or, if we have to find any power or root of a number, we may, by taking its logarithm out of the tables and multiplying or dividing it by the index of the power or root, and then finding in the tables the number whose logarithm is equal to the product or quotient, determine the power or root of the proposed number.

15. We thus see, that by the aid of a *Table of Logarithms*, the arithmetical operations of multiplication and division of all numbers within the limits of the tables, may be replaced by the addition and subtraction of their logarithms; and those involving the powers and roots of numbers, by the single operation of multiplying or dividing their logarithms by the index of the root or power, as the case may be.

16. These advantages of logarithms in effecting numerical calculations, are especially seen in the case of large numbers; and they bring within the range of computation many arithmetical operations which, on account of their intricacy and the labor they involve, would be otherwise impracticable.

Properties of Logarithms to the Base 10, *or Common System.*

17. If we make successively $x = 1, 2, 3, 4$, &c., in the equation $a^x = y$, when $a = 10$, we shall have for the corresponding values of y,

$10^1 = 10, (10)^2 = 100, (10)^3 = 1000, (10)^4 = 10000, (10)^5 = 100000$, &c.
And for *negative* values of x, $-1, -2, -3, -4, -5$, &c., we have
$(10)^{-1} = \frac{1}{10} = .1, (10)^{-2} = \frac{1}{100} = .01, (10)^{-3} = \frac{1}{1000} = .001, (10)^{-4} = \frac{1}{10000} = .0001$, &c.

Hence, in the *common* system of logarithms,

$$\begin{array}{llll} \log 10 = 1 & \text{and} & \log 0.1 & = -1 \\ \log 100 = 2 & \text{"} & \log 0.01 & = -2 \\ \log 1000 = 3 & \text{"} & \log 0.001 & = -3 \\ \log 10000 = 4 & \text{"} & \log 0.0001 & = -4 \\ \text{\&c.} & & \text{\&c.} & \end{array}$$

From which we see, that those numbers only which correspond to the exact powers of 10 have entire numbers for their logarithms; that the logarithm of any number between 1 and 10 is some number between 0 and 1, that is a fraction; the logarithm of any number between 10 and 100 is some number between 1 and 2, or 1 *plus* a fraction; the logarithm of any number between 100 and 1000 is some number between 2 and 3, or 2 *plus* a fraction: and so on.

Again, for numbers less than 1, we see that the logarithm of any number between 1 and .1 is some number between 0 and —1; the logarithm of any number between .1 and .01 is some number between —1 and —2; the logarithm of any number between .01 and .001 is some number between —2 and —3; and so on: which show that if the number be *less* than 1, its logarithm, both decimal part and integral part, will be *negative.*

18. The entire part of a logarithm is called the *characteristic* of the logarithm.

19. Hence, from the above table we see, that the characteristic of the logarithm of a number >1, which is expressed by a single figure, is 0; if the number be expressed by 2 figures, the characteristic is 1; if it be expressed by 3 figures, the characteristic is 2; and in general, if it be expressed by m figures, the characteristic is $m-1$; that is, *the characteristic of the logarithmic of a number* >1 *is one less than the number of significant figures in the number.*

20. The characteristic of the logarithm of a number <1 is always negative. If the fraction be expressed decimally, *the characteristic of its logarithm will be always* 1 *added to the number of ciphers following the decimal point.* Thus, .00462 lies between .001 and .01; its logarithm must be some number between —3 and —2, and is therefore —3 *plus* a fraction. The characteristic is —3, the number 3 being composed of 1+2, or 1 plus the number of ciphers after the decimal point.

21. A Logarithm will therefore usually consist of a whole number, followed by a decimal part less than 1; and if the number to which it corresponds be less than 1, its entire value, both decimal and integral part, will be *negative.* In the *Tables,* however, it is usual to print positive decimals only. We proceed now to show how, nevertheless, the logarithms of all numbers whatever, both the logarithms of numbers less than 1, and whose values are consequently negative, and the

logarithms of numbers greater than 10 and whose values consequently have an integral as well as a decimal part, can be included in a table where only positive decimals are printed.

22. *A negative logarithm may be expressed so that its characteristic only shall be negative.*

Let $-(n+d)$ be such a logarithm, n being its characteristic, and d its decimal part: thus,

$$-(n+d)=-(n+1)+1-d=-(n+1)+d'$$

where the new decimal part d' is positive. Hence we see, that the transformation of a logarithm entirely negative, into one whose characteristic only is negative, is effected by increasing the characteristic by 1, and substituting for the decimal part, unity *minus* the decimal part. This difference may be readily formed by subtracting the last digit on the right of the decimal part from 10, and the rest from 9. And conversely, a logarithm whose characteristic only is negative may be made entirely negative by performing on the decimal part the operation just described, and diminishing the characteristic by 1. Thus,

$$-(2.2346899)=-3+(1-0.2346899)=-3+.7653101.$$
$$=\bar{3}.7653101.$$

where we place the sign — *above* and not *before* the characteristic, to show that the sign affects only the characteristic: and the whole is used as an abbreviated mode of writing $-3+0.7653101$. If, again, we wish to convert this into an expression entirely negative, we have,

$$\bar{3}.7653101=-3+.7653101=-2-(1-.7653101)=$$
$$-(2.2346899).$$

23. In practice, a negative logarithm is always expressed so that its characteristic only is negative: when negative logarithms are expressed in this manner, there are certain peculiarities in multiplying or dividing them by any quantity, which must be attended to, and will be understood by the following illustration.

24. If a logarithm whose characteristic only is negative, is to be divided by any number, we must make the characteristic divisible by the number, and correct the expression by a corresponding addition. Thus, if

$\overline{7}.3295642$ is to be divided by 3, we have

$$\overline{7}.3295642 = -9 + 2.3295642,$$

and the quotient by 3, is $\overline{3}.7765214$.

25. In the common system, the same decimal part serves for the logarithms of all numbers which differ from one another only in the position of the place of units relative to significant digits.

Let N be any whole number, having n for the characteristic and d for the decimal part of the logarithm, so that

$$\log N = n + d.$$

First, let $N \times (10)^m$ be a whole number having the same significant digits as N, but having its place of units removed m places to the *right:* then,

$$\log (N \times (10)^m) = \log (10)^m + \log N, = (m + n) + d,$$

which has $m + n$ for its characteristic, and the same decimal part as log N.

Next, let $N \div (10)^m$ be a decimal having the same significant digits as N, but having its units' place removed m places to the *left;* then

$$\log (N \div (10)^m) = \log (10)^{-m} + \log N = -m + n + d;$$

now, if the proposed decimal be greater than 1, or $N > (10)^m$, and therefore $\log N > m$, or n not less than m,

$$\log (N \div 10)^m) = (n - m) + d;$$

which has a positive characteristic $n - m$, and the same decimal part as log N: but if the proposed decimal be less than 1, or $N < (10)^m$, and therefore $\log N < m$, or n less than m, then

$$\log (N \div (10)^m) = -(m - n) + d,$$

so that in this case also, provided the logarithm be expressed so that the characteristic only is negative, the decimal part is the same as that of log N.

26. From the preceding observations we collect the principal advantages of Tables of Logarithms calculated to a base the same as the base of the system of notation in numbers, 10, to be; that since the characteristics of all numbers whole or fractional are known by inspection, they

need not be recorded in the Tables; and the characteristic being omitted, the same record in the Tables will serve for all numbers which have the same succession of significant digits, and differ only in the position of the place of units relative to these digits: Thus, the record .5386617, which in reality expresses the logarithm of 3.4567, can be made to express the logarithms of

345670, 34567, 3456.7, 345.67, 34.567, 3.4567, .34567, .034567,

or any number formed by adding ciphers to the end of the former, or to beginning of the latter immediately after the decimal point; so that every logarithm taken out of the Tables for a particular number, becomes, by simply altering its characteristic, the logarithm of an infinite variety of other numbers, that is, of all that are expressed by the same succession of significant digits.

Mode of using Tables of Logarithms for Numbers.

27. Tables of Logarithms contain all whole numbers, from 1 up to a certain limit, with their logarithms, in which the characteristic is usually suppressed.

The tables in common use contain the logarithms of all numbers from 1 to 10,000.

In making use of tables of logarithms to effect numerical calculations, the two main problems which arise are

1st. *Any number being given, to find, by means of the Tables, its logarithm.*

2d. *A logarithm being given, to find, by means of the Tables, the corresponding number.*

28. 1st. *Any number N being given, to find by the Tables its logarithm.*

The *characteristic,* whether N be a whole number or a decimal, is known by inspection.

When N, leaving the decimal point out of consideration, exceeds the limits of the Tables, we must transpose the decimal point, so that the integral part may be contained in the Tables, and may be the greatest possible. Let n be this integral part, and d the remaining decimal part of the proposed number; l the decimal part of the logarithm of n, and δ the difference between the logarithms of the numbers n

and $n+1$, (this difference is always set down in the Tables,) we have

$$\log(n+1)-\log n = \log\left(\frac{n+1}{n}\right) = \log\left(1+\frac{1}{n}\right) = 0, \text{ when } n = \infty$$

Hence, *the difference of the logarithms of two consecutive numbers is less in proportion as the numbers themselves are greater.* We may therefore assume that the difference of the logarithms will always be proportioned to the differences of the numbers; or that

$$\frac{1}{d} = \frac{\delta}{x}, \text{ hence } x = d \times \delta,$$

the value of $x, = d \times \delta$, is what must be added to l to get the decimal part of the logarithm of $n+d$, which is also the decimal part of the logarithm of the proposed number.

29. This property, that the increment of the logarithm may be assumed to be proportional to the increment of the number, is extremely important; and the application of it deserves the closest attention, as it furnishes a ready means of greatly extending the range of the Tables. We see also that the number n should be as great as possible, for the error in the above proportion is diminished as n increases, and disappears when $n = \infty$.

30. We give specimens of the usual logarithmic Tables for numbers.

Table A has the natural numbers arranged in order in the column marked N, and is limited to numbers embracing not more than three digits. If the logarithm of a number is required which contains not more than three digits, it will be found in the first column. To save the needless repetition of figures, the two first decimals of the logarithm are only written at the head of each break, and must be regarded as belonging equally to all logarithms which succeed up to the next break.

If the number contain four digits, seek the number in column N corresponding to the first three digits, and then look along the top of the pages among the numbers 0, 1, 2, 3, 4, &c., for the number that corresponds to the fourth digit, the logarithm of the given number will be found at the intersection of the horizontal lines passing through the numbers in the column N, and the vertical line in which the fourth digit is found.

The differences of the logarithms are placed in column D.

If the number contain more than 4 digits, place a decimal point after the fourth figure from the left hand, thus converting the given number into a whole number and decimal. Find the logarithm of the entire part as just explained, and then take the corresponding difference in the column D, and multiply it by the decimal part; add the product thus found to the logarithm, and the sum will be the whole logarithm sought.

The characteristic is determined by known rules, in each case.

A.

A TABLE OF LOGARITHMS FROM 1 TO 10,000.

N.	0	1	2	3	4	5	6	7	8	9	D.
280	447158	7313	7468	7623	7778	7933	8088	8242	8397	8552	155
281	8706	8861	9015	9170	9324	9478	9633	9787	9941	°°95	154
282	450249	0403	0557	0711	0865	1018	1172	1326	1479	1633	154
283	1786	1940	2093	2247	2400	2553	2706	2859	3012	3165	153
284	3318	3471	3624	3777	3930	4082	4235	4387	4540	4692	153
285	4845	4997	5150	5302	5454	5606	5758	5910	6062	6214	152
286	6366	6518	6670	6821	6973	7125	7276	7428	7579	7731	152
287	7882	8033	8184	8336	8487	8638	8789	8940	9091	9242	151
288	9892	9543	9694	9845	9995	°146	°296	°447	°597	°748	151
289	460898	1048	1198	1348	1499	1649	1799	1948	2098	2248	150
290	462398	2548	2697	2847	2997	3146	3296	3445	3594	3744	150
291	3893	4042	4191	4340	4490	4639	4788	4936	5085	5234	149
292	5383	5532	5680	5829	5977	6126	6274	6423	6571	6719	149
293	6868	7016	7164	7312	7460	7608	7756	7904	8052	8200	148
294	8347	8495	8643	8790	8938	9085	9233	9380	9527	9675	148
295	9822	9969	°116	°263	°410	°557	°704	°851	°998	1145	147
296	471292	1438	1585	1732	1878	2025	2171	2318	2464	2610	146
297	2756	2903	3049	3195	3341	3487	3633	3779	3925	4071	146
298	4216	4362	4508	4653	4799	4944	5090	5235	5381	5526	146
299	5671	5816	5962	6107	6252	6397	6542	6687	6832	6976	145

31. In Table B, the column N has four digits in it, and hence this table may be used to find directly and without the use of the column of differences, the logarithms of all numbers as high as those which contain five digits, the fifth digit of the given number being found as before in the top line above the logarithms. When the first three figures of the decimal part of the logarithm are common, they are omitted

in the tables except in the first logarithm of each break, but are always to be regarded as constituting the first figures of the logarithms immediately following in the same break.

B.

OF NUMBERS.

N.	0	1	2	3	4	5	6	7	8	9	D.	Pro.
2550	4065402	5572	5742	5913	6083	6253	6424	6599	6764	6934		
51	7105	7275	7445	7615	7786	7956	8126	8296	8466	8637		170
52	8807	8977	9147	9317	9487	9658	9828	9998	$\bar{0}168$	$\bar{0}838$		1 17
53	4070508	0678	0848	1018	1189	1359	1529	1694	1869	2039		2 34
54	2209	2379	2549	2719	2889	3059	3229	3399	3569	3739	170	3 51 4 68
55	3909	4079	4249	4419	4580	4759	4929	5099	5268	5489		5 85 6 102
56	5608	5778	5948	6118	6288	6458	6628	6798	6968	7137		7 119
57	7307	7477	7647	7817	7987	8156	8326	8496	8666	8836		8 136 9 153
58	9005	9175	9345	9515	9684	9854	$\bar{0}024$	$\bar{0}194$	$\bar{0}363$	$\bar{0}533$		
59	4080703	0873	1042	1212	1382	1551	1721	1891	2060	2230		
2560	2400	2569	2739	2909	3078	3248	3417	3587	3757	3926		
61	4096	4265	4435	4604	4774	4944	5113	5283	5452	5622		
62	5791	5961	6130	6300	6469	6639	6808	6978	7147	7317		169
63	7486	7656	7825	7994	8164	8333	8503	8672	8841	9011		
64	9180	9350	9519	9688	9858	$\bar{0}027$	$\bar{0}196$	$\bar{0}360$	$\bar{0}535$	$\bar{0}704$		1 17 2 34
65	4090874	1043	1212	1382	1551	1720	1889	2059	2228	2397		3 51 4 68
66	2567	2736	2905	3074	3243	3413	3582	3751	3920	4089		5 85 6 101
67	4259	4428	4597	4766	4935	5105	5274	5443	5612	5781		7 118
68	5950	6119	6288	6458	6627	6796	6965	7134	7303	7472	169	8 135
69	7641	7810	7979	8148	8317	8486	8655	8824	8993	9162		9 152

Example 1.—Seek the logarithm of 2967.

In tablo A, find in column N, 296, then run along the top line of the table until you come to the number 7, the logarithm 472318 is the decimal part of the logarithm of the given number, and prefixing the characteristic, the logarithm is 2.472318.

Example 2.—Find the logarithm of 296789.

Regard all the figures on the right of the *fourth* figure as decimal part. Find the logarithm of the entire part 2967, as before. We have

log 2967, (omitting characteristic) = 472318.

The column D gives 146 for the difference. Multiplying this by .89, the decimal part of the given number, we have $146 \times .89 = 146.85$, or neglecting the decimal, and since it exceeds .5, let 1 be added to the entire part, and we have 147, to be added to the logarithm as before found. Thus,

log	.2967 =	472318
"	.89	147
"	296789	472465

Now, introducing the characteristic, the whole logarithm becomes 5.472465.

32. The trouble of multiplying the difference by the decimal part may be avoided in table B, by the column of proportional parts, which contains the product of the tabular difference by $\frac{1}{10}$, $\frac{2}{10}$, $\frac{3}{10}$, &c., or of one-tenth of this tabular difference by 1, 2, 3, 4 . . . . 9.

Thus, find the logarithm of 2568946.

log 25689		= 4097472
log .4	(16.9 × 4)	= 68
log .06	(1.69 × 6)	= 10
log 2568946		= 6.4097550

33. 2d. *A logarithm being given, to find, by means of the Table, the corresponding number.*

When the decimal part l of the logarithm L is positive, and found exactly in the tables, we have only to take out the corresponding number; and the given characteristic will determine the position of the units' place.

When the decimal part l of the given logarithm is not found exactly in the tables, let n be the integer whose logarithm has a decimal part l' immediately inferior to l, d, the difference between the decimal part l' and that which immediately follows it in the tables corresponding to $n + 1$, and d' the difference $l - l'$; making the proportion.

$$\frac{d}{d'} = \frac{1}{x}$$

we find the fourth term $x = \frac{d'}{d}$, which we must reduce to a decimal and add it to the number n: we thus find a number whose logarithm has for a decimal part l, and from which we can deduce the required number, having regard to the characteristic, as in the preceding case.

When the given logarithm is entirely negative, we must transform it so as to have the decimal part positive, as has been explained, and then proceed as above.

Ex. 1.—Let the given logarithm be 5.5386617.

In the column marked 0, we seek the logarithm which is next less than the decimal part L, which we find to be 5385737, opposite to the number 3456 in the column marked N: then advancing in the horizontal line to the column marked 7, we find the four last decimals 6617 of L. Hence, affixing the digit 7 to 3456, we have 34567, and since the given characteristic is 5, the required number is 345670.

Ex. 2.—Let the given logarithm be 2.5386729.

In seeking for the decimal part in the tables as above, we find that it lies between .5386617 and .5386743: if it were exactly equal to the former, the corresponding number would be 34567, but as it surpasses .5386617 by 112, there will be an increase in 34567; and as the difference of the logarithms corresponding to an increase of unity in 34567 is 126, the required increment x, admitting the common principle of proportional parts, will be found by the proportion

$$\frac{126}{112} = \frac{1}{x} \therefore x = \frac{112}{126} = 0.89.$$

Therefore, the required number, neglecting the decimal point, is 3456789, and since the given characteristic is 2, the required number is 345.6789.

By means of the table of proportional parts of the tabular differences, we may avoid the trouble of dividing 112 by 126. In the table of proportional parts of the difference 126, the part next less than 112 is 101, which corresponds to .8 and there remains 11. If we put zero to the right of 11, we have 110, which differs very little from the part 113, which has 9 opposite to it: we conclude therefore, that 110 corresponds to .9, and 11 to .09; consequently the required number is composed of the digits 3456789; and taking account of the characteristic, the number is 345.6789, as before.

34. In actual calculations it is often necessary to add several logarithms together, and from this sum to subtract

other logarithms. In such cases it is convenient to reduce all the operations to a single addition by the use of the *arithmetical complement* of the logarithms to be subtracted.

35. *The arithmetical complement* of a logarithm is the number which remains after subtracting the logarithm from 10.

If c be the arithmetical complement of b, we shall have $c = 10 - b$. Now, if we propose to subtract the logarithm b from the logarithm a, we have, evidently, since $-b = c - 10$,

$$a - b = a + c - 10,$$

which shows that the difference between two logarithms may be found, by adding to the first logarithm the arithmetical complement of the logarithm to be subtracted, and then diminishing this sum by 10.

The arithmetical complements of logarithms may be all formed by subtracting the last digit of each from 10, and all the other digits from 9.

TRIGONOMETRY.

CHAPTER I.

THEORY OF TRIGONOMETRICAL LINES.

Object of Trigonometry. The Method of Representing the Sides and Angles of Triangles by Numbers.

1. In every triangle, plane and spherical, there are six parts: three angles and three sides. To determine all the parts of a triangle, it is generally sufficient that three of these be known; but it is further necessary, when the triangle is rectilinear or plane, that at least one of these parts shall be a side. For, with three given angles an infinite number of triangles can be formed, which will be *similar* to each other, but not *equal*. It is always understood that the sum of the three angles is equal to two right angles.

2. *Geometry* furnishes very simple constructions for determining a triangle, by means of some of its parts, as known; but these constructions, like all *graphic* processes, are attended by the inconvenience of giving an inexact and therefore insufficient approximation, on account of the imperfections of the most accurate instruments.

These constructions have been replaced by numerical calculations, which admit of any degree of precision that may be required.

3. *The special object of Trigonometry is to give the methods of calculating all the parts of a triangle, when a sufficient number of parts is given. This is called the resolution of the triangle.*

Plane Trigonometry gives the various methods of solving *plane triangles*, and *Spherical Trigonometry* treats of the resolution of *spherical triangles*.

4. To express the dimensions of right lines by *numbers*,

they are referred to a *common unit of measure*, as a *foot*, a *yard*, &c.: and thus, the sides of a plane triangle are represented by a certain number of *feet* or *yards*, &c.

5. *Angles* are designated by the *arcs* which measure them. For this purpose, the circumference, whatever be the radius, is divided into a certain number of equal parts, called *degrees*, and then an angle or an arc is expressed by a number of *degrees.*

6. Formerly, geometricians agreed to divide the circumference into 360 *degrees*, the degree into 60 *minutes*, the minute into 60 *seconds*, &c. By this division, the fourth part of the circumference, or *quadrant*, which is the measure of a *right angle*, contains 90 degrees. To avoid the embarrassment from complex numbers, it has been proposed to subject angles to the *decimal* division, thus dividing the quadrant into 100 *degrees*, the degree into 100 *minutes*, the minute into 100 *seconds*, &c.

7. *Degrees, minutes*, and *seconds*, are denoted by the symbols $^\circ$, $'$, $''$. Thus, to represent 44 degrees, 9 minutes, 37 seconds, we write $44^\circ\ 9'\ 37''$. In the decimal division, if we wish to refer this arc to the quadrant, taken as the *unit of measure*, it would be expressed thus, 0.440937: for, in this division, degrees are *hundredths* of the quadrant, minutes are *ten thousandths*, and seconds are *millionths*.

8. The old division is adopted in this treatise, the circumference being divided into 360 degrees. In the formulæ which are given, the letter π is often used to represent the semi-circumference 180°.

Definitions of Trigonometrical Lines. The Manner of using the Signs + and — to indicate the Positions of Lines when their Direction is opposed.

9. Geometricians were for a long time embarrassed by the difficulty of establishing the relations which exist between the angles and the sides of triangles; and yet this is essential to the resolution of the triangles. For, the sides being given in terms of a *linear* unit, and the angles in terms of the arc of a circumference, all the parts of a triangle would not be expressed in *homogenous* terms, and could not therefore be combined in the calculations. The use of *Trigonomet-*

rical Lines, which are *linear* functions of the angles or arcs considered, enables us to substitute *right lines* for the angles or arcs, and thus to establish the *homogeneity* between all the parts of a triangle, which is indispensable to its resolution. We will now proceed to define these trigonometrical lines.*

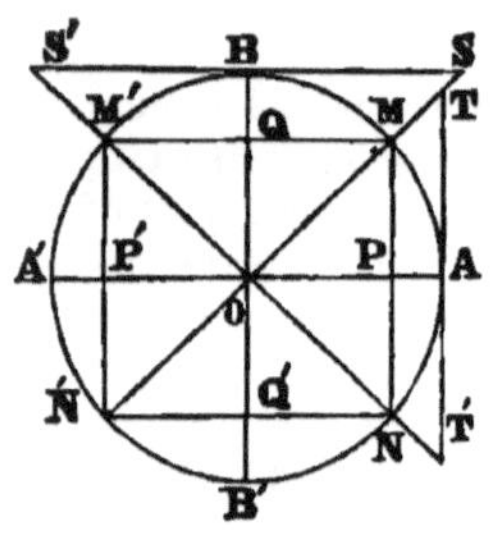

10. The sine of the arc A M, is the perpendicular M P, let fall from one extremity of the arc, upon the diameter which passes through the other extremity.

The Tangent of the arc A M, is the distance A T, intercepted on the tangent drawn at one extremity of the arc, between that extremity and the prolongation of the radius O M, which passes through the other extremity.

The Secant of an arc A M, *is the part* O T, *of the radius produced, comprised between the centre and the tangent:*

Representing the arc A M by x, the sine, tangent, and secant, are designated as follows,—

$$MP = \sin x, \; AT = \text{tang } x, \; OT = \sec x.$$

11. If M P be produced until it meets the circumference at N, the chord M N will be double of M P, and the arc M A N double of A M, hence, *the sine of an arc is half the chord which subtends double the arc.*

12. If we designate by r the radius of the circle, the side of the inscribed square will be $r\sqrt{2}$: the subtended arc to the side of the square is 90°, hence, $\sin 45° = \frac{1}{2} r \sqrt{2}$.

---

* The principle of *homogeneity* between the parts of a triangle, to which reference is made in the text, is an important one, and will be frequently referred to in this treatise. It is a general principle, not restricted to trigonometry, but necessarily applied in every branch of analytical mathematics. It is founded upon the essential relations which must exist between the two members of every Algebraic equation, to make it the consistent expression of equality between equal magnitudes; and serves also to point out the modifications which any equation not homogeneous in *form*, must have, that the relations expressed by the equations may be consistent with established facts or truths.

13. In like manner, the side of the inscribed hexagon being equal to the radius, and the subtended arc being 60°, we shall have sin 30° = $\frac{1}{2}r$.

14. *The complement* of an arc or angle is the difference between that arc or angle and 90°. When the arc is greater than 90°, its complement is *negative;* thus, the complement of 127° is —37°. The two acute angles of a right-angled triangle are complements of each other.

15. We designate by the terms *cosine, cotangent,* and *cosecant* of an arc, the sine, the tangent, and the secant of the complement of that arc: and to express these new lines, we employ the abbreviations *cos, cot, cosec;* thus, by the definitions, we have $\cos x = \sin (90° - x)$, $\cot x = \text{tang} (90° - x)$, $\text{cosec } x = \sec (90° - x)$.

17. If we draw the radius O B perpendicular to O A, and also M Q, and B S, perpendicular to O B, the arc B M will have M Q for its *sine*, B S for its *tangent*, O S for its *secant*: but the arc B M is the complement of A M; hence, designating always A M by x, we have

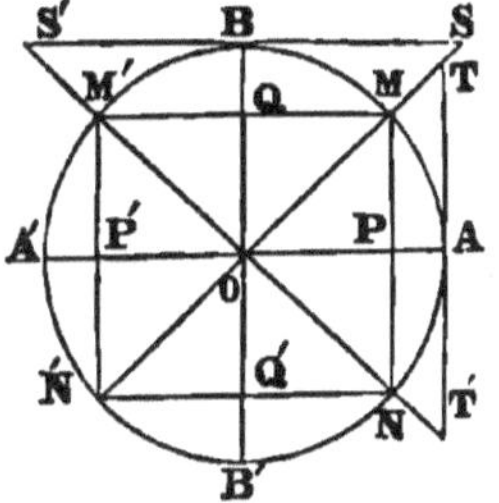

M Q = $\cos x$, B S = $\cot x$, O S = $\text{cosec } x$.

We notice that M Q = O P; which shows, *that the cosine of an arc is equal to the part of the radius comprised between the centre and the foot of the sine.*

18. The distance A P, comprised between the *origin* of the arc and the foot of the sine, is called the *verse-sine;* and the distance B Q, the *co-verse-sine.*

19. In giving to the point M every possible position on the circumference, the trigonometrical lines take situations altogether different from those which they have, when the arc A M is less than 90°. For example, if we have the arc A M′, whose complement is B M′, and *negative*, its cosine Q M′ or O P′, is found situated on the *left* of the point O, whilst before, it was on the right; such changes in the positions of the lines introduce into the calculations difficulties, which the following proposition will explain.

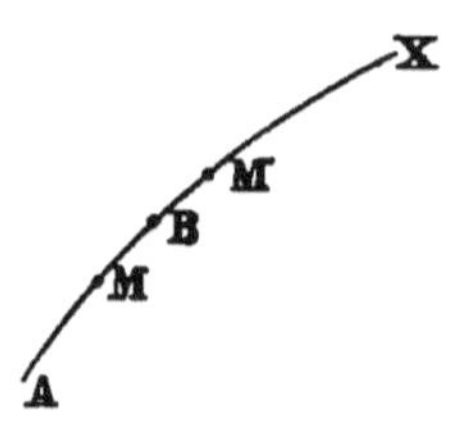

Let A B X be any line whatever, on which there are two points A and B, separated a distance $AB=a$. Assume, as known, the distance x of the point B from any point, as M, of the line A B X, and let it be required to find the distance from the point A to the point M. Representing this distance by z, it is evident we shall have

$$z=a+x, \text{ or, } z=a-x,$$

according as the point M is between B and X, or between B and A: so that two different formulæ are required for these two positions of the point M. But this inconvenience is avoided, and a single formula made to suffice, by giving different Algebraic signs to distances which have contrary positions with respect to the point B. Thus, if, in the first formula $z=a+x$, we make successively $x=+BM$, and $x=-BM$, we have, under the first supposition, $z=a+BM$, and under the second, $z=a-BM$, as should be. We thus see, that the first formula $z=a+x$, will correspond to all positions of the point M, and the second is unnecessary.

We might also take x positive between B and A, and negative between B and X: in this case, the second formula $z=a-x$, would be used. It would be easy to multiply illustrations, but what precedes is sufficient to make apparent the importance of the following rule of *Descartes.*

20. *If we consider on any line whatever, straight or curved, different distances, measured from a common origin, fixed on that line, we may introduce into the calculation distances which have opposite positions with respect to this origin, by giving those lying in one direction the sign* +, *and those lying in the opposite direction, the sign* —.

21. The rule by which *positive* distances are estimated is altogether arbitrary, but once fixed, *negative* distances must be reckoned in an opposite direction. With trigonometrical lines, the usage is, to consider them as positive in the positions they occupy when the arc is less than 90°.

22. We shall soon have occasion to make numerous applica-

tions of this rule; but before proceeding further, the student must be guarded against a very common error, which consists, in assimilating the principle in question to a theorem susceptible of an *à priori* demonstration. It is far otherwise, and whatever considerations may have been adduced in its support, it is to be regarded as a simple *agreement* which must not be violated, and the utility of which is made apparent by its applications.

Progressive Changes in Trigonometrical Lines. The Method of reducing them to the First Quadrant.

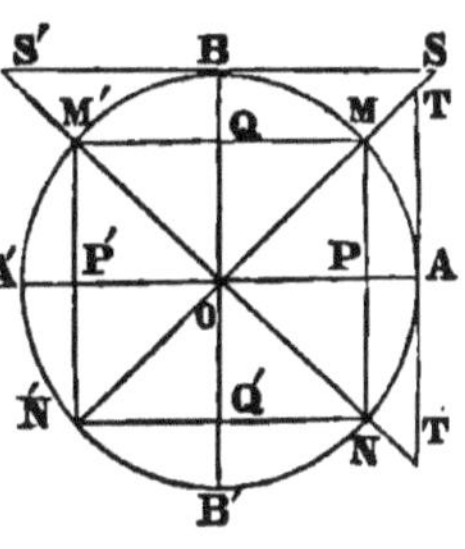

23. When the radius O M coincides with O A, the arc A M is o, the sine is o, the tangent is o, and the secant is equal to O A: at the same time, the cotangent and cosecant are infinite, since the lines B S and O S increase more and more as O M approaches O A, and become infinite, when they coincide. Thus, denoting the radius by r, we have

$$\sin o = o, \quad \text{tang}\, o = o, \quad \sec o = r$$
$$\cos o = r, \quad \cot o = \infty, \quad \text{cosec}\, o = \infty.$$

24. If the radius O M approaches the position O B, it is easy to see that the sine, tangent, and secant increase, while the cosine, cotangent, and cosecant diminish. When the point M is at the middle point of A B, the arc A M is 45°, the triangle O P M is isosceles, and the sine is equal to the cosine. Now this triangle gives $2\,\overline{MP}^2 = r^2$, hence, $MP = \frac{1}{2} r\sqrt{2}$, and therefore

$$\sin 45° = \cos 45° = \tfrac{1}{2} r \sqrt{2}.$$

25. The triangles O A T, O B S, being also isosceles and equal to each other, the tangent and cotangent are equal to the radius, hence

$$\text{tang}\, 45° = \cot 45° = r.$$

26. Finally, the secant and cosecant are also equal, and the triangle O A T giving $\overline{OT}^2 = 2r^2$, we have $OT = r\sqrt{2}$, and hence

$$\sec 45° = \text{cosec } 45° = r\sqrt{2}.$$

27. When the point M reaches B, the sine is equal to B O, the tangent and secant are infinite, the cosine M Q is o, the cotangent B S is also o, and the cosecant becomes equal to O B. We have then,

$$\sin 90° = r, \quad \text{tang } 90° = \infty, \quad \sec 90° = \infty$$
$$\cos 90° = o, \quad \cot 90° = o, \quad \text{cosec } 90° = r.$$

These values are indeed consequences from those which were found when the arc was o; for, the arcs o and 90° being complements of each other, we ought to have sin 90° = cos o, tang 90° = cot o, sec 90° = cosec o, cos 90° = sin o, cot 90° = tang o, cosec 90° = sec o.

28. Continuing the rotation of the radius O M, let us suppose that it has reached the point M′: the arc is then A M′, and its sine M′ P′. Draw M′ M parallel to A′A, and construct all the trigonometrical lines of the arc A M as indicated in the figure, we see at once that the sines M P and M′P′ will be equal, and therefore sin A M′ = sin A M.

29. To get the tangent, produce the radius O M′ below the diameter A A′, we see that the tangent, which is here A T′, is found in a position opposite to that which it had at first: consequently, it is *negative*. But the triangles O A T, O A T′, being equal, give A T′ = A T, hence tang A M′ = — tang A M.

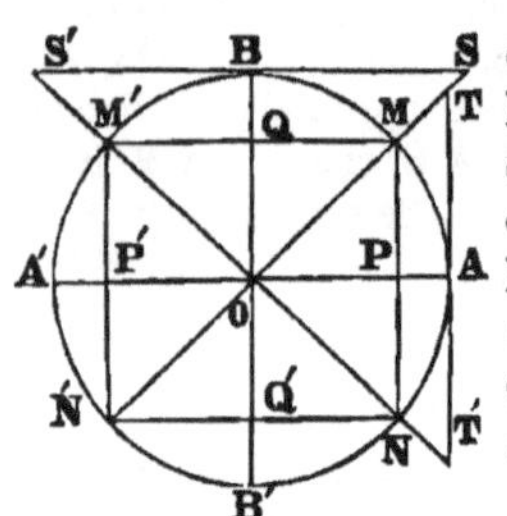

30. From the definition, the secant of the arc A M′ is O T′. This line is no longer laid off on the radius O M′ on the *same side* of the centre with the point M′, but it lies in the *opposite* direction, and is, consequently, *negative;* and since O T′ = O T, it follows that sec A M′ = — sec A M.

31. The cosine, cotangent, and cosecant present remarkable analogies. Since the arc A M′ exceeds 90°, its complement is negative; and, further, as the cosine Q M′ or O P′ is situated on the left of the point O, we also consider this cosine as negative. The same reasoning applies to the cotangent

B S′. As to the cosecant O S′, its sign is still positive, since it is on O M′, on the same side of the centre as the point M′. The triangle O Q M and O Q M′ being equal, as well as the triangles O B S, O B S′, we have Q M′ = Q M, B S′ = B S; hence, cos A M′ = —cos A M; cot A M′ = —cot A M; cosec A M′ = cosec A M.

32. *The supplement* of an arc or angle is the difference between that arc or angle and 180°: thus, A′ M′, or its equal A M, is the supplement of A M′, and we may enunciate the properties found above by saying, *that two supplementary arcs have their trigonometrical lines equal and with contrary signs, with the exception of the sine and cosecant, which are also equal, but have the same algebraic signs.*

33. To express these properties by equations, designate A M′ by x, we shall have A M = A′ M′ = $180° - x$, and we may write

$$\left.\begin{array}{ll} \sin x = \sin(180° - x), & \cos x = -\cos(180° - x), \\ \text{tang}\, x = -\text{tang}(180° - x), & \cot x = -\cot(180° - x), \\ \sec x = -\sec(180° - x), & \text{cosec}\, x = \text{cosec}(180° - x), \end{array}\right\} (1)$$

34. It is also evident, that from 90° to 180°, the sine, tangent, and secant decrease; whilst, on the contrary, the cosine, cotangent, and cosecant increase. If O M coincides with O A′, we have, for an arc of 180°,

$$\sin 180° = o, \quad \text{tang}\, 180° = o, \quad \sec 180° = -r,$$
$$\cos 180° = -r, \quad \cot 180° = -\infty, \quad \text{cosec}\, 180° = \infty.$$

All these values may be deduced from the relations (1), by making $x = 180°$. Thus, $\cos x = -\cos(180° - x)$, becomes $\cos 180° = -\cos o$; but $\cos o = r$; therefore, $\cos 180° = -r$.

35. The applications of analysis to geometry frequently involve the consideration of arcs which contain many semi-circumferences. It is necessary then to give formulæ for reducing all such arcs to the first quadrant. To abridge, we will consider specially the sine and cosine, which are the trigonometrical lines most used; and since every arc greater than a semi-circumference is composed of an arc less than 180°, increased by an arc containing 180° one or more times, we will, in the first place, determine the sine and cosine for the arc $180° + x$, x being $< 180°$.

Let A M be the arc designated by x, which may be taken,

3*

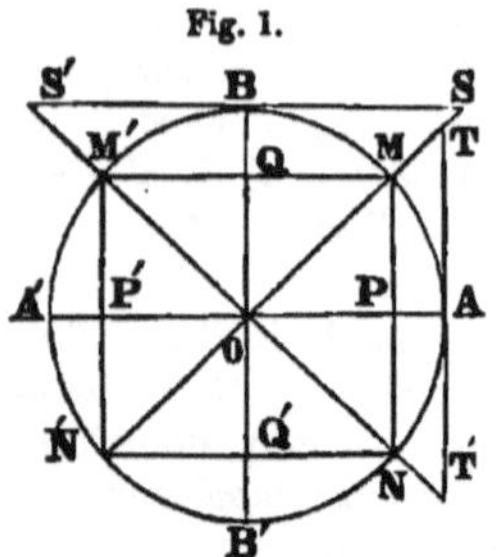

Fig. 1.

at pleasure, between o and 180°; adding to A M the semi-circumference M A' N', the new arc A M A' N' will be equal to $180° + x$. The two arcs have equal sines, viz., M P and N' P', or, what is the same thing, O Q and O Q'; but since these lines have opposite positions, we should, in accordance with the agreement established (Art. 20), give to them contrary algebraic signs. The cosines O P and O P' are also equal, and must, in like manner, have different algebraic signs: hence

$$\sin (180° + x) = -\sin x, \ \cos (180° + x) = -\cos x. \quad (2)$$

36. In the next place, add 360° to A M, it is evident we will return to the same point M of the circumference, and that consequently all the trigonometrical lines become the same. Hence we have,

$$\sin (360° + x) = \sin x, \ \cos (360 + x) = \cos x. \quad (3)$$

37. In general, whatever value we attribute to the arc x, if we add to x, 180°, or an *odd* number of semi-circumferences, its extremity will be moved from the vertex of one diameter to the opposite vertex; and hence, it is evident, that the algebraic signs of the sine and cosine are changed. But, if we add to x, 360°, or an *even* number of semi-circumferences, since we return to the same point of the circumference, no trigonometrical line should change its sign.

38. It remains now that we consider *negative* arcs, that is, arcs which are described when the radius, which at first coincided with O A, moves in the direction A B' A', contrary to that which it before followed.

Let A M and A N be two equal and opposite arcs, designated by x, and $-x$. It is evident that their sines M P and N P are also equal and inversely disposed. To obtain the cosines, we observe, that the complements of these arcs, $90° - x$, and $90° + x$, are represented by the arcs B M and B M N, of which the sines M Q and N Q' are equal and similarly situated: hence we have

$$\sin (-x) = -\sin x, \ \cos (-x) = \cos x. \quad (4)$$

39. Although in the figure the arcs A M and A N are less than 90°, these formulæ are nevertheless general. For, it is evident that by augmenting the two arcs at will, provided they continue equal, the sines M P and N P do not cease to be equal and opposite, we shall therefore always have $\sin(-x) = -\sin x$. With regard to the cosine, substitute in the second formula (4) arcs greater than 90°, A B M′ and A B′ N′, for example; and make $x =$ A B M′, and $-x = -$A B′ N′. The complement $90° - x$ of the first arc is negative, and is represented in the figure by the arc B M′, situated on the left of the point B, whilst the complement $90° + x$ of the other arc is equal to B A N′, and always on the right of the point B. But the sines M′ Q and N′ Q′ of these complementary arcs are equal, and are similarly disposed with respect to the diameter B B′; hence, we have always $\cos(-x) = \cos x$. Formulæ (4) are therefore general.

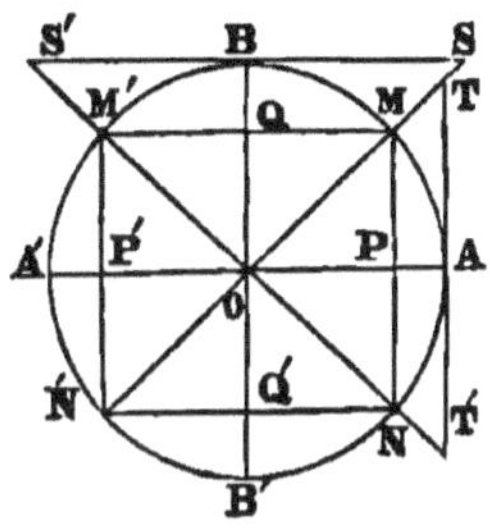

40. *It is well to note that the cosine of any arc whatever, positive or negative, is always represented in magnitude and position, by the distance from the centre to the foot of the sine.*

41. Formulæ (1), (2), (3), (4), may be applied to all possible arcs, positive and negative. For brevity, let us consider only the sine and cosine.

1st. Resuming the two formulæ $\sin x = \sin(180° - x)$, $\cos x = -\cos(180° - x)$ (Art. 33), which were demonstrated for positive arcs comprised between o and 180°, they become, when x is changed into $180° + x$,

$$\sin(180° + x) = \sin(-x), \quad \cos(180° + x) = -\cos(-x),$$

which results are evident by virtue of the relations (2) and (4). We may increase the arc by additions of 180°, to infinity, and the formulæ will remain unchanged. If we substitute $-x$ for $+x$, we see, in like manner, that the two formulæ are still true. They therefore suit all possible arcs.

2d. Formulæ (2) which have been demonstrated for all positive arcs, are equally applicable to negative arcs. For, changing x into $-x$, they become

$$\sin(180^\circ - x) = -\sin(-x,) = \sin x$$
$$\cos(180^\circ - x) = -\cos(-x,) = -\cos x.$$

thus reproducing formulæ (1).

3d. Since the addition of 180° to any arc whatever $+x$ or $-x$, only changes the algebraic sign of the sine or cosine, it follows, that the addition of 360° should produce no change: hence, formulæ (3) apply also to negative arcs.

4th. Formulæ (4) require no special demonstration, since it is evident that in them x may be replaced by $-x$.

42. Nothing is now easier than to reduce to the first quadrant the trigonometrical lines of any arc whatever. Suppose $x = 1029^\circ$, to be an arc whose *sine* is required. Take 360° from this arc as many times as possible, and there remains 309°. Then by formula (3), $\sin x = \sin 309^\circ$. Now take 180° from 309°: by formula (2), we have $\sin(180^\circ + x) = -\sin x$, hence, $\sin x = -\sin 129^\circ$. Finally, $\sin x = -\sin 51^\circ$, since 51°, is the supplement of 129°. We might carry the reduction farther, by observing $\sin 57^\circ, = \cos(90^\circ - 51^\circ)$, and hence $\sin x = -\sin 51^\circ = -\cos 39^\circ$.

43. If the given arc had been $x = -1029^\circ$, the algebraic sign of the sine would have been contrary to the preceding, and we should have $\sin x = \cos 39^\circ$.

Of Arcs which correspond to a given Sine, Cosine, &c.

44. The preceding discussion has shown, that there exists an infinite number of arcs which have the same trigonometrical lines. Let us now suppose one of these lines given, and let it be required to determine the arcs which correspond to it.

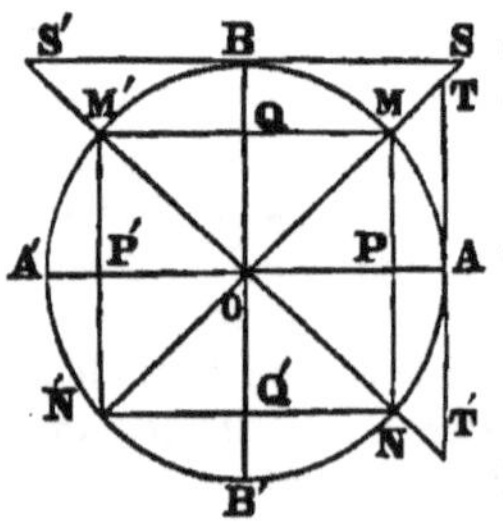

Assume $\sin x = a$. On the radius O B, perpendicular to O A, lay off $OQ = a$, and through the point Q draw M M′ parallel to O A. It is evident that we must take for values of x all arcs which terminate at the points M and M′. Designate the arc A M by a, and 180° by π; A M′ will be equal to $\pi - a$, and the positive arcs terminating in M and M′, will be comprised in the two series.

$$a,\ 2\pi + a,\ 4\pi + a,\ 6\pi + a, \text{ \&c.}$$
$$\pi - a,\ 3\pi - a,\ 5\pi - a,\ 7\pi - a, \text{ \&c.}$$

We have $AB'A'M = 2\pi - a$, and $AB'A'M' = \pi + a$. If we add to these arcs any number of circumferences, and take the resulting arcs negatively, we shall have all the negative arcs which correspond to the given sine, viz.,

$$-2\pi + a,\ -4\pi + a,\ -6\pi + a, \text{ \&c.}$$
$$-\pi - a,\ -3\pi - a,\ -5\pi - a, \text{ \&c.}$$

The arcs of these four series may be embraced in two very simple formulæ. For, we observe, that in two of these series the arc a is added to all the *even* multiples of π, negative as well as positive; and that in the other two, it is subtracted from the *odd* multiples of π; if, therefore, we designate by n, any whole number, positive or negative, and which may also be zero, then, all the required arcs may be expressed by the formulæ

$$x = 2n\pi + a,\ x = (2n + 1)\pi - a. \qquad (5)$$

45. We have supposed a positive: if $\sin x = -a$, we must lay off a on OB', by making $OQ' = a$. Then the arcs which are terminated in N and N′, are the values of x. Make $ABN' = a$, it is easy to see that we shall have $ABN = 3\pi - a$, $AB'N' = 2\pi - a$, and $AN = a - \pi$. Hence, the values of x, positive and negative, which correspond to the sine OQ', are

$$a,\ 2\pi + a,\ 4\pi + a, \text{ \&c.};\ 3\pi - a,\ 5\pi - a,\ 7\pi - a, \text{ \&c.};$$
$$-2\pi + a,\ -4\pi + a,\ 6\pi + a, \text{ \&c.};\ \pi - a,\ -\pi - a,\ -3\pi - a, \text{ \&c.};$$

and it is evident that they are still comprised in formulæ (5).

46. In all the cases, if a exceeds the radius of the circle, r, the arc x will be imaginary, for the greatest positive sine is $+r$, and the greatest negative sine is $-r$.

47. Assume $\cos x = a$. If a is positive, take on OA, $OP = a$, and at the point P erect the perpendicular MN; the values of x are the various arcs, positive and negative, terminating in M and N. Making $AM = a$, it is easy to see, that all these arcs will be embraced in the four series

$$a,\ 2\pi + a,\ 4\pi + a, \text{ \&c.};\ 2\pi - a,\ 4\pi - a,\ 6\pi - a, \text{ \&c.};$$
$$-a,\ -2\pi - a,\ -4\pi - a, \text{ \&c.};\ -2\pi + a,\ -4\pi + a,\ -6\pi + a, \text{ \&c.};$$

and designating by k any whole number whatever, positive or negative, all these arcs may be comprised in the two formulæ

$$x=2k\pi+\alpha, \qquad x=2k\pi-\alpha. \qquad (6)$$

48. If cos $x=-a$, lay off a on O A′, then designating by α the arc A M M′, no change is necessary in what precedes. If a is greater than r, the arc x becomes imaginary.

49. Assume now tang $x=a$, and let a be positive. Take the tangent A T$=a$, above O A, and draw the right line T M N′, passing through the centre, and meeting the circumference in M and N′. The values of x are the positive or negative arcs which terminate in M and N′. Make the arc AM$=\alpha$, we shall have AMN′$=\pi+\alpha$, AN′M$=2\pi-\alpha$, ANN′$=\pi-\alpha$; and the required arcs will be that of the series

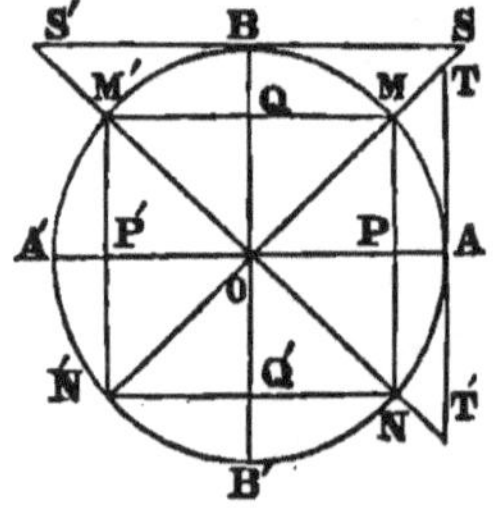

$$\begin{array}{llll} \alpha, & 2\pi+\alpha, & 4\pi+\alpha, & \&c., \\ \pi+\alpha, & 3\pi+\alpha, & 5\pi+\alpha, & \&c., \\ -2\pi+\alpha, & -4\pi+\alpha, & -6\pi+\alpha, & \&c., \\ -\pi+\alpha, & -3\pi+\alpha, & -5\pi+\alpha, & \&c. \end{array}$$

In these four series the arc α is found added to all the multiples of π, positive and negative; hence the general formula for the required arcs is

$$x=k\pi+\alpha. \qquad (7)$$

50. When the given tangent is negative, we lay it off on AT′ below OA; α represents, in this case, an arc comprised between 90° and 180°, such as A B M′. It is also evident that the tangent may have any magnitude we please.

51. We will not examine the cases in which the arc is made known by the other trigonometrical lines. But we readily see that arcs which have the same sine, the same cosine, or the same tangent, have also the same cosecant, or the same secant, or the same cotangent; and this will be made more apparent when we have established the relations between the trigonometrical lines themselves.

52. We must not forget that, in these formulæ, a always represents the smallest arc between o and 180°, corresponding to the given trigonometrical line, π the semi-circumference, and k any whole number whatever, positive or negative, and which may also be zero.

How the Sines, Cosines, &c. may be reduced to Simple Ratios.

53. In trigonometry, an arc being only employed as the measure of an angle, it is not its absolute magnitude which is considered, but its ratio to the circumference of which it forms a part. Now, it is precisely this ratio which is indicated by the number of degrees in the arc; and it is evident that this suffices to determine an angle; for, all the arcs comprised in the same angle, and described from its vertex as a centre, contain an equal number of degrees, whatever be the radii of these arcs.

54. The ratios which exist between the trigonometrical lines of these arcs, and the radii of the circles to which they belong, depend, in like manner, upon the number of degrees in the arcs; thus, MP, M′P′, M″P″, &c. being the sines of similar arcs, we have

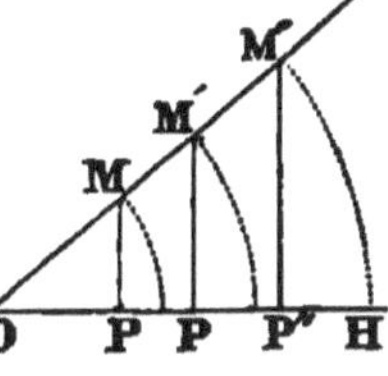

$$\frac{MP}{OM}=\frac{M'P'}{OM'}=\frac{M''P''}{OM''}, \text{ \&c.}$$

Hence, when an angle is given, the sine itself is not determined, but the ratio of the sine to the radius, as is expressed in the above ratios.

55. The same remark applies to cosines, tangents, &c. Hence, in all trigonometrical calculations, it is not the *absolute* lengths of the trigonometrical lines, but their ratios to the radius of the circle to which they belong, that must be considered; and these ratios only are introduced into the operations. For this purpose the method is simple. It is only necessary to take as *unity* the radius of the circle in which the trigonometrical lines are considered; then the numerical values of these lines will not differ from the ratios themselves. We sometimes give to these ratios the designation of *natural sines, natural cosines*, &c.

56. The simple form of the ratios which the trigonometrical lines assume when the radius is 1, tends to simplify very much the trigonometrical calculations. In the *fundamental* formulæ no hypothesis will be made upon the value of the radius, but it will be always designated in these formulæ by r.

57. When results are obtained in which the radius is taken as 1, we may readily modify them for any value of the radius. For, from what has been said, it is evident that ratios of the sines, cosines, &c., to a radius r, must be equal to the sines, cosines, &c., when the radius is unity; hence, in these results, it is only necessary to change sin a, tang b &c., into $\frac{\sin a}{r}$, $\frac{\text{tang}\, b}{r}$, &c.

For example, suppose that we have found between the arcs a and b, the relation

$$\text{tang}\, b = \frac{1 - \cos a}{1 + \sin a},$$

making the substitutions, we have

$$\frac{\text{tang}\, b}{r} = \frac{1 - \frac{\cos a}{r}}{1 + \frac{\sin a}{r}}$$

and reducing, we shall have, without making any hypothesis upon the value of the radius r,

$$\text{tang}\, b = \frac{r\,(r - \cos a)}{r + \sin a}.$$

58. We must be careful not to suppose that there is a determinate length which corresponds to the radius 1, and another corresponding to the radius r, just as there are distances equal to 1 yard, 2 yards, &c. In fact, each trigonometrical line of a given angle is expressed by different numbers, according to the hypothesis made upon the radius, but these numbers have always a constant ratio to that which represents the radius, and it is the ratio which alone enters into the calculations.

Relations of the Trigonometrical Lines to each other.

59. The triangles of the figure make known the relations of the six trigonometrical lines to each other.

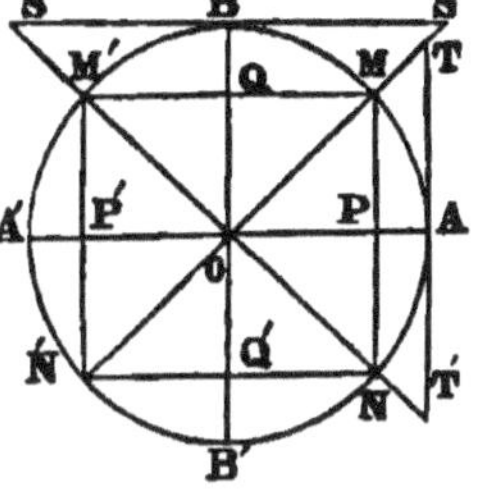

1st. In the first place, the right-angled triangle O M P gives

$$\overline{MP}^2 + \overline{OP}^2 = \overline{OM}^2.$$

And if we make the arc $AM = a$, the radius $OM = r$, and substitute for the several lines their trigonometrical designations, viz., $MP = \sin a$, $OP = \cos a$, &c., we have for the first relation

$$\sin^2 a + \cos^2 a = r^2. \qquad (8)$$

2d. The similar triangles O M P, O T A, give the proportions

$$\frac{AT}{MP} = \frac{OA}{OP}; \quad \frac{OT}{OM} = \frac{OA}{OP};$$

and the similar triangles O M Q, O S B, give the proportions

$$\frac{BS}{MQ} = \frac{OB}{OQ}; \quad \frac{OS}{OM} = \frac{OB}{OQ}.$$

Making the necessary substitutions, we have the following relations,

$$\frac{\text{tang } a}{\sin a} = \frac{r}{\cos a}, \text{ or, tang } a = \frac{r \sin a}{\cos a} \qquad (9)$$

$$\frac{\sec a}{r} = \frac{r}{\cos a}, \text{ or, } \sec a = \frac{r^2}{\cos a} \qquad (10)$$

$$\frac{\cot a}{\cos a} = \frac{r}{\sin a}, \text{ or, } \cot a = \frac{r \cos a}{\sin a} \qquad (11)$$

$$\frac{\text{cosec } a}{r} = \frac{r}{\sin a}, \text{ or, cosec } a = \frac{r^2}{\sin a} \qquad (12)$$

60. Formula (8) serves to determine the sine in terms of the cosine, and reciprocally. If we assume the sin a as known, we shall have

$$\cos a = \pm \sqrt{r^2 - \sin^2 a}$$

4

We obtain two equal values for the cosine, with contrary signs, because for the same sine, O Q, for example, there are two cosines O P and O P′, which are equal and opposite to each other.

61. Formulæ (9), (10), (11), (12), make known the values of the tangent, secant, cotangent, and cosecant, when we know the sine and cosine.

62. To make an application of these formulæ, let us take the value of sin 30° = ½r, which has been found (Art. 13). With this value, and by the aid of the above formulæ, we may readily calculate the cos 30°, and then, the tangent 30°, the sec 30°, &c. Observing that the complement of 30° is 60°, we form the following table:

$$\sin\ 30° = \cos 60° = \frac{r}{2}, \qquad \cos 30° = \sin 60° = \frac{r\sqrt{3}}{2}.$$

$$\text{tang}\ 30° = \cot 60° = \frac{r\sqrt{3}}{3}, \quad \cot 30° = \text{tang}\ 60° = r\sqrt{3}.$$

$$\sec\ 30° = \text{cosec}\ 60 = \frac{2r\sqrt{3}}{3}, \quad \text{cosec}\ 30° = \sec 60° = 2r.$$

63. Although deduced from a figure in which the arc a is $<90°$, the formulæ of (Art. 59) are not the less general. This will be evident, if we only consider the absolute values of the trigonometrical lines; for, these lines always form right-angled similar triangles, from which we may deduce results corresponding with those found in the article cited. With respect to the algebraic signs of the lines, it is manifest, that formula (8) still holds good, since it contains *squares* only; it remains then to examine whether the other formulæ always give for the tangent, secant, &c., the algebraic signs which correspond to their positions.

64. In the first quadrant, that is, from 0 to 90°, the sine and the cosine being positive, the four formulæ give positive values, as they should do. In the second quadrant, the sine is positive, and the cosine negative, and therefore the values of the tangent, secant, and cotangent are negative, while that of the cosecant remains positive; and the figure shows that these are the signs which these lines should have. In the third quadrant, the sine and cosine are negative; hence, the

values of the tangent and cotangent are positive, whilst the values of the secant and cosecant are negative, and these signs correspond with the positions which the four lines take. In the fourth quadrant, the sine is negative, and the cosine positive; the values of the tangent, cotangent, and cosecant are negative, and that of the secant is positive, still corresponding with the indications from the figure.

Beyond 360°, the values of the sine and cosine correspond in magnitude and in sign, for any arc $360° + a$, with those of the arc a; and these results are confirmed by the four formulæ. And, it will also follow, that the tangent, secant, &c., must have the same values for an arc of $360° + a$, as for an arc a.

65. Let us now suppose the arcs negative. Since $\sin(-a) = -\sin a$, and $\cos(-a) = \cos a$, (Art 38); it follows, that in changing the sign of the arc, the values given by the formulæ, for the tangent, cotangent, and cosecant, take contrary signs, without changing their magnitude, while the secant remains the same. These results also correspond with the indications of the figure.

66. Finally, it might be apprehended that the formulæ would fail for the arc 0, 90°, 180°, &c., because, in these cases, the triangles cease to exist. But it is easy to see that they still give results which correspond to these values of the arcs. For example, if we make $a = 90°$, we shall have $\sin 90° = r$, $\cos 90° = 0$; hence, $\text{tang } 90° = \infty$, $\sec 90° = \infty$, $\cot 90° = 0$, $\text{cosec } 90° = r$: values which we ought to have. It is proper to observe, that the value $\text{tang } 90° = \infty$, should be taken with the doubtful sign $\pm$; for, it is, at the same time, the limit of positive tangents, which we obtain by increasing the arc from 0 to 90°, and the limit of the negative tangents, which we obtain, by decreasing the arc from 180° to 90°. The same remark applies to other trigonometrical lines capable of becoming infinite.

67. We conclude therefore that the five formulæ we have been discussing are general, and are not limited by any exceptions.

68. Having demonstrated the generality of formulæ (9) and (10), we might have concluded from them that of (11) and (12), without further proof; for the latter can be deduced from the former by substituting $90° - a$, for a.

69. In general, whenever a relation between trigonometrical lines has been demonstrated for all possible values of arcs, we may substitute for these arcs, their complements, which only requires that we should change the sines, tangents, and secants, into cosines, cotangents, and cosecants, and reciprocally.

70. The five relations first discussed may be used to find others. We proceed to exhibit the most remarkable.

1st. Multiplying together formulæ (9) and (11), we have

$$\text{tang } a \times \cot a = r^2 \qquad (13)$$

which shows, *that the radius is a mean proportional between the tangent and cotangent.* This could be at once deduced from the similar triangles O T A, O S B.

2d. Formula (9), gives

$$r^2 + \text{tang}^2 a = r^2 + \frac{r^2 \sin^2 a}{\cos^2 a} = \frac{r^2 (\sin^2 a + \cos^2 a)}{\cos^2 a}$$

But $\sin^2 a + \cos^2 a = r^2$, and $\sec^2 a = \frac{r^4}{\cos^2 a}$, hence

$$r^2 + \text{tang}^2 a = \sec^2 a, \qquad (14)$$

a result which is evident from the right-angled triangle O T A.

We deduce, by an analogous process, the other formula

$$r^2 + \cot^2 a = \text{cosec}^2 a, \qquad (15)$$

which may also be obtained from formula (14), by substituting for a, $90° - a$.

3d. Formulæ (10) and (12) give

$$\frac{1}{\sec a} = \frac{\cos a}{r^2}, \quad \frac{1}{\text{cosec } a} = \frac{\sin a}{r^2}.$$

Squaring these equations and adding the results, member by member, and observing that $\sin^2 a + \cos^2 a = r^2$, we have

$$\frac{1}{\sec^2 a} + \frac{1}{\text{cosec}^2 a} = \frac{1}{r^2} \qquad (16)$$

71. In general, one of the six trigonometrical lines being given, the relations (8), (9), (10), (11), (12), will make known the five others; and the operation becomes a simple resolution of equations.

For example, if we wish to find the sine and cosine by means of the tangent, we make use of equations (8) and (9)

$$\sin^2 a + \cos^2 a = r^2, \text{ tang } a = \frac{r \sin a}{\cos a}$$

and deduce from them the values of sin a and cos a. The second gives $r^2 \sin^2 a = \text{tang}^2 a \cos^2 a$; then, by means of the first, we readily obtain

$$\sin a = \frac{\pm r \text{ tang } a}{\sqrt{r^2 + \text{tang}^2 a}}; \cos a = \frac{\pm r^2}{\sqrt{r^2 + \text{tang}^2 a}} \quad (17)$$

The double sign $\pm$ indicates that there are two sines and two cosines, equal and oppositely situated, which correspond to the same tangent; as is also shown by the figure. Care must be taken that the superior signs are taken together, and the inferior signs together, otherwise we should not have

$$\frac{r \sin a}{\cos a} = \text{tang } a.$$

Formulæ to find the sin (a + b) *and cos* (a + b), *also sin* (a — b) *and cos* (a — b).

72. *Knowing the sines and cosines of two arcs* a *and* b, *it is required to find the sine and cosine of the sum of these arcs and of the difference of these arcs.*

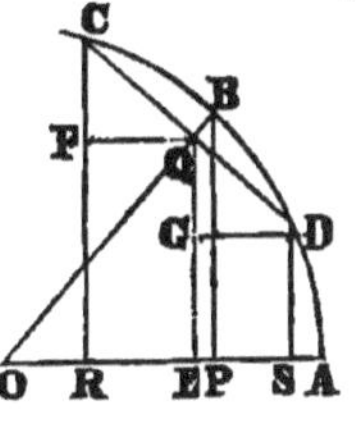

Let A B = a, and B C = B D = b, be the two given arcs. Draw the chord C D, and the radius O B perpendicular to the chord at the middle point Q; draw also the radius O A, and the perpendiculars B P, C R, D S. We shall have

B P = sin a, O P = cos a, C Q = sin b, O Q = cos b,
A C = $a + b$, C R = sin $(a + b)$, O R = cos $(a + b)$,
A D = $a - b$, D S = sin $(a - b)$, O S = cos $(a - b)$.

Now, draw Q E perpendicular to O A, and Q F, and D G, parallel to O A. The triangles C Q F, Q D G, will be equal, having the angles equal, and the side Q C = D Q: hence, D G = Q F, and G Q = C F, and we have

$$\sin(a+b) = \text{CR} = \text{FR} + \text{CF} = \text{QE} + \text{CF},$$
$$\cos(a+b) = \text{OR} = \text{OE} - \text{ER} = \text{OE} - \text{QF},$$
$$\sin(a-b) = \text{DS} = \text{QE} - \text{QG} = \text{QE} - \text{CF},$$
$$\cos(a-b) = \text{OS} = \text{OE} + \text{DG} = \text{OE} + \text{QF}.$$

The triangles O B P, and O Q E, are similar, since B P and Q E are parallel; and O B P is also similar to C Q F, because of the perpendicularity of their sides: hence we have

$$\frac{\text{QE}}{\text{BP}} = \frac{\text{OQ}}{\text{OB}}, \quad \frac{\text{OE}}{\text{OP}} = \frac{\text{OQ}}{\text{OB}},$$
$$\frac{\text{CF}}{\text{OP}} = \frac{\text{CQ}}{\text{OB}}, \quad \frac{\text{QF}}{\text{BP}} = \frac{\text{CQ}}{\text{OB}}.$$

From these equalities, we deduce

$$\text{QE} = \frac{\sin a \cos b}{r}, \quad \text{OE} = \frac{\cos a \cos b}{r},$$
$$\text{CF} = \frac{\cos a \sin b}{r}, \quad \text{QF} = \frac{\sin a \sin b}{r}.$$

Substituting these values in $\sin(a+b)$, $\cos(a+b)$, &c., we get

$$\sin(a+b) = \frac{\sin a \cos b + \cos a \sin b}{r}, \qquad (18)$$

$$\cos(a+b) = \frac{\cos a \cos b - \sin a \sin b}{r}, \qquad (19)$$

$$\sin(a-b) = \frac{\sin a \cos b - \cos a \sin b}{r}, \qquad (20)$$

$$\cos(a-b) = \frac{\cos a \cos b + \sin a \sin b}{r}. \qquad (21)$$

73. The figure employed seems to attach certain restrictions upon these formulæ, since it is assumed that a and b are *positive* arcs, and $a + b < 90°$, and that $a > b$ in the expressions for $\sin(a-b)$ and $\cos(a-b)$. We may readily modify the constructions to correspond with these various cases, but as these cases are numerous, the following process will be found preferable.

1st. The restriction $a > b$ may be removed from the formulæ (20) and (21). For when $a < b$, we know (Art. 38), that we have

$$\sin(a-b) = -\sin(b-a), \text{ and } \cos(a-b) = \cos(b-a).$$

But, b being greater than a, we may obtain sin $(b-a)$ and cos $(b-a)$ from the formulæ (20) and (21), by changing a into b, and b into a: but this substitution will only cause the sign of the first to be changed, while the second will remain the same: we shall, therefore, have, for sin $(a-b)$ and cos $(a-b)$, the same formulæ as in the case of $a > b$. Hence, these formulæ apply to all cases in which a and b are positive, and $a + b < 90°$, and we may therefore give to a and b any values we please between 0 and 45°.

2d. Since the formulæ which embrace $(a-b)$ may be deduced from those which express sin $(a+b)$, and cos $(a+b)$, by changing b into $-b$, it follows, that formulæ (18) and (19) are true for all values of a between 0 and 45°, and for all values of b between $-45°$ and $+45°$. Further, these formulæ also suit negative values of a, taken from 0 to $-45°$. For, representing by α an arc $< 45°$, and making $a = -\alpha$, we shall have

$$\sin(a+b) = \sin(-\alpha+b) = -\sin(\alpha-b),$$
$$\cos(a+b) = \cos(-\alpha+b) = \cos(\alpha-b).$$

The arcs α and $-b$ are within the limits for which the formulæ (18) and (19) have been demonstrated: therefore,

$$\sin(a+b) = -\sin(\alpha-b) = \frac{-\sin\alpha\cos b + \cos\alpha\sin b}{r},$$

$$\cos(a+b) = \cos(\alpha-b) = \frac{\cos\alpha\cos b + \sin\alpha\sin b}{r}$$

Since $a = -\alpha$, we have (Art. 38) $\sin\alpha = -\sin a$, $\cos\alpha = \cos a$, and, hence, these formulæ reduce to (18) and (19).

3d. We will now prove that in formulæ (18) and (19), we may extend indefinitely the positive and negative limits of a and b. Make $a = 90° + \alpha$, α being any arc whatever between $-45°$ and $+45°$: we shall have, on taking the complements,

$$\sin(a+b) = \sin(90°+\alpha+b) = \cos(-\alpha-b) = \cos(\alpha+b) = \frac{\cos\alpha\cos b - \sin\alpha\sin b}{r},$$

$$\cos(a+b) = \cos(90°+\alpha+b) = \sin(-\alpha-b) = -\sin(\alpha+b) = \frac{-\sin\alpha\cos b - \cos\alpha\sin b}{r}$$

But since $a = 90^\circ + \alpha$, we have, by known reductions,

$$\sin a = \sin (90^\circ + \alpha) = \cos (-\alpha) = \cos \alpha,$$
$$\cos a = \cos (90^\circ + \alpha) = \sin (-\alpha) = -\sin \alpha.$$

Therefore, we may substitute for $\cos \alpha$, the $\sin a$, and for $\sin \alpha$, $\cos a$, which again reduces to the formulæ (18) and (19). Now, taking α between -45° and $+45^\circ$, the arc $90^\circ + \alpha$, or a, passes through all magnitudes from 45° to 135°, and the positive limit of a is extended to 135°. Repeating the same reasoning, it is evident that this limit may be again extended from 90° to infinity.

74. The demonstration given (2d), to show that if the formulæ (18) and (19) are true for positive values of $a < 45^\circ$, they are also true for the same values taken negatively, may be applied to the case in which the positive limit of a is greater than 45°. Therefore, since we have seen that they are true for all positive values of a, they are also for all negative values.

75. As to the arc b, it is evident it is subject to the same reasoning that has been applied to a, and we may also extend to infinity each of its limits. Therefore, formulæ (18) and (19), as well as formulæ (20) and (21), are proved for all values of the arcs a and b.

Formulæ for the Multiplication and Division of Arcs.

76. Hereafter, the radius will always be considered as unity, so that we shall have $r = 1$, and under this hypothesis formulæ (8), (9), (10), (11), (12), (18), (19), (20), (21) become

$$\sin^2 a + \cos^2 a = 1,\ \text{tang}\ a = \frac{\sin a}{\cos a},\ \cot a = \frac{\cos a}{\sin a},\ \sec a = \frac{1}{\cos a},$$

$$\text{cosec}\ a = \frac{1}{\cos a},\ \sin (a \pm b) = \sin a \cos b \pm \cos a \sin b,$$

$$\cos (a \pm b) = \cos a \cos b \mp \sin a \sin b.$$

77. If now we make $b = a$ in the expressions for $\sin (a + b)$ and $\cos (a + b)$, we have

$$\sin 2a = 2 \sin a \cos a, \qquad (1)$$
$$\cos 2a = \cos^2 a - \sin^2 a, \qquad (2)$$

which enable us to determine the sine and cosine of double of an arc, when we know the sine and cosine of that arc.

78. Making $b=2a$, the same expressions give

$$\sin 3a = \sin a \cos 2a + \cos a \sin 2a,$$
$$\cos 3a = \cos a \cos 2a - \sin a \sin 2a;$$

and substituting for $\sin 2a$ and $\cos 2a$, their values, and remembering that $\sin^2 a + \cos^2 a = 1$, we have

$$\sin 3a = 3\sin a - 4\sin^3 a, \quad (3)$$
$$\cos 3a = 4\cos^3 a - 3\cos a. \quad (4)$$

By the same process we may deduce the expressions for $\sin 4a$, $\sin 5a$, &c., and for $\cos 4a$, $\cos 5a$, &c., or of any other *multiple* of a.

79. The formulæ for the *division* of arcs are readily deduced. Thus, to find the sine and cosine of *half* an arc, make $a = \frac{1}{2}a$ in formulæ (1) and (2); we have

$$2\sin\tfrac{1}{2}a\cos\tfrac{1}{2}a = \sin a, \quad (5)$$
$$\cos^2\tfrac{1}{2}a - \sin^2\tfrac{1}{2}a = \cos a, \quad (6)$$

and we have also

$$\cos^2\tfrac{1}{2}a + \sin^2\tfrac{1}{2}a = 1. \quad (7)$$

If we suppose the value of $\cos a$ to be known, we have only to resolve the equations (6) and (7) to get the values of $\sin\frac{1}{2}a$ and $\cos\frac{1}{2}a$. We get

$$\sin\tfrac{1}{2}a = \sqrt{\frac{1-\cos a}{2}}, \qquad \cos\tfrac{1}{2}a = \sqrt{\frac{1+\cos a}{2}}. \quad (8)$$

80. It is proper to note that these values of $\sin\frac{1}{2}a$ and $\cos\frac{1}{2}a$ should be affected with the double sign ± 1. We may readily see why these two equal values with contrary signs should exist, for we observe that it is not the arc a itself which enters into these values, but its *cosine*, so that the formulæ should make known, at the same time, the sine and cosine of the half arc for all arcs which have the same cosine. These arcs are made known by the formula, Art. 44,

$$x = 2k\pi \pm \alpha,$$

in which α denotes the smallest positive arc corresponding to the given cosine, π the semi-circumference, and k any whole number, positive or negative. We ought then to find for $\sin\frac{1}{2}a$ and $\cos\frac{1}{2}a$, the values included in

$$\sin(k\pi \pm \tfrac{1}{2}\alpha) \text{ and } \cos(k\pi \pm \tfrac{1}{2}\alpha).$$

If k is *even*, $k\pi$ will be a multiple of 360°, and we may suppress it without altering the value of the sine or cosine, and there will result

$$\sin(\pm \tfrac{1}{2} a) = \pm \sin \tfrac{1}{2} a, \text{ and } \cos(\pm \tfrac{1}{2} a) = \cos \tfrac{1}{2} a.$$

If k is *odd*, we may still suppress $k\pi$, but the signs of the sine and cosine must be changed (Art. 35), and we have

$$-\sin(\pm \tfrac{1}{2} a) = \mp \sin \tfrac{1}{2} a, \text{ and } -\cos(\pm \tfrac{1}{2} a) = -\cos \tfrac{1}{2} a.$$

We see, therefore, that we have in fact two equal values for $\sin \frac{1}{2} a$ and $\cos \frac{1}{2} a$, affected with contrary signs.

81. If we have the *sine* given instead of the *cosine*, it will be only necessary to substitute in formulæ (8) for $\cos a$, its value, $\sqrt{1 - \sin^2 a}$. Since this new radical also has the double sign $\pm$, we should have four values for each of the unknown quantities $\sin \frac{1}{2} a$, and $\cos \frac{1}{2} a$.

But we may obtain these values under another form. Resuming equations (5) and (7),

$$2 \sin \tfrac{1}{2} a \cos \tfrac{1}{2} a = \sin a,$$
$$\cos^2 \tfrac{1}{2} a + \sin^2 \tfrac{1}{2} a = 1,$$

we deduce from them the values of $\sin \frac{1}{2} a$ and $\cos \frac{1}{2} a$. Adding the first equation to the second, and afterwards subtracting the first from the second, we have, when we extract the square root of both numbers of the resulting equations,

$$\cos \tfrac{1}{2} a + \sin \tfrac{1}{2} a = \sqrt{1 + \sin a},$$
$$\cos \tfrac{1}{2} a - \sin \tfrac{1}{2} a = \sqrt{1 - \sin a};$$

from which we readily deduce the values sought:

$$\sin \tfrac{1}{2} a = \tfrac{1}{2} \sqrt{1 + \sin a} - \tfrac{1}{2} \sqrt{1 - \sin a}, \quad (9)$$
$$\cos \tfrac{1}{2} a = \tfrac{1}{2} \sqrt{1 + \sin a} + \tfrac{1}{2} \sqrt{1 - \sin a}. \quad (10)$$

Since there are two radicals, each of these expressions has four values, as might have been anticipated. For, these expressions should make known the sine and cosine of the half arc for all arcs which have the same sine.

But these arcs are made known by the formulæ, (Art. 44,)

$$x = 2k\pi + a, \quad x = (2k + 1)\pi - a;$$

hence, the expressions for $\sin \frac{1}{2} a$ and $\cos \frac{1}{2} a$, must give the sine and cosine for all arcs represented by

$$k\pi + \tfrac{1}{2}a, \text{ and } (k + \tfrac{1}{2})\pi - \tfrac{1}{2}a.$$

But we may suppress $k\pi$, taking care to retain or change the signs of the sine and cosine, according as k is even or odd. Hence, we should have for $\sin \frac{1}{2} a$ and also for $\cos \frac{1}{2} a$, four values, viz.,

$$\sin \tfrac{1}{2} a = \pm \sin \tfrac{1}{2} a, \text{ and } \sin \tfrac{1}{2} a = \pm \sin (\tfrac{1}{2}\pi - \tfrac{1}{2} a),$$
$$\cos \tfrac{1}{2} a = \pm \cos \tfrac{1}{2} a, \text{ and } \cos \tfrac{1}{2} a = \pm \cos (\tfrac{1}{2}\pi - \tfrac{1}{2} a).$$

We see, further, that they are equal, two and two, with contrary signs. If $a = 90°$, we have $\frac{1}{2}a = 45°$, $\frac{1}{2}\pi - \frac{1}{2}a = 45°$; and the four values are reduced to two.

82. Another observation suggests itself. Sine $\pi = 180°$, it follows that the two arcs $\frac{1}{2}a$, and $\frac{1}{2}\pi - \frac{1}{2}a$, are complements of each other. Hence, the preceding values may also be presented thus:

$$\sin \tfrac{1}{2} a = \pm \sin \tfrac{1}{2} a, \quad \sin \tfrac{1}{2} a = \pm \cos \tfrac{1}{2} a,$$
$$\cos \tfrac{1}{2} a = \pm \cos \tfrac{1}{2} a, \quad \cos \tfrac{1}{2} a = \pm \sin \tfrac{1}{2} a.$$

That is, the values of $\sin \frac{1}{2} a$ and $\cos \frac{1}{2} a$ are the same, and this is indicated by formulæ (9) and (10).

83. One difficulty remains to be explained. When the arc a and its sine are known, to ascertain which of the four solutions must be selected for $\sin \frac{1}{2} a$ or for $\cos \frac{1}{2} a$. For brevity, let us consider $\sin \frac{1}{2} a$ only. Taking the radicals with their different signs, the four values of $\sin \frac{1}{2} a$ may be written,

$$\sin \tfrac{1}{2} a = \pm \tfrac{1}{2} (\sqrt{1 + \sin a} - \sqrt{1 - \sin a}),$$
$$\sin \tfrac{1}{2} a = \pm \tfrac{1}{2} (\sqrt{1 + \sin a} + \sqrt{1 - \sin a}).$$

In the first place, it is evident, that the two first are equal with contrary signs, and the same for the two last. If we square the first, the result is $< \frac{1}{2}$, while the square of the other gives a result $> \frac{1}{2}$. But we know that $\sin 45° = \cos 45° = \frac{1}{2}$; hence, neglecting the signs, the two first values are less than the sin 45°, while the two last are greater than sin 45°.

But, when the arc is given, we may readily determine *à priori* whether the sine of the half arc is positive or negative, and whether it should be greater or less than 45°. Thus, all indetermination vanishes. The same reasoning applies to the cosine. If, for example, we take $a < 90°$:

sin $\frac{1}{2}a$ must be positive and less than sin 45°: cos $\frac{1}{2}a$ must also be positive, but greater than 45°: and those values of (9) and (10) must be taken which fulfil these conditions.

84. Let us now consider the *trisection* of arcs. Substituting for a, $\frac{1}{3}a$, formulæ (3) and (4) become

$$\sin a = 3 \sin \tfrac{1}{3}a - 4 \sin^3 \tfrac{1}{3}a,$$
$$\cos a = 4 \cos^3 \tfrac{1}{3}a - 3 \cos \tfrac{1}{3}a.$$

If now the cos a be given, and we wish to find the value of cos $\frac{1}{3}a$, make cos $a = b$, cos $\frac{1}{3}a = z$, and the second of these equations becomes

$$z^3 - \tfrac{3}{4}z + \tfrac{1}{4}b = 0 \qquad (11)$$

The resolution of this equation determines the value of cos $\frac{1}{3}a$. Without entering into the algebraic details, we will show *à priori* that the three values of z in this equation are real.

Since the cosine is given, the formula of the arcs corresponding to this cosine is $2k\pi \pm a$ (Art. 44); hence, equation (11) has for its values all those which are embraced in the expression

$$z = \cos\frac{2k\pi \pm a}{3}.$$

The whole number k, can only have one of the three forms $3n$, $3n+1$, $3n-1$, (n being also a whole number). Make $k = 3n$, $k = 3n+1$, $k = 3n-1$, successively; we shall have, suppressing useless circumferences,

$$z = \cos\frac{3n.2\pi \pm a}{3} = \cos(2n\pi \pm \frac{a}{3}) = \cos(\pm\frac{a}{3}) = \cos\frac{a}{3};$$

$$z = \cos\frac{(3n+1).2\pi \pm a}{3} = \cos(2n\pi + \frac{2\pi}{3} \pm \frac{a}{3}) = \cos(\frac{2\pi}{3} \pm \frac{a}{3});$$

$$z = \cos\frac{(2n-1).2\pi \pm a}{3} = \cos(-\frac{2\pi}{3} \pm \frac{a}{3}) = \cos(\frac{2\pi}{3} \pm \frac{a}{3});$$

the two last values are the same as the preceding; then there are in all three different values, as follows:

$$z = \cos\frac{a}{3}, \quad z = \cos(\frac{2\pi}{3} + \frac{a}{3}), \quad 2 = \cos(\frac{2\pi}{3} - \frac{a}{3}).$$

It sometimes happens that two of these values are equal to each other: thus, the first is equal to the third, when $a = \pi$.

Formulæ relative to Tangents.

85. Let us now find *the tangent of the sum and difference of two arcs, when the tangents of these arcs are known.*

According to the relations established (Art. 59), we have in the first place,

$$\text{tang}\,(a+b) = \frac{\sin\,(a+b)}{\cos\,(a+b)}.$$

Substituting for $\sin\,(a+b)$ and $\cos\,(a+b)$ their values (Art. 72), we have

$$\text{tang}\,(a+b) = \frac{\sin a \cos b + \cos a \sin b}{\cos a \cos b + \sin a \sin b}.$$

To express this value of $\text{tang}\,(a+b)$ in terms of tangents only, divide the numerator and denominator by $\cos a \cos b$, there results

$$\text{tang}\,(a+b) = \frac{\dfrac{\sin a}{\cos a} + \dfrac{\sin b}{\cos b},}{1 - \dfrac{\sin a \sin b}{\cos a \cos b}.}$$

But $\dfrac{\sin a}{\cos a} = \text{tang}\,a$, and $\dfrac{\sin b}{\cos b} = \text{tang}\,b$; substituting these values, we have

$$\text{tang}\,(a+b) = \frac{\text{tang}\,a + \text{tang}\,b}{1 - \text{tang}\,a\,\text{tang}\,b}. \qquad (1)$$

By a like process, we find for $\text{tang}\,(a-b)$

$$\text{tang}\,(a-b) = \frac{\text{tang}\,a - \text{tang}\,b}{1 - \text{tang}\,a\,\text{tang}\,b}. \qquad (2)$$

86. Let $b = a$, we shall have for the tangent of *double* arcs,

$$\text{tang}\,2a = \frac{2\,\text{tang}\,a}{1 - \text{tang}^2 a}. \qquad (3)$$

87. To determine tangent of $\frac{1}{2}a$, in terms of $\text{tang}\,a$, substitute in (3) $\frac{1}{2}a$ for a, we get

$$\frac{2\,\text{tang}\,\frac{1}{2}a}{1 - \text{tang}^2 \frac{1}{2}a} = \text{tang}\,a,$$

which reduces to this equation of the second degree

$$\text{tang}^2 \tfrac{1}{2} a + \frac{2}{\text{tang } a} \text{tang } \tfrac{1}{2} a - 1 = 0, \qquad (4)$$

from which we deduce

$$\text{tang } \tfrac{1}{2} a = \frac{1}{\text{tang } a} (-1 \pm \sqrt{1 + \text{tang}^2 a}).$$

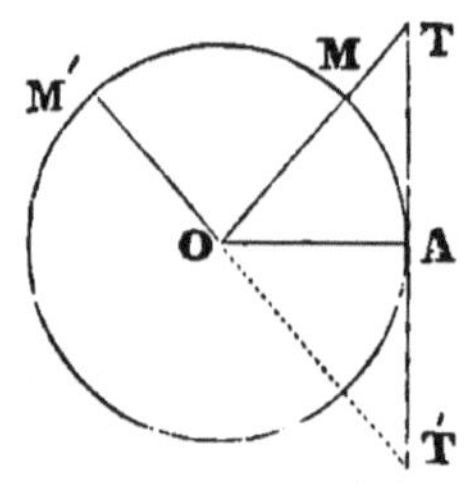

Equation (4) having -1 for its last term, we know, without solving it, that the product of the two values of tang $\frac{1}{2} a$ is equal to -1. Hence, if A T and A T' are those values, in the situation corresponding to their signs, we must have $AT \times AT' = \overline{OA}^2$, which shows that O A is a mean proportional between A T and A T', and hence the angle T O T' is a right angle; or the arc M M' is equal to 90°.

88. We frequently meet with the following formulæ,

$$\text{tang } \tfrac{1}{2} a = \sqrt{\frac{1 - \cos a}{1 + \cos a}}, \qquad (5)$$

$$\text{tang } \tfrac{1}{2} a = \frac{\sin a}{1 + \cos a}, \qquad (6)$$

$$\text{tang } \tfrac{1}{2} a = \frac{1 - \cos a}{\sin a}. \qquad (7)$$

These formulæ are easily deduced from those already known. It is evident that we have, (Art. 76,) (Art. 79,)

$$\text{tang } \tfrac{1}{2} a = \frac{\sin \frac{1}{2} a}{\cos \frac{1}{2} a} = \sqrt{\frac{1 - \cos a}{1 + \cos a}},$$

$$\text{tang } \tfrac{1}{2} a = \frac{\sin \frac{1}{2} a \cos \frac{1}{2} a}{\cos^2 \frac{1}{2} a} = \frac{\sin a}{1 + \cos a},$$

$$\text{tang } \tfrac{1}{2} a = \frac{\sin^2 \frac{1}{2} a}{\sin \frac{1}{2} a \cos \frac{1}{2} a} = \frac{1 - \cos a}{\sin a},$$

Formulæ frequently used.

89. The formulæ for $\sin(a \pm b)$ and $\cos(a \pm b)$, (Art. 72,) lead to a great many formulæ which are often used by astronomers. We will now refer to the most important of these.

Combining formulæ for $\sin(a+b)$ and $\sin(a-b)$ (Art. 72), and those for $\cos(a+b)$ and $\cos(a-b)$, by addition and subtraction, we have

$$2\sin a \cos a = \sin(a+b) + \sin(a-b),$$
$$2\cos a \sin b = \sin(a+b) - \sin(a-b),$$
$$2\cos a \cos b = \cos(a-b) + \cos(a+b),$$
$$2\sin a \sin b = \cos(a-b) - \cos(a+b).$$

These formulæ may be used to transform the product of a sine and cosine, or of two cosines, or of two sines, into a sum or difference of two trigonometrical lines.

90. Let p and q denote any two arcs whatever; make $a+b=p$, and $a-b=q$; we shall have $a = \frac{1}{2}(p+q)$, $b = \frac{1}{2}(p-q)$. Substituting these values in the preceding formulæ, we get

$$\sin p + \sin q = 2\sin \tfrac{1}{2}(p+q) \cos \tfrac{1}{2}(p-q),$$
$$\sin p - \sin q = 2\cos \tfrac{1}{2}(p+q) \sin \tfrac{1}{2}(p-q),$$
$$\cos p + \cos q = 2\cos \tfrac{1}{2}(p+q) \cos \tfrac{1}{2}(p-q),$$
$$\cos q - \cos p = 2\sin \tfrac{1}{2}(p+q) \sin \tfrac{1}{2}(p-q).$$

These formulæ are of frequent use in logarithmic calculations, to change a sum or difference into a product.

91. Finally, by division, and remembering that, in general, $\frac{\sin A}{\cos A} = \text{tang } A = \frac{1}{\cos A}$ the last formulæ lead to the following,

$$\frac{\sin p + \sin q}{\sin p - \sin q} = \frac{\sin \frac{1}{2}(p+q) \cos \frac{1}{2}(p-q)}{\cos \frac{1}{2}(p+q) \sin \frac{1}{2}(p-q)} = \frac{\text{tang } \frac{1}{2}(p+q)}{\text{tang } \frac{1}{2}(p-q)}$$

$$\frac{\sin p + \sin q}{\cos p + \cos q} = \frac{\sin \frac{1}{2}(p+q)}{\cos \frac{1}{2}(p+q)} = \text{tang } \tfrac{1}{2}(p+q)$$

$$\frac{\sin p + \sin q}{\cos p - \cos q} = \frac{\cos \frac{1}{2}(p-q)}{\sin \frac{1}{2}(p-q)} = \cot \tfrac{1}{2}(p-q)$$

$$\frac{\sin p - \sin q}{\cos p + \cos q} = \frac{\sin \frac{1}{2}(p-q)}{\cos \frac{1}{2}(p-q)} = \text{tang } \tfrac{1}{2}(p-q)$$

$$\frac{\sin p-\sin q}{\cos q-\cos p}=\frac{\cos \frac{1}{2}(p+q)}{\sin \frac{1}{2}(p+q)}=\cot \tfrac{1}{2}(p+q)$$

$$\frac{\cos p+\cos q}{\cos q-\cos p}=\frac{\cos \frac{1}{2}(p+q)\cos \frac{1}{2}(p-q)}{\sin \frac{1}{2}(p+q)\sin \frac{1}{2}(p-q)}=\frac{\cot \frac{1}{2}(p+q)}{\operatorname{tang} \frac{1}{2}(p-q)}.$$

Among these formulæ, we observe that the first may be enunciated as follows: *The sum of the sines of two arcs is to the difference of the same sines, as the tangent of half the sum of the arcs is to the tangent of half their difference.*

92. We sometimes meet with trigonometrical transformations, the origin of which we may not be able to trace. We may assure ourselves of their accuracy, by verifying them. Thus, to verify the relation

$$\sin(a+b)\sin(a-b)=\sin^2 a-\sin^2 b,$$

we substitute for $\sin(a+b)$ and $\sin(a-b)$ their values (Art. 72), and we have

$$\sin(a+b)\sin(a-b)=\sin^2 a\cos^2 b-\cos^2 a\sin^2 b;$$

then putting for $\cos^2 a$ and $\cos^2 b$, their values $1-\sin^2 a$, $1-\sin^2 b$, we may readily deduce the proposed equality.

To verify the relation $\cos a=\dfrac{1-\operatorname{tang}^2 \frac{1}{2}a}{\cos^2 a+\operatorname{tang}^2 \frac{1}{2}a}$, substitute for $\operatorname{tang}\frac{1}{2}a$ its value $\dfrac{\sin \frac{1}{2}a}{\cos \frac{1}{2}a}$, and the second member becomes

$$\frac{\cos^2 \frac{1}{2}a-\sin^2 \frac{1}{2}a}{\cos^2 \frac{1}{2}a+\sin^2 \frac{1}{2}a.}$$

But since $\cos^2 \frac{1}{2}a+\sin^2 \frac{1}{2}a=1$, and $\cos^2 \frac{1}{2}a-\sin^2 \frac{1}{2}a=\cos a$, (Art. 59 and 79,) these substitutions reduce the second member to $\cos a$, as was required.

As a further exercise, the student may verify the following transformations,

$$\cos(a+b)\cos(a-b)=\cos^2 a-\sin^2 b.$$

$$\operatorname{tang}(45°+a)=\frac{1+\operatorname{tang} a}{1-\operatorname{tang} a}$$

$$\cos a=\frac{1}{1+\operatorname{tang} a\operatorname{tang}\frac{1}{2}a}$$

$$\text{tang } a + \text{tang } b = \frac{\sin (a+b)}{\cos a \cos b}.$$

$$\text{tang } a + \text{tang } b + \text{tang } c = \text{tang } a \text{ tang } b \text{ tang } c.$$

The last formula supposes that $a+b+c=180°$, and it proves *that there may be an infinite number of combinations of three quantities, such, that their sum shall be equal to their continued product.*

Geometrical Demonstrations of Formulæ Already Found.

93. It has been observed, that, after establishing *geometrically* the *formulæ* which express sin $(a+b)$ and sin $(a-b)$, we have deduced the other formulæ by the ordinary processes of algebra. From this the conclusion is drawn, that the first being true for all possible arcs, the last are not less general, and in this consists the essence of all analysis. On the contrary, in geometrical constructions, the conclusions drawn from each figure are necessarily limited to the particular construction which is made. Still, as geometry serves to make the truth more apparent, and constitutes a most valuable means of mental discipline for the student, we proceed now to demonstrate in this way the principal results already obtained.

94. *The sine and cosine of an arc being given, to find the sine and cosine of the double arc.*

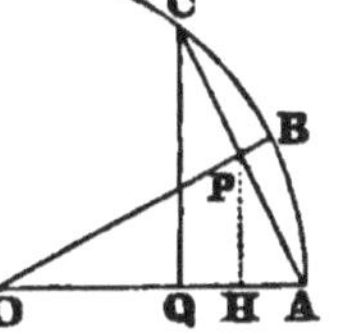

Let arc $AB = BC = a$, and make the constructions indicated in the figure, we have

$\sin a = AP$, $\cos a = OP$, $\sin 2a = CQ = 2PH$.

$\cos 2a = OQ = OH - QH = OH - AH$.

The right-angled triangle OPA gives

$$PH = \frac{AP \times OP}{OA},\ OH = \frac{\overline{OP}^2}{OA},\ AH = \frac{\overline{AP}^2}{OA};$$

consequently, substituting for the lines their trigonometrical designations, and making radius $OA = 1$, we find again the formulæ (1), (2), art. 78, as follows,

$$\sin 2a = 2\,PH = 2 \sin a \cos a,$$
$$\cos 2a = OH - AH = \cos^2 a - \sin^2 a.$$

95. *Having given the cosine of an arc, to find the sine and cosine of the half arc.*

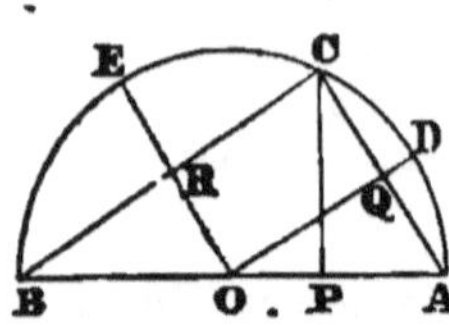

Make the arc $AC = a$, draw CP perpendicular to the diameter AB, and the chords AC and BC, and the radii OD and OE, intersecting the chords at their middle points. Assuming the radius $OA = 1$, we shall have $OP = \cos a$, $AP = 1 - \cos a$, $BP = 1 + \cos a$, $AC = 2 \sin \frac{1}{2} a$, $BC = \cos \frac{1}{2} a$. Since each chord is a mean proportional between the diameter and the adjacent segment, we have

$$\overline{AC}^2 = AB \times AP, \text{ or, } 4 \sin^2 \tfrac{1}{2} a = 2\,(1 - \cos a)$$
$$\overline{BC}^2 = AB \times BP, \text{ or, } 4 \cos^2 \tfrac{1}{2} a = 2\,(1 + \cos a).$$

From these results we deduce the formulæ previously established for $\sin \frac{1}{2} a$, $\cos \frac{1}{2} a$,

$$\sin \tfrac{1}{2} a = \sqrt{\frac{1 - \cos a}{2}},\ \cos \tfrac{1}{2} a = \sqrt{\frac{1 + \cos a}{2}}.$$

96. *The sine and cosine of an arc being given, to find the sine and cosine of three times the given arc.*

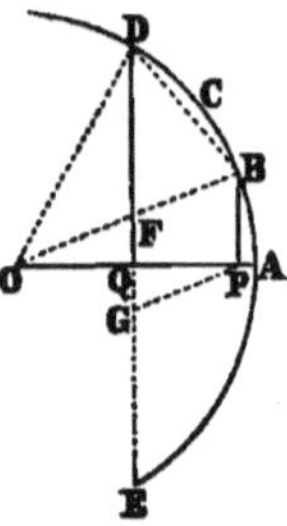

Let the radius $OB = 1$, the arc $AB = BC = CD = a$. The isosceles triangle BOD is similar to BDF, for the angle OBD is common, and the angle BDF, which is measured by the half of BAE or BCD, is equal to BOD. We have, therefore,

$$\frac{BF}{BD} = \frac{BD}{OB}, \text{ hence, } BF = 4 \sin^2 a.$$

Draw PG parallel to BF, we have $PG = BF = 4 \sin^2 a$, and the similar triangles QGP, OBP, give

$$\frac{QG}{BP} = \frac{PG}{OB}, \text{ hence, } QG = 4 \sin^3 a,$$

$$\frac{PQ}{OP} = \frac{PG}{OB}, \text{ hence, } PQ = 4 \sin^2 a \cos a.$$

But we have $\sin 3a = DQ = DF + FG - QG = BD + BP - QG = 3 \sin a - QG$,

$$\cos 3a = OQ = OP - PQ = \cos a - PQ.$$

Substituting the values just found for QG and PQ, we have

$$\sin 3a = 3 \sin a - 4 \sin^3 a$$
$$\cos 3a = \cos a - 4 \sin^2 \cos a.$$

The first of these results corresponds with formula (3), Art. 78, and substituting for $\sin^2 a$ its equivalent $1 - \cos^2 a$, the second reduces to formula (4).

97. *Having given the tangents of two arcs, to find the tangent of their sum, and the tangent of their difference.*

Let the radius $OA = 1$, $AB = a$, $BC = b$. At the extremities of the radii OA and OB draw the tangents AT and BS, and let fall the perpendicular SH on OA. By hypothesis we have given $BR = \text{tang}\, a$, $BS = \text{tang}\, b$; and it is required to find $AT = \text{tang}\,(a + b)$.

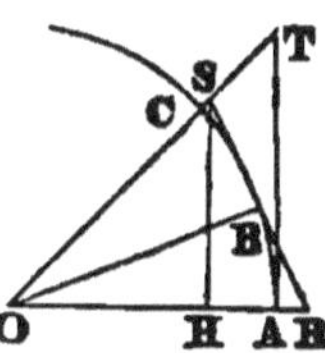

Since the triangles OAT, OHS are similar, we have

$$\frac{AT}{OA} = \frac{SH}{OH}, \text{ hence, } \text{tang}\,(a+b) = \frac{SH}{OH}.$$

Since the triangles SHR and OBR are similar, we deduce SH from the proportion,

$$\frac{SH}{SR} = \frac{OB}{OR}, \text{ or, } SH = \frac{\text{tang}\, a + \text{tang}\, b}{OR}.$$

To find OR, we know, from an established theorem, that we have

$$\overline{SR}^2 = \overline{OR}^2 + \overline{OS}^2 - 2\, OR \times OH.$$

But, $\quad \overline{SR}^2 = (BR + BS)^2 = \overline{BR}^2 + \overline{BS}^2 + 2 BR \times BS.$

Hence, $2\, OR \times OH = \overline{OR}^2 - \overline{BR}^2 + \overline{OS}^2 - \overline{BS}^2 - 2 BR \times BS$

$$= 2\overline{OB}^2 - 2 BR \times BS = 2 - 2\, \text{tang}\, a\, \text{tang}\, b.$$

Consequently,

$$OH = \frac{1 - \text{tang}\, a\, \text{tang}\, b}{OR}.$$

Substituting in tang $(a+b)$, the values of SH and OH, we have the formula already known, Art. 85,

$$\text{tang}\,(a+b)=\frac{\text{tang}\,a+\text{tang}\,b}{1-\text{tang}\,a\ \text{tang}\,b.}$$

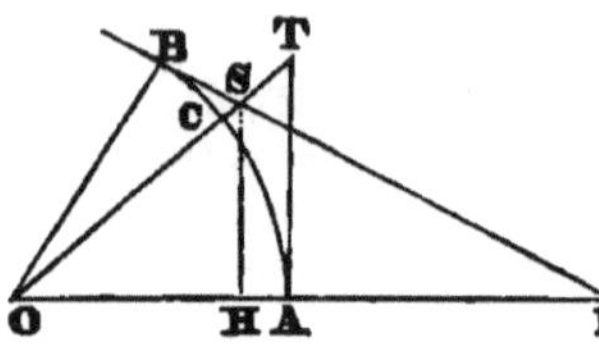

We may readily determine, by a like process, the value of tang $(a-b)$. In this case, $AC=a-b$, and no difficulty exists in immediately deducing the expression. Since RS is equal to tang a — tang b, the second term of the numerators of SH and OH have a change of sign in this case, and we get, as before,

$$\text{tang}\,(a-b)=\frac{\text{tang}\,a-\text{tang}\,b}{1+\text{tang}\,a\ \text{tang}\,b.}$$

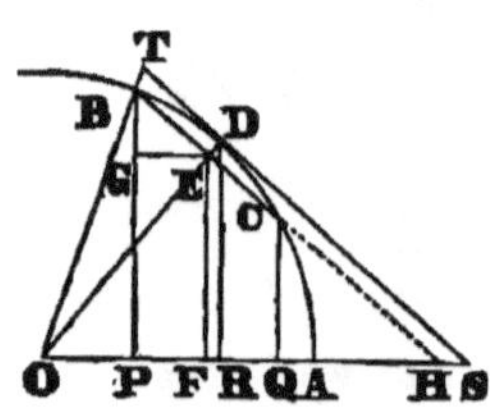

98. To demonstrate geometrically the formulæ

$\sin p+\sin q=2\sin\frac{1}{2}(p+q)\cos\frac{1}{2}(p-q)$

$\sin p-\sin q=2\cos\frac{1}{2}(p+q)\sin\frac{1}{2}(p-q)$.

Take $AB=p$, $AC=q$, draw the chord BC, and the radius OD cutting the chord perpendicularly at its middle point: let fall on OA, the perpendiculars BP, CQ, DR, EF, and draw EG parallel to OA. From the construction we have

$$BP=\sin p,\ CQ=\sin q,\ EF=\frac{\sin p+\sin q}{2},\ BG=\frac{\sin p-\sin q}{2}.$$

$$AD=\tfrac{1}{2}(p+q),\ DR=\sin\tfrac{1}{2}(p+q),\ OR=\cos\tfrac{1}{2}(p+q).$$

$$BD=\tfrac{1}{2}(p-q),\ BE=\sin\tfrac{1}{2}(p-q),\ OE=\cos\tfrac{1}{2}(p-q).$$

But the similar triangles ODR, OEF, BGE, give

$$\frac{EF}{DR}=\frac{OE}{OD},\ \text{and}\ \frac{BG}{OR}=\frac{BE}{OD},$$

hence, $EF=\dfrac{DR\times OE}{OD}$, and $BG=\dfrac{OR\times BE}{OD}$.

Substituting for the different lines their values, doubling these expressions, and making the radius O D = 1, we obtain the required formulæ.

99. *To demonstrate geometrically that the sum of the sines of two arcs is to the difference of these sines, as the tangent of half the sum of the two arcs is to the tangent of half the difference of the two arcs.*

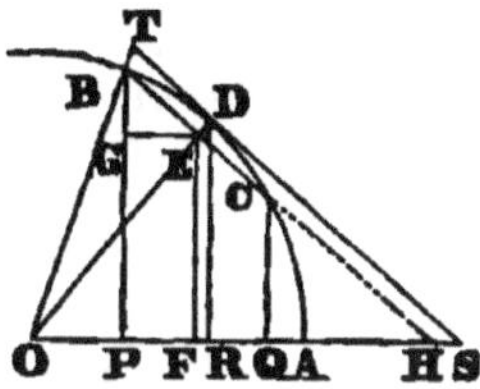

Make the same construction as in the preceding; further, draw the tangent S T at the point D, and terminated in S and T, on the radii O A and O B produced: produce also B C to H. On account of the parallels, we have

$$\frac{\text{EF}}{\text{BG}}=\frac{\text{EH}}{\text{EB}}=\frac{\text{DS}}{\text{DT}}.$$

But we have $2\,\text{EF} = \sin p + \sin q$, $2\,\text{BG} = \sin p - \sin q$, $\text{DS} = \tan \tfrac{1}{2}(p+q)$, $\text{DT} = \text{tang DB} = \text{tang}\ \tfrac{1}{2}(p-q)$; hence

$$\frac{\sin p + \sin q}{\sin p - \sin q} = \frac{\text{tang}\ \frac{1}{2}(p+q)}{\text{tang}\ \frac{1}{2}(p-q)}$$

which was to have been demonstrated.

CHAPTER II.

Trigonometrical Tables. Solution of Triangles.

Construction of Trigonometrical Tables.

100. To realize the advantage of substituting trigonometrical lines for the corresponding angles and arcs, we must have some ready means of knowing the numerical values of these trigonometrical lines, when the angle or arc is given, and reciprocally. For this purpose *tables* are formed, in which the numerical values of the trigonometrical lines are written down opposite to the arcs to which they correspond. We propose now to explain the manner of calculating the numerical values of the sine, cosine, &c., for all arcs, differing from each other by 10″. This is the law of variation in the arcs as calculated in the *Tables of Callet.*

We begin by calculating the value of sin 10″. For this purpose, we remark that the ratio of the circumference to the diameter is $\pi = 3.14159\ 26535\ 89793\ \ldots$. When the radius is taken for *unity*, the semi-circumference is equal to π; and since there are 648000 seconds in 180°, we shall have, in terms of the radius,

$$\text{arc } 10'' = \frac{\pi}{648000} = 0{,}00004\ 84813\ 68110. \text{ - - - } (1)$$

But, for a very small arc, the sine and arc are so nearly equal, we may consider the value just found for an arc of 10″ as a near approximate value for the sin 10″. To explain this and to show the degree of approximation made by substituting the value of the arc of 10″ for that of the sin 10″, we shall demonstrate some preliminary principles.

101. In the first place, we shall prove, *that in the first quadrant, an arc is greater than its sine and less than its tangent.*

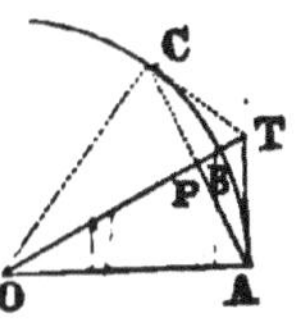

Let A P be the sine of the arc A B, and A T its tangent. Revolve the figure around O T, and suppose the point A falls on C. We shall have arc A C > chord A C, and consequently arc A B > A P; hence, the arc is greater than the sine. We have also arc A C < A T + C T; therefore, arc A B < A T; or, the arc is less than the tangent.

From this, it follows that if $\frac{\text{tang } a}{\sin a}$ differs very little from 1, the ratio of $\frac{a}{\sin a}$ will differ from it still less.

102. In the second place, it can be shown, *that if an arc decreases to zero, the ratio of this arc to its sine will differ from unity by less than any assignable quantity; that is, the limit of this ratio will be unity.*

The formula tang $a = \frac{\sin a}{\cos a}$, gives $\frac{\text{tang } a}{\sin a} = \frac{1}{\cos a}$. But, in diminishing the arc a, (which is supposed to be $< 90°$,) the cosine increases and may approach as near as we please to unity: hence, the ratio $\frac{1}{\cos a}$, or its equal $\frac{\text{tang } a}{\sin a}$, diminishes continually, and has *unity* for its limit.

The arc being > sine and < the tangent, the ratio $\frac{a}{\sin a}$ can neither be < 1 nor $> \frac{\text{tang } a}{\sin a}$. But the ratio $\frac{\text{tang } a}{\sin a}$ may approach unity as near as we please; hence, the ratio $\frac{a}{\sin a}$ has *unity* also for its *limit*. Hence, in taking the value of the arc of 10″ for that of the sin 10″, we have only done what this demonstration shows we had the right to do, by way of a near approximation.

103. To determine the degree of approximation, that we may reject useless decimals. We have $\sin a = 2 \sin \frac{1}{2} a \cos \frac{1}{2} a$. But the inequality tang $\frac{1}{2} a > \frac{1}{2} a$ reduces to $\frac{\sin \frac{1}{2} a}{\cos \frac{1}{2} a} > \frac{1}{2} a$, and gives $2 \sin \frac{1}{2} a > a \cos \frac{1}{2} a$; hence, $\sin a > a \cos^2 \frac{1}{2} a$. But $\cos^2 \frac{1}{2} a = 1 - \sin^2 \frac{1}{2} a$; therefore, $\cos^2 \frac{1}{2} a > 1 - (\frac{1}{2} a)^2$; and hence, finally,

$$\sin a > a - \frac{a^3}{4}.$$

To apply the result to arc of 10″, we shall have, from the value (1) when the 5th decimal figure is increased by 1, arc $10'' < 0.00005$; hence, $\frac{1}{4}$ (arc 10″)^3 $< 0.00000\ 00000\ 00032$. From which we have

$$\sin 10'' > \text{arc } 10'' - \frac{(\text{arc } 10'')^3}{4} > 0{,}00004\ \ 84813\ 68110\ldots -$$
$$0.00000\ 00000\ 00032,$$

$$\text{and } \sin 10'' > 0{,}00004\ 84843\ 68078.$$

We see that the sin 10″ does not commence to differ from the arc 10″ until the 13th decimal; and even then, the 13th decimal in the value of the arc is greater only by 1. Hence, we may take

$$\sin 10'' = 0{,}00004\ 84813\ 681,$$

and may be assured that the error is less than a unit of the 13th order. Indeed, it is evident that the preceding value becomes too small if we take 1 from its last figure, and too large, if we add 1 to it, for then it would exceed the arc.

Placing the value of sin 10″ under the radical $\sqrt{1-\sin^2 10''}$, we find the value of cos 10″ to be

$$\cos 10'' = 0{,}99999\ 99988\ 248.$$

We may now readily determine the sines and cosines of 20″, 30″, 40″, &c., to 45°, by the aid of the known formulæ,

$$\sin(a+b) = \sin a \cos b + \cos a \sin b$$
$$\cos(a+b) = \cos a \cos b - \sin a \sin b.$$

104. The following process which is borrowed from an English Geometrician, *Thomas Simpson*, gives the most rapid mode of making the required calculations.

The formulæ, Art. 89, give

$$\sin(a+b) = 2\cos b \sin a - \sin(a-b)$$
$$\cos(a+b) = 2\cos b \cos a - \cos(a-b).$$

We may regard the arcs $a-b$, a, $a+b$, as three consecutive terms of an arithmetical progression whose common ratio is b: hence, if we represent by t, t', t'', these three terms, we have

$$\sin t'' = 2 \cos b \sin t' - \sin t$$
$$\cos t'' = 2 \cos b \cos t' - \cos t.$$

The first formula shows that, having calculated the values of two consecutive sines, we will find the next sine following, by multiplying the last by $2 \cos b$, and the one before the last by -1, and adding the two products together. The same applies to the cosines.

Thus, to find the values of the sines and cosines for all arcs, differing from each other by $10''$ successively, we make $b = 10''$; and representing by α and β the known value of sin $10''$ and cos $10''$, we shall have

$\sin 0 = 0$	$\cos 0 = 1$
$\sin 10'' = \alpha$	$\cos 10'' = \beta$
$\sin 20'' = 2\beta \sin 10''$	$\cos 20'' = 2\beta \cos 10'' - 1$
$\sin 30'' = 2\beta \sin 20'' - \sin 10''$	$\cos 30'' = 2\beta \cos 20'' - \cos 10''$
$\sin 40'' = 2\beta \sin 30'' - \sin 20''$	$\cos 40'' = 2\beta \cos 30'' - \cos 20''$
&c., &c., &c.	&c., &c., &c.

But since 2β differs very little from two units, the calculation may be further abridged. Designating by k the difference $2 - 2\beta$, we shall have $k = 0{,}00000\ 00023\ 504$, and $2\beta = 2 - k$. Hence, the value of $\sin t''$ becomes

$$\sin t'' = 2 \sin t' - k \sin t' - \sin t$$

and therefore, $\sin t'' - \sin t' = (\sin t' - \sin t) - k \sin t'$.

When the difference $\sin t'' - \sin t'$ has been calculated, we may determine the $\sin t''$, by adding $\sin t'$ to this difference. But the last formula shows, that the difference $\sin t'' - \sin t'$ is equal to the difference $\sin t' - \sin t$ (a difference already calculated before arriving at the arc t'',) *minus* the product $k \sin t'$. Hence, the only tedious operation, which is repeated for each sine, consists in multiplying the last sine by the constant number $k = 0{,}00000\ 00023\ 504$. But the operation may itself be abridged by forming in advance the products of 23504 by the natural numbers 1, 2, 3 —— to 9. We shall then have immediately the partial products which compose each product, such as $k \sin t$, and it only remains to add them. By calculations, nearly similar, the values of the cosines are determined.

6

In so long a series of operations, errors are multiplied, since it is impossible to preserve thirteen decimals exact throughout. To ascertain the degree of precision on which we must rely, we shall find, presently, by methods which give a certain approximation, the values of many sines and cosines; and then the number of decimals common to those results and those which are given by the calculation just explained, will indicate, with sufficient certainty, the decimals which may be regarded as exact in the intermediate results.

If we find that the approximation is insufficient, we select as the basis of calculation an arc less than 10″, 1″ for example, and recommence the operations.

105. In practice it is more convenient to have the logarithms of the numbers corresponding to the trigonometrical lines rather than the trigonometrical lines themselves. But in supposing the radius to be unity, the sines and cosines will be fractions, and their logarithms will be negative. To make these logarithms positive, we make $r = 10^{10}$, which is equivalent to dividing the radius into 10,000000000 equal parts; and then the logarithms of the sines and cosines will only be negative for angles which differ so little from 0 or 90°, that this difference may be neglected.

It is easy to transform all the results which have been determined under the hypothesis of $r = 1$, to the condition $r = 10^{10}$, by multiplying them by 10^{10}, or by adding 10 to their logarithms. Indeed, when $r = 1$, we find the ratio of the sines and cosines to the radius, and it is evident that if we divide the radius into m equal parts, we must multiply all of these ratios by m, to know the number of parts contained in the sines and cosines.

106. The logarithms of the tangents are made known by the formula $\text{tang}\, a = \dfrac{r \sin a}{\cos a}$, which gives

$$\log \text{tang}\, a = \log \sin a + (10 - \log \cos a);$$

that is, we must add to the logarithm of the sine the arithmetical complement of the logarithm of the cosine.

The logarithms of the cotangents are found by means of the relation $\text{tang}\, a \cot a = r^2$, which gives

$$\log \cot a = 10 + (10 - \log \tan a).$$

Some tables do not contain the cotangent, but this result shows that we may easily supply them by adding 10 to the arithmetical complement of the logarithm of the tangent.

Nor do the tables usually contain secants and cosecants, but their logarithms are readily calculated from those of the sines and cosines.

The Tables are not usually constructed beyond 45°. For arcs greater than 45°, the logarithms of the sines and tangents, are obtained from the cosines and cotangents of the complement, and reciprocally. Thus, if $a > 45°$, we shall have, $\sin a = \cos (90° - a)$.

The tables are so arranged as to avoid the necessity of calculating this complement.

Calculation of Sines and Cosines for every 9°, to verify the Tables.

107. To effect the verifications referred to, Art. 104, we propose now to calculate the sines and cosines for arcs varying by 9°.

Let $\sin 18° = x$; $2x$ will be the chord of 36°, or the side of the regular inscribed decagon. But this side is equal to the greater segment of the radius, when it is divided into extreme and mean ratio: hence, the radius being 1, we have $4x^2 = 1 \times (1 - 2x)$.

From this we get

$$x^2 + \tfrac{1}{2}x = \tfrac{1}{4},$$

and solving the equation, and neglecting the negative value of x as useless, we have

$$x = \sin 18° = \cos 72° = \tfrac{1}{4}(-1 + \sqrt{5}).$$

With this value we readily find

$$\sqrt{1 - x^2} = \cos 18° = \sin 72° = \tfrac{1}{4}\sqrt{10 + 2\sqrt{5}}.$$

Substituting these values of sin 18° and cos 18° for $\sin a$ and $\cos a$ in the formulæ for $\sin 2a$ and $\cos 2a$, Art. 77, we have

$$\sin 36° = \cos 54° = \tfrac{1}{4}\sqrt{10 - 2\sqrt{5}}$$

$$\cos 36° = \sin 54° = \tfrac{1}{4}(1 + \sqrt{5})$$

Substituting now the same value of the sin 18° in the formulæ which give the values of sin $\frac{1}{2}a$ and cos $\frac{1}{2}a$ in terms of the sine, Art. 81, we obtain

$$\sin 9° = \cos 81° = \tfrac{1}{4}\sqrt{3+\sqrt{5}} - \tfrac{1}{4}\sqrt{5-\sqrt{5}}$$

$$\cos 9° = \sin 81° = \tfrac{1}{4}\sqrt{3+\sqrt{5}} + \tfrac{1}{4}\sqrt{5-\sqrt{5}}.$$

Finally, if we substitute in the same formulæ for sin a the value $\sin 54° = \frac{1}{4}(1+\sqrt{5})$, we deduce

$$\sin 27° = \cos 63° = \tfrac{1}{4}\sqrt{5+\sqrt{5}} - \tfrac{1}{4}\sqrt{3-\sqrt{5}}$$

$$\cos 27° = \sin 63° = \tfrac{1}{4}\sqrt{5+\sqrt{5}} + \tfrac{1}{4}\sqrt{3-\sqrt{5}}.$$

Further, remembering that $\sin 45° = \cos 45° = \frac{1}{2}\sqrt{2}$, we deduce the following table:

$$\sin 0 = \cos 90° = 0$$

$$\sin 9° = \cos 81° = \tfrac{1}{4}\sqrt{3+\sqrt{5}} - \tfrac{1}{4}\sqrt{5-\sqrt{5}}$$

$$\sin 18° = \cos 72° = \tfrac{1}{4}(-1+\sqrt{5})$$

$$\sin 27° = \cos 63° = \tfrac{1}{4}\sqrt{5+\sqrt{5}} - \tfrac{1}{4}\sqrt{3-\sqrt{5}}$$

$$\sin 36° = \cos 54° = \tfrac{1}{4}\sqrt{10-2\sqrt{5}}$$

$$\sin 45° = \cos 45° = \tfrac{1}{2}\sqrt{2}$$

$$\sin 54° = \cos 36° = \tfrac{1}{4}(1+\sqrt{5})$$

$$\sin 63° = \cos 27° = \tfrac{1}{4}\sqrt{5+\sqrt{5}} + \tfrac{1}{4}\sqrt{3-\sqrt{5}}$$

$$\sin 72° = \cos 18° = \tfrac{1}{4}\sqrt{10+2\sqrt{5}}$$

$$\sin 81° = \cos 9° = \tfrac{1}{4}\sqrt{3+\sqrt{5}} + \tfrac{1}{4}\sqrt{5-\sqrt{5}}$$

$$\sin 90° = \cos 0° = 1.$$

These expressions being very simple and involving only the square root, may be calculated decimally to any degree of approximation desired. These are the values which are used to verify the calculations explained in Art. 114.

By *bisecting* the arcs by the aid of the established formulæ, we might deduce the values of sine and cosine of arcs of 4° 30′, and 2° 15′, and the successive multiples of 2° 15′ would supply new verifications.

Arrangement and Manner of using the Tables of Callet.

107. The Tables of *Callet* are the best for the old division of the circumference, and those of *Borda* for the new or *decimal* division. In the Tables of *Callet* three tables are especially to be observed. The first contains the logarithms of numbers to 108000, and the application of this table has been explained in the introduction. The second contains the logarithms of the sines, tangent, and cosine for all arcs, differing by one minute, according to the new division; and the third contains the logarithms of the sine, cosine, tangent, and cotangent, for every 10″, by the old division. Inasmuch as the instruments used, and astronomical calculations made are referred to the sexigesimal division of the circumference, the third table only will be referred to in this place.

108. This Table gives the logarithm of the sine and tangent for every second up to 5°, and consequently the logarithms of the cosine and cotangent of all angles above 85°, and when the arcs are within these limits we find here the corresponding logarithms. Afterwards, we have the logarithms of the sine, cosine, tangent, and cotangent for every 10″. These are written in the columns headed *sine, cosine,* &c. When the tang or cotang is greater than the radius, its logarithm exceeds 10. In the table the *tens* are suppressed, but we must not forget to restore it in practice.

109. If we only consider the degrees which are at the head of each page, we might suppose that the tables only extended to 45°; but if we observe that the columns marked at the top *sin, cos,* &c., are marked at the bottom *cos, sin,* &c., we see that by taking the degrees and titles at the bottom of each page, at the same time *ascending* the columns placed on the right for minutes and seconds, we shall find the logarithms of sines and cosines from 45° to 90°.

From what has been said, we find immediately

L. sin 6° 32′ 30″ = 9,0566218

L. cot 81° 46′ 20″ = 9,1601596.

110. When the given angle contains seconds and fractions of a second, it is necessary to make use of the tabular *differences*, and to make calculations similar to those which have been explained in finding the logarithms of numbers. In doing this, the differences of *log sin*, *log cos*, &c., are regarded as proportional to those of the arcs; and this proportion, although not altogether exact, gives nevertheless a sufficient approximation. We remark that the same *differences* are common to *log tang* and *log cotang*.

The following examples are given for practice in the use of the Tables.

1st. Find the log. sin 6° 32′ 37″, 8.

log sin 6° 32′ 30″ (dif. 1836)	9,0566218
log for 7″	1285.2
log for 0.8	146.88
log sin 6° 32′ 37″, 8	= 9,056765

2d. Find the log cos 83° 27′ 22″, 2

log cos 83° 27′ 30″ (dif. 1836)	9,0566218
log for —7″	1285.2
log for —0.8	146.88
log cos 83° 27′ 22″, 2	= 9,0567650

3d. Find the log tang 8° 13′ 52″, 76

log tang 8° 13′ 50″ (dif. 1486)	9,1603083
log for 2″	297.2
log for 0,7	104.02
log for 0,06	8.916
log tang 8° 13′ 52″, 76	= 9,1603493

4th. Find the log cot 81° 46′ 17″, 24

log cot 81° 46′ 10″ (dif. 1486)	9,1603083
log for —2″	297.2
log for —0,7	104.02
log for —0,06	8.916
log cot 81° 46′ 7″, 24	= 9,1603493.

111. Let it now be required to determine the angle when we know the logarithm of a sine, a cosine, &c. Thus, let log sin $x = 9{,}0567650$, find x.

The next less logarithm given in the tables for logarithms of sines is 9,0566218, and this logarithm corresponds to 6° 32′ 30″. The difference between this logarithm and the given logarithm is 1432, and the tabular difference, corresponding to 10″, is 1836. Hence, dividing 1432 by 1836, and adding the *tenths* of the quotient as seconds, we find 7″, 8. Therefore, the required arc is $x = 6°32' 37'', 8$. The following calculations will explain this example, and others of an analogous character.

1st. Find the angle corresponding to log sin 9,0567650.

log sin $x = 9{,}0567650$		
The next less log. 9,0566218 (tab. dif. 1836)		6° 32′ 30″
1st remainder	14320	7
2d remainder	14680	0.8
		$x = 6°\ 32'\ 37'', 8$

2d. Find the angle corresponding to log cos 9,0567650.

log cos $x = 9{,}0567650$		
The next log. 9,0568054 (tab. dif. 1836)		83° 26′ 20″
1st remainder	4040	2
2d remainder	3680	0.2
		$x = 83°\ 27'\ 22'', 2$

3d. Find the angle corresponding to log tang 9,1603493.

log tang $x = 9{,}1603493$		
For 9,1603083 (tab. dif. 1486)		8° 13′ 50″
1st remainder	4100	2
2d remainder	11280	0.7
3d remainder	8780	0.06
		$x = 8°\ 13'\ 52'', 76$

4th. Find the angle corresponding to log cot 9,1603493.

log cot x = 9,1603493		
For	9,1604569 (tab. dif. 1486)	
		81° 46′ 0″
1st remainder	10760	7″
2d remainder	2580	0.2
3d remainder	6080	0.04
		x = 81° 46′ 7″,24

112. The formulæ which contain *sines*, *cosines*, &c., assume almost always that the radius has been regarded as unity. To apply the tables to them, two methods may be used.

The first consists in introducing the radius r in the formulæ as in Art. 57, and then using the logarithms of the tables, always remembering that log $r = 10$.

In the second process, we do not change the formulæ, but we retain the hypothesis of $r = 1$, and subtract 10 from every logarithm taken from the trigonometrical table. It answers to effect this subtraction on the characteristic only, which may, by this operation, become negative, but it is better to employ such logarithms as the table gives, and to take no account of the 10 until the end of the operation. This method of correction will be always easy, for, in the calculations, the logarithms are always added or subtracted, and it is evident that each additive logarithm will give one *ten* too much in the result, while each subtractive logarithm will have one *ten* too little.

113. The use of the *arithmetical complement* enables us to substitute addition for subtraction of logarithms, as has been heretofore explained. In this case, the *ten* which must be subtracted from this logarithm to reduce it to the hypothesis of $r = 1$, is compensated by that which we have to add in using the arithmetical complement. Besides, the error of a *ten* in a logarithm is so considerable, that it could not fail to be noticed. For illustration, the following examples are given.

1st. Let $x = 419 \times \sin^2 40°$, hence, $\log x = \log 419 + 2 \log \sin 40°$. Taking log sin 40° from the tables, log x contains *two* tens too much, which must be taken from the result.

$$\begin{aligned} \log 419 &= 2{,}6222140 \\ 2 \log \sin 40^\circ &= 19{,}6161350 \\ \hline \log x &= 2{,}2383490 \end{aligned}$$

If we wish to obtain x within $\frac{1}{100}$, add 2 to the characteristic, and we find $x = 173{,}12$.

2d. Let $\sin x = \dfrac{314 \times \sin 30^\circ}{411 \times \cos^2 45^\circ}$,

we get

$$\log \sin x = \log 314 - \log 411 + \log \sin 30^\circ - 2 \log \cos 15.$$

Operating by the complements, the two tens which must be taken from 2 log cos 15°, are compensated by the two tens for which we take the complements. The log sin 30° and the complement of the log 411 introduce two tens too much; but as the angle must be sought in the tables, we only take away one ten from the result, as is shown below. The arithmetical complements are designated by Lo'g.

$$\begin{aligned} \log 314 &= 2{,}49692965 \\ \text{lo'g}\, 411 &\quad 7{,}38615818 \\ \log \sin 30^\circ &\quad 9{,}6989700 \\ 2\ \text{lo'g} \cos 15^\circ &\quad 0{,}0301127 \\ \hline \log \sin x &= 9{,}6121702 \end{aligned}$$

This logarithm is prepared to suit the tables. We find $x = 24^\circ\ 10'\ 7''$.

Resolution of Triangles. Relations between the Sides and Angles of a Plane Triangle.

114. Hereafter we will designate the angles of a triangle by the large letters A, B, C, placed at their vertices, while the sides opposite to these angles will be respectively designated by the small letters a, b, c. If a triangle be right-angled, the letter A will always correspond to the right angle, and the letter a to the hypothenuse.

115. The solution of all plane triangles depends upon the following theorems which are here enunciated and demonstrated.

116. *Theorem* I. *In every right-angled plane triangle, each side about the right angle is equal to the hypothenuse multiplied by the sine of the angle opposite to this side.*

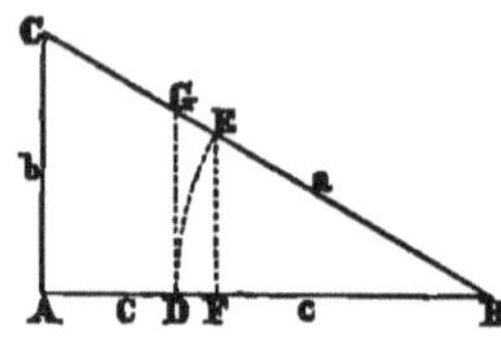

Let A B C be a triangle right-angled at A. From the point B as a centre, with any radius, describe the arc D E, and let fall the perpendicular E F. The sine of the angle B is the ratio of E F to the radius B E (Art. 54). The similar triangles B C A, B E F, give

$$\frac{AC}{BC}=\frac{EF}{BE}, \text{ and we have } \frac{b}{a}=\sin B,$$

hence, $b = a \sin B,$ (1)

which was to be proved.

117. The angle B is the complement of the angle C; hence, sin B = cos C, and we conclude, *that in every right-angled plane triangle, each side about the right angle is equal to the hypothenuse multiplied by the cosine of the adjacent angle.*

118. *Theorem* II. *In every right-angled plane triangle, each side about the right angle is equal to the other side multiplied by the tangent of the angle opposite to the first side.*

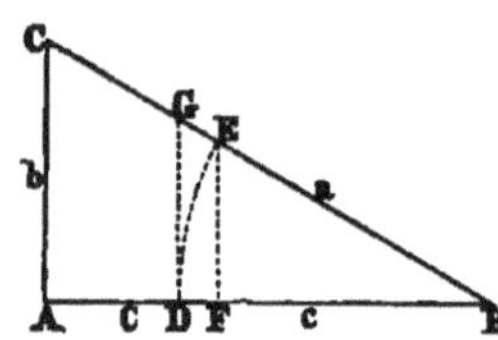

For, in the triangle A B C, describe the arc D E, from B as a centre, and with any radius; erect D G perpendicular to A B. The ratio of D G to B D is the tangent of the angle B, (Art. 54.) But the similar triangles give

$$\frac{AC}{AB}=\frac{AG}{BD}, \text{ or } \frac{b}{c}=\text{tang } B,$$

therefore, $b = c \text{ tang } B,$ (2)

which was to be proved.

119. We may deduce this result directly from Theorem I. Indeed, applying this theorem to each side b and c, and observing that sin C = cos B, we have

$$b = a \sin B \text{ and } c = a \cos B;$$

hence, $\frac{b}{c} = \frac{\sin B}{\cos B} = \text{tang } B$ or $b = c \text{ tang } B$.

120. *Theorem* III. *In every plane triangle the sides are proportional to the sines of the opposite angles.*

Let A and B be any two angles of the triangle A B C. Let fall the perpendicular C D from the vertex C on the side A B. If the perpendicular falls within the triangle A B C, the two right-angled triangles, A C D and B C D, will give

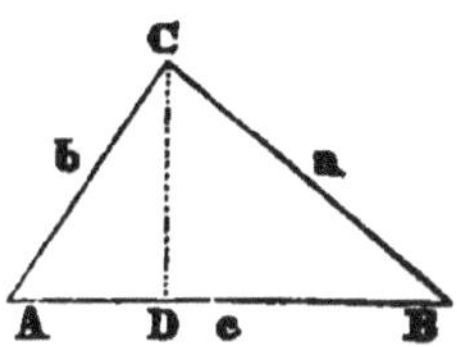

$$CD = b \sin A, \text{ and } CD = a \sin B;$$

hence, $b \sin A = a \sin B$;

and we have $$\frac{a}{b} = \frac{\sin A}{\sin B} \qquad (3)$$

If the perpendicular falls on the prolongation of B A, the triangle A C D gives C D = b sin C A D. But the angle C A D being the supplement of C A B, we have

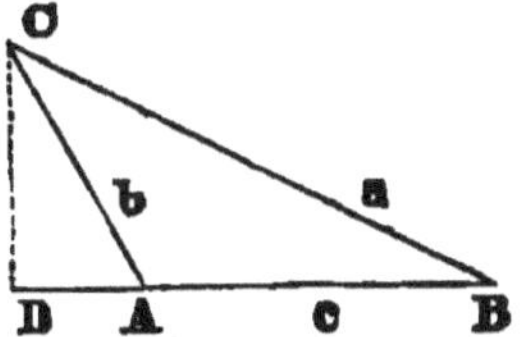

$$\sin CAD = \sin CAB = \sin A,$$

and we have still the same proportion before found.

121. *Theorem* IV. *In every plane triangle the square of either side is equal to the sum of the squares of the other two sides, minus twice twice the rectangle of these two sides, multiplied by the cosine of their included angle.* That is, we have

$$a^2 = b^2 + c^2 - 2\,b\,c \cos A. \qquad (4)$$

Let A B C be the given triangle; let fall the perpendicular C D on A B. When the angle A is acute, we have, from a known theorem,

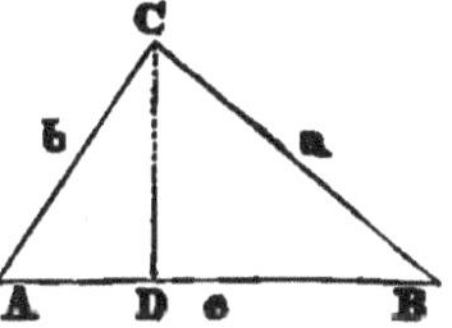

$$\overline{CB}^2 = \overline{AC}^2 + \overline{AB}^2 - 2\,AB \times AD,$$

or, $a^2 = b^2 + c^2 - 2\,c \times AD$,

But the triangle A C D gives A D $= b \cos A$, and substituting this value of A D, we have formula (4).

When the angle A is obtuse, we have

$$a^2 = b^2 + c^2 - 2c \times \mathrm{AD}$$

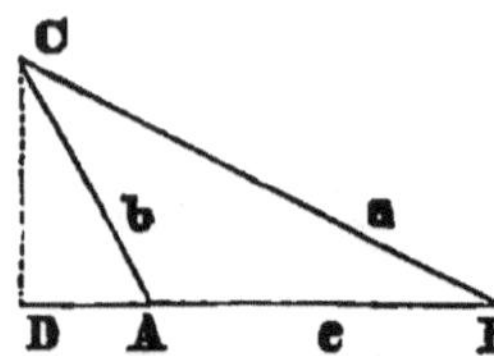

And the triangle ACD gives AD $= b$ cos CAD. But CAD being the supplement of CAB or A, we have cos CAD $= -$ cos CAB $= -$ cos A; hence, AD $= -b$ cos A; and substituting this value in that of a^2, we have the equation before found.

122. Theorem IV is sufficient of itself to solve every plane triangle. For, it is evident we have, for each side, the three equations

$$a^2 = b^2 + c^2 - 2bc \cos A, \qquad (5)$$
$$b^2 = a^2 + c^2 - 2ac \cos B, \qquad (6)$$
$$c^2 = a^2 + b^2 - 2ab \cos C, \qquad (7)$$

by which we may always determine three of the six parts of a triangle, when the other three are known, except in the cases when the triangle is impossible, or when the three angles are given.

123. Theorem III., since it expresses a relation between two sides and their opposite angles, must be a consequence from the three last equations. It may be deduced as follows:

The first equation gives $\cos A = \dfrac{b^2 + c^2 - a^2}{2bc}$; hence,

$$\sin^2 A = 1 - \cos^2 A = \frac{4b^2c^2 - (b^2 + c^2 - a^2)^2}{4b^2c^2}$$

$$= \frac{2a^2b^2 + 2a^2c^2 + 2b^2c^2 - a^4 - b^4 - c^4}{4b^2c^2}$$

and consequently,

$$\frac{\sin A}{a} = \frac{\sqrt{2a^2b^2 + 2a^2c^2 + 2b^2c^2 - a^4 - b^4 - c^4}}{2abc}$$

The other two equations give, in like manner, the ratios $\dfrac{\sin B}{b}$ and $\dfrac{\sin C}{c}$; but we may deduce them at once by changing, in the second member of the preceding equation, first a into b and b into a, and afterwards a into c and c into a. Since the second member is a symmetrical function of a, b, c, it will remain the same when the changes are made upon these letters, and we have therefore, in accordance with Theorem III.,

$$\frac{\sin A}{a} = \frac{\sin B}{b} = \frac{\sin C}{c}.$$

RESOLUTION OF RIGHT-ANGLED PLANE TRIANGLES.

124. *First case. Having given the hypothenuse* a *and an acute angle* B, *to find the angle* C, *and the two sides* b *and* c.

We have $C = 90° - B$. We may then determine b and c by means of Theorem I., which gives

$$b = a \sin B, \quad c = a \cos B.$$

The values of b and c, will of course be calculated by means of logarithms.

125. *Second case. The side* b *of the right angle, and the acute angle* B *being given, to find* C, a *and* c.

We have again $C = 90° - B$. Theorem I. gives a, by the relation

$$b = a \sin B, \quad \text{hence, } a = \frac{b}{\sin B},$$

and Theorem II. gives c, since we have

$$c = b \operatorname{tang} C, \text{ or } c = b \cot B.,$$

126. *Third case. The hypothenuse* a, *and a side* b, *being given, to find the other side* c, *and the angles* B *and* C.

From the known property of the right-angled triangle, we have $c^2 = a^2 - b^2$, or $c = \sqrt{(a+b)\ (a-b)}$, which makes known the value of c very readily by means of logarithms. We determine B by the relation $b = a \sin B$ (Art. 116), which gives $\sin B = \frac{b}{a}$, and finally we have $C = 90° - B$.

If we determine the angles first, we may still find c by the relation $c = a \sin C$.

127. *Fourth case. Having given the two sides* b *and* c *of the right angle, to find the hypothenuse* a *and the angles* B *and* C.

We first determine B, by the relation $b = c$ tang B (Theo. II.), and this gives afterwards $C = 90° - B$. The hypothenuse a is made known by the relation (Theo. I.) $b = a \sin B$.

We might find a directly from the formula $a = \sqrt{b^2 + c^2}$. But since the binomial $b^2 + c^2$ cannot be decomposed into factors, it is not suitable for logarithmic calculation, and it is better to find the angle B in the first place, and then make use of it to find a.

Resolution of Oblique Plane Triangles.

128 *Case* I. *Having given a side,* a, *and two angles of a triangle, to find the three other parts.*

If the sum of the two given angles be taken from 180°, we shall have the third angle. The sides b and c are then determined by Theorem III, by means of the proportions

$$\frac{b}{a} = \frac{\sin B}{\sin A}, \quad \frac{c}{a} = \frac{\sin C}{\sin A}.$$

129. *Case* II. *Having given two sides,* a *and* b, *and the angle* A *opposite one of them, to find the side* c *and the two other angles* B *and* C.

The simplest solution is, to find first the angle B, opposite the side b, by the proportion $\frac{\sin B}{\sin A} = \frac{b}{a}$.

The angles A and B being known, we have $C = 180° - (A + B)$; and the side c is found by the relation $\frac{c}{a} = \frac{\sin C}{\sin A}$.

130. This solution requires some development. The first proportion makes known sin B, thus,

$$\sin B = \frac{b \sin A}{a}:$$

and from the tables we find for the value of B an acute angle. But the same sine corresponds also to a supplementary angle, which is obtuse: so that, calling M the angle of the tables, we shall have for B the two values $B = M$, $B = 180° - M$, which seem to indicate two triangles, and suggest the following important observations.

1st. When the known angle A is obtuse or a right angle, the two other angles must be acute; and in this case, we take B = M only; and in order that the triangle may be possible, it is further necessary that we have $a > b$. This condition is sufficient.

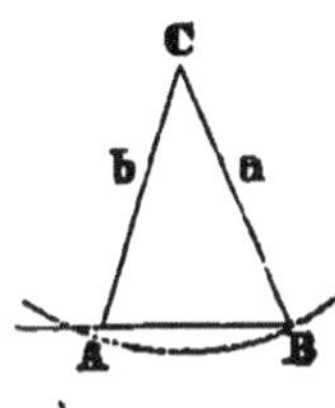

2d. When A is acute and $a > b$, we must have A > B, and it is still necessary to reject the value of B = 180° — M. In this case the triangle is always possible.

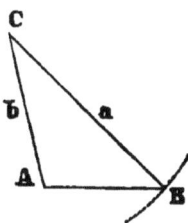

3d. But when we have A acute and $a < b$, we may take B indifferently equal to M, or to 180° — M. For, let the angle B A C = A, and A C = b; the circle described from C as a centre and with a radius a will, in certain cases, cut A B in two points, B and B′; and we shall then have two triangles A C B, A C B′, which will be constructed with the given parts, and in which the angles A B C, A B′ C are supplements of each other. The condition for two solutions is, that the side a which is supposed less than b, be greater than the perpendicular C D let fall on A B. If the side a is equal to C D, the circle is tangent to A B, and the two solutions reduce to a single right-angled triangle A C D. Finally, if the side $a <$ C D, the triangles are impossible; and we will now show that this impossibility is indicated by the value of the sin B.

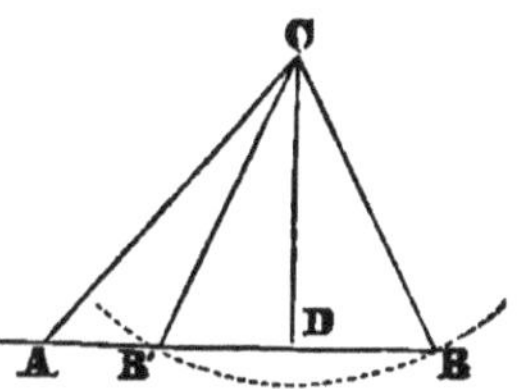

The right-angled triangle A C D gives C D = b sin A. But, by the hypothesis, $a <$ C D. Hence, we have

$$a < b \sin A, \text{ and } \frac{b \sin A}{a} > 1; \text{ hence, } \sin B > 1.$$

Thus, the value of sin B is greater than unity; but no sine is greater than unity, hence the triangle is impossible.

131. We have undertaken to find the side c after determining the angle B. We may, however, find c at once when a, b, and A are given; for, by Theo. IV., we have $a^2 = b^2 + c^2 - 2bc \cos A$; hence,

$$c^2 - 2b \cos A . c = a^2 - b^2.$$

This equation is of the second degree, and gives

$$c = b \cos A \pm \sqrt{a^2 - b^2 + b^2 \cos^2 A.} = b \cos A \pm \sqrt{a^2 - b^2 \sin^2 A.}$$

Since the side of a triangle must always be a real and positive quantity, we have only to inquire what relations between a, b, A, will give for c one or two values which fulfil these conditions: and this inquiry may be made by the student without any difficulty.

132. The preceding values of c being inconvenient for logarithmic calculations, are of no use in trigonometry. Still, as we sometimes meet with like expressions, we will here explain the transformations adopted by astronomers to facilitate the use of such results.

Placing these values under the form

$$c = b \cos A \pm a \sqrt{1 - \frac{b^2 \sin^2 A}{a^2}}$$

As these values are supposed to be real, the quantity $\dfrac{b \sin A}{a}$ is less than 1, and we may consider it as the sine of an angle ϕ, which we may determine in placing

$$\sin \phi = \frac{b \sin A}{a}.$$

We have then

$$b = \frac{a \sin \phi}{\sin A}, \quad \sqrt{1 - \frac{b^2 \sin^2 A}{a^2}} = \cos \phi; \text{ and consequently,}$$

$$c = \frac{a (\sin \phi \cos A \pm \sin A \cos \phi)}{\sin A} = \frac{a \sin (\phi \pm A)}{\sin A},$$

results which may be readily calculated by logarithms.

This solution exactly corresponds with the first, for the auxiliary angle ϕ is no other than the angle B.

133. *Case* III. *Having given in a triangle the two sides* a *and* b *and the included angle* C. *to find the side* c, *and the angles* A *and* B.

By Theorem III., we have $\frac{a}{b}=\frac{\sin A}{\sin B}$. This proportion contains two unknown quantities A and B: but we may deduce from it, in the first place,

$$\frac{a+b}{a-b}=\frac{\sin A+\sin B}{\sin A-\sin B}.$$

And we know, further, that

$$\frac{\sin A+\sin B}{\sin A-\sin B}=\frac{\text{tang}\,\frac{1}{2}(A+B)}{\text{tang}\,\frac{1}{2}(A-B)}$$

Therefore,

$$\frac{a+b}{a-b}=\frac{\text{tang}\,\frac{1}{2}(A+B)}{\text{tang}\,\frac{1}{2}(A-B)} \qquad (1)$$

which shows that, *in any triangle, the sum of two sides is to their difference as the tangent of half the sum of the angles opposite to those sides is to the tangent of half their difference.*

We have $A+B=180°-C$; hence $\frac{1}{2}(A+B)=90°-\frac{1}{2}C$: therefore, the three first terms of the above proportion are known, and we may find from them the value of $\frac{1}{2}(A-B)$. Knowing now the values of the *half sum* and *half difference* of the angles A and B, each may be determined; for, by *adding* the half sum to the half difference, we obtain the greater angle, and by subtracting the half difference from the half sum we get the less angle. Thus,

$$A=\frac{A+B}{2}+\frac{A-B}{2}, \text{ and } B=\frac{A+B}{2}-\frac{A-B}{2},$$

A and B being found, we obtain the side c, by the proportion

$$\frac{c}{a}=\frac{\sin C}{\sin A} \qquad (2)$$

134. This proportion requires that we find three new logarithms, viz.: those of a, sin A, sin C. We may abridge the calculation by the following process:

Since we have $\frac{\sin A}{a}=\frac{\sin B}{b}=\frac{\sin C}{c}$; we must also have

$$\frac{\sin A+\sin B}{\sin C}=\frac{a+b}{c}; \text{ hence, } c=\frac{(a+b)\sin C}{\sin A+\sin B}.$$

By known formulæ we have

$$\sin A + \sin B = 2 \sin \tfrac{1}{2}(A + B) \cos \tfrac{1}{2}(A - B)$$
$$\sin C = 2 \sin \tfrac{1}{2} C \cos \tfrac{1}{2} C.$$

And we have also, $\sin \frac{1}{2}(A + B) = \sin (90° - \frac{1}{2} C) = \cos \frac{1}{2} C$.

Substituting these values in c, and making the necessary reductions, we have

$$c = \frac{(a + b) \sin \frac{1}{2} C}{\cos \frac{1}{2}(A - B)} \qquad (3)$$

Since the logarithm of $(a + b)$ is already known, this value of c may be used in preference to that given by the proportion (2), and we thus save the calculation of one logarithm.

135. We have determined the value of c after finding those of A and B. To find c directly, we make use of Theorem IV., which gives

$$c = \sqrt{a^2 + b^2 - 2ab \cos C}.$$

But inasmuch as logarithms cannot readily be applied to this formula, we make use of an auxiliary angle.

We have $\cos^2 \frac{1}{2} C + \sin^2 \frac{1}{2} C = 1$, and $\cos^2 \frac{1}{2} C - \sin^2 \frac{1}{2} C = \cos C$. Hence

$$c = \sqrt{(a^2 + b^2)(\cos^2 \tfrac{1}{2} C + \sin^2 \tfrac{1}{2} C) - 2ab(\cos^2 \tfrac{1}{2} C - \sin^2 \tfrac{1}{2} C)}$$
$$= \sqrt{(a + b)^2 \sin^2 \tfrac{1}{2} C + (a - b)^2 \cos^2 \tfrac{1}{2} C.}$$
$$= (a + b) \sin \tfrac{1}{2} C \sqrt{1 + \frac{(a - b)^2 \cot^2 \frac{1}{2} C}{(a + b)^2}}$$

Since the magnitude of a tangent is unlimited, we will make

$$\text{tang}\, \phi = \frac{(a - b) \cot \frac{1}{2} C}{a + b}.$$

Then the radical becomes

$$\sqrt{1 + \text{tang}^2 \phi} = \sqrt{1 + \frac{\sin^2 \phi}{\cos^2 \phi}} = \frac{1}{\cos \phi}.$$

Hence, $$c = \frac{(a + b) \sin \frac{1}{2} C.}{\cos \phi}$$

Thus, we will find successively the auxiliary angle ϕ and the side c by means of two formulæ which may be readily calculated by the tables.

This solution differs from the preceding only in appearance: for, tang $\frac{1}{2}$(A + B) being equal to cot $\frac{1}{2}$ C, the value of tang ϕ is identical with that of tang $\frac{1}{2}$(A — B) deduced from equation (1), and hence the last value of c is found also to be identical with the formula (3).

136. In practice, it often happens that the sides are known by their logarithms. Let us suppose that the sides a and b are thus known, and that, further, C is given, and that A and B are the only quantities to be determined. To find $\frac{1}{2}$ (A — B), by means of the proportion (1), it would first be necessary to find a and b in the tables; but this may be avoided by making use of an auxiliary angle.

Let ψ be this auxiliary angle, given by the assumed relation tang $\psi = \frac{b}{a}$. We have, from known formulæ,

$$\text{tang}\,(45^\circ - \psi) = \frac{\text{tang}\,45^\circ - \text{tang}\,\psi}{1 + \text{tang}\,45^\circ\,\text{tang}\,\psi} = \frac{1 - \text{tang}\,\psi}{1 + \text{tang}\,\psi}$$

Substituting for tang ψ its value, we have

$$\text{tang}\,(45^\circ - \psi) = \frac{a-b}{a+b}.$$

But we have, Art. 133,

$$\text{tang}\,\tfrac{1}{2}\,(\text{A} - \text{B}) = \frac{a-b}{a+b}\,\text{tang}\,\tfrac{1}{2}\,(\text{A} + \text{B})$$

Hence,

$$\text{tang}\,\tfrac{1}{2}\,(\text{A} - \text{B}) = \text{tang}\,(45^\circ - \psi)\,\text{tang}\,\tfrac{1}{2}\,(\text{A} + \text{B}).$$

And since ψ is known, we readily find $\frac{1}{2}$(A — B). By this process we have two logarithms less to calculate than if we had determined the sides a and b.

137. *Case* IV. *Having given the three sides* a, b, c, *to find the three angles.*

By Theorem IV. we have $a^2 = b^2 + c^2 - 2\,b\,c \cos \text{A}$; hence

$$\cos \text{A} = \frac{b^2 + c^2 - a^2}{2\,b\,c}$$

In like manner, we may determine B and C. But it is necessary to find a formula more suitable for logarithmic calculations.

Art. 78 gives the formula $2\sin^2 \frac{1}{2}A = 1 - \cos A$, and substituting in it the value of $\cos A$, we have successively

$$2\sin^2 \tfrac{1}{2}A = 1 - \frac{b^2 + c^2 - a^2}{2bc} = \frac{a^2 - b^2 - c^2 + 2bc}{2bc}$$

$$= \frac{a^2 - (b-c)^2}{2bc} = \frac{(a+b-c)\ (a-b+c)}{2bc}$$

Hence, $\sin \frac{1}{2}A = \sqrt{\dfrac{(a+b-c)\ (a-b+c)}{4bc}}$

To simplify this formula, make the perimeter $a+b+c=2s$, we deduce $a+b-c = 2s-2c = 2(s-c)$, and $a-b+c = 2s-2b = 2(s-b)$;

hence, $\sin \frac{1}{2}A = \sqrt{\dfrac{(s-b)\ (s-c)}{bc}}$

From which we form the following important *rule, that the sine of half an angle of a plane triangle is equal to the square root of half the sum of the three sides minus one of the adjacent sides, multiplied by the half sum of the three sides minus the other adjacent side, divided by the rectangle of the adjacent sides.*

Although the angle $\frac{1}{2}A$ is determined by its sine, there is no ambiguity in the result, because A being an angle of a triangle, we must always have $A < 180°$, and $\frac{1}{2}A < 90°$.

138. We may determine with equal facility the formulæ which determine $\cos \frac{1}{2}A$ and $\text{tang} \frac{1}{2}A$. Resuming the known relation $2\cos^2 \frac{1}{2}A = 1 + \cos A$, we deduce, by similar transformations to those just employed,

$$\cos \tfrac{1}{2}A = \sqrt{\frac{s(s-a)}{bc}}$$

If we divide the value of $\sin \frac{1}{2}A$, by that of $\cos \frac{1}{2}A$, we find for $\text{tang} \frac{1}{2}A$,

$$\text{tang} \tfrac{1}{2}A = \sqrt{\frac{(s-b)\ (s-c)}{s(s-a)}}$$

139. We know that it is not always possible to form a triangle with three sides taken at will. This impossibility is indicated by the calculation. For, if we make use of the formula

$$\sin \tfrac{1}{2}A = \sqrt{\frac{(s-b)\ (s-c)}{bc}}$$

when the triangle is possible, this expression must give a real value for sin $\frac{1}{2}$ A, which must be less than 1. But if the triangle is impossible, this formula will lead to an imaginary value for sin $\frac{1}{2}$ A, or to a value >1. In order that the impossibility shall exist, it is necessary that one side be greater than the sum of the other two: let us now examine the results which the formula gives under this condition.

1st. Let $b>a+c$: we shall have $2b>a+b+c$: hence $2b>2s$ and $s-b<0$. But, further, we have plainly $a+b>c$, hence $a+b+c$ or $2s$ will be $>2c$, and $s-c>0$. Thus, the value of sin $\frac{1}{2}$ A is imaginary.

2d. Let $c>a+b$: then we shall have $s-c<0$, and $s-b>0$, and sin $\frac{1}{2}$ A will be imaginary.

3d. Let $a>b+c$; we shall have $a+b+c$ or $2s>2b+2c$: hence, $s>b+c$, and $s-b>c$, $s-c>b$, and $(s-b)(s-c)>bc$: consequently the value of sin $\frac{1}{2}$ A would be greater than 1, which would be impossible for any angle.

Application to Examples.

140. Extended trigonometrical operations require the use of various instruments, which it is not proposed to describe in this treatise. A few remarks will suffice to make intelligible the several examples which are proposed for solution.

To trace a right line upon the ground—stakes are used, and they are placed from point to point, at convenient distances, and so located that when the eye rests on the first stake, all the others will be covered by it.

An angle is traced on paper by means of an instrument called a *protractor*. This is a semi-circle divided into degrees and parts of degrees.

There are many different kinds of instruments used for measuring angles either on the ground, or in space. The chief of these are the *compass*, the *theodolite*, the *graphometer*, &c. In general, they consist of a circle, or sector of a circle, on which is marked a fixed radius, which serves as an origin for the subdivisions, whilst a second radius movable around the centre, may take any position. The plane of the circle may itself turn around its centre. To measure the angle included between two right lines which connect a given point with two other points, place the centre of the instrument at the first point, and direct the two radii upon the other two points; then read on the graduated circumference the number of degrees included between the radii, and this will be the required angle.

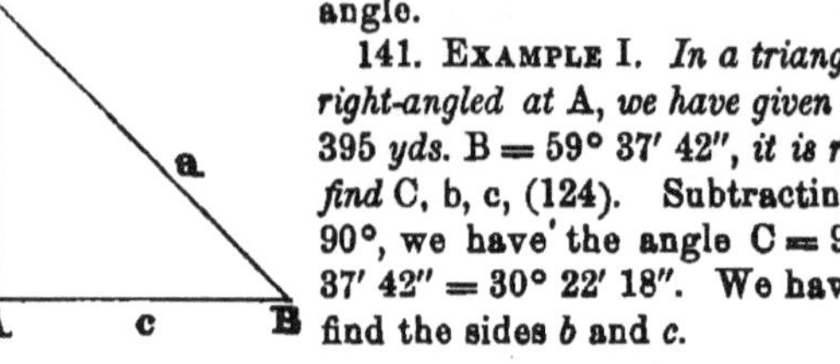

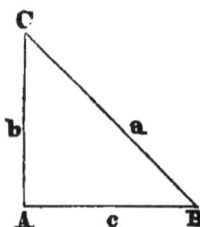

141. Example I. *In a triangle* A B C, *right-angled at* A, *we have given* a = 1785,395 *yds.* B = 59° 37′ 42″, *it is required to find* C, b, c, (124). Subtracting B from 90°, we have the angle C = 90° — 59° 37′ 42″ = 30° 22′ 18″. We have now to find the sides *b* and *c*.

Calculation of side b.		*Calculation of side* c.	
$b = a \sin B$		$c = a \cos B$	
log sin 59° 37′ 42″	9,9358919	log cos 59° 37′ 42″	9,7038132
log 1785,395	3,2517343	log 1785,395	3,2517343
log *b*	3,1876262	log *c*	2,9555475
b = 1540,374 yards.		*c* = 902,708 yards.	

142. Example II. *In a triangle* A B C, *we have given the side* a = 2597,843 *yards, the side* b = 3084,327 *yards, and the angle* A = 56° 12′ 47″,

it is required to find B, C, *and* c.

Calculation for angle B.			
$\frac{\sin B}{\sin A} = \frac{b}{a}$			
log sin 56° 12′ 47″	9,9196592	*There are two solutions.*	
log 3084,327	3,4891604		
log 2597,845 (Arith. Comp.)	6,5853868	In 1st solution B = 80° 39′ 43″	
		In 2d solution B = 99° 20′ 17″	
log sin B	9,9942064		
1st *solution:* B = 80° 39′ 43″		2d *solution:* B = 99° 20′ 17″	

Calculation for angle C.

C = 180° — A — B.

	180°
A =	56° 12′ 47″
B =	80° 39′ 43″
C =	43° 7′ 30″

To find the side C.

$$\frac{c}{a} = \frac{\sin C}{\sin A}.$$

log 2597,845	3,4146132
log sin 43° 7′ 30″	9,8347972
log sin 56° 12′ 47″ A.C.	0,0803408
log c	3,3297512

c = 2136,737 yards.

Calculation for angle C.

C = 180° — A — B

	180°
A =	56° 12′ 47″
B =	99° 20′ 17″
C =	24° 26′ 56″

To find the side C.

$$\frac{c}{a} = \frac{\sin C}{\sin A}.$$

log 2597,845	3,4146132
log sin 24° 26′ 56″	9,6168759
log sin 56° 12′ 47″ A.C.	0,0803408
log c	3,11182899

c = 1293,689 yards.

143. EXAMPLE III. *Having measured the distances* a *and* b *of the point* C *from two known points* A *and* B, *to find the points* C.

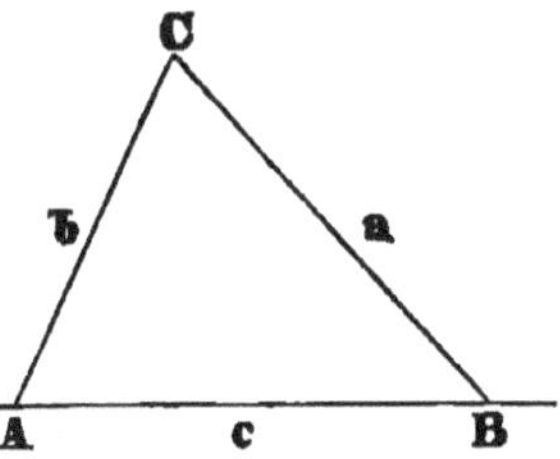

If the triangle A B C is small, we may describe two arcs of circles from the points A and B as centres, with the given distances as radii. But the operation becomes impracticable when the distances are considerable. In this case, we measure the distance A B, which will enable us, knowing the three sides of the triangle A B C, to calculate the angle A, (Art. 137.) The direction of the side A C is thus determined, and it only remains to lay off, on this direction, the given distance A C = b, and the point C is made known.

Let a = 9459,31 yards, b = 8032,29 yards, c = 8242,58, we shall have $2s$ = 25734,18, s = 12867,09, $s - b$ = 4834,80, $s - c$ = 4624,51.

$$\sin \tfrac{1}{2} A = \sqrt{\frac{(s-b)\ (s-c)}{bc}}$$

log $(s-b)$	3,6843785
log $(s-c)$	3,6650657
log b (Ar. Comp.)	6,0951606
log c (Ar. Comp.)	6,0839368
2 log sin $\frac{1}{2}$ A	19,5285416
log sin $\frac{1}{2}$ A	9,7642708

$\frac{1}{2}$ A = 35° 31′ 47″

A = 71° 3′ 34″

144. Example IV. *To find the height* A B *of a tower, the base of which is accessible.*

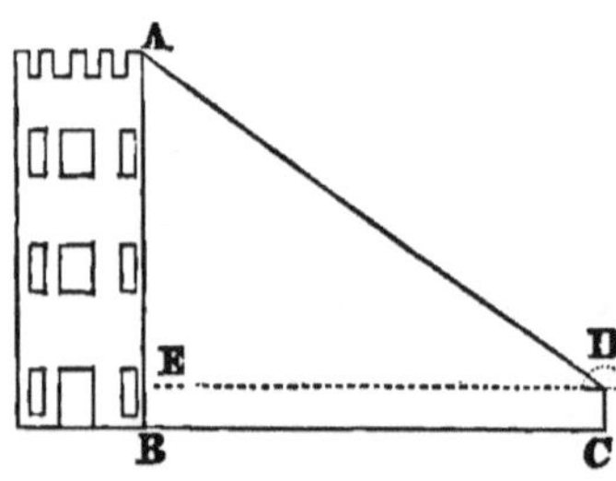

Lay off on the ground, supposed level, a base line B C, from the foot of the tower. Place the instrument at C, and measure the angle E D A formed by D A with the horizontal line D E parallel to C B. Then, in the right-angled triangle A E D, we know the side D E, and the angle D, and we may calculate A E. Adding C D to A E, we have the height required.

Let C D = 1,10 yard, D E = 61,28 yards, D = 41° 31′ 25″, we shall have

A E = 61,28 × tang 41° 31′ 25″

log tang 41° 31′ 25″	9,9471690
log 61,28	1,7873188
log A E	1,7344878

A E = 54,261 yards, A B = 55,361 yards.

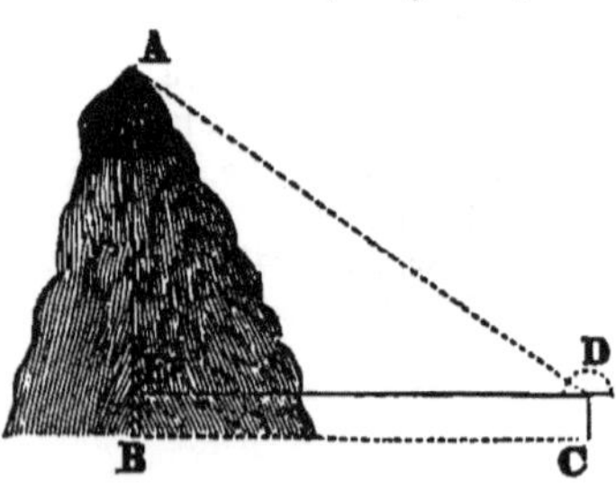

When the base of the tower is inaccessible, as when A B is the elevation of a hill above the surrounding ground, the foot of the perpendicular is unknown, and we are therefore unable to measure the distance B C. We may, nevertheless, measure the angle A D E; for, without seeing the line A B, it is still possible to make the plane of the circle with which the angles are measured, to pass through the vertical A B. Moreover, we will determine A D, as will be explained in the following example; and then knowing the hypothenuse A D and the angle D, we may find A E.

145. EXAMPLE V. *Find the distance of an observer, situated at* A, *from a point* P, *which is inaccessible, but which can be seen by the observer.*

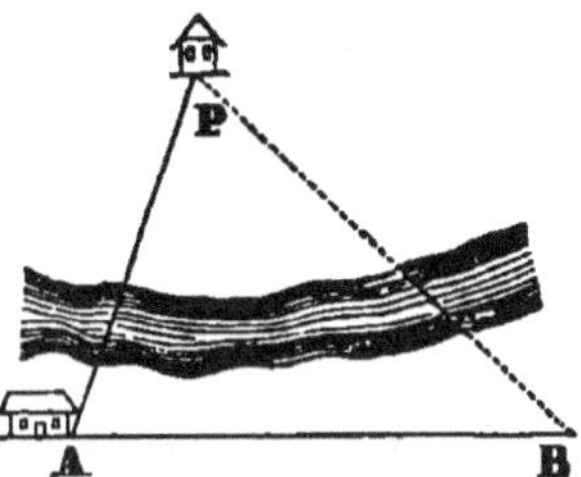

Measure a base A B, and the angles P A B, P B A, we may then readily determine A P.

Let A B = 247,49 yards, A = 62° 41′, B = 59° 42′. We deduce at once the angle P = 57° 37′; and then calculate A P, as follows.

$$\frac{AP}{AB} = \frac{\sin B}{\sin P}.$$

log A B	2,3935577
log sin B	9,9362098
log sin P (Ar. Comp.)	0,0734087
log A P	2,4031762

A P = 253,032 yards.

146. EXAMPLE VI. *To find the distance* P Q, *between two inaccessible points, which may be seen by the observer.*

Measure the base line A B, and the angles B A P, B A Q, A B P, A B Q; we may then determine, as above, the side A P of the triangle A B Q. Besides, the angle P A Q is known; for, if the four points A, B, P, Q, are in the same plane, we have P A Q = B A P — B A Q: and in every case we may determine that angle by direct measurement. Thus, in the triangle PAQ,

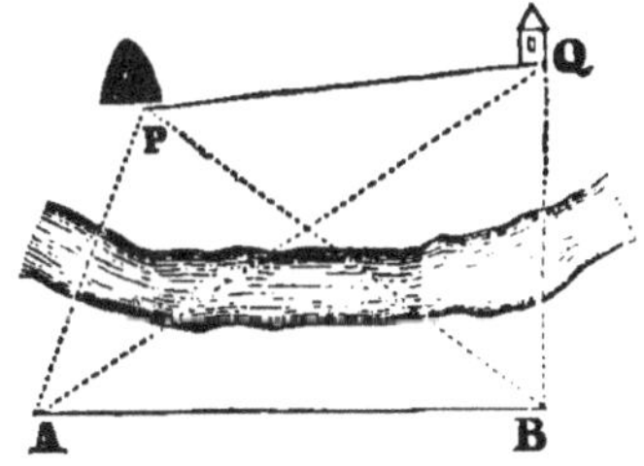

we shall know two sides and the included angle, and we may readily determine the side P Q.

Let A B = 345,29 yards, B A P = 69° 26′, B A Q = 44° 31′, P A Q = 25° 41′, A B P = 48° 15′, A B Q = 102° 14′. We deduce directly A P B = 62° 19′, A Q B = 33° 15′, and we may calculate A P and A Q as follows:

1st. *To find* A P.

$$\frac{\text{A P}}{\text{A B}} = \frac{\sin \text{A B P}}{\sin \text{A P B}}$$

log A B	2,5381840
log sin A B P	9,8727722
log sin A P B (A. C.)	0,0527973
log A P	2,4637535

A P = 290,907 yards.

2d. *To find* A Q.

$$\frac{\text{A Q}}{\text{A B}} = \frac{\sin \text{A B Q}}{\sin \text{A Q B}}$$

log A B	2,5381840
log sin A B Q	9,9900247
log sin A Q B (A. C.)	0,2609871
log A Q	2,7891958

A Q = 615,454 yards.

3d. *To find the angles* P *and* Q.

Let A Q = p, A P = q, A P Q = P, A Q P = Q. We shall have $p+q$ = 906,361, $p-q$ = 324,547. $\frac{1}{2}$ (P + Q) = 77° 9′ 30″: then, we may place

$$\frac{p+q}{p-q} = \frac{\text{tang } \frac{1}{2}(\text{P}+\text{Q})}{\text{tang } \frac{1}{2}(\text{P}-\text{Q})}$$

log tang $\frac{1}{2}$ (P + Q)	10,6421427
log ($p-q$)	2,5112776
log ($p+q$) (Ar. Com.)	7,0426988
log tang $\frac{1}{2}$ (P — Q)	10,1961191

$\frac{1}{2}$ (P — Q) = 57° 31′ 6″

P = 134° 40′ 36″ Q = 19° 38′ 24″.

4th. *To find* P Q.

$$\frac{\text{P Q}}{q} = \frac{\sin \text{P A Q}}{\sin \text{Q}}.$$

log q	2,4637535
log sin P A Q	9,6368859
log sin Q (Ar. Comp.)	0,4735196
log P Q	2,5741590

P Q = 375,110 yards.

147. *Another solution.* We have obtained the distances A P and A Q, by passing through their logarithms. We may avoid this, by using the auxiliary angle ψ, as indicated in Art. 136. Then, having found log A P, and log A Q, we obtain the angles P and Q, as follows:

To find ψ.		*To find* P *and* Q.	
$\tang \psi = \frac{AP}{AQ}$		$\tang \tfrac{1}{2}(P-Q) = \tang \tfrac{1}{2}(P+Q) + \tang (45^\circ - \psi)$	
log A P	2,4637535	log tang $\tfrac{1}{2}$ (P + Q)	10,6421427
log A Q, (Ar. Comp.)	7,2108042	log tang (45° — ψ)	9,5539790
log tang ψ	9,6745577	log tang $\tfrac{1}{2}$ (P — Q)	10,1961217
$\psi = 25^\circ\ 17'\ 50''$,		$\tfrac{1}{2}(P - Q) = 57^\circ\ 31'\ 6''$.	
$45^\circ - \psi = 19^\circ\ 42'\ 5''$			

The remainder of the solution is determined as above.

148. EXAMPLE VII. *Three points,* A, B, C, *are situated on level ground. It is required to find the point* M, *from which the distances* A B *and* A C *are seen under known angles.*

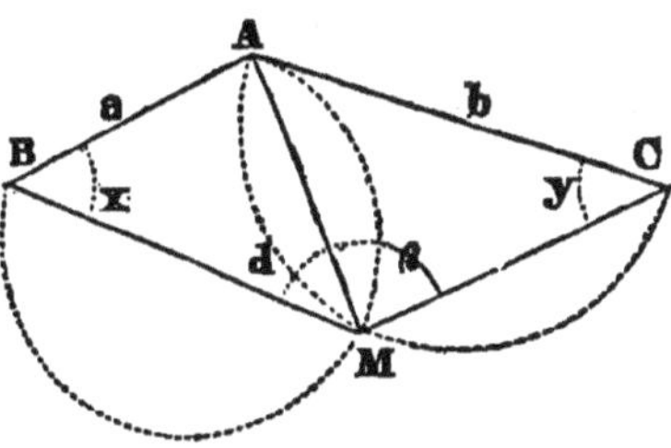

From the enunciation, A M B and A M C are known. On A B, describe a segment of a circle capable of containing the angle A M B, and on A C, a segment that will contain the angle A M C: these arcs will intersect each other in A and M, and the point M will be determined. But this construction being impracticable on the ground, we may obtain, by calculation, the angle B A M, and the distance A M. The following method by which we find first the angles A B M and A C M, seems to be the simplest.

Make the given parts $AB = a$, $AC = b$, $BAC = A$, $AMB = \alpha$, $AMC = \beta$; and let the unknown parts be represented as follows, $ABM = x$, $ACM = y$.

In the quadrilateral A B M C we have

$$x + y = 360^\circ - (A + \alpha + \beta)$$

thus the sum of the angles x and y is known. Let us now find their difference. The triangles A B M and A C M give

$$\frac{AM}{a} = \frac{\sin x}{\sin \alpha},\quad \frac{AM}{b} = \frac{\sin y}{\sin \beta}. \qquad (1)$$

Placing the values of A M drawn from these two proportions equal, we have

$$\frac{a \sin x}{\sin \alpha} = \frac{b \sin y}{\sin \beta}, \text{ or } \frac{b \sin \alpha}{a \sin \beta} = \frac{\sin x}{\sin y}.$$

Placing $a' = \dfrac{a \sin \beta}{\sin \alpha}$, the quantity a' may be readily calculated by logarithms. Hence, we have

$$\frac{b}{a} = \frac{\sin x}{\sin y}, \text{ from which we get } \frac{b + a'}{b - a'} = \frac{\sin x + \sin y}{\sin x - \sin y}$$

and therefore,

$$\frac{b + a'}{b - a'} = \frac{\operatorname{tang} \frac{1}{2}(x + y)}{\operatorname{tang} \frac{1}{2}(x - y)}$$

Since $x + y$ is known, we shall have the difference $x - y$, by this last equation, and we may determine x and y. We may then find the angle B A M $= 180° - (\alpha + x)$, and the distance A M will be made known by one of the proportions (1).

Relations between the Angles and Sides of Spherical Triangles.

149. *Fundamental Formula.* The parts of a spherical triangle, traced on the surface of a given sphere, are known, when we know the number of degrees which each of them contains. The solution of questions relating to spherical triangles depends then upon the relations which exist between the number of degrees, or, in other words, between the corresponding trigonometrical numbers, *sine*, *cosine*, &c. Consequently, we propose, in the first place, to establish the formula which connects any angle whatever with the three sides of the given triangle, and we will afterwards show how we may deduce from this formula the solution of all cases which may be presented in spherical triangles. Angles will always be designated by the large letters A, B, C, and the opposite sides by the small letters a, b, c.

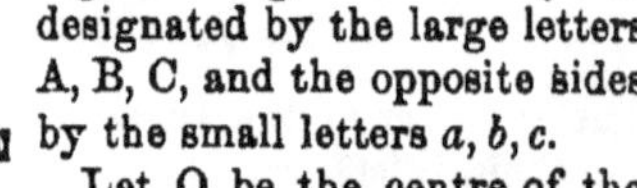

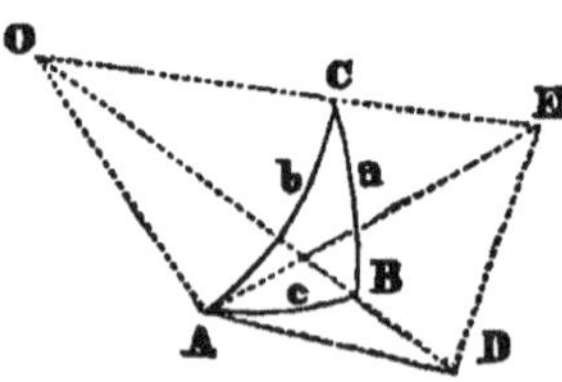

Let O be the centre of the sphere on which is situated the triangle A B C; draw the radii O A, O B, O C: erect on O A the perpendiculars AD and A E, one in the plane O A B, the other in the plane O A C, and let them meet in D and E the prolongation of the radii O B and O C. The angle D A E is equal to the angle A of the spherical triangle; and taking O A for unity, we have $AD = \text{tang}\, c$, $OD = \sec c$, $AE = \text{tang}\, b$, $OE = \sec b$.

The triangles D A E, D O E give

$$\overline{AD}^2 + \overline{AE}^2 - 2AD \times AE . \cos A = \overline{DE}^2$$
$$\overline{OD}^2 + \overline{OE}^2 - 2OD \times OE . \cos a = \overline{DE}^2$$

Taking the first equation from the second, and observing that $\overline{OD}^2 - \overline{AD}^2 = \overline{OE}^2 - \overline{AE}^2 = 1$; and substituting for the lines their trigonometrical designations, and dividing by 2, we have

$$1 - \sec b \sec c \cos a + \text{tang}\, b \,\text{tang}\, c \cos A = 0.$$

But $\sec b = \frac{1}{\cos b}$, $\text{tang}\, b = \frac{\sin b}{\cos b}$, $\sec c = \frac{1}{\cos c}$, $\text{tang}\, c = \frac{\sin c}{\cos c}$; we have therefore,

$$1 - \frac{\cos a}{\cos b \cos c} + \frac{\sin b \sin c \cos A}{\cos b \cos c} = 0.$$

Getting rid of the denominator, and transposing $\cos a$ to the first member, we have

$$\cos a = \cos b \cos c + \sin b \sin c \cos A, \qquad (1)$$

which is the fundamental formula of spherical Trigonometry.

150. In the figure the sides b and c are less than 90°, but it is easy to see this formula (1) is general. Suppose one of the sides, as b or A C, exceeds 90°. Complete the semi-circumferences CAC′, CBC′, forming thus the triangle A B C′, in which the sides a' and b', or B C′ and A C′, are supplements of a and b, and the angle B A C′ is the supplement of A. Since the sides b' and c are less than 90°, the formula (1) is applicable to the triangle A B C′, and gives

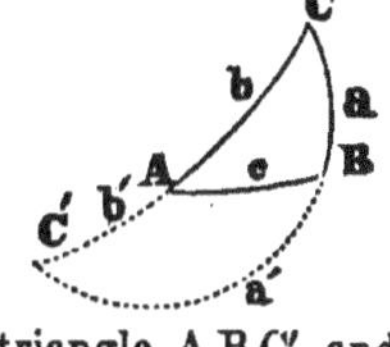

$$\cos a' = \cos b' \cos c + \sin b' \sin c \cos BAC'.$$

But $a' = 180° - a$, $b' = 180° - b$, $BAC' = 180° - A$. Substituting these values, and changing the signs of both members, we reproduce formula (1), hence it applies to cases in which $b > 90°$.

If the two sides b and c exceed 90°, produce them until they meet in A′ and form the triangle B C A′, in which the angle A′ is equal to A, and the sides b' and c' are equal to the supplements of b and c. Formula (1) applied to this triangle must hold good when $180° - b$ and $180° - c$ are substituted for b and c. But these substitutions produce no change in this formula, and it is again verified.

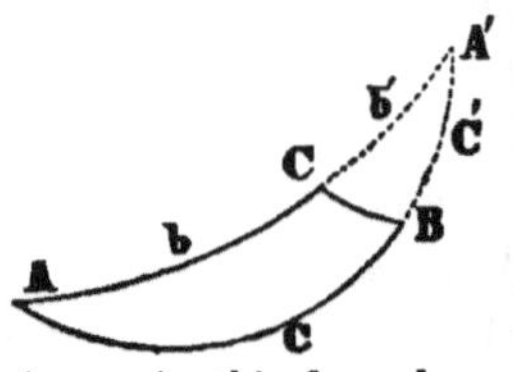

Finally, we may verify the formula in the case in which $b = 90°$ and $c = 90°$, either separately or together. But since the formula is true for values of b and c very near 90°, the formula must hold good in these particular cases.

151. If the angle $A = 90°$: we have $\cos A = 0$, and the general formula becomes $\cos a = \cos b \cos c$, that is, *in a right-angled spherical triangle, the cosine of the hypothenuse is equal to the product of the cosines of the two sides containing the right angle.*

152. We may apply formula (1) to each of the sides of the triangle, and we shall thus have three equations by means of which we may always determine any three parts of a spherical triangle when the other three parts are known. But in practice, it is necessary to have, separately, the various relations which exist between four parts of the triangle, taken in every possible combination. There are only altogether four distinct combinations—and we propose to examine these successively.

153. 1st. *Relations between the three sides and an angle.* Formula (1) gives, by permutation of the letters, the following relations:

$$\cos a = \cos b \cos c + \sin b \sin c \cos A \qquad (1)$$

$$\cos b = \cos a \cos c + \sin a \sin c \cos B \qquad (2)$$

$$\cos c = \cos a \cos b + \sin a \sin b \cos C \qquad (3)$$

To calculate A, for example, by means of the three sides, we make use of the first equation, and we draw from it

$$\cos A = \frac{\cos a - \cos b \cos c}{\sin b \sin c}$$

Then, with this value, we find from it others better adapted to logarithms, by obtaining the values of sin $\frac{1}{2}$ A, cos $\frac{1}{2}$ A, tang $\frac{1}{2}$ A, as we have done in plane triangles. Let us take the formula $2\sin^2\frac{1}{2}A = 1 - \cos A$, and substitute for cos A, the value deduced above. We have

$$2\sin^2\tfrac{1}{2}A = \frac{\cos b\cos c + \sin b\sin c - \cos a}{\sin b\sin c} = \frac{\cos(b-c) - \cos a}{\sin b\sin c}.$$

In the known formula $\cos q - \cos p = 2\sin\frac{1}{2}(p+q)\sin\frac{1}{2}(p-q)$, make $p = a$, $q = b - c$; we have $\cos(b-c) - \cos a = 2\sin\frac{1}{2}(a+b-c)\sin\frac{1}{2}(a-b+c)$; hence,

$$\sin\tfrac{1}{2}A = \sqrt{\frac{\sin\frac{1}{2}(a+b-c)\sin\frac{1}{2}(a-b+c)}{\sin b\sin c}}.$$

If we take the formula $2\cos\frac{1}{2}A = 1 + \cos A$, we shall find, by a like operation, the value of cos $\frac{1}{2}$ A,

$$\cos\tfrac{1}{2}A = \sqrt{\frac{\sin\frac{1}{2}(a+b+c)\sin\frac{1}{2}(b+c-a)}{\sin b\sin c}}.$$

Then, knowing the values of sin $\frac{1}{2}$ A and cos $\frac{1}{2}$ A, we may deduce tang $\frac{1}{2}$ A.

Place $a + b + c = 2s$, we shall have $a + b - c = 2(s-c)$, $a - b + c = 2(s-b)$, $b + c - a = 2(s-a)$, and the three formulæ may be written,

$$\sin\tfrac{1}{2}A = \sqrt{\frac{\sin(s-b)\sin(s-c)}{\sin b\sin c}}$$

$$\cos\tfrac{1}{2}A = \sqrt{\frac{\sin s\sin(s-a)}{\sin b\sin c}}$$

$$\text{tang}\tfrac{1}{2}A = \sqrt{\frac{\sin(s-b)\sin(s-c)}{\sin s\sin(s-a)}}$$

By a simple change in the letters, we may deduce analogous formulæ for $\frac{1}{2}$ B and $\frac{1}{2}$ C.

154. It is shown in geometry that the sum of the three sides of a spherical triangle is less than a circumference of a great circle. Calling δ the difference, we shall have $s = 180° - \frac{1}{2}\delta$, $s - a = 180° - (a + \frac{1}{2}\delta)$, &c.; and since the sine of an arc is equal to the sine of its supplement, we may give the preceding formulæ in the following form:

$$\sin \tfrac{1}{2} A = \sqrt{\frac{\sin (b + \tfrac{1}{2}\delta) \sin (c + \tfrac{1}{2}\delta)}{\sin b \sin c}}$$

$$\cos \tfrac{1}{2} A = \sqrt{\frac{\sin \tfrac{1}{2}\delta \sin (a + \tfrac{1}{2}\delta)}{\sin b \sin c}}$$

$$\operatorname{tang} \tfrac{1}{2} A = \sqrt{\frac{\sin (b + \tfrac{1}{2}\delta) \sin (c + \tfrac{1}{2}\delta)}{\sin \tfrac{1}{2}\delta \sin (a + \tfrac{1}{2}\delta)}}$$

155. 2d. *Relation between two sides and the two opposite angles.*

To find that which corresponds to the combination a, b, A, B, it is necessary to eliminate c between (1) and (2). The method which at once presents itself, is to deduce the values of $\sin c$ and $\cos c$, and substitute them in equation $\sin^2 c + \cos c^2 = 1$. But the following process, analogous to that adopted Art. 123, is the most simple.

Equation (1) gives

$$\cos A = \frac{\cos a - \cos b \cos c}{\sin b \sin c}$$

Hence, we have

$$\sin^2 A = 1 - \cos^2 A = 1 - \frac{(\cos a - \cos b \cos c)^2}{\sin^2 b \sin^2 c}$$

$$= \frac{(1 - \cos^2 b)(1 - \cos^2 c) - (\cos a - \cos b \cos c)^2}{\sin^2 b \sin^2 c}$$

$$= \frac{1 - \cos^2 a - \cos^2 b - \cos^2 c + 2 \cos a \cos b \cos c}{\sin^2 b \sin^2 c}$$

Therefore,

$$\frac{\sin A}{\sin a} = \sqrt{\frac{1 - \cos^2 a - \cos^2 b - \cos^2 c + 2 \cos a \cos b \cos c}{\sin a \sin b \sin c}}$$

There is no ambiguity here in the sign of the radical, inasmuch as the angles and sides being less than 90°, their sines will be positive.

Since the second member remains constant when we change A and a into B and b, and reciprocally, or into C and c, and the reverse, we conclude, that

$$\frac{\sin A}{\sin a} = \frac{\sin B}{\sin b} = \frac{\sin C}{\sin c} \qquad (4)$$

Hence, *in any spherical triangle, the sines of the angles are proportionate to the sines of the opposite sides.*

156. 3d. *Relation between two sides, the angle included by them, and the angle opposite one of them.*

Considering the combination a, b, A, C, eliminate $\cos c$, between (1) and (3), and we have

$$\cos a = \cos a \cos^2 b + \cos b \sin a \sin b \cos C + \sin b \sin c \cos A.$$

Transposing $\cos a \cos^2 b$, observing that $\cos a - \cos a \cos^2 b = \cos a \sin^2 b$, and dividing by $\sin b \sin a$, we have

$$\frac{\cos a \sin b}{\sin a} = \cos b \cos C + \frac{\sin c \cos A}{\sin a}$$

But $\frac{\sin c}{\sin a} = \frac{\sin C}{\sin A}$, hence, we have for the required relation

$$\cot a \sin b = \cos b \cos C + \sin C \cot A.$$

By making the different permutations in the letters, we may deduce the following six equations:

$$\cot a \sin b = \cos b \cos C + \sin C \cot A \qquad (5)$$
$$\cot b \sin a = \cos a \cos C + \sin C \cot B \qquad (6)$$
$$\cot a \sin c = \cos c \cos B + \sin B \cot A \qquad (7)$$
$$\cot c \sin a = \cos a \cos B + \sin B \cot C \qquad (8)$$
$$\cot b \sin c = \cos c \cos A + \sin A \cot B \qquad (9)$$
$$\cot c \sin b = \cos b \cos A + \sin A \cot C. \qquad (10)$$

157. 4th. *Relation between a side and the three angles.* This is the last relation to be determined. Eliminate b and c, between the equations (1), (2), and (3). For this purpose, substitute in equation (1) the value of $\cos c$ deduced from (3), and their results:

$$\frac{\cos a \sin b}{\sin a} = \cos b \cos C + \frac{\sin c \cos A}{\sin a}$$

and this relation, by means of the equations

$$\frac{\sin b}{\sin a} = \frac{\sin B}{\sin A} \text{ and } \frac{\sin c}{\sin a} = \frac{\sin C}{\sin A}$$

becomes $\cos a \sin B = \cos b \sin A \cos B + \cos A \sin C$.

Performing the same operations on equation (2), or rather changing, in this last equation, a and A into b and B, and *vice versa*, we have

$$\cos b \sin A = \cos a \sin B \cos C + \cos B \sin C$$

We have now only to eliminate $\cos b$ between the two last equations. This gives us, after making the necessary reductions, the required relation between A, B, C and a; and if we apply this relation to each of the three angles in succession, we deduce the three following equations:

$$\cos A = -\cos B \cos C + \sin B \sin C \cos a \qquad (11)$$
$$\cos B = -\cos A \cos C + \sin A \sin C \cos b \qquad (12)$$
$$\cos C = -\cos A \cos B + \sin A \sin B \cos c \qquad (13)$$

158. It is easy to express $\sin \frac{1}{2} a$, $\cos \frac{1}{2} a$, $\text{tang} \frac{1}{2} a$, in functions of the angles A, B, C. We deduce from formula (11),

$$\cos a = \frac{\cos A + \cos B \cos C}{\sin B \sin C.}$$

By operations analogous to those in Art. 153, we readily deduce the formulæ required. Placing $A + B + C = 180° + \epsilon$, these formulæ become

$$\sin \tfrac{1}{2} a = \sqrt{\frac{\sin \frac{1}{2} \epsilon \sin (A - \frac{1}{2}\epsilon)}{\sin B \sin C.}}$$

$$\cos \tfrac{1}{2} a = \sqrt{\frac{\sin (B - \frac{1}{2}\epsilon) \sin (C - \frac{1}{2}\epsilon)}{\sin B \sin C.}}$$

$$\text{tang} \tfrac{1}{2} a = \sqrt{\frac{\sin \frac{1}{2} \epsilon \sin (A - \frac{1}{2}\epsilon)}{\sin (B - \frac{1}{2}\epsilon) \sin (C - \frac{1}{2}\epsilon)}}$$

The quantity ϵ is called in geometry *the spherical excess.*

159. The analogy between equations (11), (12), (13) and the fundamental formula (1) is striking, and leads to a remarkable consequence. Let us conceive a spherical triangle A′ B′ C′, the sides of which, a', b', c', are the supplements of the angles A, B, C. By virtue of formula (1) we shall have $\cos a' = \cos b' \cos c' + \sin b' \sin c' \cos A'$. But $\sin a' = \sin A$, $\cos a' = -\cos A$, $\sin b' = \sin B$, &c. Hence

$$-\cos A = \cos B \cos C + \sin B \sin C \cos A'$$

From this equation we get for $\cos A'$ a value equal and with a contrary sign to that which equation (11) gives for $\cos a'$: hence, $a = 180° - A'$. In like manner, $b = 180° - B'$, and $c = 180° - C'$. Therefore, *a spherical triangle being given, if we form a second triangle from it, so that the sides of this new triangle shall be supplements of the angles of the first triangle, the sides of the first triangle will be also the supplements of the angles of the second triangle.*

160. These triangles are said to be *supplementary.* Since each of these triangles may be described by taking as *poles* the three vertices of the other, as is shown in geometry, this property leads to a designation of these triangles as *polar* one to the other.

161. By the *polar* triangle, the formulæ found above for $\frac{1}{2}a$ may be deduced from those which are known for $\frac{1}{2}A$ (Art. 153). Indeed, we can apply these formulæ to the polar triangle itself, and for this purpose it is only necessary to replace the angles by the supplements of the sides and the sides by the supplements of the angles. We thus reproduce the values of $\sin\frac{1}{2}a$, $\cos\frac{1}{2}a$, $\text{tang}\,\frac{1}{2}a$.

162. *Proportion of the four tangents.* We proceed now to demonstrate some formulæ which are of frequent use by astronomers. Taking the proportion of the sines $\frac{\sin A}{\sin B}=\frac{\sin a}{\sin b}$, we deduce, in the first place,

$$\frac{\sin A+\sin B}{\sin A-\sin B}=\frac{\sin a+\sin b}{\sin a-\sin b}.$$

Replacing now each ratio, by a ratio of tangents, from known relations, we have

$$\frac{\text{tang}\,\frac{1}{2}(A+B)}{\text{tang}\,\frac{1}{2}(A-B)}=\frac{\text{tang}\,\frac{1}{2}(a+b)}{\text{tang}\,\frac{1}{2}(a-b)}$$

This formula is called the *proportion of the four tangents.*

163. *Proportion of the four cosines.* The student may readily construct the figure for himself. From the vertex C of any triangle draw the arc of a great circle perpendicular to the side A B. Designate the perpendicular by p, and by α and β the segments adjacent to the sides a and b. The two right-angled triangles will give $\cos a=\cos p\cos\alpha$, $\cos b=\cos p\cos\beta$ (Art. 150): and from these results, we deduce the proportion of the *four cosines.*

$$\frac{\cos a}{\cos b}=\frac{\cos\alpha}{\cos\beta}.$$

164. *Formulæ of Delambre.* These formulæ which bear the name of the distinguished astronomer who first published them are:

$$\sin\tfrac{1}{2}(A+B)=\cos\tfrac{1}{2}C\,\frac{\cos\tfrac{1}{2}(a-b)}{\cos\tfrac{1}{2}c}$$

$$\sin\tfrac{1}{2}(A-B)=\cos\tfrac{1}{2}C\,\frac{\sin\tfrac{1}{2}(a-b)}{\sin\tfrac{1}{2}c}$$

$$\cos\tfrac{1}{2}(A+B)=\sin\tfrac{1}{2}C\,\frac{\cos\tfrac{1}{2}(a+b)}{\cos\tfrac{1}{2}c}$$

$$\cos\tfrac{1}{2}(A-B)=\sin\tfrac{1}{2}C\,\frac{\sin\tfrac{1}{2}(a+b)}{\sin\tfrac{1}{2}c}.$$

They are readily deduced by finding the expressions for the sines and cosines of the angles $\tfrac{1}{2}(A+B)$, $\tfrac{1}{2}(A-B)$ in functions of the sides a, b, c.

For example, let us find $\sin\tfrac{1}{2}(A+B)$. In the first place, we have $\sin\tfrac{1}{2}(A+B)=\sin\tfrac{1}{2}A\cos\tfrac{1}{2}B+\sin\tfrac{1}{2}B\cos\tfrac{1}{2}A$. Substituting for the sines and cosines, in the second member, their values expressed in terms of the sides, we have

$$\sin\tfrac{1}{2}(A+B)=\begin{cases}\sqrt{\dfrac{\sin(s-b)\sin(s-c)}{\sin b\sin c}}\sqrt{\dfrac{\sin s\sin(s-b)}{\sin a\sin c}}\\[2ex] +\sqrt{\dfrac{\sin(s-a)\sin(s-c)}{\sin a\sin c}}\sqrt{\dfrac{\sin s\sin(s-a)}{\sin b\sin c}}\end{cases}$$

$$=\frac{\sin(s-b)+\sin(s-a)}{\sin c}\sqrt{\frac{\sin s\sin(s-c)}{\sin a\sin b}}$$

The last radical is equal to $\cos\tfrac{1}{2}C$. The numerator $\sin(s-b)+(s-c)$ is a sum which may be replaced by a product, and the denominator $\sin c$ is equal to $2\sin\tfrac{1}{2}c\cos\tfrac{1}{2}c$: we have, therefore,

$$\frac{\sin(s-b)+\sin(s-a)}{\sin c}=\frac{2\sin\dfrac{s-a+s-b}{2}\cos\dfrac{s-a-s+b}{2}}{2\sin\tfrac{1}{2}c\cos\tfrac{1}{2}c.}$$

$$=\frac{\sin\tfrac{1}{2}(2s-a-b)\cos\tfrac{1}{2}(a-b)}{\sin\tfrac{1}{2}c\cos\tfrac{1}{2}c.}$$

Recollecting that $2s=a+b+c$, we have $2s-a-b=c$, and the preceding fraction is reduced to $\dfrac{\cos\tfrac{1}{2}(a-b)}{\cos\tfrac{1}{2}c.}$. We have therefore

$$\sin \tfrac{1}{2}(\mathrm{A}+\mathrm{B}) = \cos \tfrac{1}{2}\mathrm{C}\,\frac{\cos \tfrac{1}{2}(a-b)}{\cos \tfrac{1}{2}c},$$

which is the first of *Delambre's* formulæ. The three other formulæ may be obtained by analogous operations.

165. It might be supposed that in applying these formulæ to the *polar* triangle we should obtain new formulæ. But this is not the case. For example: replacing A, B, C, in the 1st by the supplements of the sides, and reciprocally, it becomes $\sin \tfrac{1}{2}(a+b) = \sin \tfrac{1}{2}c\,\dfrac{\cos \tfrac{1}{2}(\mathrm{A}-\mathrm{B})}{\sin \tfrac{1}{2}\mathrm{C}}$, and this is no other than the 4th formula, under a little different form. Reciprocally, the 4th formula will reproduce the 1st. As to the 2d, it will be made to reduce to itself, with a slight change in form: and so also with the 3d.

166. *Napier's Analogies.* These remarkable *analogies* or proportions are embraced in the following formulæ:

$$\tan \tfrac{1}{2}(\mathrm{A}+\mathrm{B}) = \cot \tfrac{1}{2}\mathrm{C}\,\frac{\cos \tfrac{1}{2}(a-b)}{\cos \tfrac{1}{2}(a+b)}$$

$$\tan \tfrac{1}{2}(\mathrm{A}-\mathrm{B}) = \cot \tfrac{1}{2}\mathrm{C}\,\frac{\sin \tfrac{1}{2}(a-b)}{\sin \tfrac{1}{2}(a+b)}$$

$$\tan \tfrac{1}{2}(a+b) = \tan \tfrac{1}{2}c\,\frac{\cos \tfrac{1}{2}(\mathrm{A}-\mathrm{B})}{\cos \tfrac{1}{2}(\mathrm{A}+\mathrm{B})}$$

$$\tan \tfrac{1}{2}(a-b) = \tan \tfrac{1}{2}c\,\frac{\sin \tfrac{1}{2}(\mathrm{A}-\mathrm{B})}{\sin \tfrac{1}{2}(\mathrm{A}+\mathrm{B})}$$

To demonstrate these formulæ directly, we should have to go over the same calculations which have just been made for the formulæ of *Delambre.* It will therefore be more simple to deduce them at once from these formulæ.

If we divide $\sin \tfrac{1}{2}(\mathrm{A}+\mathrm{B})$ by $\cos \tfrac{1}{2}(\mathrm{A}+\mathrm{B})$, and $\sin \tfrac{1}{2}(\mathrm{A}-\mathrm{B})$ by $\cos \tfrac{1}{2}(\mathrm{A}-\mathrm{B})$ in *Delambre's* formulæ, we at once deduce the two first analogies. In like manner, by dividing the 4th formula of *Delambre* by the 3d, and the 2d by the 1st, we obtain the two last analogies. We may also deduce the two last analogies from the two first by means of the polar triangle.

167. *Napier's Analogies* serve to simplify some cases in the resolution of spherical triangles. We make use of the two first when we have two sides and the included angle given;

and the two last, when we know two angles and the included side.

168. *Relations between the parts of a right-angled spherical triangle.*

To deduce the formulæ which correspond to the particular case of the right-angled triangle, it is only necessary to make $A = 90°$ in the relations heretofore found, which contain this angle. In this manner, we find the following formulæ:

Art. 153, gives $\cos a = \cos b \cos c$ (a)

Art. 155, " $\sin b = \sin a \sin B, \sin c = \sin a \sin C$ (b)

Art. 156, " $\text{tang}\, b = \text{tang}\, a \cos C, \text{tang}\, c = \text{tang}\, a \cos B,$ (c)

Art. 156, " $\text{tang}\, b = \sin c\, \text{tang}\, B, \text{tang}\, c = \sin b\, \text{tang}\, C,$ (d)

Art. 157, " $\cos B = \sin C \cos b, \quad \cos C = \sin B \cos c,$ (e)

Art. 157, " $\cos a = \cot B \cot C.$ (f)

We have thus six distinct formulæ equally adapted to logarithmic calculation. The first gives a relation between the hypothenuse and the two sides of the right angle; the second, between the hypothenuse, a side and the angle opposite; the third, between the hypothenuse, a side and the adjacent angle; the fourth, between the two sides and the angle opposite one of them; the fifth, between a side and and the two oblique angles; and the sixth, between the hypothenuse and the oblique angles. Thus, any two of the five parts being known, we have a formula to determine all the other parts.

169. Certain properties of right-angled triangles must be noticed here.

1st. Formula (a) requires that $\cos a$ have the sign of the product $\cos b \cos c$: but, for this purpose, it is necessary that the three cosines be positive, or only one positive. Hence, *in a right-angled spherical triangle, the three sides are less than* 90°; *or else two of the sides are greater than* 90°, *and the third less.*

2d. The formulæ (d) show that $\text{tang}\, b$ has the same sign as $\text{tang}\, B$, and $\text{tang}\, c$ the same sign as $\text{tang}\, C$. Hence, *each side of the right angle is of the same species as the opposite angle: that is, the angle and the side are both less than* 90°, *or both greater.*

170. *Napier's Rules.* The formulæ given in Art. 168 are embodied with great elegance and convenience in *Napier's Rules*, the enunciation and proof of which are as follows.

The right angle being left out of consideration, the two sides including the right angle, and the complements of the hypothenuse and of the two other angles, are called the five *circular parts.* Any one of these being taken as the *middle* part, the two circular parts which are immediately contiguous to it, in position, are called the *adjacent* parts; and the other two parts, considered with reference to the same part as the *middle* part, are called the *opposite* parts. Then, whatever the middle part be, whether a side, or the complement of the hypothenuse, or the complement of an angle, we have always

1st. *The sine of the middle part equal to the product of the tangents of the adjacent parts.*

2d. *The sine of the middle part equal to the product of the cosines of the opposite parts.*

1st. Let the complement of the hypothenuse, $90° - a$, be the *middle* part, then the complements of the two oblique angles, viz. $90° - \text{B}$, $90° - \text{C}$, are the *adjacent* parts, and the two sides b and c, the *opposite* parts; and formula (11), Art. 157, and formula (1), Art. 149,

$$\cos \text{A} = -\cos \text{B} \cos \text{C} + \sin \text{B} \sin \text{C} \cos a$$
$$\cos a = \cos b \cos c + \sin b \sin c \cos \text{A},$$

become, since $\text{A} = 90°$, and therefore $\cos \text{A} = 0$ and $\sin \text{A} = 1$,

$\cos a = \cot \text{B} \cot \text{C}$, or, $\sin (90° - a) = \text{tang}(90° - \text{B}) \times \text{tang} (90° - \text{C})$,

$\cos a = \cos b \cos c$, or, $\sin (90° - a) = \cos b \cos c$,

which prove the rules.

2d. Let the complement of the angle C, $90° - \text{C}$, be the *middle* part; then the complement of the hypothenuse, $90° - a$, and the side b, are the *adjacent* parts; and the complement of the angle B, $90° - \text{B}$, and the side c, are the *opposite* parts: and the formulæ,

$\cot a \sin b = \cos b \cos \text{C} + \sin \text{C} \cot \text{A}$ (5), Art. 156.

$\cos \text{C} = -\cos \text{A} \cos \text{B} + \sin \text{A} \sin \text{B} \cos c$ (13), Art. 157.

become, since $\text{A} = 90°$,

$\cos C = \cot a \tan b$, or, $\sin (90^\circ - C) = \tan (90^\circ - a) \tan b$,
$\cos C = \sin B \cos c$, or, $\sin (90^\circ - C) = \cos (90^\circ - B) \cos c$.

which prove the rules again, with reference to the part assumed as middle part.

If $90^\circ - B$, be taken for the *middle* part, the rules may be proved exactly in the same way, and other two similar relations found.

3d. Let the side b be the *middle* part, then $90^\circ - C$ and the side c will be the *adjacent* parts, and $90^\circ - a$, and $90^\circ - B$, the *opposite* parts; and the formulæ

$$\cot c \sin b = \cos b \cos A + \sin A \cot C, \quad (10), \text{ Art. } 156,$$
$$\sin A \sin b = \sin a \sin B. \quad (4), \text{ Art. } 155,$$

become, when $A = 90^\circ$,

$\sin b = \cot C \tan c$, or, $\sin b = \tan (90^\circ - C) \tan c$,
$\sin b = \sin a \sin B$, or, $\sin b = \cos (90^\circ - a) \cos (90^\circ - B)$,

which show that the rules hold good in this case also. If c be the middle part, the rules may be proved in the same way, and other two similar relations found.

171. We are thus furnished with *ten* relations amongst the five parts of a right-angled triangle, each being a different combination of three of the parts; but five parts, taken three at a time, can be combined only in 10 different ways; consequently, the above *Rules*, when any two parts whatever are given, will supply us with formulæ in which each of the remaining parts is *separately* combined with these two given parts, and in a form adapted to the immediate application of logarithms.

172. There can evidently be only *six* distinct cases, viz.: I and II, when the hypothenuse is given together with an angle, or with one of the sides containing the right angle; III and IV, when one of the sides of the right angle is given together with the angle adjacent to it, or with the angle opposite to it; V and VI, when the two sides containing the right angle are given, or when the two angles are given. In applying *Napier's Rules*, to obtain the three unknown parts from two given ones, it is sometimes requisite to take for a middle part one of the given parts, and sometimes one of those that are sought, the sole object being

to separately combine each of the unknown parts with the given parts.

173. A spherical triangle may be *bi-rectangular* or *tri-rectangular*, that is, two of its angles may be right angles, or all three may be right. In the last case, the three sides are quadrants; in the first, the sides opposite to the two right angles are also quadrants; and the third angle having for a measure the third side, must be expressed by the same number of degrees as this side. Thus, these two cases not giving rise to any particular investigation, it will be only necessary to refer to the spherical triangle which contains one right angle.

174. *Case* I. *Having given the hypothenuse* a, *and an angle* B, *to find* b, c, *and* C.

Taking successively b, $90° - B$, and $90° - a$ for the *middle* part, we get,

$$\sin b = \sin a \sin B,\ \cos B = \cot a \operatorname{tang} c,\ \cos a = \cot B \cot C:$$

C and c are determined without ambiguity; b and B must be both greater or both less than 90°.

175. *Case* II. *Having given the hypothenuse* a, *and a side* b, *to find* c, B, *and* C.

Taking successively $90° - a$, b, and $90° - C$ for the middle part, we get

$$\cos a = \cos b \cos c,\ \sin b = \sin a \sin B,\ \cos C = \cot a \operatorname{tang} b.$$

By these formulæ, c and C are determined without ambiguity, for there is only one angle less than 180° corresponding to a given cosine; and B, determined by its sine, is not ambiguous since it must be of the same species as b.

176. *Case* III. *Having given one of the sides* b *containing the right angle, and the angle* C *adjacent to it, to find* a, c, *and* B.

Taking successively $90° - C$, b, and $90° - B$ for the middle part, we get

$$\cos C = \operatorname{tang} b \cot a,\ \sin b = \operatorname{tang} c \cot C,\ \cos B = \cos b \sin C,$$

which determine a, c, and B, without ambiguity.

177. *Case* IV. *Having given one of the sides*, b, *of the right angle, and the angle* B, *opposite to it, to find* a, c, *and* C.

Taking b, c, $90° - B$, successively for the middle part, we get

$$\sin b = \sin B \sin a,\ \sin c = \cot B \operatorname{tang} b,\ \cos B = \sin C \cos b.$$

Here, since all the unknown parts are determined by their sines, and since there are always two angles less than 180° corresponding to a given sine, an ambiguity presents itself, and it is easily seen that such ought to be the case. For if in the triangle B A C, right-angled at A, we produce B A and B C, until they meet in D, then taking D A′ = B A, and D C′ = B C, the triangles B A C, D A′ C′, will be equal in all their parts; hence, the angle A′ is right, and C′ A′ = C A = b. The triangle B A′ C′ is thus rectangular, and contains also the two given parts B and b. We may then take, at will, $a < 90°$ or $a > 90°$, but when the choice is made, the species of c will be given by the relation $\cos a = \cos b \cos c$, and the species will be the same as that of C.

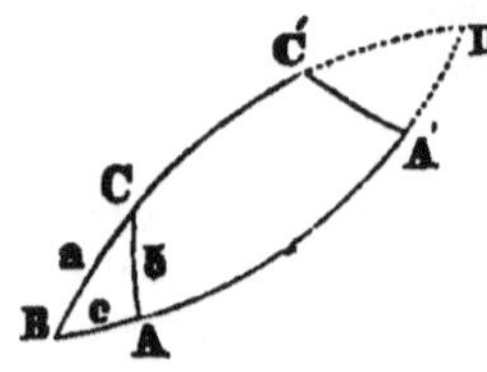

178. *Case* V. *Having given the two sides* b *and* c *containing the right angle, to find* a, B, C.

Taking successively $90° - a$, c, and b for the *middle* part, we get

$$\cos a = \cos b \cos c, \quad \sin c = \cot B \operatorname{tang} b, \quad \sin b = \cot C \operatorname{tang} c,$$

which determine a, B, C, without ambiguity.

179. *Case* VI. *Having given the two oblique angles* B *and* C, *to find* a, b, c.

Taking successively $90° - a$, $90° - B$, $90° - C$, for the *middle* part, we get

$$\cos a = \cot B \cot C, \quad \cos B = \cos b \sin C, \quad \cos C = \cos c \sin B.$$

These formulæ give no ambiguity, and if the triangle be impossible they will show it.

180. *Observations.* The solution of many cases of oblique-angled spherical triangles may be reduced to that of right-angled triangles.

1st. If in an oblique-angled spherical triangle, three parts be given, one of which is a side equal to 90°, the corresponding angle in the *polar* triangle will be a right angle. Besides, we shall know two of the five parts of this triangle, hence it may be solved by rules just given; and it is evident, when the parts of the polar triangle are known, we may readily determine those of the given triangle.

2d. When a spherical triangle is isosceles, the two equal sides are considered as only one element, and the two equal angles opposite as only one element, and in this case any two parts of the triangle being given, we may determine the triangle. For this purpose we have only to draw from the vertex to the middle point of the base the arc of a great circle, and the given triangle will be divided into two right-angled triangles equal in all respects, and in each of which two parts will be known, besides the right angle, and thus all the parts of the given triangle are determined by the Rules for the solution of right-angled triangles.

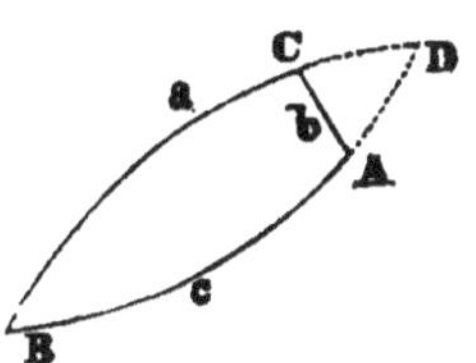

3d. Let A B C be a spherical triangle, in which we have $a+b=180°$. Producing a and c until they meet in D, we shall have $a+\mathrm{CD}=180°$; hence, $\mathrm{CD}=b$. But each known element of the triangle ABC makes known one element in the isosceles triangle AC D, and reciprocally; hence the resolution of a triangle in which the sum of two sides is equal to 180° reduces to that of an isosceles triangle, and consequently to that of a right-angled triangle.

4th. The same remark applies to a spherical triangle in which two angles are supplements of each other; for we cannot have $a+b=180°$, without having at the same time $\mathrm{A}+\mathrm{B}=180°$, and *vice versa.* Indeed, in the isosceles triangle A C D, the angle $\mathrm{CAD}=\mathrm{D}=\mathrm{B}$: but $\mathrm{CAD}+\mathrm{CAB}=180°$; hence, in the triangle A B C, we have also $\mathrm{A}+\mathrm{B}=180°$.

RESOLUTION OF OBLIQUE SPHERICAL TRIANGLES.

181. *Case* I. *Having given the three sides* a, b, c, *to find the three angles* A, B, C.

Equation (1), Art. 153, gives

$$\cos \mathrm{A}=\frac{\cos a-\cos b \cos c}{\sin b \sin c};$$

but as this formula is not adapted to logarithmic calculation, we make use of the formulæ for $\sin \frac{1}{2} A$, $\cos \frac{1}{2} A$, $\operatorname{tang} \frac{1}{2} A$, given in Art. 153.

$$\sin \tfrac{1}{2} A = \sqrt{\frac{\sin(s-b)\sin(s-c)}{\sin b \sin c}}$$

$$\cos \tfrac{1}{2} A = \sqrt{\frac{\sin s \sin(s-a)}{\sin b \sin c}}$$

$$\operatorname{tang} \tfrac{1}{2} A = \sqrt{\frac{\sin(s-b)\sin(s-c)}{\sin s \sin(s-a)}}$$

182. *Case* II. *Having given two sides* a *and* b, *and the angle* A *opposite one of them, to find* c, B, *and* C.

We obtain the angle B, opposite the side b, immediately, by the proportion

$$\frac{\sin B}{\sin b} = \frac{\sin A}{\sin a}, \text{ hence, } \sin B = \frac{\sin A \sin b}{\sin a.}$$

To find c and C, we make use of *Napier's analogies*, Art. 166, which give

$$\operatorname{tang} \tfrac{1}{2} c = \operatorname{tang} \tfrac{1}{2}(a-b)\frac{\sin \frac{1}{2}(A+B)}{\sin \frac{1}{2}(A-B)}$$

$$\cot \tfrac{1}{2} C = \operatorname{tang} \tfrac{1}{2}(A-B)\frac{\sin \frac{1}{2}(a+b)}{\sin \frac{1}{2}(a-b)}$$

The angle B being determined by its sine, may be either acute or obtuse. Nevertheless, for certain values of the given parts a, b, A, only one triangle exists. We will resume the consideration of this case hereafter, presenting as it does an analogous discussion to that which has been made in Case II of plane triangles.

We may also determine C directly by making use of the formula (5), Art. 156,

$$\cot A \sin C + \cos b \cot C = \cot a \sin b. \qquad (5)$$

For this purpose, we will first determine an auxiliary angle φ, by making $\cot A = \cos b \cot \varphi$, which gives

$$\cot \varphi = \frac{\cot A}{\cos b}.$$

Then, substituting the value of $\cot A = \cos b \cot \varphi = \frac{\cos b \cos \varphi}{\sin \varphi}$ in equation (5), this equation becomes

$$\cos b\,(\sin C \cos\phi + \cos C \sin\phi) = \cot a \sin b \sin\phi,$$

from which we deduce

$$\sin(C+\phi) = \frac{\operatorname{tang} b \sin\phi}{\operatorname{tang} a}$$

which makes known $c+\phi$. Put $c+\phi = m$, we have $C = m - \phi$.

Having found C, we obtain the side c, by the rule that the sines of the angles are proportional to the sines of the opposite sides. But if we wish to calculate c directly, we make use of formula (1) Art. 153,

$$\cos b \cos c + \cos A \sin b \sin c = \cos a.$$

We reduce, as above, the first member to a single term, by means of the auxiliary angle ϕ, by placing $\cos A \sin b = \cos b \cot\phi$, which gives

$$\cot\phi = \cos A \operatorname{tang} b.$$

Consequently, the equation becomes

$$\cos b\,(\sin\phi \cos c + \cos\phi \sin c) = \cos a \sin\phi,$$

hence,
$$\sin(c+\varphi) = \frac{\cos a \sin\phi}{\cos b}.$$

Knowing now the value of ϕ, we readily determine c.

183. *Case* III. *Having given two sides*, a *and* b, *with the included angle* C, *to find* A, B, *and* c.

Formulæ (5) and (6), Art. 156, give for A and B,

$$\cot A = \frac{\cot a \sin b - \cos b \cos C}{\sin C}$$

$$\cot B = \frac{\cot b \sin a - \cos a \cos C}{\sin C}.$$

By making use of auxiliary angles, we may readily reduce each numerator to a single term. But Napier's Analogies furnish a more simple solution. We have

$$\operatorname{tang} \tfrac{1}{2}(A+B) = \cot \tfrac{1}{2} C \frac{\cos \frac{1}{2}(a-b)}{\cos \frac{1}{2}(a+b)}$$

$$\operatorname{tang} \tfrac{1}{2}(A-B) = \cot \tfrac{1}{2} C \frac{\sin \frac{1}{2}(a-b)}{\sin \frac{1}{2}(a+b)}$$

These analogies make known $\frac{1}{2}(A+B)$ and $\frac{1}{2}(A-B)$, from which we readily deduce the values of A and B.

When A and B are known, c is found by the proportion of the sines. If we want to determine c directly, we take the well-known formula, Art. 153,

$$\cos c = \cos a \cos b + \sin a \sin b \cos C,$$

in which we make

$$\sin b \cos C = \frac{\cos b \cos \phi}{\sin \phi} = \cos b \cot \phi.$$

We have, without any ambiguity,

$$\cot \phi = \operatorname{tang} b \cos C, \quad \cos c = \frac{\cos b \sin (a + \phi)}{\sin \phi}.$$

184. *Case* IV. *Having given two angles,* A *and* B, *with the included side* c, *to find* a, b, C.

We may determine a and b by formulæ (7) and (9), Art. 156,

$$\cot a = \frac{\cot A \sin B + \cos B \cos c}{\sin c}$$

$$\cot b = \frac{\cot B \sin A + \cos A \cos c}{\sin c};$$

or, better still, by *Napier's Analogies:*

$$\operatorname{tang} \tfrac{1}{2}(a + b) = \operatorname{tang} \tfrac{1}{2} c \frac{\cos \frac{1}{2}(A - B)}{\cos \frac{1}{2}(A + B)}$$

$$\operatorname{tang} \tfrac{1}{2}(a - b) = \operatorname{tang} \tfrac{1}{2} c \frac{\sin \frac{1}{2}(A - B)}{\sin \frac{1}{2}(A + B)}$$

The angle C is found by the proportion of the sines. To find C directly, we take formula (13), Art. 157,

$$\cos C = \sin A \sin B \cos c - \cos A \cos B,$$

and placing $\sin B \cos c = \cos B \cot \phi$, we have

$$\cot \phi = \operatorname{tang} B \cos c, \quad \cos C = \frac{\cos B \sin (A - \phi)}{\sin \phi}$$

This case is analogous to Case III., and presents no ambiguity.

185. *Case* V. *Having given two angles* A *and* B, *and the side* a *opposite one of them, to find* b, c, *and* C.

This case is precisely analogous to Case II. It is discussed in the same way, and presents like ambiguities.

The side b is deduced from the proportion $\frac{\sin b}{\sin a} = \frac{\sin B}{\sin A}$, and c and C are determined by the formulæ employed in Case II.

$$\text{tang } \tfrac{1}{2} c = \text{tang } \tfrac{1}{2}(a - b) \frac{\sin \frac{1}{2}(A + B)}{\sin \frac{1}{2}(A - B)}$$

$$\cot \tfrac{1}{2} C = \text{tang } \tfrac{1}{2}(A - B) \frac{\sin \frac{1}{2}(a + b)}{\sin \frac{1}{2}(a - b)}$$

The side c is also given by the formula (7),

$$\cot a \sin c - \cos B \cos c = \cot A \sin B.$$

In which, if we make $\cot a = \cos B \cot \phi$, we have

$$\cot \phi = \frac{\cot a}{\cos B}, \sin (c - \phi) = \frac{\text{tang } B \sin \phi}{\text{tang } A.}$$

Finally, the angle C is found either by the formula that the sines of the angles are proportional to the sines of the opposite sides, or by means of the equation

$$\cos a \sin B \sin C - \cos B \cos C = \cos A.$$

We may reduce the first member to a single term by placing

$$\cos a \sin B = \cos B \cot \phi,$$

hence, $\cot \phi = \cos a \text{ tang } B, \sin (C - \phi) = \frac{\cos A \sin \phi}{\cos B.}.$

These values make known ϕ, $C - \phi$, and consequently the angle C.

186. *Case* VI. *Having given the three angles* A, B, C, *to find the three sides* a, b, c.

This case is solved by analogous calculations to those in Case I. Thus, to find a, we make use of the formulæ, Art. 158,

$$\sin \tfrac{1}{2} a = \sqrt{\frac{\sin \frac{1}{2} \varepsilon \sin (A - \frac{1}{2} \varepsilon)}{\sin B \sin C}}$$

$$\cos \tfrac{1}{2} a = \sqrt{\frac{\sin (B - \frac{1}{2} \varepsilon) \sin (C - \frac{1}{2} \varepsilon)}{\sin B \sin C}}$$

$$\text{tang } \tfrac{1}{2} a = \sqrt{\frac{\sin \frac{1}{2} \varepsilon \sin (A - \frac{1}{2} \varepsilon)}{\sin (B - \frac{1}{2} \varepsilon) \sin (C - \frac{1}{2} \varepsilon)}}$$

We remark that the three last cases bear a strong analogy to the three first. Indeed, by the properties of the *polar* triangle, they may be deduced, one from the other.

Ambiguous Cases in Spherical Triangles.

187. The only cases in which ambiguity arises, as to the values of the computed elements, are the II. and V., and it becomes necessary to investigate the conditions to which the given elements are respectively subject, when they correspond to two triangles, or to only one, or to none at all.

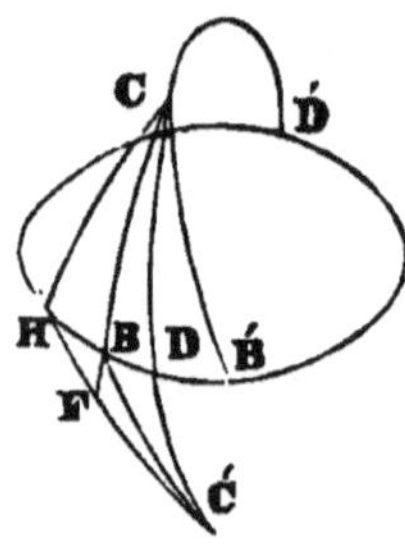

Let DHD′, DCD′, be two great circles at right angles to each other, take $CD < 90°$, and from the point C, draw the arcs of great circles to the different points of the circumference DHD′. Produce CD until C′D is equal to CD, and join C′B. The triangles CDB, C′DB are equal, since they have sides including a right equal, each to each: hence, $CB = C'B$. But $CDC' < CB + BC'$; hence, $CD < CB$. Therefore, 1st, *the arc* CD *perpendicular to the circumferences* DHD′ *is the shortest arc that can be drawn from the point* C *to this circumference: while* CD′ *is the longest arc that can be drawn to the same circumference.*

Let $DB' = DB$; the triangles CDB, CDB′ are equal, since they have two sides and the included right angle equal, each to each; hence $CB' = CB$. *Therefore,* 2d, *the oblique arcs equally distant from* CD *or* CD′ *are equal.*

Finally, let $DH > DB$; draw C′H, and produce CB until it meets C′H in I. Since the arc CC′ is less than a semicircumference, it must be intersected by the prolongation of CB beyond the point C′, which makes it necessary that the point I should be between H and C′. We have then $C'B < C'I + IB$, and consequently $C'B + BC < C'I + IC$. But we have $IC < IH + HC$, and consequently $C'I + IC < C'H + HC$; therefore, *à fortiori*, $C'B + BC < C'H + HC$. But $C'B = BC$, and $C'H = HC$; hence we have $BC < HC$.

Therefore, 3d, *of oblique arcs that is the longest which is at the greatest distance from* CD, *or the least distance from* CD′.

188. Let us now suppose that we have to construct a spherical triangle with the two sides a and b, and the angle A opposite to one of them given.

In the first place, it is to be observed that there are certain cases of impossibility that are indicated by the calculation itself. Make the angle CAB=A, and AC=b, produce AC and AB, until they meet in E, then draw CD perpendicular to AE.

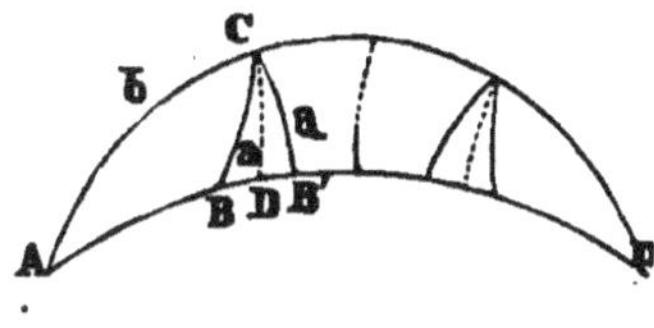

The arc CD must be of the same species as A, that is, be less than 90°, if A<90°, or greater than 90° if A>90°: hence, if A is acute, CD is the shortest distance from the point C to the semi-circumference, and it is the greatest if A is oblique (187.)

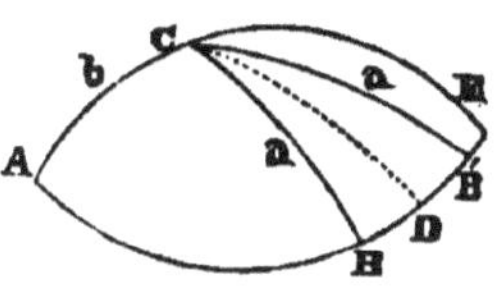

In the first hypothesis, the triangle will be impossible if we have $a<\text{CD}$, which gives $\sin a<\sin \text{CD}$; and in the second case, it will be impossible, if we have $a>\text{CD}$, which gives still $\sin a<\sin \text{CD}$. But in the right-angled triangle ACD, we have $\sin \text{CD}=\sin b \sin \text{A}$; hence, in the two hypotheses, we should have $\sin a<\sin b \sin \text{A}$. On the other hand, when the angle B of the triangle ACB is sought, we have

$$\frac{\sin \text{B}}{\sin \text{A}}=\frac{\sin b}{\sin a}, \text{ which gives } \sin \text{B}=\frac{\sin b \sin \text{A}}{\sin a}.$$

Hence, this value of sin B would be greater than 1, which indicates an evident impossibility.

If $a=\text{CD}$, we should have only the right-angled triangle ACD which would be possible; and this is indicated by the value of sin B, which becomes $\sin \text{B}=1$. It is assumed that the angle A is not equal to 90°.

189. Passing by these cases, let us examine the various relations which the given parts a, b, A may present, with respect to magnitude.

Let A<90°, and $b<90°$. Since A and $b<90°$, AD is also <90°; hence AD<DE. This being established, if we have also $a<b$, it is evident we may place one arc

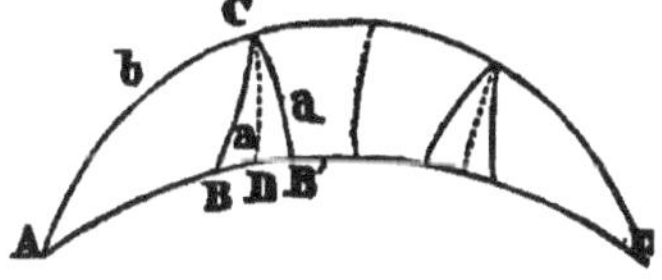

10

$CB = a$ between CA and CD, and that on the other side, we may place another arc $CB' = CB = a$, between CD and CE: that is, there will be two triangles ACB and ACB', constructed with the same given parts a, b, A. When $a = b$, the triangle ACB disappears, and ACB' only remains. When we have $a + b = 180°$, or $a + b > 180°$, the point B' falls on E, or passes beyond it, and there is no longer any triangle.

We may discuss in the same manner the other hypotheses. The results are comprised in the following table.

A	b	Condition	Result
$A<90°$	$b<90°$	$a<b$	*two solutions*
		$a>b$, or $a=b$, provided $a+b$ is not 180° or greater than 180°,—	*one solution*
		$a+b=180°$, or $a+b>180°$	*no solution*
	$b>90°$	$a+b<180°$,	*two solutions*
		$a+b>180°$, or $a+b=180°$, provided we have not $a=b$, or $a>b$,	*one solution*
		$a=b$, or $a>b$	*no solution*
	$b=90°$	$a<b$	*two solutions*
		$a=b$, or $a>b$	*no solution*
$A>90°$	$b<90°$	$a+b>180°$	*two solutions*
		$a+b=180°$, or $a+b<180°$, provided a is not $=b$, or $<b$	*one solution*
		$a=b$, or $a<b$	*no solution*
	$b>90°$	$a>b$	*two solutions*
		$a=b$, or $a<b$, if $a+b$ is not $=180°$, or $<180°$	*one solution*
		$a+b=180°$, or $a+b<180°$	*no solution*
	$b=90°$	$a>b$	*two solutions*
		$a=b$, or $a<b$	*no solution*
$A=90°$	$b<90°$	$a>b$, if we have not $a+b=180°$, or $>180°$	*one solution*
		$a=b$, or $a<b$	*no solution*
		$a+b=180°$, or $a+b>180°$	*no solution*
	$b>90°$	$a<b$, if $a+b$ is not $=180°$, or $<180°$	*one solution*
		$a=b$, or $a>b$	*no solution*
		$a+b=180°$, or $180<180°$	*no solution*
	$b=90°$	$a=90°$	*infinite number of solutions*
		$a<90°$, or $a>90°$	*no solution*

190. The property of the polar triangle enables us to apply these results to a triangle in which A, B, and a are given, as in Case V. It is only necessary to change a, b, and A, into A, B, and a, the sign $>$ into $<$, and the sign $<$ into $>$.

When the conditions which are given lead to one of the cases in which there is only one solution, the calculation always indicates two. To ascertain which must be preserved, it will be only necessary to note that the greater angles must lie opposite to the greater sides, and reciprocally.

Suppose that we have given $A = 112°$, $a = 102°$, $b = 106°$. In the preceding table, among the cases which correspond to $A > 90°$, we consider those in which $b > 90°$, and note in these, that one in which we have $a = b$, or $a < b$, we observe, further, that we have $a + b = 208°$; hence, $a + b > 180°$. We conclude, therefore, from the table, that we have only one solution, and since $b > a$, the angle $B > A$, hence B is obtuse.

Surface of the Spherical Triangle.

191. *The surface of a spherical triangle is equal to the excess of the sum of its three angles above two right angles.* This *excess*, which is also called the *spherical excess*, may be found from the sides, by the aid of very simple formulæ, which will now be explained. Designating, as before, the spherical excess by ε, we shall have

$$\sin \tfrac{1}{2}\varepsilon = \sin(\tfrac{1}{2}(A + B + C) - 90°) = -\cos \tfrac{1}{2}(A + B + C).$$

It is required to find $\cos \tfrac{1}{2}(A + B + C)$.

In the first place we have

$$\cos \tfrac{1}{2}(A + B) = \cos \tfrac{1}{2}A \cos \tfrac{1}{2}B - \sin \tfrac{1}{2}A \sin \tfrac{1}{2}B.$$

Now putting $B + C$ for B, we readily get

$$\cos \tfrac{1}{2}(A+B+C) = \cos \tfrac{1}{2}A \cos \tfrac{1}{2}B \cos \tfrac{1}{2}C - \cos \tfrac{1}{2}A \sin \tfrac{1}{2}B \sin \tfrac{1}{2}C - \cos \tfrac{1}{2}B \sin \tfrac{1}{2}A \sin \tfrac{1}{2}C - \cos \tfrac{1}{2}C \sin \tfrac{1}{2}A \sin \tfrac{1}{2}B.$$

Substituting for the sines and cosines, their values in terms of the sides (Art. 153); and considering, in the first place, the 1st term which contains three cosines, we shall have

$$\cos \tfrac{1}{2}A \cos \tfrac{1}{2}B \cos \tfrac{1}{2}C = \sqrt{\frac{\sin s \sin (s-a)}{\sin b \sin c} \times \frac{\sin s \sin (s-b)}{\sin a \sin c} \times \frac{\sin s \sin (s-c)}{\sin a \sin b}}$$

$$= \sin s \sqrt{\frac{\sin s \sin (s-a) \sin (s-b) \sin (o-c)}{\sin a \sin b \sin c}}$$

Designating the radical which occurs in the second member by R, the expression assumes the form

$$\cos \tfrac{1}{2} A \cos \tfrac{1}{2} B \cos \tfrac{1}{2} C = \frac{R \sin s}{\sin a \sin b \sin c.}$$

If we now consider the 2d term, we find

$$\cos \tfrac{1}{2} A \sin \tfrac{1}{2} B \sin \tfrac{1}{2} C = \sqrt{\frac{\sin s \sin (s-a)}{\sin b \sin c} \times \frac{\sin (s-a) \sin (s-c)}{\sin a \sin c} \times \frac{\sin (s-a) \sin (s-b)}{\sin a \sin b}}$$

or, simplifying,

$$\cos \tfrac{1}{2} A \sin \tfrac{1}{2} B \sin \tfrac{1}{2} C = \frac{R \sin (s-a)}{\sin a \sin b \sin c.}$$

By a single change in the letters, we readily find analogous values of the two other terms, viz.:

$$\cos \tfrac{1}{2} B \sin \tfrac{1}{2} A \sin \tfrac{1}{2} C = \frac{R \sin (s-b)}{\sin a \sin b \sin c}$$

$$\cos \tfrac{1}{2} C \sin \tfrac{1}{2} A \sin \tfrac{1}{2} B = \frac{R \sin (s-c)}{\sin a \sin b \sin c}$$

By means of these transformations we get

$$\cos \tfrac{1}{2} (A+B+C) = \frac{R(\sin s - \sin (s-a) - \sin (s-b) - \sin (s-c))}{\sin a \sin b \sin c}$$

In this value the multiplier of R may be reduced to a single term, as follows.

The two first terms being a difference of sines, we may make use of the formula

$$\sin p - \sin q = 2 \sin \tfrac{1}{2} (p-q) \cos \tfrac{1}{2} (p+q)$$

and we have

$$\sin s - \sin (s-a) = 2 \sin \tfrac{1}{2} a \cos \tfrac{1}{2} (b+c)$$

For the two others, we make use of the formula

$$\sin p + \sin q = 2 \sin \tfrac{1}{2} (p \times q) - \cos \tfrac{1}{2} (p-q)$$

and we find

$$\sin (s-b) + \sin (s-c) = 2 \sin \tfrac{1}{2} a \cos \tfrac{1}{2} (b-c)$$

Hence, the multiplier of R will be equal to

$$2 \sin \tfrac{1}{2} a (\cos \tfrac{1}{2} (b+c) - \cos \tfrac{1}{2} (b-c))$$

and this last expression may itself be reduced to $-4 \sin \frac{1}{2} a \sin \frac{1}{2} b \sin \frac{1}{2} c$.

As to the denominator, we remark that

$$\sin a \sin b \sin c = 8 \sin \tfrac{1}{2} a \cos \tfrac{1}{2} a \sin \tfrac{1}{2} b \cos \tfrac{1}{2} b \sin \tfrac{1}{2} c \cos \tfrac{1}{2} c.$$

Consequently, after making all the reductions, and restoring the radical designated by R, we shall have $\cos \frac{1}{2}(A+B+C)$. Then changing the sign of the result, we have the required formula,

$$\sin \tfrac{1}{2} \varepsilon = \frac{\sqrt{\sin s \sin (s-a) \sin (s-b) \sin (s-c)}}{2 \cos \frac{1}{2} a \cos \frac{1}{2} b \cos \frac{1}{2} c.}$$

If we find the value of $\cos \frac{1}{2} \varepsilon$, in the place of $\sin \frac{1}{2} \varepsilon$, we obtain the following formula.

$$\cos \tfrac{1}{2} \varepsilon = \frac{1 + \cos a + \cos b + \cos c}{4 \cos \frac{1}{2} a \cos \frac{1}{2} b \cos \frac{1}{2} c,}$$

which is not convenient to be used in logarithmic calculations. The transformations required are complicated, and it would be useless to stop here to make them.

Finally, we introduce, in this place, without developing the calculations, the very elegant formulæ of *Simon Lhuillier*. By means of $\cos \frac{1}{2} \varepsilon$, we obtain $\sin \frac{1}{4} \varepsilon$, $\cos \frac{1}{4} \varepsilon$, $\text{tang} \frac{1}{4} \varepsilon$, and we get the formulæ referred to.

$$\sin \tfrac{1}{4} \varepsilon = \sqrt{\frac{\sin \frac{1}{2} s \sin \frac{1}{2} (s-a) \sin \frac{1}{2} (s-b) \sin \frac{1}{2} (s-c)}{\cos \frac{1}{2} a \cos \frac{1}{2} b \cos \frac{1}{2} c}}$$

$$\cos \tfrac{1}{4} \varepsilon = \sqrt{\frac{\cos \frac{1}{2} s \cos \frac{1}{2} (s-a) \cos \frac{1}{2} (s-b) \cos \frac{1}{2} (s-c)}{\cos \frac{1}{2} a \cos \frac{1}{2} b \cos \frac{1}{2} c}}$$

$$\text{tang} \tfrac{1}{4} \varepsilon = \sqrt{\text{tang} \tfrac{1}{2} s \,\text{tang} \tfrac{1}{2} (s-a) \,\text{tang} \tfrac{1}{2} (s-b) \,\text{tang} \tfrac{1}{2} (s-c)}$$

Application in Spherical Trigonometry.

192. Example 1. *To reduce an angle to the horizon.*

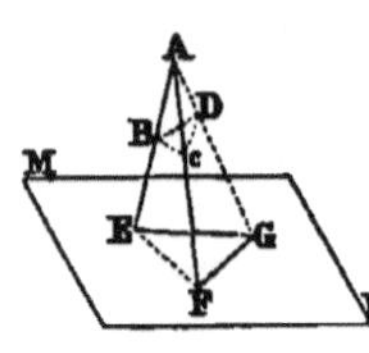

Let B A C be an angle situated in a plane inclined to the horizon, and A D the vertical passing through the vertex A. Draw at will the horizontal plane M N, meeting the lines A B, A C, A D, in E, F, G: the angle E G F is *the horizontal projection* of the angle B A C, or, in other words, it is the angle B A C, *reduced to the horizon.*

It is required to calculate the angle E G F, supposing the angles B A C, B A D, C A D, as known, being determined by instrumental measurements.

The graphic solution would be easy, for, the line A G being arbitrary, we should have sufficient elements given to construct, in the first place, the right-angled triangles E A G, and F A G, then the triangle E A F, and finally the triangle E G F.

The calculation of the angle E G F is easily made. If we describe a sphere from A, as a centre, with any radius, the right lines A B, A C, A D, determine, by their intersection with the surface of the sphere, a spherical triangle B C D, the sides of which are known in degrees by means of the given angles, and the angle B C D equal to E G F is the required angle. The solution is readily made by using the well-known formula for sin ½ A in terms of the three sides.

$$\sin \tfrac{1}{2} A = \sqrt{\frac{\sin (s-b) \sin (s-b)}{\sin b \sin c,}}$$

in which we make $a = $ B A C, $b = $ B A D, $c = $ C A D, $s = \frac{1}{2}(a + b + c)$.

Let $a = 47° \ 45' \ 39''$, $b = 69° \ 49' \ 19''$, $c = 80° \ 17' \ 36''$, we shall have $2s, = 197° \ 52' \ 34''$, $s = 98° \ 56' \ 17''$, $s - b = 29° \ 6' \ 58''$, $s - c = 18° \ 38' \ 41''$, and we may make the following calculation.

Log sin $(s-b)$	9,6871552
Log sin $(s-c)$	9,5047412
Log sin b (Arith. Compl.) .	0,0275078
Log sin c (Arith. Compl.) .	0,0062623
2 Log sin $\frac{1}{2}$ A	19,2256665
Log sin $\frac{1}{2}$ A	9,6128332

$\frac{1}{2}$ A = 24° 12′ 27″,9

A = 48° 24′ 56″

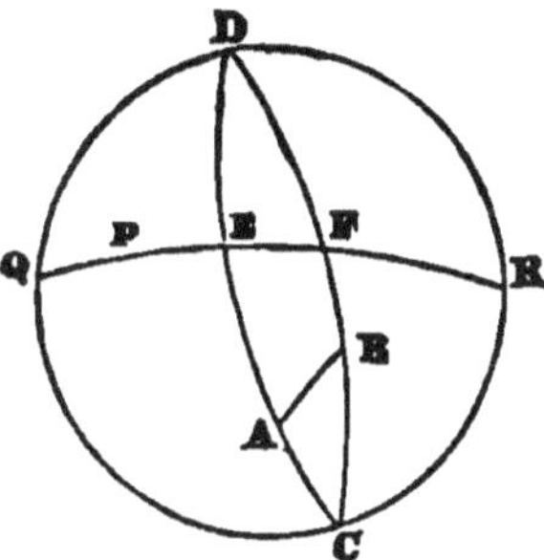

193. Example II. *Having given the latitudes and longitudes of two points on the surface of the earth, to find the distance between these points.*

Let A and B be the two points. Let Q R be the circle of the *equator*, C, the *north pole*, and C E D, C F D, the meridians passing through the points A and B. Finally, suppose that the longitudes are reckoned from the point P in the direction P E F.

The difference of longitudes, P F — P E, is equal to the arc E F, or to the angle C included between the two meridians; and the arcs A C, B C, are the complements of the given latitudes A E, B F. Thus, in the spherical triangle A B C, we know the angle C, as well as the sides which contain this angle, and it is required to calculate the side A B. Case III., in which two sides and the included angle are given, give the formulæ,

$$\cot\phi = \operatorname{tang} b \cos C, \quad \cos c = \frac{\cos b \sin(a+\phi)}{\sin\phi}.$$

Suppose, for example, that we wish to calculate the distance from *Brest* to *Cayenne*.

The longitude of Brest = 6° 49′ Latitude = 48° 23′ 14″.
The longitude of Cayenne = 54° 35′ Latitude = 4° 36′ 15″.

Both longitudes are *west* and are reckoned from the meridian of Paris; both latitudes are north.

With these data, we find, in the first place,

$$C = 54° 35' - 6° 49' = 47° 46'$$
$$a = 90° - 48° 23' 14'' = 41° 36' 46''$$
$$b = 90° - 4° 51' 15'' = 85° 3' 45''$$

We may now calculate c as follows.

Calculation of the auxiliary angle ϕ.		*Calculation of the side* c.	
$\log \cos C =$	9,8274671	$\log \cos b$	8,9348468
$\log \text{tang}\, b$	11,0635386	$\log \sin (a + \phi)$	9,8773621
		$\log \sin \phi$ (Arith. Comp.)	0,8945642
$\log \cot \phi$	10,8910057		
$\phi = 7° 19' 26''$		$\log \cos c$	9,7067731
$a + \phi = 48° 56' 12.$		$c = 59° 23' 54'' 38.$	

Thus, the arc which measures the distance between Brest and Cayenne is 59° 23′ 54″, 38. To estimate this arc in myriametres, we must remember that the fourth of the terrestrial meridian is 10,000,000 metres, or 1000 myriametres.

Hence, we have the proportion

$$\frac{x}{1000} = \frac{59° 23' 54'' 38}{90°}$$

and reducing the arcs to seconds, we find the distance x to be

$$x = \frac{213834 \times 1000}{324,000} = 65,983 \text{ myriametres.}$$

This last operation could have been easier, if the arc c had been expressed in *centesimal* degrees. For example, take in this division an arc 37° 45′ 69″: referring it to the quadrant, it will be expressed by 0,374569, and multiplying the number by the value of the quadrant, in terms of myriametres, we find at once, by the simple removal of the decimal point, 374,569 myriametres.

CHAPTER III.

FORMULÆ USED IN THE HIGHER MATHEMATICS. SINES AND COSINES IN SERIES. RESOLUTION OF BINOMIAL EQUATIONS AND OF THE EQUATION OF THE 3D DEGREE.

FORMULA OF DE MOIVRE.

195. This formula, which bears the name of the French Geometer who discovered it, is as follows:

$$(\cos\phi + \sqrt{-1}\sin\phi)^n = \cos n\phi + \sqrt{-1}\sin n\phi \qquad (A)$$

It expresses that to find the value of the binomial $\cos\phi + \sqrt{-1}\sin\phi$, raised to any power, we have only to multiply the arc ϕ by the exponent of this power. Either sign + or — may be placed before $\sqrt{-1}$, for this would involve only the change of $+\phi$ into $-\phi$.

196. *Case* 1st. n *positive and entire.* We have, by multiplication,

$$(\cos\phi + \sqrt{-1}\sin\phi)(\cos\psi + \sqrt{-1}\sin\psi) = \cos\phi\cos\psi - \sin\phi\sin\psi + \sqrt{-1}(\sin\phi\cos\psi + \cos\phi\sin\psi).$$

But, from known formulæ, the real part of this product is equal to $\cos(\phi+\psi)$, and the imaginary part is equal to $\sqrt{-1}\sin(\phi+\psi)$; hence,

$$(\cos\phi + \sqrt{-1}\sin\phi)(\cos\psi + \sqrt{-1}\sin\psi) = \cos(\phi+\psi) + \sqrt{-1}(\sin\phi+\psi).$$

That is to say, when we multiply two expressions of the form $\cos\phi + \sqrt{-1}\sin\phi$, we obtain a like expression, in which the two arcs are added together. If now we wish to multiply this product by a new factor of the same form, we have only to add the new arc to the two others, and so on, for any number of factors. Hence, if there should be n factors, all equal to $\cos\phi + \sqrt{-1}\sin\phi$, we have

$$(\cos\phi + \sqrt{-1}\sin\phi)^n = \cos n\phi + \sqrt{-1}\sin n\phi \qquad (1)$$

(111)

197. *Case* 2d. n *positive and fractional.* Substituting $\frac{\phi}{n}$ for ϕ in formula (1), we have,

$$\left(\cos\frac{\phi}{n}+\sqrt{-1}\sin\frac{\phi}{n}\right)^n=\cos\phi+\sqrt{-1}\sin\phi.$$

Now, extracting the nth root, and putting a fractional exponent in place of the radical, formula (1) is found proved for an exponent $\frac{1}{n}$, for, we shall have

$$(\cos\phi+\sqrt{-1}\sin\phi)^{\frac{1}{n}}=\cos\frac{\phi}{n}+\sqrt{-1}\sin\frac{\phi}{n}. \qquad (2)$$

which proves formula (1) for an exponent $\frac{1}{n}$.

In general, the expression $A^{\frac{m}{n}}$ signifies that the mth power of A is taken, and then, the nth root of this result. Hence, if we raise $\cos\phi+\sqrt{-1}\sin\phi$ to the mth power by formula (1), and afterwards extract the nth root by formula (2), we shall have

$$(\cos\phi+\sqrt{-1}\sin\phi)^{\frac{m}{n}}=\cos\frac{m\phi}{n}+\sqrt{-1}\sin\frac{m\phi}{n} \qquad (3)$$

which proves formula (1) for any positive fractional exponent $\frac{m}{n}$.

198. *Case* 3d. n *negative.* In this case, we observe that

$$(\cos n\phi+\sqrt{-1}\sin n\phi)(\cos n\phi-\sqrt{-1}\sin n\phi)=\cos^2 n\phi+\sin^2 n\phi=1$$

from which we get

$$\frac{1}{\cos n\phi+\sqrt{-1}\sin n\phi}=\cos n\phi-\sqrt{-1}\sin n\phi$$

or, the equivalent expression

$$(\cos\phi+\sqrt{-1}\sin\phi)^{-n}=\cos(-n\phi)+\sqrt{-1}\sin(-n\phi), \qquad (4)$$

hence, formula (A) is verified when the exponent n is positive or negative, entire or fractional; and, in general, $\cos\frac{m}{n}\phi+\sqrt{-1}\sin\frac{m}{n}\phi$ is one of the n different values which the expression $(\cos\phi+\sqrt{-1}\sin\phi)^{\frac{m}{n}}$, according to the principles of Algebra, admits of.

199. There is no difficulty in assigning the remaining values; for, from what has been proved, since ϕ is any arc whatever, and may therefore be replaced by the general value of all arcs which have the same sine and cosine as ϕ, viz., $2r\pi+\phi$, r being any whole number, positive or negative, it follows that

$$\cos\frac{m}{n}(2r\pi+\phi)+\sqrt{-1}\sin\frac{m}{n}(2r\pi+\phi) \quad \text{(B)}$$

is a value of $\quad (\cos 2r\pi+\phi+\sqrt{-1}\sin(2r\pi+\phi))^{\frac{m}{n}}$,

or of $\quad (\cos\phi+\sqrt{-1}\sin\phi)^{\frac{m}{n}}$.

We will now show that the expression (B) admits of n different values, and no more.

If we make $r=0, 1, 2,$ &c. to $n-1$, we get n different values; for, if two of them were alike, as $r=p$, and $r=q$, it would be necessary that the arcs $\frac{m}{n}(2p\pi+\phi)$, $\frac{m}{n}(2q\pi+\phi)$, should differ by a multiple of 2π, or that $\frac{m(p-q)\pi}{n}$ should be a multiple of π, which is impossible, since p and q are both less than n, and m not divisible by n. Also, if we take for r some number beyond the limits 0 and $n-1$, we shall get no new value; for, if $r=\lambda n+r'$, in which λ is any positive or negative number, and r' positive and less than n, so that r may represent any positive or negative numbers whatever; then the above expression becomes

$$\cos\left(2\lambda m\pi+\frac{m}{n}(2r'\pi+\phi)\right)+\sqrt{-1}\sin\left(2\lambda m\pi+\frac{m}{n}(2r'\pi+\phi)\right),$$

or, suppressing the multiple of 2π

$$\cos\frac{m}{n}(2r'\pi+\phi)+\sqrt{-1}\sin\frac{m}{n}(2r'\pi+\phi),$$

which, since r' is positive and less than n, is comprised among the values obtained in making $r=0, 1, 2, ----- n-1$.

Consequently, the complete value of $(\cos\phi+\sqrt{-1}\sin\phi)^{\frac{m}{n}}$ is given by the formula

$$(\cos\phi+\sqrt{-1}\sin\phi)^{\frac{m}{n}}=\cos\frac{m}{n}(2r\pi+\phi)+\sqrt{-1}\sin\frac{m}{n}(2r\pi+\phi),$$

r being any whole number, positive or negative, not excluding zero, and the n values of the first member are found from the second, by taking n from 0 to $n-1$. We see, also, by assuming $n\phi = 2\pi r + \alpha$, in which r is any whole number, and α is any angle between 0 and π, we obtain n different values of ϕ, every one of which furnishes a different value of $\cos\phi + \sqrt{-1}\sin\phi$, but the same value of $\cos n\phi + \sqrt{-1}\sin n\phi$.

FORMULÆ FOR $\sin n\phi$ AND $\cos n\phi$ IN TERMS OF SINE AND COSINE OF THE SIMPLE ARC.

200. Resuming *De Moivre's* formula

$$\cos n\phi + \sqrt{-1}\sin n\phi = (\cos\phi + \sqrt{-1}\sin\phi)^n \quad (1)$$

in which n is considered entire and positive, we have, when ϕ is changed into $-\phi$

$$\cos n\phi - \sqrt{-1}\sin n\phi = (\cos\phi - \sqrt{-1}\sin\phi)^n. \quad (2)$$

Adding and subtracting these equations, we find these values,

$$\cos n\phi = \frac{(\cos\phi + \sqrt{-1}\sin\phi)^n + (\cos\phi - \sqrt{-1}\sin\phi)^n}{2}, \quad (3)$$

$$\sin n\phi = \frac{(\cos\phi + \sqrt{-1}\sin\phi)^n - (\cos -\phi\sqrt{-1}\sin\phi)^n}{2\sqrt{-1}}. \quad (4)$$

201. Expanding the powers in the second member by the binomial formula, and suppressing terms which destroy each other, we obtain

$$\cos n\phi = (\cos\phi)^n - \frac{n(n-1)}{1,2}(\cos\phi)^{n-2}(\sin\phi)^2 + \frac{n(n-1)(n-2)(n-3)}{1,2,3,4}(\cos\phi)^{n-4}(\sin\phi)^4, \text{ \&c.} \quad (5)$$

$$\sin n\phi = \frac{n}{1}(\cos\phi)^{n-1}\sin\phi - \frac{n(n-1)(n-2)}{1,2,3}(\cos\phi)^{n-3}(\sin\phi)^3 + \frac{n(n-1)(n-2)(n-3)(n-4)}{1,2,3,4,5}(\cos\phi)^{n-5}(\sin\phi)^5 - \text{ \&c.} \quad (6)$$

These formulæ express the values of the sine and cosine of the multiple arc $n\phi$ in terms of the sine and cosine of the simple arc ϕ. Their law is evident, and, like the binomial theorem from which they are deduced, each is to be continued until we arrive at a term zero.

FORMULÆ FOR $(\sin\phi)^n$, $(\cos\phi)^n$ IN TERMS OF SINE AND COSINE OF THE MULTIPLE ARCS.

202. In the higher mathematics, it is often necessary to express $(\sin\phi)^n$, and $(\cos\phi)^n$ in terms of the sines and cosines of the multiple arcs $(n\,\phi)$.

When n is a positive whole number, which is the ordinary case, it may be effected as follows.

Assume $\cos\phi+\sqrt{-1}\sin\phi=u$, $\cos\phi-\sqrt{-1}\sin\phi=v$,

we shall have $2\cos\phi=u+v$, $2\sqrt{-1}\sin\phi=u-v$; hence,

$$2^n(\cos\phi)^n=(u+v)^n,\ (2\sqrt{-1})^n(\sin\phi)^n=(u-v)^n;$$

and developing $(u+v)^n$, $(u-v)^n$, we get

$$2^n(\cos\phi)^n=u^n+\frac{n}{1}u^{n-1}v+\frac{n(n-1)}{1,2}u^{n-2}v^2+\ \&c. \quad (7)$$

$$(2\sqrt{-1})^n(\sin\phi)^n=u^n-\frac{n}{1}u^{n-1}v+\frac{n(n-1)}{1,2}u^{n-2}v^2-\&c. \quad (8)$$

If n be odd, there will be an even number of terms $n+1$ in the expansions, and grouping together the terms which are equally distant from the extremes, we have for formula (7),

$$2^n(\cos\phi)^n=(u^n+v^n)+\frac{n}{1}uv(u^{n-2}+v^{n-2})+\frac{n(n-1)}{1,2}u^2v^2(u^{n-4}+v^{n-4})\cdots\cdots\frac{n(n-1)(n-2)\cdots(n+2)}{1,2,3,4\cdots\cdots n}u^nv^n(u+v).$$

But De Moivre's formula gives

$$u^n+v^n=(\cos\phi+\sqrt{-1}\sin\phi)^n+(\cos\phi-\sqrt{-1}\sin\phi)^n=+\cos n\phi+\sqrt{-1}\sin n\phi+\cos n\phi-\sqrt{-1}\sin n\phi=2\cos n\phi.$$

Besides, the powers of uv are equal to 1, for we have

$$uv=(\cos\phi+\sqrt{-1}\sin\phi)(\cos\phi-\sqrt{-1}\sin\phi)=\cos^2\phi+\sin^2\phi=1.$$

We see, consequently, that the expression for $2^n(\cos\phi)^n$ will be simplified by these reductions; and dividing all the terms by 2, it becomes

$$2^{n-1}(\cos\phi)^n=\cos n\phi+\frac{n}{1}\cos(n-2)\phi+\frac{n(n-1)}{1,2}\cos(n-4)\phi\cdots\cdots\frac{n(n-1)(n-2)\cdots(n+2)}{1,2,3\cdots\cdots\cdots n}\cos\phi. \quad (9)$$

If n be *even*, the number of terms in the development will be odd, and there will be only one middle term; and if we divide the middle term into two others, each equal to half of the middle term, we shall find, by similar reductions to the preceding,

$$2^{n-1}(\cos\phi)^n = \cos n\phi + \frac{n}{1}\cos(n-2)\phi + \frac{n(n-1)}{1,2}\cos(n-4)\phi \cdots\cdots \tfrac{1}{2}\left(\frac{n(n-1)(n-2)\cdots(n+1)}{1,2,3\cdots\cdots\cdots n}\right). \qquad (10)$$

If now we consider formula (8), the terms being alternately + and —, if n be odd, the coefficients of the terms equally distant from the extremes will be equal, but will have contrary signs. We shall have, therefore, by the same process used in formula (7),

$$(2\sqrt{-1})^n(\sin\phi)^n = u^n - v^n - \frac{n}{1}uv(u^{n-2}-v^{n-2}) + \frac{n(n-1)}{1,2}, u^2v^2(u^{n-4}-v^{n-4})\cdots\cdots \pm \frac{n(n-1)(n-2)\cdots(n+2)}{1,2,3\cdots\cdots\cdots n}u^n v^n(u-v).$$

To make the necessary reductions, we remark that $uv=1$, and in general $u^n - v^n = 2\sqrt{-1}\sin n\phi$. The two members will then be divisible by $2\sqrt{-1}$, and we shall have

$$(2\sqrt{-1})^{n-1}(\sin\phi)^n = \sin(n\phi) - \frac{n}{1}\sin(u-v)\phi + \frac{n(n-1)}{1,2}\sin(n-4)\phi\cdots\cdots \pm \frac{n(n-1)(n-2)\cdots(n+2)}{1,2,3\cdots\cdots\cdots n}\sin\phi. \quad (11)$$

If n is an even number, the terms equally distant from the extremes in formula (8) have equal coefficients with the same sign, and by analogous reasoning to that used for formula (7) we find

$$\tfrac{1}{2}(2\sqrt{-1})^n(\sin\phi)^n = \cos n\phi - \frac{n}{1}\cos(n-2)\phi + \frac{n(n-1)}{1,2}\cos(n-4)\phi\cdots\cdots \pm \tfrac{1}{2}\left(\frac{n(n-1)(n-2)\cdots\cdots(n-1)}{1,2,3\cdots\cdots n}\right). \quad (12)$$

Formulæ (9), (10), (11), (12) are the required formulæ; and they enable us to convert the powers of a cosine or of a sine into a series of terms, each of which contains the sine or cosine of a multiple arc, of the first degree only.

It will be observed that in the last two formulæ, the imaginary factor $\sqrt{-1}$ being raised to an even power, must produce a real result equal to $+1$, or -1, according as the index of the power is even or odd. Also, for the convenience of the calculation, we must stop at and exclude the first negative arc; and we must take only half of the last term when it involves the arc zero.

Development of the Sine and Cosine into Series.

203. We shall now show how *Euler* deduces from formulæ (5) and (6), Art. 201, the series which give the values of the sine and cosine in functions of the arc. We may, without departing from the hypothesis of n being a whole number, assume ϕ, so that $n\phi$ shall be equal to any given arc a. Let, therefore, $n\phi=a$, we shall have $n=\frac{a}{\phi}$, and with this value the formulæ become

$$\cos a=(\cos\phi)^n-\frac{a(a-\phi)}{1,2}(\cos\phi)^{n-2}\left(\frac{\sin\phi}{\phi}\right)^2+\frac{a(a-\phi)(a-2\phi)(a-3\phi)}{1,2,3,4}$$

$$(\cos\phi)^{n-4}\left(\frac{\sin\phi}{\phi}\right)^4-\&c., \&c. \qquad (1)$$

$$\sin a=\frac{a}{1}(\cos\phi)^{n-1}\left(\frac{\sin\phi}{\phi}\right)-\frac{a(a-\phi)(a-2\phi)}{1,2,3}(\cos\phi)^{n-3}\left(\frac{\sin\phi}{\phi}\right)^3$$

$$+\frac{a(a-\phi)(a-2\phi)(a-3\phi)(a-4\phi)}{1,2,3,4,5}(\cos\phi)^{n-5}\left(\frac{\sin\phi}{\phi}\right)^5-\&c. \quad (2)$$

Now suppose that ϕ diminishes to zero, the number n must increase to infinity; then, these formulæ will contain no traces of ϕ and of n, and will contain a only. For, when $\phi=0$, $\cos\phi=1$, and $\frac{\sin\phi}{\phi}=1$; and for this value of ϕ, we may also admit that the powers of $\cos\phi$ and of $\frac{\sin\phi}{\phi}$ are equal to 1, however great the indices of the powers. Consequently, the above formulæ become

$$\cos a=1-\frac{a^2}{1,2}+\frac{a^4}{1,2,3,4}-\frac{a^6}{1,2,3,4,5,6}+\&c. \quad (3)$$

$$\sin a=a-\frac{a^3}{1,2,3}+\frac{a^5}{1,2,3,4,5}-\frac{a^7}{1,2,3,4,5,6,7}+\&c. \quad (4)$$

As the number n has become infinite, these series do not terminate; but they are no less convenient to give approximate values of the sine and cosine when a is a small fraction, which is the case in practice, although the formulæ, being convergent, may be applied for any value of a.

202. By dividing series (4) by series (3), we obtain the expansion of tang a, the first four terms of which are

$$\text{tang}\, a = a + \frac{a^3}{3} + \frac{2a^5}{3.5} + \frac{17a^7}{3.3.5.7} \ . + \&c. \qquad (5)$$

In like manner, by dividing 1 by each of the series (3), (4), and (5), we may obtain the expansions for sec a, cosec a, cotang a.

Resolution of Binomial Equations, &c.

203. *De Moivre's* formula affords a ready means of resolving binomial equations of the form $y^n = \pm A$. Let a represent one of the n roots of A, and make $y = ax$. The binomial equation becomes $x^n = \pm 1$.

Consider, in the first place, the case

$$x^n = +1. \qquad (1)$$

Placing $x = \cos\phi + \sqrt{-1}\sin\phi$, the formula of *De Moivre* gives

$$x^n = \cos n\phi + \sqrt{-1}\sin n\phi.$$

Consequently, all the values of ϕ are made known by the equation

$$\cos n\phi + \sqrt{-1}\sin n\phi = +1$$

and will give values of x corresponding to equation (1).

That the last equation may be satisfied, it is necessary that the imaginary factor $\sqrt{-1}\sin n\phi$ disappear. This requires that $n\phi$ should be a multiple of the semi-circumference. It is also necessary that we have $\cos n\phi = 1$, which requires that $n\phi$ should be an even multiple of the semi-circumference. Hence, designating the semi-circumference by π, and by $2k$ any even number whatever, we shall have $n\phi = 2k\pi$, from which we get $\phi = \frac{2k\pi}{n}$. Consequently

$$x = \cos\frac{2k\pi}{n} + \sqrt{-1}\sin\frac{2k\pi}{n}.$$

The reasoning will be the same for $-\sqrt{-1}$. Thus, we shall find all the values of x in the equation $x^n = 1$, by taking all the values of x comprised in the equation

$$x = \cos\frac{2k\pi}{n} \pm \sqrt{-1}\sin\frac{2k\pi}{n} \qquad (a)$$

Equation (1) has n values for x, and we know that all these values are unequal. But in formula (a) we may give to k every possible entire value, positive and negative; and we will now prove that in this way we may obtain n different values of x, from equation (a), and no more; and these values will therefore be the values of x in equation (1).

In the first place, it is useless to give to k negative values: for, if we put $-k$ for $+k$, the two values of formula (a) reduce to the same form exactly.

In the second place, it is unnecessary to take $k = n$ or $> n$, for we may take from k the greatest multiple of n, which it contains; this is equivalent to taking from the arc $\frac{2k\pi}{n}$ one or more circumferences, which changes neither the cosine nor the sine.

Finally, if we take numbers k' and $n-k'$ between 0 and n, equally distant from the extremes, the corresponding values of x will be the same. For, let $k = n - k'$, we shall have

$$x = \cos\frac{2(n-k')\pi}{n} \pm \sqrt{-1}\sin\frac{2(n-k')\pi}{n}$$

$$= \cos\frac{-2k'\pi}{n} \pm \sqrt{-1}\sin\frac{-2k'\pi}{n}$$

$$= \cos\frac{2k'\pi}{n} \pm \sqrt{-1}\sin\frac{2k'\pi}{n}$$

and these values are the same as those which correspond to $k = k'$. It is therefore useless to give to k values greater than $\frac{1}{2}n$; and hence, whether we suppose n equal to an even number $2p$, or to an odd number $2p+1$, we shall be limited in giving values to k, to the values $k = 0, 1, 2, 3 \ldots p$.

It remains now that we show that formula (a) gives all the values of x belonging to equation (1).

If the equation is of the form $x^{2p} = 1$, formula (a) becomes

$$x = \cos \frac{kp}{p} \pm \sqrt{-1} \sin \frac{k\pi}{p}$$

For the extreme numbers $k = 0$, and $k = p$, we find $x = +1$, and $x = -1$. For the intermediate numbers $1, 2, 3, ---$ $p-1$, the arc is comprised between 0 and 180°; then the sine which is a multiplier of $\sqrt{-1}$ does not become zero: hence, these values of x are imaginary. Further, among these imaginary values, none are repeated; for, in the two roots which form part of the same couple, that is, which result from the same value of k, the imaginary factor $\sqrt{-1}$ has contrary signs; and in the couples which result from different values of k, the real parts are different, whereas they are the cosines of arcs which increase continuously from 0 to 180°. There are therefore $2p-2$ different imaginary values, and two real values; in all, $2p$ values for the equation $x^{2p} = 1$, as there should be.

If the equation (1) has the form $x^{2p+1} = 1$, formula (a) becomes

$$x = \cos \frac{2k\pi}{2p+1} \pm \sqrt{-1} \sin \frac{2k\pi}{2p+1}$$

In this form, there will be only one real value for x, $x = +1$, which corresponds to $k = 0$, the other values are imaginary, and further, the total number of values is equal to the exponent $2p+1$.

Let us now consider the equation

$$x^n = -1. \qquad (2)$$

If we make $x = \cos\phi \pm \sqrt{-1} \sin\phi$, we shall have $x^n = \cos n\phi \pm \sqrt{-1} \sin n\phi$.

Thus, the values of x will be made known, by determining the values of ϕ from the equation

$$\cos n\phi \pm \sqrt{-1} \sin n\phi = -1.$$

But this condition requires that we shall have separately $\sin n\phi = 0$ and $\cos n\phi = -1$.

Hence, the arc $n\phi$ must be an odd multiple of π. Therefore, making

$$n\phi = (2k+1)\pi, \text{ we deduce } \phi = \frac{(2k+1)\pi}{n}$$

and we shall have

$$x = \cos\frac{(2k+1)\pi}{n} \pm \sqrt{-1}\sin\frac{(2k+1)\pi}{n}. \qquad (\beta)$$

We need not take negative multiple values of π, for there would result the same values of x as when these multiple values are positive. Nor will be necessary to take $k > n$ or even equal to n: for, in taking from k the multiple of n which it contains, we diminish the arc $\frac{(2k+1)\pi}{n}$ one or more entire circumferences, which does not change the values of x in equation (β). The number $k = n - 1$ gives

$$x = \cos\frac{(2n-1)\pi}{n} \pm \sqrt{-1}\sin\frac{(2n-1)\pi}{n}$$

$$= \cos\frac{-\pi}{n} \pm \sqrt{-1}\sin\frac{-\pi}{n} = \cos\frac{\pi}{n} \pm \sqrt{-1}\sin\frac{\pi}{n}.$$

These values are the same as those obtained by the hypothesis $k = 0$; and, in general, the values of k equally distant from 0 and $n-1$ give the same values of x. Thus, if we make $k = n - 1 - k'$, we have

$$x = \cos\frac{(2n-2k'-1)\pi}{n} \pm \sqrt{-1}\sin\frac{(2n-2k'-1)\pi}{n}$$

$$x = \cos\frac{(2n+1)\pi}{n} \mp \sqrt{-1}\sin\frac{(2k'+1)\pi}{n}$$

a result which corresponds with that which we have when we make $k = k'$. Hence, it follows, that we shall obtain all the values of x by giving to k values which do not exceed $\frac{1}{2}(n-1)$. If n is an even number, $2p$, we make $k = 0, 1, 2, 3, \cdots p - 1$; and if n is an odd number, $2p + 1$, we make $k = 0, 1, 2, 3 \cdots p$.

In the case in which $n = 2p$, the equation to be resolved is $x^{2p} = -1$, and the numbers $k = 0, 1, 2, 3, \cdots p - 1$ give in the formula (β) the increasing arcs

$$\frac{\pi}{2p}, \ \frac{3\pi}{2p}, \ \frac{5\pi}{2p}, \qquad \frac{(2p-1)\pi}{2p},$$

which are all included between 0 and 180°. Consequently the series of no one of them is equal to 0, and their cosines are all unequal. Each arc thus gives two imaginary values for x which differ from each other by the sign of the factor $\sqrt{-1}$, and which therefore cannot be repeated. Hence x will have $2p$ different values as it should have.

When $n = 2p + 1$, the equation is $x^{2p+1} = -1$, and the values of $k = 0, 1, 2 \text{---} p$, give from formula (β) the following arcs

$$\frac{\pi}{2p+1},\ \frac{3\pi}{2p+1},\qquad \frac{(2p+1)\pi}{2p+1}\ \text{or}\ \pi.$$

The last arc being equal to π, gives $x = -1$. For each of the other arcs, x has two imaginary values, and it is easy to see, that, among all these values, none are repeated. Hence x has $2p + 1$ different values.

204. When we know the values of x from the equations $x^n = +1$, and $x^n = -1$, it is easy to form the real divisors of the second degree of the binomials $x^n - 1$, and $x^n + 1$.

In the first place, formula (α) gives for the factors of the first degree of the binomial $x^n - 1$, the two expressions

$$x - \cos\frac{2k\pi}{n} - \sqrt{-1}\sin\frac{2k\pi}{n}$$

$$x - \cos\frac{2k\pi}{n} + \sqrt{-1}\sin\frac{2k\pi}{n}$$

and, multiplying them together, we have

$$x^2 - 2x\cos\frac{2k\pi}{n} + 1 \qquad (\alpha')$$

This formula contains all the real divisors of the 2d degree of the binomial $x^n - 1$. To deduce these divisors, it is only necessary to substitute for k the positive numbers from 0 to $\frac{1}{2}n$.

In like manner, we find for the divisors of the 2d degree of the binomial $x^n + 1$, the formula

$$x^2 - 2x\cos\frac{(2k+1)\pi}{n} + 1 \qquad (\beta')$$

in which, by giving to k all entire positive values from 0 to $\frac{1}{2}(n-1)$, we shall find the divisors of the 2d degree.

Since the formulæ (α) and (β) comprise the real values of x from the equations $x^n = 1$, and $x^n = -1$, it follows that the two last formulæ, (α') and (β'), must also include the real factors of the 1st degree of the binomials x^n-1, and x^n+1. But it is to be observed that in these formulæ, they are presented under the form of being raised to the 2d power. Thus, if $k=0$, in formula (a'), it becomes x^2-2k+1, or $(x-1)^2$. The factor required is therefore $x-1$.

IRREDUCIBLE CASE OF CUBIC EQUATIONS.

205. *De Moivre's* formula furnishes a very direct and simple process of converting the *irreducible case* of cubic equations to a finite real form.

The solution of a cubic equation by the method of *Cardan* is as follows. Assuming a cubic equation, which has been deprived of its second term by known rule, we have

$$x^3+3px+2q=0. \qquad (1)$$

Place x equal to the sum of two other unknown quantities; that is, put

$$x=a+b.$$

We shall then have

$$x^3=a^3+b^3+3ab(a+b);$$

that is, replacing $a+b$ by x, and transposing,

$$x^3-3abx-a^3-b^3=0;$$

and, in order that this may be identical with the proposed equation, we must determine a and b so as to satisfy these conditions, viz.:

$$ab=-p, \quad a^3+b^3=-2q.$$

The problem is thus reduced to the determination of a and b from these two equations. From the first we have

$$a^3b^3=-p^3;$$

hence, combining this with the second, we have the sum of two quantities, a^3+b^3, and their product a^3b^3 given, to determine these quantities; a problem which may be readily solved by the quadratic equation

$$v^2+qv-\frac{p^3}{27}=0, \qquad (2)$$

by making $v=b^3$, in which the two values of v will be the expressions for a^3 and b^3. Hence, solving the equation, and separating these two values, we have

$$a^3 = -q + \sqrt{q^2 + p^3} \qquad (3)$$

$$b^3 = -q - \sqrt{q^2 + p^3} \qquad (4)$$

and, since $x = a + b$, there results the following general expression for the values of x in the proposed equation, viz.,

$$x = \left(-q + \sqrt{q^2 + p^3}\right)^{\frac{1}{3}} + \left(-q - \sqrt{q^2 + p^3}\right)^{\frac{1}{3}},$$

which is *Cardan's formula.*

Since the cube root of a^3 may be represented indifferently by either of the three expressions

$$a, \quad \frac{-1 + \sqrt{-3}}{2}\,a, \quad \frac{-1 - \sqrt{-3}}{2}\,a,$$

and the cube root of b^3 by either of the expressions

$$b, \quad \frac{-1 + \sqrt{-3}}{2}\,b, \quad \frac{-1 - \sqrt{-3}}{2}\,b,$$

it would appear that the proposed equation admits of *nine* values. It must be remembered, however, that in all cases where we assign the root of any expression, no reference is made to the mode in which the given power was formed, the root being in fact the expression from which the given power has been actually produced. When we speak of a proposed power having a multiplicity of roots, we merely refer to the various expressions, from either of which that power *might* be produced; and as many of these as prove inconsistent with the conditions involved in the production of the power, are of course to be rejected. Now, one of the conditions in virtue of which a^3 and b^3 have been produced, is

$$a\,b = -p;$$

but of the nine products which the preceding expressions for a and b furnish, six are imaginary; such a combination of values must therefore be rejected, as inconsistent with the conditions to be fulfilled; the other three products are possible. Hence, the only admissible solutions are the three following, where the product of a and b fulfils the above condition.

$$a+b,$$

$$\frac{-1+\sqrt{-3}}{2}a+\frac{-1-\sqrt{-3}}{2}b,$$

$$\frac{-1-\sqrt{-3}}{2}a+\frac{-1+\sqrt{-3}}{2}b.$$

If the values of v in equation (2), as exhibited in (3) and (4), are real, the formula for x will consist of the cube roots of two real quantities; but, if imaginary, then the expression for x will be the cube roots of two imaginary quantities; and, consequently, such value must itself be imaginary; that is, if the relation between p and q be such that

$$q^2+p^3<v,$$

the expression for x will consist of two parts, each of which is imaginary.

But the sum of these parts, that is, the complete expressions for x, must be real; because the preceding relation between p and q is that which necessitates the reality of all the values of the equation. The calculation, therefore, must furnish the means of effecting a reduction of the imaginary symbols. This difficulty, which has taxed the ingenuity of analysts, has been called the *irreducible case* of cubic equations, and is usually solved by developing each part separately, and then by combining the two, representing the aggregate by an infinite series, from which the imaginary symbol has disappeared.

De Moivre's formula furnishes a ready means of effecting the required reduction. For this purpose the two cubic radicals must be first reduced to the form

$$\sqrt[n]{\cos\phi+\sqrt{-1}\sin\phi}.$$

Since we have the condition $q^2+p^3<v$, p must be negative. Placing $-p$ for p in the given equation, it becomes

$$x^3-3px+2q=v. \qquad (1)$$

Then we shall have $q^2-p^3<v$, or $q^2<p^3$, and the values of a must be given by the formula

$$a=\sqrt[3]{-q+\sqrt{q^2-p^3}},$$

and those of x by the formula

$$x = a + \frac{p}{a} = \sqrt{p}\left\{\frac{a}{\sqrt{p}} + \frac{\sqrt{p}}{a}\right\}$$

in which we must substitute the different values of a.

The general value of a may then be placed under the form

$$\frac{a}{\sqrt{p}} = \sqrt[3]{-\frac{q}{\sqrt{p^3}} + \sqrt{1 - \frac{q^2}{p^3}}\sqrt{-1}.}$$

And since we assume $q^2 < p^3$, we may determine an arc ϕ by means of the equation

$$\cos\phi = \frac{-q}{\sqrt{p^3}}$$

There exists an infinite number of arcs corresponding to a given cosine; but we will limit ourselves here to that which is $< 180°$.

By *De Moivre's* formula we have

$$(\cos\tfrac{1}{3}\phi + \sqrt{-1}\sin\tfrac{1}{3}\phi)^3 = \cos\phi + \sqrt{-1}\sin\phi. \quad (2)$$

Hence, reciprocally, we must have

$$\sqrt[3]{\cos\phi + \sqrt{-1}\sin\phi} = \cos\tfrac{1}{3}\phi + \sqrt{-1}\sin\tfrac{1}{3}\phi.$$

and consequently,

$$\frac{a}{\sqrt{p}} = \sqrt[3]{\cos\phi + \sqrt{-1}\sin\phi} = \cos\tfrac{1}{3}\phi + \sqrt{-1}\sin\tfrac{1}{3}\phi.$$

$$\frac{\sqrt{p}}{a} = \frac{1}{\cos\tfrac{1}{3}\phi + \sqrt{-1}\sin\tfrac{1}{3}\phi} = \cos\tfrac{1}{3}\phi - \sqrt{-1}\sin\tfrac{1}{3}\phi$$

$$\frac{a}{\sqrt{p}} + \frac{\sqrt{p}}{a} = 2\cos\tfrac{1}{3}\phi.$$

We observe that the second member of formula (2) does not undergo any change when we add to ϕ any number of circumferences. Hence, designating 180° by π, and by k any whole number, the values of x would embrace all the values comprised in the formula

$$x = 2\sqrt{p}\cos\tfrac{1}{3}(\phi + 2k\pi)$$

However, there will appear but three distinct values for x; for, after making $k = 0$, 1, and 2, all the other substitutions reduce to the same results, and the only values for x will be the following:

$$x = 2\sqrt{p}\cos\tfrac{1}{3}\phi$$
$$x = 2\sqrt{p}\cos\tfrac{1}{3}(\phi + 2\pi)$$
$$x = 2\sqrt{p}\cos\tfrac{1}{3}(\phi + 4\pi).$$

CHAPTER IV.

Promiscuous Examples.

Plane Trigonometry.

1. Having measured a distance 200 feet in a direct horizontal line from the bottom of a tower, the angle of elevation of its top, taken at that distance, was found to be 47° 30′; what is the height of the tower? *Ans.* 218.26 feet.

2. What is the perpendicular height of a balloon, when its angles of elevation as taken by two observers, at the same time, both on the same side of it, and in the same vertical plane, were 35° and 64°, their distance apart being 880 yards? *Ans.* 935.757 yards.

3. Wanting to know the distance between two inaccessible objects from the top of a tower, 120 feet high, which lay in the same right line with the two objects, the angles formed by the vertical line of the tower with lines conceived to be drawn to the bottom of each of the objects, are found to be 33° and 64° 30′, what is the distance between the objects?

Ans. 173.656 feet.

4. From the edge of a ditch 36 feet wide, surrounding a fort, the angle of elevation of the top of the wall is measured and found equal to 62° 40′, what is the height of the wall, and what is the length of a ladder from the point on the ditch to the top of the wall?

Ans. Height of wall, 69.64 feet.
Length of ladder, 78.4 feet.

5. A ladder, 40 feet long, can be so planted, that it shall reach a window 33 feet from the ground, on one side of the street; and by turning it over, without moving the foot, it will do the same by a window 21 feet high, on the other side; what is the breadth of the street? *Ans.* 56.649 feet.

6. From the top of a tower, by the sea-side, 143 feet high, the angle of depression of a ship's bottom was observed to be 35°, what was the ship's distance from the bottom of the tower? *Ans.* 204.22 feet.

7. What is the perpendicular height of a hill; its angle of elevation, taken at the bottom of it, being 46°, and 200 yards farther off, on a level with the bottom of it, the angle of elevation was 31°? *Ans.* 286.28 yards.

8. Wanting to know the height of an inaccessible tower; at the least distance from it, on the same horizontal plane, its angle of elevation is found to be 58°; then going 300 feet directly from it, the angle of elevation is found to be 32°. Required its height, and the distance from it to the 1st station. *Ans.* Height, 307.53.
Distance, 192.15.

9. Being on a horizontal plane, and wanting to know the height of a tower on the top of an inaccessible hill; the angle of elevation of the top of the hill is found to be 40°, and of the top of the tower 51°; and then measuring in a line directly from it to the distance of 200 feet farther, the angle to the top of the tower is found to be 33° 45′; what is the height of the tower? *Ans.* 93.33148 feet.

10. From a window near the bottom of a house, on a level with the bottom of a steeple, the angle of elevation of the top of the steeple is found to be 40°; then from another window 18 feet directly above the former, the like angle was 37° 30′; what is the height of the steeple, and what its distance?
Ans. Height, 210.44 feet.
Distance, 250.79 "

11. Two ships-of-war, intending to cannonade a fort, are, by the shallowness of the water, kept so far from it, they suspect that their guns cannot have effect. In order to measure the distance, they separate from each other 440 yards; then each ship measures the angles the other ship and the fort subtend, which angles are 83° 45′ and 85° 15′. What was the distance between each ship, and what the distance to the fort? *Ans.* 2292.26 yards.
2298.05 "

12. Wanting to know the breadth of a river, a base line of 500 yards is measured close by one side of it; and at each

end of this line the angles subtended by the other end and a tree close by the bank on the other side of the river are found to be 58° and 79° 12′. What is the perpendicular breadth of the river? *Ans.* 529.48.

13. A point of land was observed by a ship at sea, to bear east-by-south; and after sailing north-east 12 miles, it was found to bear southeast-by-east. It is required to determine the place of that headland and the ship's distance from it at the last observation. *Ans.* 26.0728 miles.

14. Wanting to know the distance between a house and a mill, which was seen at a distance on the other side of a river, a base line of 600 yards is measured, and at each end of it, the angles subtended by the other end and the house and mill are observed, and are as follows, at one end, 58° 20′; and at the other end, 53° 30′ and 98° 45′. What was the distance between the house and the mill?

Ans. 959.5866 yards.

15. Wanting to know the distance from the observer to an inaccessible object O, on the other side of a river; and having no instruments for taking angles, and only a chain for measuring distances; two stations, A and B, are taken, 500 yards apart, and from each station in a direct line to the object O, distances of 100 yards are measured, viz., A C = 100 yards, and B D = 100 yards; and the diagonal B D is also measured 550 yards, and the diagonal B C, 560 yards. What is the distance of the object O from the stations A and B?

Ans. A O = 536.25 yards.
B O = 500.09 "

16. From a convenient station P, where could be seen three objects, A, B, and C, whose distances from each are known to be A B = 800 yards, A C = 600 yards, B C = 400 yards, the horizontal angles A P C = 33° 45′, and B P C = 22° 30′, are measured. Required to determine the distances P A, P B, P C. *Ans.* P A = 710.193 yards.
P C = 1042.522 "
P B = 934.291 "

Spherical Trigonometry.

1. In the right-angled triangle A B C, the hypothenuse A B = 65° 5′, the angle A = 48° 12′; find the sides A C, C B, and the angle B.

Ans. A C = 55° 7′ 22″.
B C = 42° 32′ 19″.
B = 64° 46′ 14″.

2. In the oblique-angled triangle A B C, given A B = 76° 20′, B C = 119° 17′, and B = 52° 5′, find A C, and angles B and C.

Ans. A C = 66° 5′ 36″.
A = 131° 10′ 42″.
C = 56° 58′ 58″.

3. In an oblique-angled triangle, the three sides are, a = 81° 17′, b = 114° 3′, c = 59° 12′; required the angles A, B, C.

Ans. A = 62° 39′ 42″.
B = 124° 60′ 50″.
C = 50° 31′ 42″.

4. In the right-angled triangle A B C, a = 115° 25′, c = 60° 59′; required the remaining parts.

Ans. B = 148° 56′ 45″.
C = 75° 30′ 33″.
b = 152° 13′ 50″.

5. In the right-angled triangle A B, c = 116° 30′ 43″, and b = 29° 41′ 32″; required the other parts.

Ans. C = 103° 52′ 46″.
B = 33° 30′ 22″.
a = 112° 48′ 58″.

6. In the oblique-angled triangle A B C, b = 91° 03′ 25″, a = 40° 36′ 37″; required the other parts.

Ans. B = 115° 35′ 41″.
C = 58° 30′ 57″.
c = 70° 58′ 52″.

7. In the oblique-angled triangle A B C, A = 103° 59′ 57″, B = 46° 18′ 07″, and a = 42° 68′ 48″; find the remaining parts.

Ans. b = 30°
C = 36° 07′ 54″.
c = 24° 03′ 56″.

8. In the oblique-angled triangle A B C, a = 40° 18′ 29″,

$b = 67° 14' 28''$, and $c = 89° 47' 06''$; find the remaining parts.

Ans. A = 34° 22′ 16″.
B = 53° 35′ 16″.
C = 119° 13′ 32″.

9. In the oblique-angled triangle A B C, A = 109° 55′ 42″, B = 116° 38′ 33″, C = 120° 43′ 37″; find the remaining parts.

Ans. a = 98° 21′ 40.
b = 109° 50′ 22.
c = 115° 13′ 26.

10. In the oblique-angled triangle A B C, b = 83° 19′ 42″, c = 23° 27′ 46″, A = 20° 39′ 48″; find the other parts.

Ans. B = 156° 30′ 16″.
C = 9° 11′ 48″.
a = 61° 32′ 12″.

11. In the oblique-angled triangle A B C, A = 34° 15′ 03″, B = 42° 15′ 13″, c = 76° 35′ 36″; find the other parts.

Ans. a = 40° 00′ 10″.
b = 50° 10′ 30″.
C = 121° 36′ 19″.

12.* Knowing the longitude of the sun E S = 67° 19′ 20″,

---

* The following definitions will aid the general student in the solution of this and the following examples: —

1. A plane through centre of the sun, and parallel to the earth's equator, cuts from the celestial sphere a great circle, called the *equinoctial.*

2. The earth revolves about the sun in a plane which cuts the celestial sphere in a great circle, called the *ecliptic.*

3. The planes of the equinoctial and ecliptic intersect each other in a right line through the sun. The points in which the line pierces the celestial sphere are called the *equinoxes.* One is the *vernal equinox*, the other is *the autumnal equinox.*

4. A great circle through the poles of the equinoctial is called a *declination circle.*

5. The *declination* of a star is its angular distance from the equinoctial measured upon the declination circle through the star.

6. The *right ascension* of a star is the angular distance from the vernal equinox to the intersection of the declination circle through the star with the equinoctial. It is measured on the equinoctial, eastward to 360°.

7. A great circle through the poles of the ecliptic is called a *circle of latitude.*

8. The *latitude of a star* is its angular distance from the ecliptic, measured on the circle of latitude passing through the star.

9. The *longitude of a star* is the angular distance from the vernal equinox

and the obliquity of the ecliptic SES' = 23° 28' 30"; find the right ascension of the sun E S' and his declination, S S'.

Ans. R. A. = 65° 30' 28".7.
Decl. = 21° 33' 52".9.

13. Knowing the right ascension of the sun E S' = 65° 30' 28".7, and the obliquity of the ecliptic S E S' = 23° 28' 30"; find the longitude of the sun, E S, and his declination, S S'.

Ans. Long. = 67° 19' 20".
Decl. = 21° 33' 52".

14. Knowing the declination of the sun, S S' = 21° 23' 52"; find his longitude E S and right ascension E S', the obliquity of ecliptic as before. *Ans.* Long. = 67° 19' 20".
R. A. = 65° 30' 28".7.

15. Knowing the right ascension of a star E A = 225° 20', and declination A *s* = 89° 9' 15"; find the distance *s* E to one of the points in which the ecliptic cuts the equator, and the angle *s* E A, which the arc of a great circle passing through the star and the equinox makes with the equator.

Ans. Distance, = 90° 35' 40".5.
Angle, = 90° 36' 5".7.

16. Knowing E S' the longitude of the star Sirius to be 101° 21' 13", and S S' its latitude = 39° 32' 1".8, and E M' the longitude of the moon = 100° 33' 17", and M M' her latitude = 5° 13' 19".8; find their distance M S. *Ans.* 34° 19' 04".6.

P is the pole of the equinoctial.
P' is the pole of the ecliptic.
E is vernal equinox.
E Q is the equinoctial.
E C is the ecliptic.
S place of the sun.
s place of star.
E S' right ascension of *s*.
E S" longitude of *s* eastward.
s S' declination of *s*.

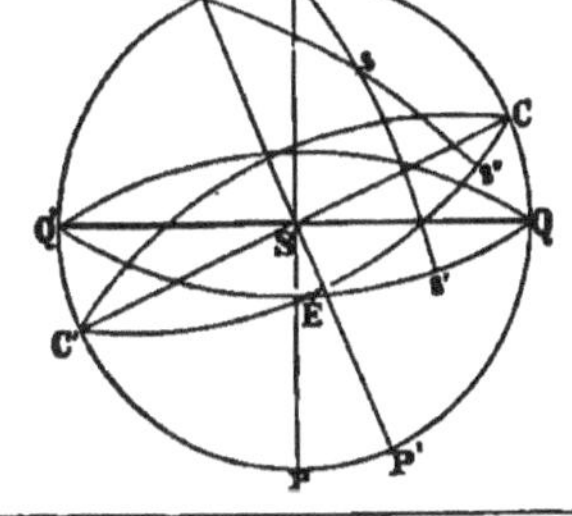

to the intersection of the circle of latitude through the star with the ecliptic. It is measured on the ecliptic, sometimes eastwardly and westwardly, each 180°; more commonly it is measured westwardly to 360°.

10. The planes of the equinoctial and of the ecliptic make an angle of 23° 28' 30".

11. The sun is always in the ecliptic. He has no latitude.

A A declination of *s*.
E A right ascension of *s*.
L *s* latitude of *s*.
E L longitude of *s*.

E vernal equinox.
M moon's place.
S place of star.
E Q ecliptic.
P pole of ecliptic.
E M′ longitude of moon.
M M′ latitude of moon.
E S longitude of star.
S S latitude of star.

Tables of Trigonometrical Formulæ.

The following Tables contain the most important formulæ used in Trigonometrical calculations. Those which have not been formally demonstrated may be readily verified, and will supply, in this way, a valuable exercise for the student. The formulæ marked with an *asterisk* should be carefully committed to memory.

Table I.

* 1. $\sin^2 a + \cos^2 a = 1$

* 2. $\frac{\sin a}{\cos a} = \text{tang}\, a$

* 3. $\frac{\cos a}{\sin a} = \cot a$

* 4. $\text{tang}\, a \cot a = 1$

* 5. $\sec a \cos a = 1$

* 6. $\text{cosec}\, a \sin a = 1$

* 7. $1 + \text{tang}^2 a = \sec^2 a$

* 8. $1 + \text{cotang}^2 a = \text{cosec}^2 a$

* 9. $\text{ver-sin}\, a = 1 - \cos a$

*10. $\text{co-ver-sin}\, a = 1 - \sin a$

TABLE II.

* $\sin 0$	$= 0$	$\sin(180^\circ + a)$	$= -\sin a$
* $\cos 0$	$= 1$	$\cos(180^\circ + a)$	$= -\cos a$
* $\mathrm{tang}\, 0$	$= 0$	$\mathrm{tang}(180^\circ + a)$	$= \mathrm{tang}\, a$
* $\mathrm{cotang}\, 0$	$= \infty$	$\mathrm{cotang}(180^\circ + a$	$= \mathrm{cotang}\, a$
* $\sec 0$	$= 1$	$\sec(180^\circ + a)$	$= -\sec a$
* $\mathrm{cosec}\, 0$	$= \infty$	$\mathrm{cosec}(180^\circ + a)$	$= -\mathrm{cosec}\, a$
* $\sin(90^\circ - a)$	$= \cos a$	$\sin(270^\circ - a)$	$= -\cos a$
* $\cos(90^\circ - a)$	$= \sin a$	$\cos(270^\circ - a)$	$= -\sin a$
* $\mathrm{tang}(90^\circ - a)$	$= \cot a$	$\mathrm{tang}(270^\circ - a)$	$= \cot a$
* $\cot(90^\circ - a)$	$= \mathrm{tang}\, a$	$\cot(270^\circ - a)$	$= \mathrm{tang}\, a$
* $\sec(90^\circ - a)$	$= \mathrm{cosec}\, a$	$\sec(270^\circ - a)$	$= -\mathrm{cosec}\, a$
* $\mathrm{cosec}(90^\circ - a)$	$= \sec a$	$\mathrm{cosec}(270^\circ - a$	$= -\sec a$
* $\sin 90^\circ$	$= 1$	$\sin 270^\circ$	$= -1$
* $\cos 90^\circ$	$= 0$	$\cos 270^\circ$	$= 0$
* $\mathrm{tang}\, 90^\circ$	$= \infty$	$\mathrm{tang}\, 270^\circ$	$= -\infty$
* $\mathrm{cotang}\, 90^\circ$	$= 0$	$\mathrm{cotang}\, 270^\circ$	$= 0$
* $\sec 90^\circ$	$= \infty$	$\sec 270^\circ$	$= \infty$
* $\mathrm{cosec}\, 90^\circ$	$= 1$	$\mathrm{cosec}\, 270^\circ$	$= -1$
* $\sin(90^\circ + a)$	$= \cos a$	$\sin(270^\circ + a)$	$= -\cos a$
* $\cos(90^\circ + a)$	$= -\sin a$	$\cos(270^\circ + a)$	$= \sin a$
* $\mathrm{tang}(90^\circ + a)$	$= -\cot a$	$\mathrm{tang}(270^\circ + a)$	$= -\cot a$
* $\cot(90^\circ + a)$	$= -\mathrm{tang}\, a$	$\cot(270^\circ + a)$	$= -\tan a$
$\sec(90^\circ + a)$	$= -\mathrm{cosec}\, a$	$\sec(270^\circ + a)$	$= \mathrm{cosec}\, a$
$\mathrm{cosec}(90^\circ + a)$	$= \sec a$	$\mathrm{cosec}(270^\circ + a)$	$= -\sec a$
* $\sin(180^\circ - a)$	$= \sin a$	$\sin(360 - a)$	$= -\sin a$
* $\cos(180^\circ - a)$	$= -\cos a$	$\cos(360^\circ - a)$	$= \cos a$
* $\mathrm{tang}(180^\circ - a)$	$= -\mathrm{tang}\, a$	$\tan(360^\circ - a)$	$= -\tan a$
* $\mathrm{cotang}(180^\circ - a)$	$= -\cot a$	$\cot(360^\circ - a)$	$= -\cot a$
$\sec(180^\circ - a)$	$= -\sec a$	$\sec(360^\circ - a)$	$= \sec a$
$\cos(180^\circ - a)$	$= \mathrm{cosec}\, a$	$\mathrm{cosec}(360^\circ - a)$	$= -\mathrm{cosec}\, a$
* $\sin 180^\circ$	$= 0$	$\sin 360^\circ$	$= 0$
* $\cos 180^\circ$	$= -1$	$\cos 360^\circ$	$= 1$
* $\mathrm{tang}\, 180^\circ$	$= 0$	$\tan 360^\circ$	$= 0$
* $\mathrm{cotang}\, 180^\circ$	$= -\infty$	$\cot 360^\circ$	$= \infty$
* $\sec 180^\circ$	$= -1$	$\sec 360^\circ$	$= 1$
* $\mathrm{cosec}\, 180^\circ$	$= 0$	$\mathrm{cosec}\, 360^\circ$	$= \infty$

TABLE III.

$$*\sin(a \pm b) = \sin a \cos b \pm \sin b \cos a$$

$$*\cos(a \pm b) = \cos a \cos b \mp \sin a \cos b$$

$$*\operatorname{tang}(a \pm b) = \frac{\tan a \pm \tan b}{1 \mp \tan a \tan b}$$

$$*\sin 2a = 2 \sin a \cos a$$

$$\cos 2a = \cos^2 a - \sin^2 a = 2\cos^2 a - 1 = 1 - 2\sin^2 a.$$

$$\operatorname{tang} 2a = \frac{2 \tan a}{1 - \tan^2 a}$$

$$\sin \frac{1}{2} a = \sqrt{\frac{1 - \cos a}{2}}$$

$$\cos \frac{1}{2} a = \sqrt{\frac{1 + \cos a}{2}}$$

$$\operatorname{tang} \frac{1}{2} a = \sqrt{\frac{1 - \cos a}{1 + \cos a}} = \frac{1 - \cos a}{\sin a} = \frac{\sin a}{1 + \cos a}$$

$$\sin a = 2 \sin \tfrac{1}{2} a \cos \tfrac{1}{2} a$$

$$= 3 \sin a - 4 \sin^3 a$$

$$\cos 3a = 4\cos^3 a - 3 \cos a$$

$$\sin(n+1)a = 2 \sin n a \cos a - \sin(n-1)a$$

$$\cos(n+1)a = 2 \cos n a \cos a - \cos(n-1)a$$

$$\sin a + \sin b = 2 \sin \tfrac{1}{2}(a+b) \cos \tfrac{1}{2}(a-b)$$

$$\sin a - \sin b = 2 \sin \tfrac{1}{2}(a-b) \cos \tfrac{1}{2}(a+b)$$

$$\cos a + \cos b = 2 \cos \tfrac{1}{2}(a+b) \cos \tfrac{1}{2}(a-b)$$

$$\cos a - \cos b = -2 \sin \tfrac{1}{2}(a+b) \sin \tfrac{1}{2}(a-b)$$

$$\frac{\sin a + \sin b}{\sin a - \sin b} = \frac{\tan \frac{1}{2}(a+b)}{\tan \frac{1}{2}(a-b)}$$

$$\sin(a+b) + \sin(a-b) = 2 \sin a \cos b$$

$$\sin(a+b) - \sin(a-b) = 2 \sin b \cos a$$

$$\cos(a+b) + \cos(a-b) = 2 \cos a \cos b$$

$$\cos(a+b) - \cos(a-b) = -2 \sin a \sin b$$

$$\sin(a+b) \cos(a+b) = \sin^2 a - \sin^2 b = \cos^2 b - \cos^2 a$$

$$\cos(a+b) \cos(a-b) = \cos^2 a - \sin^2 b = \cos^2 a + \cos^2 b - 1$$

$$*\sin 45^\circ = \cos 45^\circ = \frac{1}{\sqrt{2}}$$

$$*\tan 45^\circ = \cot 45^\circ = 1$$

$$*\sin 30° = \cos 60° = \tfrac{1}{2}$$

$$*\cos 30° = \sin 60° = \tfrac{1}{2}\sqrt{3}$$

$$*\tan 30° = \cot 60° = \frac{1}{\sqrt{3}}$$

$$*\cot 30° = \text{tang}\, 60° \quad \sqrt{3}$$

The following formulæ, although of less frequent occurrence, may be found useful, and can be readily deduced from the above.

$$\sin(45° \pm a) = (\cos 45° \mp a) = \frac{\cos a \pm \sin a}{\sqrt{2}}$$

$$\tan(45° \pm a) = \quad = \frac{1 \pm \tan a}{1 \mp \tan a}$$

$$\tan^2(45° \pm \tfrac{1}{2}a) = \frac{1 \pm \sin a}{1 \mp \sin a}$$

$$\tan(45° \pm \tfrac{1}{2}a) = \frac{1 \pm \sin a}{\cos a}$$

$$\frac{\sin(a+b)}{\sin(a-b)} = \frac{\tan a + \tan b}{\tan a - \tan b} = \frac{\cot b + \cot a}{\cot b - \cot a}$$

$$\frac{\cos(a+b)}{\cos(a-b)} = \frac{\cot b - \tan a}{\cot b + \tan a} = \frac{\cot a - \tan b}{\cot a + \tan a}$$

$$\frac{\sin a + \sin b}{\cos a + \cos b} = \tan \tfrac{1}{2}(a+b)$$

$$\frac{\sin a + \sin b}{\cos a - \cos b} = -\cot \tfrac{1}{2}(a-b)$$

$$\frac{\sin a - \sin b}{\cos a + \cos b} = \tan \tfrac{1}{2}(a-b)$$

$$\frac{\sin a - \sin b}{\cos a - \cos b} = -\cot \tfrac{1}{2}(a+b)$$

$$\frac{\cos a + \cos b}{\cos a - \cos b} = -\cot \tfrac{1}{2}(a+b) \cot \tfrac{1}{2}(a-b)$$

$$\tan a + \tan b = \frac{\sin(a+b)}{\cos a \cos b}$$

$$\cot a + \cot b = \frac{\sin(a+b)}{\sin a \sin b}$$

$$\tan a - \tan b = \frac{\sin(a-b)}{\cos a \cos b}$$

$$\cot a - \cot b = -\frac{\sin(a-b)}{\sin a \sin b}$$

$$\tan^2 a - \tan^2 b = \frac{\sin(a+b)\sin(a-b)}{\cos^2 a \cos^2 b}$$

$$\cot^2 a - \cot^2 b = -\frac{\sin(a+b)\sin(a-b)}{\sin^2 a \sin^2 b}$$

It will also be a useful exercise to verify the following values of $\sin a$, $\cos a$, $\tan a$.

TABLE OF THE MOST USEFUL VALUES OF SIN a, COS a, TANG a.

Values of sin a.	*Values of cos a.*	*Values of tan a.*
1. $\cot a \tan a$	1. $\frac{\sin a}{\tan a}$	1. $\frac{\sin a}{\cos a}$
2. $\frac{\cos a}{\cot a}$	2. $\sin a \cot a$	2. $\frac{1}{\cot a}$
3. $\sqrt{1-\cos^2 a}$	3. $\sqrt{1-\sin^2 a}$	3. $\sqrt{\frac{1}{\cos^2 a}-1}$
4. $\frac{1}{\sqrt{1+\cot^2 a}}$	4. $\frac{1}{\sqrt{1+\operatorname{tang}^2 a}}$	4. $\frac{\sin a}{\sqrt{1-\sin^2 a}}$
5. $\frac{\tan a}{\sqrt{1+\tan^2 a}}$	5. $\frac{\cot a}{\sqrt{1+\cot^2 a}}$	5. $\frac{\sqrt{1-\cos^2 a}}{\cos a}$
6. $2 \sin \frac{1}{2} a \cos \frac{1}{2} a$	6. $\cos^2 \frac{1}{2} a - \sin^2 \frac{1}{2} a$	
7. $\frac{\sqrt{1-\cos 2a}}{2}$	7. $1 - 2 \sin^2 \frac{1}{2} a$	7. $\frac{2 \tan \frac{1}{2} a}{1-\tan^2 \frac{1}{2} a}$
8. $\frac{2 \tan \frac{1}{2} a}{1+\tan^2 \frac{1}{2} a}$	8. $2 \cos^2 \frac{1}{2} a - 1$	8. $\frac{2 \cot \frac{1}{2} a}{\cot^2 \frac{1}{2} a - 1}$
9. $\frac{2}{\cot \frac{1}{2} a + \tan \frac{1}{2} a}$	9. $\sqrt{\frac{1+\cos 2a}{2}}$	9. $\frac{2}{\cot \frac{1}{2} a - \tan \frac{1}{2} a}$
10 $\frac{\sin(30^\circ + a) - \sin(30^\circ - a)}{\sqrt{3}}$	10. $\frac{1-\tan^2 \frac{1}{2} a}{1+\tan^2 \frac{1}{2} a}$	10. $\cot a - 2 \cot 2a$
11. $2 \sin^2 (45^\circ + \frac{1}{2} a) - 1$	11. $\frac{\cos \frac{1}{2} a - \operatorname{tang} \frac{1}{2} a}{\cot \frac{1}{2} a + \tan \frac{1}{2} a}$	11. $\frac{1-\cos 2a}{\sin 2a}$
12. $1 - 2 \sin^2 (45^\circ - \frac{1}{2} a)$	12. $\frac{1}{1+\tan a \tan \frac{1}{2} a}$	12. $\frac{\sin^2 a}{1+\cos 2a}$

FINIS.

TABLES

OF

LOGARITHMS OF NUMBERS

AND OF

SINES AND TANGENTS

FOR EVERY

TEN SECONDS OF THE QUADRANT,

WITH OTHER USEFUL TABLES.

CONTENTS.

PREFACE.

The accompanying tables were designed to afford the means of performing trigonometrical computations with facility and precision. The tables chiefly used in this country for purposes of education extend to six decimal places, like those in the present collection; but the precision which they are designed to furnish is only attained by a serious expenditure of labor. In the Table of Logarithms of Numbers they do not furnish the correction for a fifth figure in the natural number, and the labor of computing this correction is such that I always prefer the use of Hutton's Tables, extending to seven places, even in computations to which six-place logarithms are abundantly competent. In the present collection, the correction for a fifth figure of the natural number is introduced at the bottom of each page, and the table is thus rendered nearly as useful as one of the common kind extending to 100,000. The whole has been carefully compared with standard authors, and nearly a dozen errors have thus been detected in the common tables.

The principal table in this collection is that of Logarithmic Sines and Tangents. The common tables in this country extend only to minutes, with differences to 100″. If, in a trigonometrical computation, angles are only required to the nearest minute, tables to five places are quite sufficient; but if the computation is to be carried to seconds, these can only be obtained from the common tables by a great expenditure of time and labor. In the present collection, the sines and tangents are furnished to every ten seconds of the quadrant, and at the bottom of each page is given the correction for any number of seconds less than ten, so that the precision of seconds can be obtained with almost the same facility as that of minutes with the tables in common use. Moreover, near the limits of the quadrant, by means of an auxiliary table, sines and tangents are readily obtained, even for a fraction of a second. The method of arrangement of the sines and tangents was suggested by a table in Mackay's Longitude; but the errors of that table, amounting to several thousand, have been corrected by a careful comparison with the work of Ursinus. By comparison with the same standard, more than two hundred errors (chiefly in the final figures) have been detected in the tables in common use.

The Table of Natural Sines and Tangents is of less use than the logarithmic; nevertheless, it is often important for reference, particularly in analytical geometry and the calculus; and it is useful as a stepping-stone to assist the beginner in comprehending the nature of logarithmic sines and tangents. The Traverse Table commonly used in this country furnishes the latitude and departure to every quarter degree of the quadrant, for distances from 1 to 100, and occupies ninety pages. The accompanying table occupies but six pages, and yields ten times greater precision.

The Table of Meridional Parts extends to tenths of a mile, and great care has been taken to insure its accuracy. For this purpose, I have compared all the similar tables within my reach, and among them have found two which appeared to have been computed independently. Between them there were detected 674 discrepancies in the final figures. These cases were all recomputed, and 78 errors were detected in the best copy compared. It is probable that the numbers in this table are not in every instance true to the *nearest* tenth of a mile; but it is believed that the remaining errors are few in number, as well as minute. This table is confidently pronounced more accurate than any similar one with which I have been able to compare it.

The Table of Corrections to Middle Latitude was computed entirely anew. The corresponding table in common use, which was originally computed by Workman, contains more than four hundred errors, several of them amounting to two minutes.

On the whole, it is believed that the accompanying tables will be found more convenient to the computer than any tables of six decimal places hitherto published in this country; and that they will be pronounced sufficiently extensive for all purposes of academic and collegiate instruction, as well as for practical mechanics and surveyors.

EXPLANATION OF THE TABLES.

Table of Logarithms of Numbers, pp. 1–20.

Logarithms are numbers contrived to diminish the labor of Multiplication and Division by substituting in their stead Addition and Subtraction. All numbers are regarded as powers of some one number, which is called the *base* of the system; and the exponent of that power of the base which is equal to a given number, is called the logarithm of that number.

The base of the common system of logarithms (called, from their inventor, Briggs' logarithms) is the number 10. Hence all numbers are to be regarded as powers of 10. Thus, since

$10^0=1$,	0 is the logarithm of	1	in Briggs' system;
$10^1=10$,	1 " "	10	" "
$10^2=100$,	2 " "	100	" "
$10^3=1000$,	3 " "	1000	" "
$10^4=10,000$,	4 " "	10,000	" "
&c.,	&c.,		&c.;

whence it appears that, in Briggs' system, the logarithm of every number between 1 and 10 is some number between 0 and 1, *i. e.*, is a proper fraction. The logarithm of every number between 10 and 100 is some number between 1 and 2, *i. e.*, is 1 plus a fraction. The logarithm of every number between 100 and 1000 is some number between 2 and 3, *i. e.*, is 2 plus a fraction, and so on.

The preceding principles may be extended to fractions by means of negative exponents. Thus, since

$10^{-1}=0.1$,	—1 is the logarithm of	0.1	in Briggs' system
$10^{-2}=0.01$,	—2 " "	0.01	" "
$10^{-3}=0.001$,	—3 " "	0.001	" "
$10^{-4}=0.0001$,	—4 " "	0.0001	" "
&c.,	&c.,		&c.

Hence it appears that the logarithm of every number between 1 and 0.1 is some number between 0 and —1, or may be represented by —1 plus a fraction; the logarithm of every number between 0.1 and .01 is some number between —1 and —2, or may be represented by —2 plus a fraction; the logarithm of every number between .01 and .001 is some number between —2 and —3, or is equal to —3 plus a fraction, and so on.

The logarithms of most numbers, therefore, consist of an integer and a fraction. The integral part is called the *characteristic*, and may be known from the following

RULE.

The characteristic of the logarithm of a number greater than unity, is one less than the number of integral figures in the given number.

Thus the logarithm of 297 is 2 plus a fraction; that is, the characteristic of the logarithm of 297 is 2, which is one less than the number of integral figures. The characteristic of the logarithm of 5673.29 is 3; that of 73254.1 is 4, &c.

The characteristic of the logarithm of a decimal fraction is a negative number, and is equal to the number of places by which its first significant figure is removed from the place of units.

Thus the logarithm of .0046 is —3 plus a fraction; that is, the characteristic of the logarithm is —3, the first significant figure 4 being removed three places from units.

The accompanying table contains the logarithms of all numbers from 1 to 10,000 carried to 6 decimal places.

To find the Logarithm of any Number between 1 *and* 100.

Look on the first page of the table, along the column of numbers under N, for the given number, and against it, in the next column, will be found the logarithm, with its characteristic. Thus,

opposite 13 is 1.113943, which is the logarithm of 13;
" 65 is 1.812913, " " 65.

To find the Logarithm of any Number consisting of three Figures.

Look on one of the pages from 2 to 20, along the left-hand column marked N, for the given number, and against it, in the column headed 0, will be found the decimal part of its logarithm. To this the characteristic must be prefixed, according to the rule already given. Thus

the logarithm of 347 will be found, from page 8, to be 2.540329;
" " 871 " " " 18, " 2.940018.

As the first two figures of the decimal are the same for several successive numbers in the table, they are not repeated for each logarithm separately, but are left to be supplied. Thus the decimal part of the logarithm of 339 is .530200. The first two figures of the decimal remain the same up to 347; they are therefore omitted in the table, and are to be supplied.

To find the Logarithm of any Number consisting of four Figures.

Find the three left-hand figures in the column marked N as before and the fourth figure at the head of one of the other columns. Opposite to the first three figures, and in the column under the fourth figure, will be found four figures of the logarithm, to which two figures from the column headed 0 are to be prefixed, as in the former case. The characteristic must be supplied by the usual rule. Thus

the logarithm of 3456 is 3.538574;
" " 8765 is 3.942752.

In several of the columns headed 1, 2, 3, &c., small dots are found in the place of figures. This is to show that the two figures which are to be prefixed from the first column have changed, and they are to be taken from the horizontal line directly *below*. The place of the dots is to be supplied with ciphers. Thus

the logarithm of 2045 is 3.310693;
" " 9777 is 3.990206.

The two leading figures from the column 0 must also be taken from the horizontal line below, if any dots have been passed over on the same horizontal line. Thus

the logarithm of 1628 is 3.211654.

To find the Logarithm of any Number containing more than four Figures.

By inspecting the table, we shall find that within certain limits the logarithms are nearly proportional to their corresponding numbers. Thus

the logarithm of 7250 is 3.860338;
" " 7251 is 3.860398;
" " 7252 is 3.860458;
" " 7253 is 3.860518.

Here the difference between the successive logarithms, called the *tabular difference*, is constantly 60, corresponding to a difference of unity in the natural numbers. If, then, we suppose the logarithms to be proportional to their corresponding numbers (as they are nearly), a difference of 0.1 in the numbers should correspond to a difference of 6 in the logarithms; a difference of 0.2 in the numbers should correspond to a difference of 12 in the logarithms, &c. Hence

the logarithm of 7250.1 must be 3.860344;
" " 7250.2 " 3.860350;
" " 7250.3 " 3.860356;
&c., &c.

In order to facilitate the computation, the tabular difference is inserted on page 16 in the column headed D, and the proportional part for the fifth figure of the natural number is given at the bottom of the page. Thus, when the tabular difference is 60, the corrections for .1, .2, .3, &c., are seen to be 6, 12, 18, &c.

If the given number was 72501, the characteristic of its logarithm would be 4, but the decimal part would be the me as for 7250 1

If it were required to find the correction for a sixth figure in the natural number, it is readily obtained from the Proportional Parts in the table. Thus, if the correction for .5 is 30, the correction for .05 is obviously 3.

As the differences change rapidly in the first part of the table, it was found inconvenient to give the proportional parts for each tabular difference; accordingly, for the first seven pages they are only given for the *even* differences, but the proportional parts for the odd differences will be readily found by inspection.

Required the logarithm of 452789.

The logarithm of 452700 is 5.655810

The tabular difference is 96.

Accordingly, the correction for the fifth figure, 8, is 77, and for the sixth figure, 9, is 8.6, or 9 nearly. Adding these corrections to the number before found, we obtain 5.655896.

The preceding logarithms do not pretend to be perfectly exact, but only the nearest numbers having but six decimal places. Accordingly when the fraction which is omitted exceeds half a unit in the sixth decimal place, the last figure must be increased by unity.

Required the logarithm of 8765432.

The logarithm of 8765000 is	6.942752
Correction for the fifth figure 4,	20
" " sixth figure 3,	1.5
" " seventh figure 2,	0.1
Therefore the logarithm of 8765432 is	6.942774.

Required the logarithm of 234567.

The logarithm of 234500 is	5.370143
Correction for the fifth figure 6,	111
" " sixth figure 7,	13
Therefore the logarithm of 234567 is	5.370267.

To find the Logarithm of a Decimal Fraction.

The decimal part of the logarithm of any number is the same as that of the number multiplied or divided by 10, 100, 1000, &c. Hence, for a decimal fraction, we find the logarithm as if the figures were integers and prefix the characteristic according to the usual rule.

EXAMPLES.

The logarithm of	345.6	is 2.538574;
" "	87.65	is 1.942752;
" "	2.345	is 0.370143;
" "	.1234	is $\bar{1}$.091315;
" "	.005678	is $\bar{3}$.754195.

The minus sign is placed *over* the characteristic to show that this alone is negative, while the decimal part of the logarithm is positive.

To find the Logarithm of a Vulgar Fraction.

We may reduce the vulgar fraction to a decimal, and find its logarithm by the preceding rule; or, since the value of a fraction is equal to the quotient of the numerator divided by the denominator, we may subtract the logarithm of the denominator from that of the numerator; the difference will be the logarithm of the quotient.

Required the logarithm of $\frac{3}{16}$, or 0.1875.

From the logarithm of 3, 0.477121,
Subtract the logarithm of 16, 1.204120.

Leaves logarithm of $\frac{3}{16}$, or .1875, $\bar{1}.273001$.

In the same manner we find

the logarithm of $\frac{4}{55}$ is $\bar{2}.861697$;
" " $\frac{123}{876}$ is $\bar{1}.147401$.

To find the natural Number corresponding to any Logarithm.

Look in the table in the column headed 0 for the first two figures of the logarithm, neglecting the characteristic; the other four figures are to be looked for in the same column, or in one of the nine following columns; and if they are exactly found, the first three figures of the corresponding number will be found opposite to them in the column headed N, and the fourth figure will be found at the top of the page. This number must be made to correspond with the characteristic of the given logarithm by pointing off decimals or annexing ciphers. Thus

the natural number belonging to the logarithm 4.370143 is 23450;
" " " " 1.538574 is 34.56.

If the decimal part of the logarithm can not be exactly found in the table, look for the nearest less logarithm, and take out the four figures of the corresponding natural number as before; the additional figures may be obtained by means of the Proportional Parts at the bottom of the page.

Required the number belonging to the logarithm 4.368399.
On page 6, we find the next less logarithm .368287.

The four corresponding figures of the natural number are 2335. Their logarithm is less than the one proposed by 112. The tabular difference is 186; and, by referring to the bottom of page 6, we find that, with a difference of 186, the figure corresponding to the Proportional Part 112 is 6. Hence the five figures of the natural number are 23356; and, since the characteristic of the proposed logarithm is 4, these five figures are all integral.

Required the number belonging to logarithm 5.345678.
The next less logarithm in the table is .345570
Their difference is 108.

The first four figures of the natural number are 2216.

With the tabular difference 196, the fifth figure corresponding to 108 is seen to be 5, with a remainder of 10, which furnishes a sixth figure 5 nearly. Hence the required number is 221655.

In the same manner we find

the number corresponding to logarithm 3.538672 is 3456.78;
" " " 1.994605 is 98.7654;
" " " $\bar{1}$.647817 is .444444.

TABLE OF NATURAL SINES AND TANGENTS, pp. 116–133.

This is a table of natural sines and tangents for every degree and minute of the quadrant, carried to six places of figures. Since the radius of the circle is supposed to be unity, the sine of every arc below 90° is less than unity. These sines are expressed in decimal parts of the radius; and, although the decimal point is not written in the table, it must always be prefixed. The degrees are arranged in order at the top of the page, and the minutes in the left hand vertical column. Directly under the given number of degrees at the top of the page, and opposite to the minutes on the left, will be found the sine required. The two leading figures are repeated at intervals of ten minutes. Thus

the sine of 6° 27′ is .112336;
" " 28° 53′ is .483028.

The same number in the table is both the sine of an arc and the cosine of its complement. The degrees for the cosines must be sought at the bottom of the page, and the minutes on the right. Thus

the cosine of 62° 25′ is .463038;
" " 84° 23′ is .097872.

If a sine is required for an arc consisting of degrees, minutes, and *seconds*, it may be found by means of the line at the bottom of each page, which gives the proportional part corresponding to one second of arc

Required the sine of 8° 9′ 10″.

The sine of 8° 9′ is .141765.

By referring to the bottom of page 116, in the column under 8°, we find the correction for 1″ is 4.80; hence the correction for 10″ must be 48, which, added to the number above found, gives for the sine of 8° 9′ 10″, .141813.

In the same manner we find

the cosine of 56° 34′ 28″ is .550853.

It will be observed, that since the cosines decrease while the arcs increase, the correction for the 28″ is to be subtracted from the cosine of 56° 34′.

The arrangement of the table of natural tangents is similar to that of the table of sines. The tangents for arcs less than 45° are all less than radius, and consist wholly of decimals. For arcs above 45°, the tangents are all greater than radius and contain both integral and decimal figures. The proportional parts at the bottom of each page enable us readily to find the correction for seconds. Thus

the natural tangent of 32° 29′ 18″ is .636784;
" " 74° 35′ 55″ is 3.63014.

To find the Number of Degrees, Minutes, and Seconds belonging to a given Sine or Tangent.

If the given sine or tangent is found exactly in the table, the corresponding degrees will be found at the top of the page, and the minutes on the left hand. But when the given number is not found exactly in the table, look for the sine or tangent which is next less than the proposed one, and take out the corresponding degrees and minutes. Find, also, the difference between this tabular number and the number proposed, and divide it by the proportional part for 1″ found at the bottom of the page; the quotient will be the required number of seconds.

Required the arc whose sine is .750000.

The next less sine in the table is .749919, the arc corresponding to which is 48° 35′. The difference between this sine and that proposed is 81, which, divided by 3.21, gives 25. Hence the required arc is 48° 35′ 25″.

In the same manner we find

the arc whose tangent is 2.000000, to be 63° 26′ 6″.

TABLE OF NATURAL SECANTS, pp. 134–5.

This is a table of natural secants for every ten minutes of the quadrant carried to seven places of figures. The degrees are arranged in order in the first vertical column on the left, and the minutes at the top of the page. Thus

the secant of 21° 20′ is 1.073561;
" " 81° 50′ is 7.039622.

If a secant is required for a number of minutes not given in the table, the correction for the odd minutes may be found by means of the last vertical column on the right, which shows the proportional part for one minute.

Let it be required to find the secant of 30° 33′.

The secant of 30° 30′ is 1.160592.

The correction for 1′ is 198.9, which, multiplied by 3, gives 597. Adding this to the number before found, we obtain 1.161189.

For a cosecant, the degrees must be sought in the right-hand vertical column, and the minutes at the bottom of the page. Thus

the cosecant of 47° 40′ is 1.352742.

TABLE OF LOGARITHMIC SINES AND TANGENTS, pp. 21–115.

This is a table of the logarithms of the sines and tangents for every ten seconds of the quadrant, carried to six places of decimals. The degrees and seconds are placed at the top of the page, and the minutes in the left vertical column. After the first two degrees, the three leading figures in the table of sines are only given in the column headed 0″, and are to be prefixed to the numbers in the other columns, as in the table of logarithms of numbers. Also, where the leading figures change, this change is indicated by dots, as in the former table. The correction for any number of seconds less than 10 is given at the bottom of the page.

To find the Logarithmic Sine or Tangent of a given Arc.

Look for the degrees at the top of the page, the minutes on the left hand, and the next less tenth second at the top; then, under the seconds, and opposite to the minutes, will be found four figures, to which the three leading figures are to be prefixed from the column headed 0″; to this add the proportional part for the odd seconds from the bottom of the page.

Required the logarithmic sine of 24° 27′ 34″.

The logarithmic sine of 24° 27′ 30″ is	9.617033.
Proportional part for 4″ is	18.
Logarithmic sine of 24° 27′ 34″ is	9.617051.

This is the logarithm of .414049 found in the table of natural sines on page 120. The natural sine being less than unity, the characteristic of its logarithm is negative. To obviate this inconvenience, the characteristics in the table have all been increased by 10; or the logarithmic sines may be regarded as the logarithms of natural sines computed for a radius of 10,000,000,000.

Required the logarithmic tangent of 73° 35′ 43″.

The logarithmic tangent of 73° 35′ 40″ is	10.531031.
Proportional part for 3″ is	23.
Logarithmic tangent of 73° 35′ 43″	10.531054.

When a cosine is required, the degrees and seconds must be sought at the bottom of the page, and the minutes on the right, and the correction for the odd seconds must be subtracted from the number in the table

Required the logarithmic cosine of 59° 33′ 47″.

The logarithmic cosine of 59° 33′ 40″ is	9.704682.
Proportional part for 7″ is	25.
Logarithmic cosine of 59° 33′ 47″ is	9.704657.

So, also, the logarithmic cotangent of 37° 27′ 14″ is found to be 10.115744.

The proportional parts given at the bottom of each page correspond to the degrees at the top of the page increased by 30′, and are not strictly applicable to any other number of minutes; nevertheless, the differences of the sines change so slowly, except near the commencement of the quadrant, that the error resulting from using these numbers for every part of the page will seldom exceed a unit in the sixth decimal place. For the first two degrees, the differences change so rapidly that the proportional part for 1″ is given for each minute in the right-hand column of the page. The correction for any number of seconds less than ten will be found by multiplying the proportional part for 1″ by the given number of seconds.

Required the logarithmic sine of 1° 17′ 33″.

The logarithmic sine of 1° 17′ 30″ is 8.352991

The correction for 3″ is found by multiplying 93.4 by 3, which gives 280. Adding this to the above tabular number, we obtain

the sine of 1° 17′ 33″, 8.353271.

A similar method may be employed for several of the first degrees of the quadrant, if the proportional parts at the bottom of the page are not thought sufficiently precise. This correction may, however, be obtained pretty nearly by inspection from comparing the proportional parts for two successive degrees. Thus, on page 26, the correction for 1″, corresponding to the sine of 2° 30′, is 48; the correction for 1″, corresponding to the sine of 3° 30′, is 34. Hence the correction for 1″, corresponding to the sine of 3° 0′, must be about 41; and in the same manner we may proceed for any other part of the table.

Near the close of the quadrant, the tangents vary so rapidly, that the same arrangement of the table is adopted as for the commencement of the quadrant. For the last as well as the first two degrees of the quadrant, the proportional part to 1″ is given for each minute separately. These proportional parts are computed for the minutes placed opposite to them, increased by 30′, and are not strictly applicable to any other number of seconds; nevertheless, the differences for the most part change so slowly, that the error resulting from using these numbers for every part of the same horizontal line is quite small. When great accuracy is required, the table on page 114 may be employed for arcs near the limits of the quadrant. This table furnishes the differences between the logarithmic sines and the logarithms of the arcs expressed in seconds. Thus

the logarithmic sine of 0° 5′ is	7.162696;
the logarithm of 300″ (=5′) is	2.477121;
the difference is	4.685575.

This is the number found on page 114, under the heading *log. sine A—log. A″*, opposite 5 min.; and in a similar manner the other numbers in the same column are obtained. These numbers vary quite slowly for two degrees; and hence, to find the logarithmic sine of an arc less than two degrees, we have but to add the logarithm of the arc expressed in seconds to the appropriate number found in this table.

Required the logarithmic sine of 0° 7′ 22″.

Tabular number from page 114,	4.685575.
The logarithm of 442″ is	2.645422
Logarithmic sine of 0° 7′ 22″ is	7.330997

The logarithmic tangent of an arc less than two degrees is found in a similar manner.

Required the logarithmic tangent of 0° 27′ 36″.

Tabular number from page 114,	4.685584.
The logarithm of 1656″ is	3.219060.
Logarithmic tangent of 0° 27′ 36″ is	7.904644.

The column headed *log. cot. A+log. A″* is found by adding the logarithmic cotangent to the logarithm of the arc expressed in seconds. Hence, to find the logarithmic cotangent of an arc less than two degrees, we must subtract from the tabular number the logarithm of the arc in seconds.

Required the logarithmic cotangent of 0° 27′ 36″.

Tabular number from page 114,	15.314416.
The logarithm of 1656″ is	3.219060.
Logarithmic cotangent of 0° 27′ 36″ is	12.095356.

The same method will, of course, furnish cosines and cotangents of arcs near 90°.

The secants and cosecants are omitted in this table, since they are easily derived from the cosines and sines.

The logarithmic secant is found by subtracting the logarithmic cosine from 20 ; *and the logarithmic cosecant is found by subtracting the logarithmic sine from* 20.

Thus we have found the logarithmic sine of 24° 27′ 34″ to be 9.617051. Hence the logarithmic cosecant of 24° 27′ 34″ is 10.382949.

The logarithmic cosine of 54° 12′ 40″ is 9.767008.
Hence the logarithmic secant of 54° 12′ 40″ is 10.232992.

To find the Arc corresponding to a given Logarithmic Sine or Tangent

If the given number is found exactly in the table the corresponding degrees and seconds will be found at the top of the page, and the minutes on the left. But when the given number is not found exactly in the table, look for the sine or tangent which is next less than the proposed one, and take out the corresponding degrees, minutes, and seconds. Find, also, the difference between this tabular number and the number proposed, and, corresponding to this difference, at the bottom of the page will be found a certain number of seconds, which is to be added to the arc before found.

Required the arc corresponding to the logarithmic sine 9.750000.

The next less sine in the table is 9.749987.

The arc corresponding to which is 34° 13′ 0″.

The difference between its sine and the one proposed is 13, corresponding to which, at the bottom of the page, we find 4″ nearly. Hence the required arc is 34° 13′ 4″.

In the same manner we find the arc corresponding to logarithmic tangent 10.250000, to be 60° 38′ 57″.

When the arc falls within the first two degrees of the quadrant, the odd seconds may be found by dividing the difference between the tabular number and the one proposed, by the proportional part for 1″. We thus find the arc corresponding to logarithmic sine 8.400000, to be 1° 26′ 22″ nearly.

We may employ the same method for the last two degrees of the quadrant when a tangent is given; but near the limits of the quadrant it is better to employ the auxiliary table on page 114. If we subtract the corresponding tabular number on page 114 from the given logarithmic sine, the remainder will be the logarithm of the arc expressed in seconds.

Required the arc corresponding to logarithmic sine 7.000000.

We see, from page 22, that the arc must be nearly 3′; the corresponding tabular number on page 114 is 4.685575.

The difference is 2.314425;

which is the logarithm of 206.″265.

Hence the required arc is 3′ 26.″265.

In the same manner we find the arc corresponding to logarithmic tangent 8.184608, to be 0° 52′ 35″.

TABLE FOR THE LENGTHS OF CIRCULAR ARCS, p. 135.

This table contains the lengths of every single degree up to 60, and at intervals of ten degrees up to 180; also for every minute and second up to 20. The lengths are all expressed in decimal parts of radius.

Required the length of an arc of 57° 17′ 44.″8.

Take out from their respective columns the lengths answering to each of these numbers singly, and add them all together thus:

57°	0.9948377
17′	.0049451
40″	.0001939
4″	.0000194
0.″8	.0000039
The sum is	1.0000000.

That is, the length of an arc of 57° 17′ 44.″8 is equal to the radius of the circle.

TRAVERSE TABLE, pp. 136–141.

This table shows the difference of latitude and the departure to four decimal places for distances from 1 to 10, and for bearings from 0° to 90°, at intervals of 15′. If the bearing is less than 45°, the angle will be found on the left margin of one of the pages of the table, and the distance at the top or bottom of the page; the difference of latitude wil be found in the column headed *lat.* at the top of the page, and the departure in the column headed *dep.* If the bearing is more than 45°, the angle will be found on the right margin, and the difference of latitude will be found in the column marked *lat.* at the bottom of the page, and the departure in the other column. The latitudes and departures for different distances with the same bearing, are proportional to the distances. Therefore the distances may be reckoned as tens, hundreds, or thousands, if the place of the decimal point in each departure and difference of latitude be changed accordingly.

Required the latitude and departure for the distance 32.25, and the bearing 10° 30′.

On page 136, opposite to 10° 30′, we find the following latitudes and departures, proper attention being paid to the position of the decimal points.

Distance.	*Diff. Lat.*	*Dep.*
30	29.498	5.467
2	1.966	.364
.2	.197	.036
.05	.049	.009
32.25	31.710	5.876.

TABLE OF MERIDIONAL PARTS, pp. 142–148.

This table gives the length of the enlarged meridian on Mercator's Chart to every minute of latitude expressed in geographical miles and tenths of a mile. The degrees of latitude are arranged in order at the top of the page, and the minutes on both the right and left margins. Under the degrees and opposite to the minutes are placed the meridional parts corresponding to any latitude less than 80°. Thus

the meridional parts for latitude 12° 23′ are 748.9;
" " " 57° 42′ are 4260.5.

TABLE OF CORRECTIONS TO MIDDLE LATITUDE, p. 149.

This table is used in Navigation for correcting the middle latitude The given middle latitude is to be found either in the first or last vertical column, opposite to which, and under the given difference of latitude, is inserted the proper correction in minutes, to be added to the middle latitude to obtain the latitude in which the meridian distance is accurately equal to the departure. Thus, if the middle latitude is 41°, and the difference of latitude 14°, the correction will be found to be 25′, which, added to the middle latitude, gives the corrected middle latitude 41° 25′.

A TABLE

OF

LOGARITHMS OF NUMBERS

FROM 1 TO 10,000.

N.	Log.	N.	Log.	N.	Log.	N.	Log.
1	0.000000	26	1.414973	51	1.707570	76	1.880814
2	0.301030	27	1.431364	52	1.716003	77	1.886491
3	0.477121	28	1.447158	53	1.724276	78	1.892095
4	0.602060	29	1.462398	54	1.732394	79	1.897627
5	0.698970	30	1.477121	55	1.740363	80	1.903090
6	0.778151	31	1.491362	56	1.748188	81	1.908485
7	0.845098	32	1.505150	57	1.755875	82	1.913814
8	0.903090	33	1.518514	58	1.763428	83	1.919078
9	0.954243	34	1.531479	59	1.770852	84	1.924279
10	1.000000	35	1.544068	60	1.778151	85	1.929419
11	1.041393	36	1.556303	61	1.785330	86	1.934498
12	1.079181	37	1.568202	62	1.792392	87	1.939519
13	1.113943	38	1.579784	63	1.799341	88	1.944483
14	1.146128	39	1.591065	64	1.806180	89	1.949390
15	1.176091	40	1.602060	65	1.812913	90	1.954243
16	1.204120	41	1.612784	66	1.819544	91	1.959041
17	1.230449	42	1.623249	67	1.826075	92	1.963788
18	1.255273	43	1.633468	68	1.832509	93	1.968483
19	1.278754	44	1.643453	69	1.838849	94	1.973128
20	1.301030	45	1.653213	70	1.845098	95	1.977724
21	1.322219	46	1.662758	71	1.851258	96	1.982271
22	1.342423	47	1.672098	72	1.857332	97	1.986772
23	1.361728	48	1.681241	73	1.863323	98	1.991226
24	1.380211	49	1.690196	74	1.869232	99	1.995635
25	1.397940	50	1.698970	75	1.875061	100	2.000000

N.B. In the following table, the two leading figures in the first column of logarithms are to be prefixed to all the numbers of the same horizontal line in the next nine columns; but when a point (.) occurs, its place is to be supplied by a cipher, and the two leading figures are to be taken from the next lower line.

Logarithms of Numbers.

N	0	1	2	3	4	5	6	7	8	9	D.
100	000000	0434	0868	1301	1734	2166	2598	3029	3461	3891	432
101	4321	4751	5181	5609	6038	6466	6894	7321	7748	8174	428
102	8600	9026	9451	9876	.300	.724	1147	1570	1993	2415	424
103	012837	3259	3680	4100	4521	4940	5360	5779	6197	6616	419
104	7033	7451	7868	8284	8700	9116	9532	9947	.361	.775	416
105	021189	1603	2016	2428	2841	3252	3664	4075	4486	4896	412
106	5306	5715	6125	6533	6942	7350	7757	8164	8571	8978	408
107	9384	9789	.195	.600	1004	1408	1812	2216	2619	3021	404
108	033424	3826	4227	4628	5029	5430	5830	6230	6629	7028	400
109	7426	7825	8223	8620	9017	9414	9811	.207	.602	.998	396
110	041393	1787	2182	2576	2969	3362	3755	4148	4540	4932	393
111	5323	5714	6105	6495	6885	7275	7664	8053	8442	8830	389
112	9218	9606	9993	.380	.766	1153	1538	1924	2309	2694	386
113	053078	3463	3846	4230	4613	4996	5378	5760	6142	6524	382
114	6905	7286	7666	8046	8426	8805	9185	9563	9942	.320	379
115	060698	1075	1452	1829	2206	2582	2958	3333	3709	4083	376
116	4458	4832	5206	5580	5953	6326	6699	7071	7443	7815	373
117	8186	8557	8928	9298	9668	..38	.407	.776	1145	1514	369
118	071882	2250	2617	2985	3352	3718	4085	4451	4816	5182	366
119	5547	5912	6276	6640	7004	7368	7731	8094	8457	8819	363
120	9181	9543	9904	.266	.626	.987	1347	1707	2067	2426	360
121	082785	3144	3503	3861	4219	4576	4934	5291	5647	6004	357

N.	0	1	2	3	4	5	6	7	8	9	D.
Differences.	Proportional Parts.										
	434	43	87	130	174	217	260	304	347	391	
	432	43	86	130	173	216	259	302	346	389	
	430	43	86	129	172	215	258	301	344	387	
	428	43	86	128	171	214	257	300	342	385	
	426	43	85	128	170	213	256	298	341	383	
	424	42	85	127	170	212	254	297	339	382	
	422	42	84	127	169	211	253	295	338	380	
	420	42	84	126	168	210	252	294	336	378	
	418	42	84	125	167	209	251	293	334	376	
	416	42	83	125	166	208	250	291	333	374	
	414	41	83	124	166	207	248	290	331	373	
	412	41	82	124	165	206	247	288	330	371	
	410	41	82	123	164	205	246	287	328	369	
	408	41	82	122	163	204	245	286	326	367	
	406	41	81	122	162	203	244	284	325	365	
	404	40	81	121	162	202	242	283	323	364	
	402	40	80	121	161	201	241	281	322	362	
	400	40	80	120	160	200	240	280	320	360	
	398	40	80	119	159	199	239	279	318	358	
	396	40	79	119	158	198	238	277	317	356	
	394	39	79	118	158	197	236	276	315	355	
	392	39	78	118	157	196	235	274	314	353	
	390	39	78	117	156	195	234	273	312	351	
	388	39	78	116	155	194	233	272	310	349	
	386	39	77	116	154	193	232	270	309	347	
	384	38	77	115	154	192	230	269	307	346	
	382	38	76	115	153	191	229	267	306	344	
	380	38	76	114	152	190	228	266	304	342	
	378	38	76	113	151	189	227	265	302	340	
	376	38	75	113	150	188	226	263	301	338	
	374	37	75	112	150	187	224	262	299	337	
	372	37	74	112	149	186	223	260	298	335	
	370	37	74	111	148	185	222	259	296	333	
	368	37	74	110	147	184	221	258	294	331	
	366	37	73	110	146	183	220	256	293	329	
	364	36	73	109	146	182	218	255	291	328	
	362	36	72	109	145	181	217	253	290	326	
	360	36	72	108	144	180	216	252	288	324	

N.	0	1	2	3	4	5	6	7	8	9	D.
122	086360	6716	7071	7426	7781	8136	8490	8845	9198	9552	355
123	9905	.258	.611	.963	1315	1667	2018	2370	2721	3071	351
124	093422	3772	4122	4471	4820	5169	5518	5866	6215	6562	349
125	6910	7257	7604	7951	8298	8644	8990	9335	9681	..26	346
126	100371	0715	1059	1403	1747	2091	2434	2777	3119	3462	343
127	3804	4146	4487	4828	5169	5510	5851	6191	6531	6871	341
128	7210	7549	7888	8227	8565	8903	9241	9579	9916	.253	338
129	110590	0926	1263	1599	1934	2270	2605	2940	3275	3609	335
130	3943	4277	4611	4944	5278	5611	5943	6276	6608	6940	333
131	7271	7603	7934	8265	8595	8926	9256	9586	9915	.245	330
132	120574	0903	1231	1560	1888	2216	2544	2871	3198	3525	328
133	3852	4178	4504	4830	5156	5481	5806	6131	6456	6781	325
134	7105	7429	7753	8076	8399	8722	9045	9368	9690	..12	323
135	130334	0655	0977	1298	1619	1939	2260	2580	2900	3219	321
136	3539	3858	4177	4496	4814	5133	5451	5769	6086	6403	318
137	6721	7037	7354	7671	7987	8303	8618	8934	9249	9564	315
138	9879	.194	.508	.822	1136	1450	1763	2076	2389	2702	314
139	143015	3327	3639	3951	4263	4574	4885	5196	5507	5818	311
140	6128	6438	6748	7058	7367	7676	7985	8294	8603	8911	309
.41	9219	9527	9835	.142	.449	.756	1063	1370	1676	1982	307
142	.52288	2594	2900	3205	3510	3815	4120	4424	4728	5032	305
143	5336	5640	5943	6246	6549	6852	7154	7457	7759	8061	303
144	8362	8664	8965	9266	9567	9868	.168	.469	.769	1068	301
145	161368	1667	1967	2266	2564	2863	3161	3460	3758	4055	299
146	4353	4650	4947	5244	5541	5838	6134	6430	6726	7022	297
147	7317	7613	7908	8203	8497	8792	9086	9380	9674	9968	295
148	170262	0555	0848	1141	1434	1726	2019	2311	2603	2895	293
N.	**0**	**1**	**2**	**3**	**4**	**5**	**6**	**7**	**8**	**9**	**D.**
Differences.	358	36	72	107	143	179	215	251	286	322	
	356	36	71	107	142	178	214	249	285	320	
	354	35	71	106	142	177	212	248	283	319	
	352	35	70	106	141	176	211	246	282	317	
	350	35	70	105	140	175	210	245	280	315	
	348	35	70	104	139	174	209	244	278	313	
	346	35	69	104	138	173	208	242	277	311	
	344	34	69	103	138	172	206	241	275	310	
	342	34	68	103	137	171	205	239	274	308	
	340	34	68	102	136	170	204	238	272	306	
	338	34	68	101	135	169	203	237	270	304	
	336	34	67	101	134	168	202	235	269	302	
	334	33	67	100	134	167	200	234	267	301	
	332	33	66	100	133	166	199	232	266	299	
	330	33	66	99	132	165	198	231	264	297	
	328	33	66	98	131	164	197	230	262	295	
	326	33	65	98	130	163	196	228	261	293	
	324	32	65	97	130	162	194	227	259	292	
	322	32	64	97	129	161	193	225	258	290	
	320	32	64	96	128	160	192	224	256	288	
	318	32	64	95	127	159	191	223	254	286	
	316	32	63	95	126	158	190	221	253	284	
	314	31	63	94	126	157	188	220	251	283	
	312	31	62	94	125	156	187	218	250	281	
	310	31	62	93	124	155	186	217	248	279	
	308	31	62	92	123	154	185	216	246	277	
	306	31	61	92	122	153	184	214	245	275	
	304	30	61	91	122	152	182	213	243	274	
	302	30	60	91	121	151	181	211	242	272	
	300	30	60	90	120	150	180	210	240	270	
	298	30	60	89	119	149	179	209	238	268	
	296	30	59	89	118	148	178	207	237	266	
	294	29	59	88	118	147	176	206	235	265	

Proportional Parts.

N.	0	1	2	3	4	5	6	7	8	9	D.
149	173186	3478	3769	4060	4351	4641	4932	5222	5512	5802	291
150	6091	6381	6670	6959	7248	7536	7825	8113	8401	8689	289
151	8977	9264	9552	9839	.126	.413	.699	.986	1272	1558	287
152	181844	2129	2415	2700	2985	3270	3555	3839	4123	4407	285
153	4691	4975	5259	5542	5825	6108	6391	6674	6956	7239	283
154	7521	7803	8084	8366	8647	8928	9209	9490	9771	..51	281
155	190332	0612	0892	1171	1451	1730	2010	2289	2567	2846	279
156	3125	3403	3681	3959	4237	4514	4792	5069	5346	5623	278
157	5900	6176	6453	6729	7005	7281	7556	7832	8107	8382	276
158	8657	8932	9206	9481	9755	..29	.303	.577	.850	1124	274
159	201397	1670	1943	2216	2488	2761	3033	3305	3577	3848	272
160	4120	4391	4663	4934	5204	5475	5746	6016	6286	6556	271
161	6826	7096	7365	7634	7904	8173	8441	8710	8979	9247	269
162	9515	9783	..51	.319	.586	.853	1121	1388	1654	1921	267
163	212188	2454	2720	2986	3252	3518	3783	4049	4314	4579	266
164	4844	5109	5373	5638	5902	6166	6430	6694	6957	7221	264
165	7484	7747	8010	8273	8536	8798	9060	9323	9585	9846	262
166	220108	0370	0631	0892	1153	1414	1675	1936	2196	2456	261
167	2716	2976	3236	3496	3755	4015	4274	4533	4792	5051	259
168	5309	5568	5826	6084	6342	6600	6858	7115	7372	7630	258
169	7887	8144	8400	8657	8913	9170	9426	9682	9938	.193	256
170	230449	0704	0960	1215	1470	1724	1979	2234	2488	2742	254
171	2996	3250	3504	3757	4011	4264	4517	4770	5023	5276	253
172	5528	5781	6033	6285	6537	6789	7041	7292	7544	7795	252
173	8046	8297	8548	8799	9049	9299	9550	9800	..50	.300	250
174	240549	0799	1048	1297	1546	1795	2044	2293	2541	2790	249
175	3038	3286	3534	3782	4030	4277	4525	4772	5019	5266	248
176	5513	5759	6006	6252	6499	6745	6991	7237	7482	7728	246
177	7973	8219	8464	8709	8954	9198	9443	9687	9932	.176	245
178	250420	0664	0908	1151	1395	1638	1881	2125	2368	2610	243
179	2853	3096	3338	3580	3822	4064	4306	4548	4790	5031	242
180	5273	5514	5755	5996	6237	6477	6718	6958	7198	7439	241
181	7679	7918	8158	8398	8637	8877	9116	9355	9594	9833	239

N.	0	1	2	3	4	5	6	7	8	9	D.
Differences.	Differences.	Proportional Parts.									
	292	29	58	88	117	146	175	204	234	263	
	290	29	58	87	116	145	174	203	232	261	
	288	29	58	86	115	144	173	202	230	259	
	286	29	57	86	114	143	172	200	229	257	
	284	28	57	85	114	142	170	199	227	256	
	282	28	56	85	113	141	169	197	226	254	
	280	28	56	84	112	140	168	196	224	252	
	278	28	56	83	111	139	167	195	222	250	
	276	28	55	83	110	138	166	193	221	248	
	274	27	55	82	110	137	164	192	219	247	
	272	27	54	82	109	136	163	190	218	245	
	270	27	54	81	108	135	162	189	216	243	
	268	27	54	80	107	134	161	188	214	241	
	266	27	53	80	106	133	160	186	213	239	
	264	26	53	79	106	132	158	185	211	238	
	262	26	52	79	105	131	157	183	210	236	
	260	26	52	78	104	130	156	182	208	234	
	258	26	52	77	103	129	155	181	206	232	
	256	26	51	77	102	128	154	179	205	230	
	254	25	51	76	102	127	152	178	203	229	
	252	25	50	76	101	126	151	176	202	227	
	250	25	50	75	100	125	150	175	200	225	
	248	25	50	74	99	124	149	174	198	223	
	246	25	49	74	98	123	148	172	197	221	
	244	24	49	73	98	122	146	171	195	220	
	242	24	48	73	97	121	145	169	194	218	
	240	24	48	72	96	120	144	168	192	216	

N.	0	1	2	3	4	5	6	7	8	9	D.
182	260071	0310	0548	0787	1025	1263	1501	1739	1976	2214	238
183	2451	2688	2925	3162	3399	3636	3873	4109	4346	4582	237
184	4818	5054	5290	5525	5761	5996	6232	6467	6702	6937	235
185	7172	7406	7641	7875	8110	8344	8578	8812	9046	9279	234
186	9513	9746	9980	.213	.446	.679	.912	1144	1377	1609	233
187	271842	2074	2306	2538	2770	3001	3233	3464	3696	3927	232
188	4158	4389	4620	4850	5081	5311	5542	5772	6002	6232	230
189	6462	6692	6921	7151	7380	7609	7838	8067	8296	8525	229
190	8754	8982	9211	9439	9667	9895	.123	.351	578	.806	228
191	281033	1261	1488	1715	1942	2169	2396	2622	2849	3075	227
192	3301	3527	3753	3979	4205	4431	4656	4882	5107	5332	226
193	5557	5782	6007	6232	6456	6681	6905	7130	7354	7578	225
194	7802	8026	8249	8473	8696	8920	9143	9366	9589	9812	223
195	290035	0257	0480	0702	0925	1147	1369	1591	1813	2034	222
196	2256	2478	2699	2920	3141	3363	3584	3804	4025	4246	221
197	4466	4687	4907	5127	5347	5567	5787	6007	6226	6446	220
198	6665	6884	7104	7323	7542	7761	7979	8198	8416	8635	219
199	8853	9071	9289	9507	9725	9943	.161	.378	.595	.813	218
200	301030	1247	1464	1681	1898	2114	2331	2547	2764	2980	217
201	3196	3412	3628	3844	4059	4275	4491	4706	4921	5136	216
202	5351	5566	5781	5996	6211	6425	6639	6854	7068	7282	214
203	7496	7710	7924	8137	8351	8564	8778	8991	9204	9417	213
204	9630	9843	..56	.268	.481	.693	.906	1118	1330	1542	212
205	311754	1966	2177	2389	2600	2812	3023	3234	3445	3656	211
206	3867	4078	4289	4499	4710	4920	5130	5340	5551	5760	210
207	5970	6180	6390	6599	6809	7018	7227	7436	7646	7854	209
208	8063	8272	8481	8689	8898	9106	9314	9522	9730	9938	208
209	320146	0354	0562	0769	0977	1184	1391	1598	1805	2012	207
210	2219	2426	2633	2839	3046	3252	3458	3665	3871	4077	206
211	4282	4488	4694	4899	5105	5310	5516	5721	5926	6131	205
212	6336	6541	6745	6950	7155	7359	7563	7767	7972	8176	204
213	8380	8583	8787	8991	9194	9398	9601	9805	...8	.211	203
214	330414	0617	0819	1022	1225	1427	1630	1832	2034	2236	202
215	2438	2640	2842	3044	3246	3447	3649	3850	4051	4253	202
216	4454	4655	4856	5057	5257	5458	5658	5859	6059	6260	201
217	6460	6660	6860	7060	7260	7459	7659	7858	8058	8257	200
218	8456	8656	8855	9054	9253	9451	9650	9849	..47	.246	199
219	340444	0642	0841	1039	1237	1435	1632	1830	2028	2225	198
220	2423	2620	2817	3014	3212	3409	3606	3802	3999	4196	197
N.	0	1	2	3	4	5	6	7	8	9	D.
Differences.	238	24	48	71	95	119	143	167	190	214	
	236	24	47	71	94	118	142	165	189	212	
	234	23	47	70	94	117	140	164	187	211	
	232	23	46	70	93	116	139	162	186	209	
	230	23	46	69	92	115	138	161	184	207	
	228	23	46	68	91	114	137	160	182	205	
	226	23	45	68	90	113	136	158	181	203	
	224 Proportional Parts.	22	45	67	90	112	134	157	179	202	
	222	22	44	67	89	111	133	155	178	200	
	220	22	44	66	88	110	132	154	176	198	
	218	22	44	65	87	109	131	153	174	196	
	216	22	43	65	86	108	130	151	173	194	
	214	21	43	64	86	107	128	150	171	193	
	212	21	42	64	85	106	127	148	170	191	
	210	21	42	63	84	105	126	147	168	189	
	208	21	42	62	83	104	125	146	166	187	
	206	21	41	62	82	103	124	144	165	185	
	204	20	41	61	82	102	122	143	163	184	
	202	20	40	61	81	101	121	141	162	182	
	200	20	40	60	80	100	120	140	160	180	
	198	20	40	59	79	99	119	139	158	178	

N.	0	1	2	3	4	5	6	7	8	9	D.
221	344392	4589	4785	4981	5178	5374	5570	5766	5962	6157	196
222	6353	6549	6744	6939	7135	7330	7525	7720	7915	8110	195
223	8305	8500	8694	8889	9083	9278	9472	9666	9860	..54	194
224	350248	0442	0636	0829	1023	1216	1410	1603	1796	1989	193
225	2183	2375	2568	2761	2954	3147	3339	3532	3724	3916	
226	4108	4301	4493	4685	4876	5068	5260	5452	5643	5834	192
227	6026	6217	6408	6599	6790	6981	7172	7363	7554	7744	191
228	7935	8125	8316	8506	8696	8886	9076	9266	9456	9646	190
229	9835	..25	.215	.404	.593	.783	.972	1161	1350	1539	189
230	361728	1917	2105	2294	2482	2671	2859	3048	3236	3424	188
231	3612	3800	3988	4176	4363	4551	4739	4926	5113	5301	
232	5488	5675	5862	6049	6236	6423	6610	6796	6983	7169	187
233	7356	7542	7729	7915	8101	8287	8473	8659	8845	9030	186
234	9216	9401	9587	9772	9958	.143	.328	.513	.698	.883	185
235	371068	1253	1437	1622	1806	1991	2175	2360	2544	2728	184
236	2912	3096	3280	3464	3647	3831	4015	4198	4382	4565	
237	4748	4932	5115	5298	5481	5664	5846	6029	6212	6394	183
238	6577	6759	6942	7124	7306	7488	7670	7852	8034	82:6	182
239	8398	8580	8761	8943	9124	9306	9487	9668	9849	..30	181
240	380211	0392	0573	0754	0934	1115	1296	1476	1656	1837	
241	2017	2197	2377	2557	2737	2917	3097	3277	3456	3636	180
242	3815	3995	4174	4353	4533	4712	4891	5070	5249	5428	179
243	5606	5785	5964	6142	6321	6499	6677	6856	7034	7212	178
244	7390	7568	7746	7923	8101	8279	8456	8634	8811	8989	
245	9166	9343	9520	9698	9875	..51	.228	.405	.582	.759	177
246	390935	1112	1288	1464	1641	1817	1993	2169	2345	2521	176
247	2697	2873	3048	3224	3400	3575	3751	3926	4101	4277	
248	4452	4627	4802	4977	5152	5326	5501	5676	5850	6025	175
249	6199	6374	6548	6722	6896	7071	7245	7419	7592	7766	174
250	7940	8114	8287	8461	8634	8808	8981	9154	9328	9501	173
251	9674	9847	..20	.192	.365	.538	.711	.883	1056	1228	
252	401401	1573	1745	1917	2089	2261	2433	2605	2777	2949	172
253	3121	3292	3464	3635	3807	3978	4149	4320	4492	4663	:71
254	4834	5005	5176	5346	5517	5688	5858	6029	6199	6370	
255	6540	6710	6881	7051	7221	7391	7561	7731	7901	8070	170
256	8240	8410	8579	8749	8918	9087	9257	9426	9595	9764	169
257	9933	.102	.271	.440	.609	.777	.946	1114	1283	1451	
258	411620	1788	1956	2124	2293	2461	2629	2796	2964	3132	168
259	3300	3467	3635	3803	3970	4137	4305	4472	4639	4806	167
260	4973	5140	5307	5474	5641	5808	5974	6141	6308	6474	
261	664:	6807	6973	7139	7306	7472	7638	7804	7970	8135	166
262	830:	8467	8633	8798	8964	9129	9295	9460	9625	9791	165
263	9956	.121	.286	.451	.616	.781	.945	1110	1275	1439	

N.	0	1	2	3	4	5	6	7	8	9	D.
Differences. Proportional Parts.	196	20	39	59	78	98	118	137	157	176	
	194	19	39	58	78	97	116	136	155	175	
	192	19	38	58	77	96	115	134	154	173	
	190	19	38	57	76	95	114	133	152	171	
	188	19	38	56	75	94	113	132	150	169	
	186	19	37	56	74	93	112	130	149	167	
	184	18	37	55	74	92	110	129	147	166	
	182	18	36	55	73	91	109	127	146	164	
	180	18	36	54	72	90	108	126	144	162	
	178	18	36	53	71	89	107	125	142	160	
	176	18	35	53	70	88	106	123	141	158	
	174	17	35	52	70	87	104	122	139	157	
	172	17	34	52	69	86	103	120	138	155	
	170	17	34	51	68	85	102	119	136	153	
	168	17	34	50	67	84	101	118	134	151	
	166	17	33	50	66	83	100	116	133	149	
	164	16	33	49	66	82	98	115	131	148	

N.	0	1	2	3	4	5	6	7	8	9	D.
264	421604	1768	1933	2097	2261	2426	2590	2754	2918	3082	164
265	3246	3410	3574	3737	3901	4065	4228	4392	4555	4718	
266	4882	5045	5208	5371	5534	5697	5860	6023	6186	6349	163
267	6511	6674	6836	6999	7161	7324	7486	7648	7811	7973	162
268	8135	8297	8459	8621	8783	8944	9106	9268	9429	9591	
269	9752	9914	..75	.236	.398	.559	.720	.881	1042	1203	161
270	431864	1525	1685	1846	2007	2167	2328	2488	2649	2809	
271	2969	3130	3290	3450	3610	3770	3930	4090	4249	4409	160
272	4569	4729	4888	5048	5207	5367	5526	5685	5844	6004	159
273	6163	6322	6481	6640	6799	6957	7116	7275	7433	7592	
274	7751	7909	8067	8226	8384	8542	8701	8859	9017	9175	158
275	9333	9491	9648	9806	9964	.122	.279	.437	.594	.752	
276	440909	1066	1224	1381	1538	1695	1852	2009	2166	2323	157
277	2480	2637	2793	2950	3106	3263	3419	3576	3732	3889	
278	4045	4201	4357	4513	4669	4825	4981	5137	5293	5449	156
279	5604	5760	5915	6071	6226	6382	6537	6692	6848	7003	155
280	7158	7313	7468	7623	7778	7933	8088	8242	8397	8552	
281	8706	8861	9015	9170	9324	9478	9633	9787	9941	..95	154
282	450249	0403	0557	0711	0865	1018	1172	1326	1479	1633	
283	1786	1940	2093	2247	2400	2553	2706	2859	3012	3165	153
284	3318	3471	3624	3777	3930	4082	4235	4387	4540	4692	
285	4845	4997	5150	5302	5454	5606	5758	5910	6062	6214	152
286	6366	6518	6670	6821	6973	7125	7276	7428	7579	7731	
287	7882	8033	8184	8336	8487	8638	8789	8940	9091	9242	151
288	9392	9543	9694	9845	9995	.146	.296	.447	.597	.748	
289	460898	1048	1198	1348	1499	1649	1799	1948	2098	2248	150
290	2398	2548	2697	2847	2997	3146	3296	3445	3594	3744	
291	3893	4042	4191	4340	4490	4639	4788	4936	5085	5234	149
292	5383	5532	5680	5829	5977	6126	6274	6423	6571	6719	
293	6868	7016	7164	7312	7460	7608	7756	7904	8052	8200	148
294	8347	8495	8643	8790	8938	9085	9233	9380	9527	9675	
295	9822	9969	.116	.263	.410	.557	.704	.851	.998	1145	147
296	471292	1438	1585	1732	1878	2025	2171	2318	2464	2610	
297	2756	2903	3049	3195	3341	3487	3633	3779	3925	4071	146
298	4216	4362	4508	4653	4799	4944	5090	5235	5381	5526	
299	5671	5816	5962	6107	6252	6397	6542	6687	6832	6976	145
300	7121	7266	7411	7555	7700	7844	7989	8133	8278	8422	
301	8566	8711	8855	8999	9143	9287	9431	9575	9719	9863	144
302	480007	0151	0294	0438	0582	0725	0869	1012	1156	1299	
303	1443	1586	1729	1872	2016	2159	2302	2445	2588	2731	143
304	2874	3016	3159	3302	3445	3587	3730	3872	4015	4157	
305	4300	4442	4585	4727	4869	5011	5153	5295	5437	5579	142
306	5721	5863	6005	6147	6289	6430	6572	6714	6855	6997	
307	7138	7280	7421	7563	7704	7845	7986	8127	8269	8410	141
308	8551	8692	8833	8974	9114	9255	9396	9537	9677	9818	
309	9958	..99	.239	.380	.520	.661	.801	.941	1081	1222	140
310	491362	1502	1642	1782	1922	2062	2201	2341	2481	2621	
311	2760	2900	3040	3179	3319	3458	3597	3737	3876	4015	139
N.	0	1	2	3	4	5	6	7	8	9	D.

N.	0	1	2	3	4	5	6	7	8	9	D.
312	494155	4294	4433	4572	4711	4850	4989	5128	5267	5406	139
313	5544	5683	5822	5960	6099	6238	6376	6515	6653	6791	
314	6930	7068	7206	7344	7483	7621	7759	7897	8035	8173	138
315	8311	8448	8586	8724	8862	8999	9137	9275	9412	9550	
316	9687	9824	9962	..99	.236	.374	.511	.648	.785	.922	137
317	501059	1196	1333	1470	1607	1744	1880	2017	2154	2291	
318	2427	2564	2700	2837	2973	3109	3246	3382	3518	3655	136
319	3791	3927	4063	4199	4335	4471	4607	4743	4878	5014	
320	5150	5286	5421	5557	5693	5828	5964	6099	6234	6370	
321	6505	6640	6776	6911	7046	7181	7316	7451	7586	7721	135
322	7856	7991	8126	8260	8395	8530	8664	8799	8934	9068	
323	9203	9337	9471	9606	9740	9874	...9	.143	.277	.411	134
324	510545	0679	0813	0947	1081	1215	1349	1482	1616	1750	
325	1883	2017	2151	2284	2418	2551	2684	2818	2951	3084	133
326	3218	3351	3484	3617	3750	3883	4016	4149	4282	4415	
327	4548	4681	4813	4946	5079	5211	5344	5476	5609	5741	
328	5874	6006	6139	6271	6403	6535	6668	6800	6932	7064	132
329	7196	7328	7460	7592	7724	7855	7987	8119	8251	8382	
330	8514	8646	8777	8909	9040	9171	9303	9434	9566	9697	131
331	9828	9959	..90	.221	.353	.484	.615	.745	.876	1007	
332	521138	1269	1400	1530	1661	1792	1922	2053	2183	2314	
333	2444	2575	2705	2835	2966	3096	3226	3356	3486	3616	130
334	3746	3876	4006	4136	4266	4396	4526	4656	4785	4915	
335	5045	5174	5304	5434	5563	5693	5822	5951	6081	6210	129
336	6339	6469	6598	6727	6856	6985	7114	7243	7372	7501	
337	7630	7759	7888	8016	8145	8274	8402	8531	8660	8788	
338	8917	9045	9174	9302	9430	9559	9687	9815	9943	..72	128
339	530200	0328	0456	0584	0712	0840	0968	1096	1223	1351	
340	1479	1607	1734	1862	1990	2117	2245	2372	2500	2627	
341	2754	2882	3009	3136	3264	3391	3518	3645	3772	3899	127
342	4026	4153	4280	4407	4534	4661	4787	4914	5041	5167	
343	5294	5421	5547	5674	5800	5927	6053	6180	6306	6432	126
344	6558	6685	6811	6937	7063	7189	7315	7441	7567	7693	
345	7819	7945	8071	8197	8322	8448	8574	8699	8825	8951	
346	9076	9202	9327	9452	9578	9703	9829	9954	..79	.204	125
347	540329	0455	0580	0705	0830	0955	1080	1205	1330	1454	
348	1579	1704	1829	1953	2078	2203	2327	2452	2576	2701	
349	2825	2950	3074	3199	3323	3447	3571	3696	3820	3944	124
350	4068	4192	4316	4440	4564	4688	4812	4936	5060	5183	
351	5307	5431	5555	5678	5802	5925	6049	6172	6296	6419	
352	6543	6666	6789	6913	7036	7159	7282	7405	7529	7652	123
353	7775	7898	8021	8144	8267	8389	8512	8635	8758	8881	
354	9003	9126	9249	9371	9494	9616	9739	9861	9984	.106	
355	550228	0351	0473	0595	0717	0840	0962	1084	1206	1328	122
356	1450	1572	1694	1816	1938	2060	2181	2303	2425	2547	
357	2668	2790	2911	3033	3155	3276	3398	3519	3640	3762	121
358	3883	4004	4126	4247	4368	4489	4610	4731	4852	4973	
359	5094	5215	5336	5457	5578	5699	5820	5940	6061	6182	
360	6303	6423	6544	6664	6785	6905	7026	7146	7267	7387	120
361	7507	7627	7748	7868	7988	8108	8228	8349	8469	8589	

N.	0	1	2	3	4	5	6	7	8	9	D.
362	558709	8829	8948	9068	9188	9308	9428	9548	9667	9787	120
363	9907	..26	.146	.265	.385	.504	.624	.743	.863	.982	
364	561101	1221	1340	1459	1578	1698	1817	1936	2055	2174	119
365	2293	2412	2531	2650	2769	2887	3006	3125	3244	3362	
366	3481	3600	3718	3837	3955	4074	4192	4311	4429	4548	
367	4666	4784	4903	5021	5139	5257	5376	5494	5612	5730	118
368	5848	5966	6084	6202	6320	6437	6555	6673	6791	6909	
369	7026	7144	7262	7379	7497	7614	7732	7849	7967	8084	
370	8202	8319	8436	8554	8671	8788	8905	9023	9140	9257	117
371	9374	9491	9608	9725	9842	9959	..76	.193	.309	.426	
372	570543	0660	0776	0893	1010	1126	1243	1359	1476	1592	
373	1709	1825	1942	2058	2174	2291	2407	2523	2639	2755	116
374	2872	2988	3104	3220	3336	3452	3568	3684	3800	3915	
375	4031	4147	4263	4379	4494	4610	4726	4841	4957	5072	
376	5188	5303	5419	5534	5650	5765	5880	5996	6111	6226	115
377	6341	6457	6572	6687	6802	6917	7032	7147	7262	7377	
378	7492	7607	7722	7836	7951	8066	8181	8295	8410	8525	
379	8639	8754	8868	8983	9097	0212	9326	9441	9555	9669	114
380	9784	9898	..12	.126	.241	.355	.469	.583	.697	.811	
381	580925	1039	1153	1267	1381	1495	1608	1722	1835	1950	
382	2063	2177	2291	2404	2518	2631	2745	2858	2972	3085	
383	3199	3312	3426	3539	3652	3765	3879	3992	4105	4218	113
384	4331	4444	4557	4670	4783	4896	5009	5122	5235	5348	
385	5461	5574	5686	5799	5912	6024	6137	6250	6362	6475	
386	6587	6700	6812	6925	7037	7149	7262	7374	7486	7599	112
387	7711	7823	7935	8047	8160	8272	8384	8496	8608	8720	
388	8832	8944	9056	9167	9279	9391	9503	9615	9726	9838	
389	9950	..61	.173	.284	.396	.507	.619	.730	.842	.953	
390	591065	1176	1287	1399	1510	1621	1732	1843	1955	2066	111
391	2177	2288	2399	2510	2621	2732	2843	2954	3064	3175	
392	3286	3397	3508	3618	3729	3840	3950	4061	4171	4282	
393	4393	4503	4614	4724	4834	4945	5055	5165	5276	5386	110
394	5496	5606	5717	5827	5937	6047	6157	6267	6377	6487	
395	6597	6707	6817	6927	7037	7146	7256	7366	7476	7586	
396	7695	7805	7914	8024	8134	8243	8353	8462	8572	8681	
397	8791	8900	9009	9119	9228	9337	9446	9556	9665	9774	109
398	9883	9992	.101	.210	.319	.428	.537	.646	.755	.864	
399	600973	1082	1191	1299	1408	1517	1625	1734	1843	1951	
400	2060	2169	2277	2386	2494	2603	2711	2819	2928	3036	108
401	3144	3253	3361	3469	3577	3686	3794	3902	4010	4118	
402	4226	4334	4442	4550	4658	4766	4874	4982	5089	5197	
403	5305	5413	5521	5628	5736	5844	5951	6059	6166	6274	
404	6381	6489	6596	6704	6811	6919	7026	7133	7241	7348	107
405	7455	7562	7669	7777	7884	7991	8098	8205	8312	8419	
406	8526	8633	8740	8847	8954	9061	9167	9274	9381	9488	
407	9594	9701	9808	9914	..21	.128	.234	.341	.447	.554	
408	610660	0767	0873	0979	1086	1192	1298	1405	1511	1617	106
N.	0	1	2	3	4	5	6	7	8	9	D.
...es.	119	12	24	36	48	60	71	83	95	107	
	118	12	24	35	47	59	71	83	94	106	
	117	12	23	35	47	59	70	82	94	105	
	116	12	23	35	46	58	70	81	93	104	
	115	12	23	35	46	58	69	81	92	104	
	114	11	23	34	46	57	68	80	91	103	

...l Parts.

N.	0	1	2	3	4	5	6	7
409	611723	1829	1936	2042	2148	2254	2360	2466
410	2784	2890	2996	3102	3207	3313	3419	3525
411	3842	3947	4053	4159	4264	4370	4475	4581
412	4897	5003	5108	5213	5319	5424	5529	5634
413	5950	6055	6160	6265	6370	6476	6581	6686
414	7000	7105	7210	7315	7420	7525	7629	7734
415	8048	8153	8257	8362	8466	8571	8676	8780
416	9093	9198	9302	9406	9511	9615	9719	9824
417	620136	0240	0344	0448	0552	0656	0760	0864
418	1176	1280	1384	1488	1592	1695	1799	1903
419	2214	2318	2421	2525	2628	2732	2835	2939
420	3249	3353	3456	3559	3663	3766	3869	3973
421	4282	4385	4488	4591	4695	4798	4901	5004
422	5312	5415	5518	5621	5724	5827	5929	6032
423	6340	6443	6546	6648	6751	6853	6956	7058
424	7366	7468	7571	7673	7775	7878	7980	8082
425	8389	8491	8593	8695	8797	8900	9002	9104
426	9410	9512	9613	9715	9817	9919	..21	.123
427	630428	0530	0631	0733	0835	0936	1038	1139
428	1444	1545	1647	1748	1849	1951	2052	2153
429	2457	2559	2660	2761	2862	2963	3064	3165
430	3468	3569	3670	3771	3872	3973	4074	4175
431	4477	4578	4679	4779	4880	4981	5081	5182
432	5484	5584	5685	5785	5886	5986	6087	6187
433	6488	6588	6688	6789	6889	6989	7089	7189
434	7490	7590	7690	7790	7890	7990	8090	8190
435	8489	8589	8689	8789	8888	8988	9088	9188
436	9486	9586	9686	9785	9885	9984	..84	.183
437	640481	0581	0680	0779	0879	0978	1077	1177
438	1474	1573	1672	1771	1871	1970	2069	2168
439	2465	2563	2662	2761	2860	2959	3058	3156
440	3453	3551	3650	3749	3847	3946	4044	4143
441	4439	4537	4636	4734	4832	4931	5029	5127
442	5422	5521	5619	5717	5815	5913	6011	6110
443	6404	6502	6600	6698	6796	6894	6992	7089
444	7383	7481	7579	7676	7774	7872	7969	8067
445	8360	8458	8555	8653	8750	8848	8945	9043
446	9335	9432	9530	9627	9724	9821	9919	..16
447	650308	0405	0502	0599	0696	0793	0890	0987
448	1278	1375	1472	1569	1666	1762	1859	1956
449	2246	2343	2440	2536	2633	2730	2826	2923
450	3213	3309	3405	3502	3598	3695	3791	3888
451	4177	4273	4369	4465	4562	4658	4754	4850
452	5138	5235	5331	5427	5523	5619	5715	5810
453	6098	6194	6290	6386	6482	6577	6673	6769
454	7056	7152	7247	7343	7438	7534	7629	7725
455	8011	8107	8202	8298	8393	8488	8584	8679
456	8965	9060	9155	9250	9346	9441	9536	9631
457	9916	..11	.106	.201	.296	.391	.486	.581

N.	0	1	2	3	4	5	6	7	8	9	D.
458	660865	0960	1055	1150	1245	1339	1434	1529	1623	1718	95
459	1813	1907	2002	2096	2191	2286	2380	2475	2569	2663	
460	2758	2852	2947	3041	3135	3230	3324	3418	3512	3607	94
461	3701	3795	3889	3983	4078	4172	4266	4360	4454	4548	
462	4642	4736	4830	4924	5018	5112	5206	5299	5393	5487	
463	5581	5675	5769	5862	5956	6050	6143	6237	6331	6424	
464	6518	6612	6705	6799	6892	6986	7079	7173	7266	7360	
465	7453	7546	7640	7733	7826	7920	8013	8106	8199	8293	93
466	8386	8479	8572	8665	8759	8852	8945	9038	9131	9224	
467	9317	9410	9503	9596	9689	9782	9875	9967	..60	.153	
468	670246	0339	0431	0524	0617	0710	0802	0895	0988	1080	
469	1173	1265	1358	1451	1543	1636	1728	1821	1913	2005	
470	2098	2190	2283	2375	2467	2560	2652	2744	2836	2929	92
471	3021	3113	3205	3297	3390	3482	3574	3666	3758	3850	
472	3942	4034	4126	4218	4310	4402	4494	4586	4677	4769	
473	4861	4953	5045	5137	5228	5320	5412	5503	5595	5687	
474	5778	5870	5962	6053	6145	6236	6328	6419	6511	6602	
475	6694	6785	6876	6968	7059	7151	7242	7333	7424	7516	91
476	7607	7698	7789	7881	7972	8063	8154	8245	8336	8427	
477	8518	8609	8700	8791	8882	8973	9064	9155	9246	9337	
478	9428	9519	9610	9700	9791	9882	9973	..63	.154	.245	
479	680336	0426	0517	0607	0698	0789	0879	0970	1060	1151	
480	1241	1332	1422	1513	1603	1693	1784	1874	1964	2055	90
481	2145	2235	2326	2416	2506	2596	2686	2777	2867	2957	
482	3047	3137	3227	3317	3407	3497	3587	3677	3767	3857	
483	3947	4037	4127	4217	4307	4396	4486	4576	4666	4756	
484	4845	4935	5025	5114	5204	5294	5383	5473	5563	5652	
485	5742	5831	5921	6010	6100	6189	6279	6368	6458	6547	89
486	6636	6726	6815	6904	6994	7083	7172	7261	7351	7440	
487	7529	7618	7707	7796	7886	7975	8064	8153	8242	8331	
488	8420	8509	8598	8687	8776	8865	8953	9042	9131	9220	
489	9309	9398	9486	9575	9664	9753	9841	9930	..19	.107	
490	690196	0285	0373	0462	0550	0639	0728	0816	0905	0993	
491	1081	1170	1258	1347	1435	1524	1612	1700	1789	1877	88
492	1965	2053	2142	2230	2318	2406	2494	2583	2671	2759	
493	2847	2935	3023	3111	3199	3287	3375	3463	3551	3639	
494	3727	3815	3903	3991	4078	4166	4254	4342	4430	4517	
495	4605	4693	4781	4868	4956	5044	5131	5219	5307	5394	
496	5482	5569	5657	5744	5832	5919	6007	6094	6182	6269	87
497	6356	6444	6531	6618	6706	6793	6880	6968	7055	7142	
498	7229	7317	7404	7491	7578	7665	7752	7839	7926	8014	
499	8101	8188	8275	8362	8449	8535	8622	8709	8796	8883	
500	8970	9057	9144	9231	9317	9404	9491	9578	9664	9751	
501	9838	9924	..11	..98	.184	.271	.358	.444	.531	.617	
502	700704	0790	0877	0963	1050	1136	1222	1309	1395	1482	86
503	1568	1654	1741	1827	1913	1999	2086	2172	2258	2344	
504	2431	2517	2603	2689	2775	2861	2947	3033	3119	3205	
505	3291	3377	3463	3549	3635	3721	3807	3893	3979	4065	
506	4151	4236	4322	4408	4494	4579	4665	4751	4837	4922	
507	5008	5094	5179	5265	5350	5436	5522	5607	5693	5778	
N.	0	1	2	3	4	5	6	7	8	9	D.

Differences.	Proportional Parts.								
95	10	19	29	38	48	57	67	76	86
94	9	19	28	38	47	56	66	75	85
93	9	19	28	37	47	56	65	74	84
92	9	18	28	37	46	55	64	74	83
91	9	18	27	36	46	55	64	73	82
90	9	18	27	36	45	54	63	72	81
89	9	18	27	36	45	53	62	71	80
88	9	18	26	35	44	53	62	70	79
87	9	17	26	35	44	52	61	70	78

N.	0	1	2	3	4	5	6	7	8	9	D.
508	705864	5949	6035	6120	6206	6291	6376	6462	6547	6632	85
509	6718	6803	6888	6974	7059	7144	7229	7315	7400	7485	
510	7570	7655	7740	7826	7911	7996	8081	8166	8251	8336	
511	8421	8506	8591	8676	8761	8846	8931	9015	9100	9185	
512	9270	9355	9440	9524	9609	9694	9779	9863	9948	..33	
513	710117	0202	0287	0371	0456	0540	0625	0710	0794	0879	
514	0963	1048	1132	1217	1301	1385	1470	1554	1639	1723	84
515	1807	1892	1976	2060	2144	2229	2313	2397	2481	2566	
516	2650	2734	2818	2902	2986	3070	3154	3238	3323	3407	
517	3491	3575	3659	3742	3826	3910	3994	4078	4162	4246	
518	4330	4414	4497	4581	4665	4749	4833	4916	5000	5084	
519	5167	5251	5335	5418	5502	5586	5669	5753	5836	5920	
520	6003	6087	6170	6254	6337	6421	6504	6588	6671	6754	83
521	6838	6921	7004	7088	7171	7254	7338	7421	7504	7587	
522	7671	7754	7837	7920	8003	8086	8169	8253	8336	8419	
523	8502	8585	8668	8751	8834	8917	9000	9083	9165	9248	
524	9331	9414	9497	9580	9663	9745	9828	9911	9994	..77	
525	720159	0242	0325	0407	0490	0573	0655	0738	0821	0903	
526	0986	1068	1151	1233	1316	1398	1481	1563	1646	1728	82
527	1811	1893	1975	2058	2140	2222	2305	2387	2469	2552	
528	2634	2716	2798	2881	2963	3045	3127	3209	3291	3374	
529	3456	3538	3620	3702	3784	3866	3948	4030	4112	4194	
530	4276	4358	4440	4522	4604	4685	4767	4849	4931	5013	
531	5095	5176	5258	5340	5422	5503	5585	5667	5748	5830	
532	5912	5993	6075	6156	6238	6320	6401	6483	6564	6646	
533	6727	6809	6890	6972	7053	7134	7216	7297	7379	7460	81
534	7541	7623	7704	7785	7866	7948	8029	8110	8191	8273	
535	8354	8435	8516	8597	8678	8759	8841	8922	9003	9084	
536	9165	9246	9327	9408	9489	9570	9651	9732	9813	9893	
537	9974	..55	.136	.217	.298	.378	.459	.540	.621	.702	
538	730782	0863	0944	1024	1105	1186	1266	1347	1428	1508	
539	1589	1669	1750	1830	1911	1991	2072	2152	2233	2313	
540	2394	2474	2555	2635	2715	2796	2876	2956	3037	3117	80
541	3197	3278	3358	3438	3518	3598	3679	3759	3839	3919	
542	3999	4079	4160	4240	4320	4400	4480	4560	4640	4720	
543	4800	4880	4960	5040	5120	5200	5279	5359	5439	5519	
544	5599	5679	5759	5838	5918	5998	6078	6157	6237	6317	
545	6397	6476	6556	6635	6715	6795	6874	6954	7034	7113	
546	7193	7272	7352	7431	7511	7590	7670	7749	7829	7908	79
547	7987	8067	8146	8225	8305	8384	8463	8543	8622	8701	
548	8781	8860	8939	9018	9097	9177	9256	9335	9414	9493	
549	9572	9651	9731	9810	9889	9968	..47	.126	.205	.284	
550	740363	0442	0521	0600	0678	0757	0836	0915	0994	1073	
551	1152	1230	1309	1388	1467	1546	1624	1703	1782	1860	
552	1939	2018	2096	2175	2254	2332	2411	2489	2568	2647	
553	2725	2804	2882	2961	3039	3118	3196	3275	3353	3431	78
554	3510	3588	3667	3745	3823	3902	3980	4058	4136	4215	
555	4293	4371	4449	4528	4606	4684	4762	4840	4919	4997	
556	5075	5153	5231	5309	5387	5465	5543	5621	5699	5777	
557	5855	5933	6011	6089	6167	6245	6323	6401	6479	6556	
558	6634	6712	6790	6868	6945	7023	7101	7179	7256	7334	

N.	0	1	2	3	4	5	6	7	8	9	D.
559	747412	7489	7567	7645	7722	7800	7878	7955	8033	8110	78
560	8188	8266	8343	8421	8498	8576	8653	8731	8808	8885	77
561	8963	9040	9118	9195	9272	9350	9427	9504	9582	9659	
562	9736	9814	9891	9968	..45	.123	.200	.277	.354	.43	
563	750508	0586	0663	0740	0817	0894	0971	1048	1125	1202	
564	1279	1356	1433	1510	1587	1664	1741	1818	1895	1972	
565	2048	2125	2202	2279	2356	2433	2509	2586	2663	2740	
566	2816	2893	2970	3047	3123	3200	3277	3353	3430	3506	
567	3583	3660	3736	3813	3889	3966	4042	4119	4195	4272	
568	4348	4425	4501	4578	4654	4730	4807	4883	4960	5036	76
569	5112	5189	5265	5341	5417	5494	5570	5646	5722	5799	
570	5875	5951	6027	6103	6180	6256	6332	6408	6484	6560	
571	6636	6712	6788	6864	6940	7016	7092	7168	7244	7320	
572	7396	7472	7548	7624	7700	7775	7851	7927	8003	8079	
573	8155	8230	8306	8382	8458	8533	8609	8685	8761	8836	
574	8912	8988	9063	9139	9214	9290	9366	9441	9517	9592	
575	9668	9743	9819	9894	9970	..45	.121	.196	.272	.347	75
576	760422	0498	0573	0649	0724	0799	0875	0950	1025	1101	
577	1176	1251	1326	1402	1477	1552	1627	1702	1778	1853	
578	1928	2003	2078	2153	2228	2303	2378	2453	2529	2604	
579	2679	2754	2829	2904	2978	3053	3128	3203	3278	3353	
580	3428	3503	3578	3653	3727	3802	3877	3952	4027	4101	
581	4176	4251	4326	4400	4475	4550	4624	4699	4774	4848	
582	4923	4998	5072	5147	5221	5296	5370	5445	5520	5594	
583	5669	5743	5818	5892	5966	6041	6115	6190	6264	6338	74
584	6413	6487	6562	6636	6710	6785	6859	6933	7007	7082	
585	7156	7230	7304	7379	7453	7527	7601	7675	7749	7823	
586	7898	7972	8046	8120	8194	8268	8342	8416	8490	8564	
587	8638	8712	8786	8860	8934	9008	9082	9156	9230	9303	
588	9377	9451	9525	9599	9673	9746	9820	9894	9968	..42	
589	770115	0189	0263	0336	0410	0484	0557	0631	0705	0778	
590	0852	0926	0999	1073	1146	1220	1293	1367	1440	1514	
591	1587	1661	1734	1808	1881	1955	2028	2102	2175	2248	73
592	2322	2395	2468	2542	2615	2688	2762	2835	2908	2081	
593	3055	3128	3201	3274	3348	3421	3494	3567	3640	3713	
594	3786	3860	3933	4006	4079	4152	4225	4298	4371	4444	
595	4517	4590	4663	4736	4809	4882	4955	5028	5100	5173	
596	5246	5319	5392	5465	5538	5610	5683	5756	5829	5902	
597	5974	6047	6120	6193	6265	6338	6411	6483	6556	6629	
598	6701	6774	6846	6919	6992	7064	7137	7209	7282	7354	
599	7427	7499	7572	7644	7717	7789	7862	7934	8006	8079	72
600	8151	8224	8296	8368	8441	8513	8585	8658	8730	8802	
601	8874	8947	9019	9091	9163	9236	9308	9380	9452	9524	
602	9596	9669	9741	9813	9885	9957	..29	.101	.173	.245	
603	780317	0389	0461	0533	0605	0677	0749	0821	0893	0965	
604	1037	1109	1181	1253	1324	1396	1468	1540	1612	1684	
605	1755	1827	1899	1971	2042	2114	2186	2258	2329	2401	
606	2473	2544	2616	2688	2759	2831	2902	2974	3046	3117	
607	3189	3260	3332	3403	3475	3546	3618	3689	3761	3832	71
608	3904	3975	4046	4118	4189	4261	4332	4403	4475	4546	
609	4617	4689	4760	4831	4902	4974	5045	5116	5187	5259	
610	5330	5401	5472	5543	5615	5686	5757	5828	5899	5970	
611	6041	6112	6183	6254	6325	6396	6467	6538	6609	6680	

N.	0	1	2	3	4	5	6	7	8	9	D.
612	786751	6822	6893	6964	7035	7106	7177	7248	7319	7390	71
613	7460	7531	7602	7673	7744	7815	7885	7956	8027	8098	
614	8168	8239	8310	8381	8451	8522	8593	8663	8734	8804	
615	8875	8946	9016	9087	9157	9228	9299	9369	9440	9510	
616	9581	9651	9722	9792	9863	9933	...4	..74	.144	.215	70
617	790285	0356	0426	0496	0567	0637	0707	0778	0848	0918	
618	0988	1059	1129	1199	1269	1340	1410	1480	1550	1620	
619	1691	1761	1831	1901	1971	2041	2111	2181	2252	2322	
620	2392	2462	2532	2602	2672	2742	2812	2882	2952	3022	
621	3092	3162	3231	3301	3371	3441	3511	3581	3651	3721	
622	3790	3860	3930	4000	4070	4139	4209	4279	4349	4418	
623	4488	4558	4627	4697	4767	4836	4906	4976	5045	5115	
624	5185	5254	5324	5393	5463	5532	5602	5672	5741	5811	
625	5880	5949	6019	6088	6158	6227	6297	6366	6436	6505	69
626	6574	6644	6713	6782	6852	6921	6990	7060	7129	7198	
627	7268	7337	7406	7475	7545	7614	7683	7752	7821	7890	
628	7960	8029	8098	8167	8236	8305	8374	8443	8513	8582	
629	8651	8720	8789	8858	8927	8996	9065	9134	9203	9272	
630	9341	9409	9478	9547	9616	9685	9754	9823	9892	9961	
631	800029	0098	0167	0236	0305	0373	0442	0511	0580	0648	
632	0717	0786	0854	0923	0992	1061	1129	1198	1266	1335	
633	1404	1472	1541	1609	1678	1747	1815	1883	1952	2021	
634	2089	2158	2226	2295	2363	2432	2500	2568	2637	2705	
635	2774	2842	2910	2979	3047	3116	3184	3252	3321	3389	68
636	3457	3525	3594	3662	3730	3798	3867	3935	4003	4071	
637	4139	4208	4276	4344	4412	4480	4548	4616	4685	4753	
638	4821	4889	4957	5025	5093	5161	5229	5297	5365	5433	
639	5501	5569	5637	5705	5773	5841	5908	5976	6044	6112	
640	6180	6248	6316	6384	6451	6519	6587	6655	6723	6790	
641	6858	6926	6994	7061	7129	7197	7264	7332	7400	7467	
642	7535	7603	7670	7738	7806	7873	7941	8008	8076	8143	
643	8211	8279	8346	8414	8481	8549	8616	8684	8751	8818	67
644	8886	8953	9021	9088	9156	9223	9290	9358	9425	9492	
645	9560	9627	9694	9762	9829	9896	9964	..31	..98	.165	
646	810233	0300	0367	0434	0501	0569	0636	0703	0770	0837	
647	0904	0971	1039	1106	1173	1240	1307	1374	1441	1508	
648	1575	1642	1709	1776	1843	1910	1977	2044	2111	2178	
649	2245	2312	2379	2445	2512	2579	2646	2713	2780	2847	
650	2913	2980	3047	3114	3181	3247	3314	3381	3448	3514	
651	3581	3648	3714	3781	3848	3914	3981	4048	4114	4181	
652	4248	4314	4381	4447	4514	4581	4647	4714	4780	4847	
653	4913	4980	5046	5113	5179	5246	5312	5378	5445	5511	66
654	5578	5644	5711	5777	5843	5910	5976	6042	6109	6175	
655	6241	6308	6374	6440	6506	6573	6639	6705	6771	6838	
656	6904	6970	7036	7102	7169	7235	7301	7367	7433	7499	
657	7565	7631	7698	7764	7830	7896	7962	8028	8094	8160	
658	8226	8292	8358	8424	8490	8556	8622	8688	8754	8820	
659	8885	8951	9017	9083	9149	9215	9281	9346	9412	9478	
660	9544	9610	9676	9741	9807	9873	9939	...4	..70	.136	
661	820201	0267	0333	0399	0464	0530	0595	0661	0727	0792	
662	0858	0924	0989	1055	1120	1186	1251	1317	1382	1448	
663	1514	1579	1645	1710	1775	1841	1906	1972	2037	2103	65
664	2168	2233	2299	2364	2430	2495	2560	2626	2691	2756	

N.	0		1	2	3	4	5	6	7	8	9	D.
Differences.	71	Propor. Parts.	7	14	21	28	36	43	50	57	64	
	70		7	14	21	28	35	42	49	56	63	
	69		7	14	21	28	35	41	48	55	62	
	68		7	14	20	27	34	41	48	54	61	
	67		7	13	20	27	34	40	47	54	60	
	66		7	13	20	26	33	40	46	53	59	
	65		7	13	20	26	33	39	46	52	59	

N.	0	1	2	3	4	5	6	7	8	9
665	822822	2887	2952	3018	3083	3148	3213	3279	3344	3409
666	3474	3539	3605	3670	3735	3800	3865	3930	3996	4061
667	4126	4191	4256	4321	4386	4451	4516	4581	4646	4711
668	4776	4841	4906	4971	5036	5101	5166	5231	5296	5361
669	5426	5491	5556	5621	5686	5751	5815	5880	5945	6010
670	6075	6140	6204	6269	6334	6399	6464	6528	6593	6658
671	6723	6787	6852	6917	6981	7046	7111	7175	7240	7305
672	7369	7434	7499	7563	7628	7692	7757	7821	7886	7951
673	8015	8080	8144	8209	8273	8338	8402	8467	8531	8595
674	8660	8724	8789	8853	8918	8982	9046	9111	9175	9239
675	9304	9368	9432	9497	9561	9625	9690	9754	9818	9882
676	9947	..11	..75	.139	.204	.268	.332	.396	.460	.525
677	830589	0653	0717	0781	0845	0909	0973	1037	1102	1166
678	1230	1294	1358	1422	1486	1550	1614	1678	1742	1806
679	1870	1934	1998	2062	2126	2189	2253	2317	2381	2445
680	2509	2573	2637	2700	2764	2828	2892	2956	3020	3083
681	3147	3211	3275	3338	3402	3466	3530	3593	3657	3721
682	3784	3848	3912	3975	4039	4103	4166	4230	4294	4357
683	4421	4484	4548	4611	4675	4739	4802	4866	4929	4993
684	5056	5120	5183	5247	5310	5373	5437	5500	5564	5627
685	5691	5754	5817	5881	5944	6007	6071	6134	6197	6261
686	6324	6387	6451	6514	6577	6641	6704	6767	6830	6894
687	6957	7020	7083	7146	7210	7273	7336	7399	7462	7525
688	7588	7652	7715	7778	7841	7904	7967	8030	8093	8156
689	8219	8282	8345	8408	8471	8534	8597	8660	8723	8786
690	8849	8912	8975	9038	9101	9164	9227	9289	9352	9415
691	9478	9541	9604	9667	9729	9792	9855	9918	9981	..43
692	840106	0169	0232	0294	0357	0420	0482	0545	0608	0671
693	0733	0796	0859	0921	0984	1046	1109	1172	1234	1297
694	1359	1422	1485	1547	1610	1672	1735	1797	1860	1922
695	1985	2047	2110	2172	2235	2297	2360	2422	2484	2547
696	2609	2672	2734	2796	2859	2921	2983	3046	3108	3170
697	3233	3295	3357	3420	3482	3544	3606	3669	3731	3793
698	3855	3918	3980	4042	4104	4166	4229	4291	4353	4415
699	4477	4539	4601	4664	4726	4788	4850	4912	4974	5036
700	5098	5160	5222	5284	5346	5408	5470	5532	5594	5656
701	5718	5780	5842	5904	5966	6028	6090	6151	6213	6275
702	6337	6399	6461	6523	6585	6646	6708	6770	6832	6894
703	6955	7017	7079	7141	7202	7264	7326	7388	7449	7511
704	7573	7634	7696	7758	7819	7881	7943	8004	8066	8128
705	8189	8251	8312	8374	8435	8497	8559	8620	8682	8743
706	8805	8866	8928	8989	9051	9112	9174	9235	9297	9358
707	9419	9481	9542	9604	9665	9726	9788	9849	9911	9972
708	850033	0095	0156	0217	0279	0340	0401	0462	0524	0585
709	0646	0707	0769	0830	0891	0952	1014	1075	1136	1197
710	1258	1320	1381	1442	1503	1564	1625	1686	1747	1809
711	1870	1931	1992	2053	2114	2175	2236	2297	2358	2419
712	2480	2541	2602	2663	2724	2785	2846	2907	2968	3029
713	3090	3150	3211	3272	3333	3394	3455	3516	3577	3637
714	3698	3759	3820	3881	3941	4002	4063	4124	4185	4245
715	4306	4367	4428	4488	4549	4610	4670	4731	4792	4852
716	4913	4974	5034	5095	5156	5216	5277	5337	5398	5459
717	5519	5580	5640	5701	5761	5822	5882	5943	6003	6064
718	6124	6185	6245	6306	6366	6427	6487	6548	6608	6668
719	6729	6789	6850	6910	6970	7031	7091	7152	7212	7272

N.	0	1	2	3	4	5	6	7	8	9	D.
720	857332	7393	7453	7513	7574	7634	7694	7755	7815	7875	60
721	7935	7995	8056	8116	8176	8236	8297	8357	8417	8477	
722	8537	8597	8657	8718	8778	8838	8898	8958	9018	9078	
723	9138	9198	9258	9318	9379	9439	9499	9559	9619	9679	
724	9739	9799	9859	9918	9978	..38	..98	.158	.218	.278	
725	860338	0398	0458	0518	0578	0637	0697	0757	0817	0877	
726	0937	0996	1056	1116	1176	1236	1295	1355	1415	1475	
727	1534	1594	1654	1714	1773	1833	1893	1952	2012	2072	
728	2131	2191	2251	2310	2370	2430	2489	2549	2608	2668	
729	2728	2787	2847	2906	2966	3025	3085	3144	3204	3263	
730	3323	3382	3442	3501	3561	3620	3680	3739	3799	3858	59
731	3917	3977	4036	4096	4155	4214	4274	4333	4392	4452	
732	4511	4570	4630	4689	4748	4808	4867	4926	4985	5045	
733	5104	5163	5222	5282	5341	5400	5459	5519	5578	5637	
734	5696	5755	5814	5874	5933	5992	6051	6110	6169	6228	
735	6287	6346	6405	6465	6524	6583	6642	6701	6760	6819	
736	6878	6937	6996	7055	7114	7173	7232	7291	7350	7409	
737	7467	7526	7585	7644	7703	7762	7821	7880	7939	7998	
738	8056	8115	8174	8233	8292	8350	8409	8468	8527	8586	
739	8644	8703	8762	8821	8879	8938	8997	9056	9114	9173	
740	9232	9290	9349	9408	9466	9525	9584	9642	9701	9760	
741	9818	9877	9935	9994	..53	.111	.170	.228	.287	.345	
742	870404	0462	0521	0579	0638	0696	0755	0813	0872	0930	58
743	0989	1047	1106	1164	1223	1281	1339	1398	1456	1515	
744	1573	1631	1690	1748	1806	1865	1923	1981	2040	2098	
745	2156	2215	2273	2331	2389	2448	2506	2564	2622	2681	
746	2739	2797	2855	2913	2972	3030	3088	3146	3204	3262	
747	3321	3379	3437	3495	3553	3611	3669	3727	3785	3844	
748	3902	3960	4018	4076	4134	4192	4250	4308	4366	4424	
749	4482	4540	4598	4656	4714	4772	4830	4888	4945	5003	
750	5061	5119	5177	5235	5293	5351	5409	5466	5524	5582	
751	5640	5698	5756	5813	5871	5929	5987	6045	6102	6160	
752	6218	6276	6333	6391	6449	6507	6564	6622	6680	6737	
753	6795	6853	6910	6968	7026	7083	7141	7199	7256	7314	
754	7371	7429	7487	7544	7602	7659	7717	7774	7832	7889	
755	7947	8004	8062	8119	8177	8234	8292	8349	8407	8464	57
756	8522	8579	8637	8694	8752	8809	8866	8924	8981	9039	
757	9096	9153	9211	9268	9325	9383	9440	9497	9555	9612	
758	9669	9726	9784	9841	9898	9956	..13	..70	.127	.185	
759	880242	0299	0356	0413	0471	0528	0585	0642	0699	0756	
760	0814	0871	0928	0985	1042	1099	1156	1213	1271	1328	
761	1385	1442	1499	1556	1613	1670	1727	1784	1841	1898	
762	1955	2012	2069	2126	2183	2240	2297	2354	2411	2468	
763	2525	2581	2638	2695	2752	2809	2866	2923	2980	3037	
764	3093	3150	3207	3264	3321	3377	3434	3491	3548	3605	
765	3661	3718	3775	3832	3888	3945	4002	4059	4115	4172	
766	4229	4285	4342	4399	4455	4512	4569	4625	4682	4739	
767	4795	4852	4909	4965	5022	5078	5135	5192	5248	5305	
768	5361	5418	5474	5531	5587	5644	5700	5757	5813	5870	
769	5926	5983	6039	6096	6152	6209	6265	6321	6378	6434	56
770	6491	6547	6604	6660	6716	6773	6829	6885	6942	6998	
771	7054	7111	7167	7223	7280	7336	7392	7449	7505	7561	
772	7617	7674	7730	7786	7842	7898	7955	8011	8067	8123	
773	8179	8236	8292	8348	8404	8460	8516	8573	8629	8685	
774	8741	8797	8853	8909	8965	9021	9077	9134	9190	9246	
N.	0	1	2	3	4	5	6	7	8	9	D.
Differ.	60 Pro. Parts.	6	12	18	24	30	36	42	48	54	
	59	6	12	18	24	30	35	41	47	53	
	58	6	12	17	23	29	35	41	46	52	
	57	6	11	17	23	29	34	40	46	51	
	56	6	11	17	22	28	34	39	45	50	

N.	0	1	2	3	4	5	6	7	8	9	D.
775	889302	9358	9414	9470	9526	9582	9638	9694	9750	9806	56
776	9862	9918	9974	..30	..86	.141	.197	.253	.309	.365	
777	890421	0477	0533	0589	0645	0700	0756	0812	0868	0924	
778	0980	1035	1091	1147	1203	1259	1314	1370	1426	1482	
779	1537	1593	1649	1705	1760	1816	1872	1928	1983	2039	
780	2095	2150	2206	2262	2317	2373	2429	2484	2540	2595	
781	2651	2707	2762	2818	2873	2929	2985	3040	3096	3151	
782	3207	3262	3318	3373	3429	3484	3540	3595	3651	3706	
783	3762	3817	3873	3928	3984	4039	4094	4150	4205	4261	55
784	4316	4371	4427	4482	4538	4593	4648	4704	4759	4814	
785	4870	4925	4980	5036	5091	5146	5201	5257	5312	5367	
786	5423	5478	5533	5588	5644	5699	5754	5809	5864	5920	
787	5975	6030	6085	6140	6195	6251	6306	6361	6416	6471	
788	6526	6581	6636	6692	6747	6802	6857	6912	6967	7022	
789	7077	7132	7187	7242	7297	7352	7407	7462	7517	7572	
790	7627	7682	7737	7792	7847	7902	7957	8012	8067	8122	
791	8176	8231	8286	8341	8396	8451	8506	8561	8615	8670	
792	8725	8780	8835	8890	8944	8999	9054	9109	9164	9218	
793	9273	9328	9383	9437	9492	9547	9602	9656	9711	9766	
794	9821	9875	9930	9985	..39	..94	.149	.203	.258	.312	
795	900367	0422	0476	0531	0586	0640	0695	0749	0804	0859	
796	0913	0968	1022	1077	1131	1186	1240	1295	1349	1404	
797	1458	1513	1567	1622	1676	1731	1785	1840	1894	1948	54
798	2003	2057	2112	2166	2221	2275	2329	2384	2438	2492	
799	2547	2601	2655	2710	2764	2818	2873	2927	2981	3036	
800	3090	3144	3199	3253	3307	3361	3416	3470	3524	3578	
801	3633	3687	3741	3795	3849	3904	3958	4012	4066	4120	
802	4174	4229	4283	4337	4391	4445	4499	4553	4607	4661	
803	4716	4770	4824	4878	4932	4986	5040	5094	5148	5202	
804	5256	5310	5364	5418	5472	5526	5580	5634	5688	5742	
805	5796	5850	5904	5958	6012	6066	6119	6173	6227	6281	
806	6335	6389	6443	6497	6551	6604	6658	6712	6766	6820	
807	6874	6927	6981	7035	7089	7143	7196	7250	7304	7358	
808	7411	7465	7519	7573	7626	7680	7734	7787	7841	7895	
809	7949	8002	8056	8110	8163	8217	8270	8324	8378	8431	
810	8485	8539	8592	8646	8699	8753	8807	8860	8914	8967	
811	9021	9074	9128	9181	9235	9289	9342	9396	9449	9503	
812	9556	9610	9663	9716	9770	9823	9877	9930	9984	..37	53
813	910091	0144	0197	0251	0304	0358	0411	0464	0518	0571	
814	0624	0678	0731	0784	0838	0891	0944	0998	1051	1104	
815	1158	1211	1264	1317	1371	1424	1477	1530	1584	1637	
816	1690	1743	1797	1850	1903	1956	2009	2063	2116	2169	
817	2222	2275	2328	2381	2435	2488	2541	2594	2647	2700	
818	2753	2806	2859	2913	2966	3019	3072	3125	3178	3231	
819	3284	3337	3390	3443	3496	3549	3602	3655	3708	3761	
820	3814	3867	3920	3973	4026	4079	4132	4184	4237	4290	
821	4343	4396	4449	4502	4555	4608	4660	4713	4766	4819	
822	4872	4925	4977	5030	5083	5136	5189	5241	5294	5347	
823	5400	5453	5505	5558	5611	5664	5716	5769	5822	5875	
824	5927	5980	6033	6085	6138	6191	6243	6296	6349	6401	
825	6454	6507	6559	6612	6664	6717	6770	6822	6875	6927	
826	6980	7033	7085	7138	7190	7243	7295	7348	7400	7453	
827	7506	7558	7611	7663	7716	7768	7820	7873	7925	7978	52
828	8030	8083	8135	8188	8240	8293	8345	8397	8450	8502	
829	8555	8607	8659	8712	8764	8816	8869	8921	8973	9026	
830	9078	9130	9183	9235	9287	9340	9392	9444	9496	9549	
N.	0	1	2	3	4	5	6	7	8	9	D.
Differ.	55 P. Parts	6	11	17	22	28	33	39	44	50	
	54	5	11	16	22	27	32	38	43	49	
	53	5	11	16	21	27	32	37	42	48	
	52	5	10	16	21	26	31	36	42	47	

N.	0	1	2	3	4	5	6	7	8	9	D.
409	611723	1829	1936	2042	2148	2254	2360	2466	2572	2678	106
410	2784	2890	2996	3102	3207	3313	3419	3525	3630	3736	
411	3842	3947	4053	4159	4264	4370	4475	4581	4686	4792	
412	4897	5003	5108	5213	5319	5424	5529	5634	5740	5845	105
413	5950	6055	6160	6265	6370	6476	6581	6686	6790	6895	
414	7000	7105	7210	7315	7420	7525	7629	7734	7839	7943	
415	8048	8153	8257	8362	8466	8571	8676	8780	8884	8989	
416	9093	9198	9302	9406	9511	9615	9719	9824	9928	..32	104
417	620136	0240	0344	0448	0552	0656	0760	0864	0968	1072	
418	1176	1280	1384	1488	1592	1695	1799	1903	2007	2110	
419	2214	2318	2421	2525	2628	2732	2835	2939	3042	3146	
420	3249	3353	3456	3559	3663	3766	3869	3973	4076	4179	103
421	4282	4385	4488	4591	4695	4798	4901	5004	5107	5210	
422	5312	5415	5518	5621	5724	5827	5929	6032	6135	6238	
423	6340	6443	6546	6648	6751	6853	6956	7058	7161	7263	
424	7366	7468	7571	7673	7775	7878	7980	8082	8185	8287	102
425	8389	8491	8593	8695	8797	8900	9002	9104	9206	9308	
426	9410	9512	9613	9715	9817	9919	..21	.123	.224	.326	
427	630428	0530	0631	0733	0835	0936	1038	1139	1241	1342	
428	1444	1545	1647	1748	1849	1951	2052	2153	2255	2356	101
429	2457	2559	2660	2761	2862	2963	3064	3165	3266	3367	
430	3468	3569	3670	3771	3872	3973	4074	4175	4276	4376	
431	4477	4578	4679	4779	4880	4981	5081	5182	5283	5383	100
432	5484	5584	5685	5785	5886	5986	6087	6187	6287	6388	
433	6488	6588	6688	6789	6889	6989	7089	7189	7290	7390	
434	7490	7590	7690	7790	7890	7990	8090	8190	8290	8389	
435	8489	8589	8689	8789	8888	8988	9088	9188	9287	9387	99
436	9486	9586	9686	9785	9885	9984	..84	.183	.283	.382	
437	640481	0581	0680	0779	0879	0978	1077	1177	1276	1375	
438	1474	1573	1672	1771	1871	1970	2069	2168	2267	2366	
439	2465	2563	2662	2761	2860	2959	3058	3156	3255	3354	
440	3453	3551	3650	3749	3847	3946	4044	4143	4242	4340	98
441	4439	4537	4636	4734	4832	4931	5029	5127	5226	5324	
442	5422	5521	5619	5717	5815	5913	6011	6110	6208	6306	
443	6404	6502	6600	6698	6796	6894	6992	7089	7187	7285	
444	7383	7481	7579	7676	7774	7872	7969	8067	8165	8262	
445	8360	8458	8555	8653	8750	8848	8945	9043	9140	9237	97
446	9335	9432	9530	9627	9724	9821	9919	..16	.113	.210	
447	650308	0405	0502	0599	0696	0793	0890	0987	1084	1181	
448	1278	1375	1472	1569	1666	1762	1859	1956	2053	2150	
449	2246	2343	2440	2536	2633	2730	2826	2923	3019	3116	
450	3213	3309	3405	3502	3598	3695	3791	3888	3984	4080	96
451	4177	4273	4369	4465	4562	4658	4754	4850	4946	5042	
452	5138	5235	5331	5427	5523	5619	5715	5810	5906	6002	
453	6098	6194	6290	6386	6482	6577	6673	6769	6864	6960	
454	7056	7152	7247	7343	7438	7534	7629	7725	7820	7916	
455	8011	8107	8202	8298	8393	8488	8584	8679	8774	8870	95
456	8965	9060	9155	9250	9346	9441	9536	9631	9726	9821	
457	9916	..11	.106	.201	.296	.391	.486	.581	.676	.771	

N.	0	1	2	3	4	5	6	7	8	9	D.
Differences.	106 Proportional Parts.	11	21	32	42	53	64	74	85	95	
	105	11	21	32	42	53	63	74	84	95	
	104	10	21	31	42	52	62	73	83	94	
	103	10	21	31	41	52	62	72	82	93	
	102	10	20	31	41	51	61	71	82	92	
	101	10	20	30	40	51	61	71	81	91	
	100	10	20	30	40	50	60	70	80	90	
	99	10	20	30	40	50	59	69	79	89	
	98	10	20	29	39	49	59	69	78	88	
	97	10	19	29	39	49	58	68	78	87	
	96	10	19	29	38	48	58	67	77	86	

N.	0	1	2	3	4	5	6	7	8	9
458	660865	0960	1055	1150	1245	1339	1434	1529	1623	1718
459	1813	1907	2002	2096	2191	2286	2380	2475	2569	2663
460	2758	2852	2947	3041	3135	3230	3324	3418	3512	3607
461	3701	3795	3889	3983	4078	4172	4266	4360	4454	4548
462	4642	4736	4830	4924	5018	5112	5206	5299	5393	5487
463	5581	5675	5769	5862	5956	6050	6143	6237	6331	6424
464	6518	6612	6705	6799	6892	6986	7079	7173	7266	7360
465	7453	7546	7640	7733	7826	7920	8013	8106	8199	8293
466	8386	8479	8572	8665	8759	8852	8945	9038	9131	9224
467	9317	9410	9503	9596	9689	9782	9875	9967	..60	.153
468	670246	0339	0431	0524	0617	0710	0802	0895	0988	1080
469	1173	1265	1358	1451	1543	1636	1728	1821	1913	2005
470	2098	2190	2283	2375	2467	2560	2652	2744	2836	2929
471	3021	3113	3205	3297	3390	3482	3574	3666	3758	3850
472	3942	4034	4126	4218	4310	4402	4494	4586	4677	4769
473	4861	4953	5045	5137	5228	5320	5412	5503	5595	5687
474	5778	5870	5962	6053	6145	6236	6328	6419	6511	6602
475	6694	6785	6876	6968	7059	7151	7242	7333	7424	7516
476	7607	7698	7789	7881	7972	8063	8154	8245	8336	8427
477	8518	8609	8700	8791	8882	8973	9064	9155	9246	9337
478	9428	9519	9610	9700	9791	9882	9973	..63	.154	.245
479	680336	0426	0517	0607	0698	0789	0879	0970	1060	1151
480	1241	1332	1422	1513	1603	1693	1784	1874	1964	2055
481	2145	2235	2326	2416	2506	2596	2686	2777	2867	2957
482	3047	3137	3227	3317	3407	3497	3587	3677	3767	3857
483	3947	4037	4127	4217	4307	4396	4486	4576	4666	4756
484	4845	4935	5025	5114	5204	5294	5383	5473	5563	5652
485	5742	5831	5921	6010	6100	6189	6279	6368	6458	6547
486	6636	6726	6815	6904	6994	7083	7172	7261	7351	7440
487	7529	7618	7707	7796	7886	7975	8064	8153	8242	8331
488	8420	8509	8598	8687	8776	8865	8953	9042	9131	9220
489	9309	9398	9486	9575	9664	9753	9841	9930	..19	.107
490	690196	0285	0373	0462	0550	0639	0728	0816	0905	0993
491	1081	1170	1258	1347	1435	1524	1612	1700	1789	1877
492	1965	2053	2142	2230	2318	2406	2494	2583	2671	2759
493	2847	2935	3023	3111	3199	3287	3375	3463	3551	3639
494	3727	3815	3903	3991	4078	4166	4254	4342	4430	4517
495	4605	4693	4781	4868	4956	5044	5131	5219	5307	5394
496	5482	5569	5657	5744	5832	5919	6007	6094	6182	6269
497	6356	6444	6531	6618	6706	6793	6880	6968	7055	7142
498	7229	7317	7404	7491	7578	7665	7752	7839	7926	8014
499	8101	8188	8275	8362	8449	8535	8622	8709	8796	8883
500	8970	9057	9144	9231	9317	9404	9491	9578	9664	9751
501	9838	9924	..11	..98	.184	.271	.358	.444	.531	.617
502	700704	0790	0877	0963	1050	1136	1222	1309	1395	1482
503	1568	1654	1741	1827	1913	1999	2086	2172	2258	2344
504	2431	2517	2603	2689	2775	2861	2947	3033	3119	3205
505	3291	3377	3463	3549	3635	3721	3807	3893	3979	4065
506	4151	4236	4322	4408	4494	4579	4665	4751	4837	4922
507	5008	5094	5179	5265	5350	5436	5522	5607	5693	5778

N.	0	1	2	3	4	5	6	7	8	9	D.
508	705864	5949	6035	6120	6206	6291	6376	6462	6547	6632	85
509	6718	6803	6888	6974	7059	7144	7229	7315	7400	7485	
510	7570	7655	7740	7826	7911	7996	8081	8166	8251	8336	
511	8421	8506	8591	8676	8761	8846	8931	9015	9100	9185	
512	9270	9355	9440	9524	9609	9694	9779	9863	9948	..33	
513	710117	0202	0287	0371	0456	0540	0625	0710	0794	0879	
514	0963	1048	1132	1217	1301	1385	1470	1554	1639	1723	84
515	1807	1892	1976	2060	2144	2229	2313	2397	2481	2566	
516	2650	2734	2818	2902	2986	3070	3154	3238	3323	3407	
517	3491	3575	3659	3742	3826	3910	3994	4078	4162	4246	
518	4330	4414	4497	4581	4665	4749	4833	4916	5000	5084	
519	5167	5251	5335	5418	5502	5586	5669	5753	5836	5920	
520	6003	6087	6170	6254	6337	6421	6504	6588	6671	6754	83
521	6838	6921	7004	7088	7171	7254	7338	7421	7504	7587	
522	7671	7754	7837	7920	8003	8086	8169	8253	8336	8419	
523	8502	8585	8668	8751	8834	8917	9000	9083	9165	9248	
524	9331	9414	9497	9580	9663	9745	9828	9911	9994	..77	
525	720159	0242	0325	0407	0490	0573	0655	0738	0821	0903	
526	0986	1068	1151	1233	1316	1398	1481	1563	1646	1728	82
527	1811	1893	1975	2058	2140	2222	2305	2387	2469	2552	
528	2634	2716	2798	2881	2963	3045	3127	3209	3291	3374	
529	3456	3538	3620	3702	3784	3866	3948	4030	4112	4194	
530	4276	4358	4440	4522	4604	4685	4767	4849	4931	5013	
531	5095	5176	5258	5340	5422	5503	5585	5667	5748	5830	
532	5912	5993	6075	6156	6238	6320	6401	6483	6564	6646	
533	6727	6809	6890	6972	7053	7134	7216	7297	7379	7460	81
534	7541	7623	7704	7785	7866	7948	8029	8110	8191	8273	
535	8354	8435	8516	8597	8678	8759	8841	8922	9003	9084	
536	9165	9246	9327	9408	9489	9570	9651	9732	9813	9893	
537	9974	..55	.136	.217	.298	.378	.459	.540	.621	.702	
538	730782	0863	0944	1024	1105	1186	1266	1347	1428	1508	
539	1589	1669	1750	1830	1911	1991	2072	2152	2233	2313	
540	2394	2474	2555	2635	2715	2796	2876	2956	3037	3117	80
541	3197	3278	3358	3438	3518	3598	3679	3759	3839	3919	
542	3999	4079	4160	4240	4320	4400	4480	4560	4640	4720	
543	4800	4880	4960	5040	5120	5200	5279	5359	5439	5519	
544	5599	5679	5759	5838	5918	5998	6078	6157	6237	6317	
545	6397	6476	6556	6635	6715	6795	6874	6954	7034	7113	
546	7193	7272	7352	7431	7511	7590	7670	7749	7829	7908	79
547	7987	8067	8146	8225	8305	8384	8463	8543	8622	8701	
548	8781	8860	8939	9018	9097	9177	9256	9335	9414	9493	
549	9572	9651	9731	9810	9889	9968	..47	.126	.205	.284	
550	740363	0442	0521	0600	0678	0757	0836	0915	0994	1073	
551	1152	1230	1309	1388	1467	1546	1624	1703	1782	1860	
552	1939	2018	2096	2175	2254	2332	2411	2489	2568	2647	
553	2725	2804	2882	2961	3039	3118	3196	3275	3353	3431	78
554	3510	3588	3667	3745	3823	3902	3980	4058	4136	4215	
555	4293	4371	4449	4528	4606	4684	4762	4840	4919	4997	
556	5075	5153	5231	5309	5387	5465	5543	5621	5699	5777	
557	5855	5933	6011	6089	6167	6245	6323	6401	6479	6556	
558	6634	6712	6790	6868	6945	7023	7101	7179	7256	7334	

N.	0	1	2	3	4	5	6	7	8	9	D.
559	747412	7489	7567	7645	7722	7800	7878	7955	8033	8110	78
560	8188	8266	8343	8421	8498	8576	8653	8731	8808	8885	77
561	8963	9040	9118	9195	9272	9350	9427	9504	9582	9659	
562	9736	9814	9891	9968	..45	.123	.200	.277	.354	.43	
563	750508	0586	0663	0740	0817	0894	0971	1048	1125	1202	
564	1279	1356	1433	1510	1587	1664	1741	1818	1895	1972	
565	2048	2125	2202	2279	2356	2433	2509	2586	2663	2740	
566	2816	2893	2970	3047	3123	3200	3277	3353	3430	3506	
567	3583	3660	3736	3813	3889	3966	4042	4119	4195	4272	
568	4348	4425	4501	4578	4654	4730	4807	4883	4960	5036	76
569	5112	5189	5265	5341	5417	5494	5570	5646	5722	5799	
570	5875	5951	6027	6103	6180	6256	6332	6408	6484	6560	
571	6636	6712	6788	6864	6940	7016	7092	7168	7244	7320	
572	7396	7472	7548	7624	7700	7775	7851	7927	8003	8079	
573	8155	8230	8306	8382	8458	8533	8609	8685	8761	8836	
574	8912	8988	9063	9139	9214	9290	9366	9441	9517	9592	
575	9668	9743	9819	9894	9970	..45	.121	.196	.272	.347	75
576	760422	0498	0573	0649	0724	0799	0875	0950	1025	1101	
577	1176	1251	1326	1402	1477	1552	1627	1702	1778	1853	
578	1928	2003	2078	2153	2228	2303	2378	2453	2529	2604	
579	2679	2754	2829	2904	2978	3053	3128	3203	3278	3353	
580	3428	3503	3578	3653	3727	3802	3877	3952	4027	4101	
581	4176	4251	4326	4400	4475	4550	4624	4699	4774	4848	
582	4923	4998	5072	5147	5221	5296	5370	5445	5520	5594	
583	5669	5743	5818	5892	5966	6041	6115	6190	6264	6338	74
584	6413	6487	6562	6636	6710	6785	6859	6933	7007	7082	
585	7156	7230	7304	7379	7453	7527	7601	7675	7749	7823	
586	7898	7972	8046	8120	8194	8268	8342	8416	8490	8564	
587	8638	8712	8786	8860	8934	9008	9082	9156	9230	9303	
588	9377	9451	9525	9599	9673	9746	9820	9894	9968	..42	
589	770115	0189	0263	0336	0410	0484	0557	0631	0705	0778	
590	0852	0926	0999	1073	1146	1220	1293	1367	1440	1514	
591	1587	1661	1734	1808	1881	1955	2028	2102	2175	2248	73
592	2322	2395	2468	2542	2615	2688	2762	2835	2908	2081	
593	3055	3128	3201	3274	3348	3421	3494	3567	3640	3713	
594	3786	3860	3933	4006	4079	4152	4225	4298	4371	4444	
595	4517	4590	4663	4736	4809	4882	4955	5028	5100	5173	
596	5246	5319	5392	5465	5538	5610	5683	5756	5829	5902	
597	5974	6047	6120	6193	6265	6338	6411	6483	6556	6629	
598	6701	6774	6846	6919	6992	7064	7137	7209	7282	7354	
599	7427	7499	7572	7644	7717	7789	7862	7934	8006	8079	72
600	8151	8224	8296	8368	8441	8513	8585	8658	8730	8802	
601	8874	8947	9019	9091	9163	9236	9308	9380	9452	9524	
602	9596	9669	9741	9813	9885	9957	..29	.101	.173	.245	
603	780317	0389	0461	0533	0605	0677	0749	0821	0893	0965	
604	1037	1109	1181	1253	1324	1396	1468	1540	1612	1684	
605	1755	1827	1899	1971	2042	2114	2186	2258	2329	2401	
606	2473	2544	2616	2688	2759	2831	2902	2974	3046	3117	
607	3189	3260	3332	3403	3475	3546	3618	3689	3761	3832	71
608	3904	3975	4046	4118	4189	4261	4332	4403	4475	4546	
609	4617	4689	4760	4831	4902	4974	5045	5116	5187	5259	
610	5330	5401	5472	5543	5615	5686	5757	5828	5899	5970	
611	6041	6112	6183	6254	6325	6396	6467	6538	6609	6680	

N.	0	1	2	3	4	5	6	7	8	9	D.
Differences.	77 Propor. Parts.	8	15	23	31	39	46	54	62	69	
	76	8	15	23	30	38	46	53	61	68	
	75	8	15	23	30	38	45	53	60	68	
	74	7	15	22	30	37	44	52	59	67	
	73	7	15	22	29	37	44	51	58	66	
	72	7	14	22	29	36	43	50	58	65	
	71	7	14	21	28	36	43	50	57	64	

N.	0	1	2	3	4	5	6	7	8	9	D.
612	786751	6822	6893	6964	7035	7106	7177	7248	7319	7390	71
613	7460	7531	7602	7673	7744	7815	7885	7956	8027	8098	
614	8168	8239	8310	8381	8451	8522	8593	8663	8734	8804	
615	8875	8946	9016	9087	9157	9228	9299	9369	9440	9510	
616	9581	9651	9722	9792	9863	9933	...4	..74	.144	.215	70
617	790285	0356	0426	0496	0567	0637	0707	0778	0848	0918	
618	0988	1059	1129	1199	1269	1340	1410	1480	1550	1620	
619	1691	1761	1831	1901	1971	2041	2111	2181	2252	2322	
620	2392	2462	2532	2602	2672	2742	2812	2882	2952	3022	
621	3092	3162	3231	3301	3371	3441	3511	3581	3651	3721	
622	3790	3860	3930	4000	4070	4139	4209	4279	4349	4418	
623	4488	4558	4627	4697	4767	4836	4906	4976	5045	5115	
624	5185	5254	5324	5393	5463	5532	5602	5672	5741	5811	
625	5880	5949	6019	6088	6158	6227	6297	6366	6436	6505	69
626	6574	6644	6713	6782	6852	6921	6990	7060	7129	7198	
627	7268	7337	7406	7475	7545	7614	7683	7752	7821	7890	
628	7960	8029	8098	8167	8236	8305	8374	8443	8513	8582	
629	8651	8720	8789	8858	8927	8996	9065	9134	9203	9272	
630	9341	9409	9478	9547	9616	9685	9754	9823	9892	9961	
631	800029	0098	0167	0236	0305	0373	0442	0511	0580	0648	
632	0717	0786	0854	0923	0992	1061	1129	1198	1266	1335	
633	1404	1472	1541	1609	1678	1747	1815	1884	1952	2021	
634	2089	2158	2226	2295	2363	2432	2500	2568	2637	2705	
635	2774	2842	2910	2979	3047	3116	3184	3252	3321	3389	68
636	3457	3525	3594	3662	3730	3798	3867	3935	4003	4071	
637	4139	4208	4276	4344	4412	4480	4548	4616	4685	4753	
638	4821	4889	4957	5025	5093	5161	5229	5297	5365	5433	
639	5501	5569	5637	5705	5773	5841	5908	5976	6044	6112	
640	6180	6248	6316	6384	6451	6519	6587	6655	6723	6790	
641	6858	6926	6994	7061	7129	7197	7264	7332	7400	7467	
642	7535	7603	7670	7738	7806	7873	7941	8008	8076	8143	
643	8211	8279	8346	8414	8481	8549	8616	8684	8751	8818	67
644	8886	8953	9021	9088	9156	9223	9290	9358	9425	9492	
645	9560	9627	9694	9762	9829	9896	9964	..31	..98	.165	
646	810233	0300	0367	0434	0501	0569	0636	0703	0770	0837	
647	0904	0971	1039	1106	1173	1240	1307	1374	1441	1508	
648	1575	1642	1709	1776	1843	1910	1977	2044	2111	2178	
649	2245	2312	2379	2445	2512	2579	2646	2713	2780	2847	
650	2913	2980	3047	3114	3181	3247	3314	3381	3448	3514	
651	3581	3648	3714	3781	3848	3914	3981	4048	4114	4181	
652	4248	4314	4381	4447	4514	4581	4647	4714	4780	4847	
653	4913	4980	5046	5113	5179	5246	5312	5378	5445	5511	66
654	5578	5644	5711	5777	5843	5910	5976	6042	6109	6175	
655	6241	6308	6374	6440	6506	6573	6639	6705	6771	6838	
656	6904	6970	7036	7102	7169	7235	7301	7367	7433	7499	
657	7565	7631	7698	7764	7830	7896	7962	8028	8094	8160	
658	8226	8292	8358	8424	8490	8556	8622	8688	8754	8820	
659	8885	8951	9017	9083	9149	9215	9281	9346	9412	9478	
660	9544	9610	9676	9741	9807	9873	9939	...4	..70	.136	
661	820201	0267	0333	0399	0464	0530	0595	0661	0727	0792	
662	0858	0924	0989	1055	1120	1186	1251	1317	1382	1448	
663	1514	1579	1645	1710	1775	1841	1906	1972	2037	2103	65
664	2168	2233	2299	2364	2430	2495	2560	2626	2691	2756	

N.	0	1	2	3	4	5	6	7	8	9	D.
Differences.	71 (Propor. Parts.)	7	14	21	28	36	43	50	57	64	
	70	7	14	21	28	35	42	49	56	63	
	69	7	14	21	28	35	41	48	55	62	
	68	7	14	20	27	34	41	48	54	61	
	67	7	13	20	27	34	40	47	54	60	
	66	7	13	20	26	33	40	46	53	59	
	65	7	13	20	26	33	39	46	52	59	

N.	0	1	2	3	4	5	6	7	8	9	D.
665	822822	2887	2952	3018	3083	3148	3213	3279	3344	3409	65
666	3474	3539	3605	3670	3735	3800	3865	3930	3996	4061	
667	4126	4191	4256	4321	4386	4451	4516	4581	4646	4711	
668	4776	4841	4906	4971	5036	5101	5166	5231	5296	5361	
669	5426	5491	5556	5621	5686	5751	5815	5880	5945	6010	
670	6075	6140	6204	6269	6334	6399	6464	6528	6593	6658	
671	6723	6787	6852	6917	6981	7046	7111	7175	7240	7305	
672	7369	7434	7499	7563	7628	7692	7757	7821	7886	7951	
673	8015	8080	8144	8209	8273	8338	8402	8467	8531	8595	64
674	8660	8724	8789	8853	8918	8982	9046	9111	9175	9239	
675	9304	9368	9432	9497	9561	9625	9690	9754	9818	9882	
676	9947	..11	..75	.139	.204	.268	.332	.396	.460	.525	
677	830589	0653	0717	0781	0845	0909	0973	1037	1102	1166	
678	1230	1294	1358	1422	1486	1550	1614	1678	1742	1806	
679	1870	1934	1998	2062	2126	2189	2253	2317	2381	2445	
680	2509	2573	2637	2700	2764	2828	2892	2956	3020	3083	
681	3147	3211	3275	3338	3402	3466	3530	3593	3657	3721	
682	3784	3848	3912	3975	4039	4103	4166	4230	4294	4357	
683	4421	4484	4548	4611	4675	4739	4802	4866	4929	4993	
684	5056	5120	5183	5247	5310	5373	5437	5500	5564	5627	63
685	5691	5754	5817	5881	5944	6007	6071	6134	6197	6261	
686	6324	6387	6451	6514	6577	6641	6704	6767	6830	6894	
687	6957	7020	7083	7146	7210	7273	7336	7399	7462	7525	
688	7588	7652	7715	7778	7841	7904	7967	8030	8093	8156	
689	8219	8282	8345	8408	8471	8534	8597	8660	8723	8786	
690	8849	8912	8975	9038	9101	9164	9227	9289	9352	9415	
691	9478	9541	9604	9667	9729	9792	9855	9918	9981	..43	
692	840106	0169	0232	0294	0357	0420	0482	0545	0608	0671	
693	0733	0796	0859	0921	0984	1046	1109	1172	1234	1297	
694	1359	1422	1485	1547	1610	1672	1735	1797	1860	1922	
695	1985	2047	2110	2172	2235	2297	2360	2422	2484	2547	62
696	2609	2672	2734	2796	2859	2921	2983	3046	3108	3170	
697	3233	3295	3357	3420	3482	3544	3606	3669	3731	3793	
698	3855	3918	3980	4042	4104	4166	4229	4291	4353	4415	
699	4477	4539	4601	4664	4726	4788	4850	4912	4974	5036	
700	5098	5160	5222	5284	5346	5408	5470	5532	5594	5656	
701	5718	5780	5842	5904	5966	6028	6090	6151	6213	6275	
702	6337	6399	6461	6523	6585	6646	6708	6770	6832	6894	
703	6955	7017	7079	7141	7202	7264	7326	7388	7449	7511	
704	7573	7634	7696	7758	7819	7881	7943	8004	8066	8128	
705	8189	8251	8312	8374	8435	8497	8559	8620	8682	8743	
706	8805	8866	8928	8989	9051	9112	9174	9235	9297	9358	61
707	9419	9481	9542	9604	9665	9726	9788	9849	9911	9972	
708	850033	0095	0156	0217	0279	0340	0401	0462	0524	0585	
709	0646	0707	0769	0830	0891	0952	1014	1075	1136	1197	
710	1258	1320	1381	1442	1503	1564	1625	1686	1747	1809	
711	1870	1931	1992	2053	2114	2175	2236	2297	2358	2419	
712	2480	2541	2602	2663	2724	2785	2846	2907	2968	3029	
713	3090	3150	3211	3272	3333	3394	3455	3516	3577	3637	
714	3698	3759	3820	3881	3941	4002	4063	4124	4185	4245	
715	4306	4367	4428	4488	4549	4610	4670	4731	4792	4852	
716	4913	4974	5034	5095	5156	5216	5277	5337	5398	5459	
717	5519	5580	5640	5701	5761	5822	5882	5943	6003	6064	
718	6124	6185	6245	6306	6366	6427	6487	6548	6608	6668	60
719	6729	6789	6850	6910	6970	7031	7091	7152	7212	7272	
N.	0	1	2	3	4	5	6	7	8	9	D.
Differ.	64 Pro. Parts.	6	13	19	26	32	38	45	51	58	
	63	6	13	19	25	32	38	44	50	57	
	62	6	12	19	25	31	37	43	50	56	
	61	6	12	18	24	31	37	43	49	55	
	60	6	12	18	24	30	36	42	48	54	

N.	0	1	2	3	4	5	6	7	8	9	D.
720	857332	7393	7453	7513	7574	7634	7694	7755	7815	7875	60
721	7935	7995	8056	8116	8176	8236	8297	8357	8417	8477	
722	8537	8597	8657	8718	8778	8838	8898	8958	9018	9078	
723	9138	9198	9258	9318	9379	9439	9499	9559	9619	9679	
724	9739	9799	9859	9918	9978	..38	..98	.158	.218	.278	
725	860338	0398	0458	0518	0578	0637	0697	0757	0817	0877	
726	0937	0996	1056	1116	1176	1236	1295	1355	1415	1475	
727	1534	1594	1654	1714	1773	1833	1893	1952	2012	2072	
728	2131	2191	2251	2310	2370	2430	2489	2549	2608	2668	
729	2728	2787	2847	2906	2966	3025	3085	3144	3204	3263	
730	3323	3382	3442	3501	3561	3620	3680	3739	3799	3858	59
731	3917	3977	4036	4096	4155	4214	4274	4333	4392	4452	
732	4511	4570	4630	4689	4748	4808	4867	4926	4985	5045	
733	5104	5163	5222	5282	5341	5400	5459	5519	5578	5637	
734	5696	5755	5814	5874	5933	5992	6051	6110	6169	6228	
735	6287	6346	6405	6465	6524	6583	6642	6701	6760	6819	
736	6878	6937	6996	7055	7114	7173	7232	7291	7350	7409	
737	7467	7526	7585	7644	7703	7762	7821	7880	7939	7998	
738	8056	8115	8174	8233	8292	8350	8409	8468	8527	8586	
739	8644	8703	8762	8821	8879	8938	8997	9056	9114	9173	
740	9232	9290	9349	9408	9466	9525	9584	9642	9701	9760	
741	9818	9877	9935	9994	..53	.111	.170	.228	.287	.345	
742	870404	0462	0521	0579	0638	0696	0755	0813	0872	0930	58
743	0989	1047	1106	1164	1223	1281	1339	1398	1456	1515	
744	1573	1631	1690	1748	1806	1865	1923	1981	2040	2098	
745	2156	2215	2273	2331	2389	2448	2506	2564	2622	2681	
746	2739	2797	2855	2913	2972	3030	3088	3146	3204	3262	
747	3321	3379	3437	3495	3553	3611	3669	3727	3785	3844	
748	3902	3960	4018	4076	4134	4192	4250	4308	4366	4424	
749	4482	4540	4598	4656	4714	4772	4830	4888	4945	5003	
750	5061	5119	5177	5235	5293	5351	5409	5466	5524	5582	
751	5640	5698	5756	5813	5871	5929	5987	6045	6102	6160	
752	6218	6276	6333	6391	6449	6507	6564	6622	6680	6737	
753	6795	6853	6910	6968	7026	7083	7141	7199	7256	7314	
754	7371	7429	7487	7544	7602	7659	7717	7774	7832	7889	
755	7947	8004	8062	8119	8177	8234	8292	8349	8407	8464	57
756	8522	8579	8637	8694	8752	8809	8866	8924	8981	9039	
757	9096	9153	9211	9268	9325	9383	9440	9497	9555	9612	
758	9669	9726	9784	9841	9898	9956	..13	..70	.127	.185	
759	880242	0299	0356	0413	0471	0528	0585	0642	0699	0756	
760	0814	0871	0928	0985	1042	1099	1156	1213	1271	1328	
761	1385	1442	1499	1556	1613	1670	1727	1784	1841	1898	
762	1955	2012	2069	2126	2183	2240	2297	2354	2411	2468	
763	2525	2581	2638	2695	2752	2809	2866	2923	2980	3037	
764	3093	3150	3207	3264	3321	3377	3434	3491	3548	3605	
765	3661	3718	3775	3832	3888	3945	4002	4059	4115	4172	
766	4229	4285	4342	4399	4455	4512	4569	4625	4682	4739	
767	4795	4852	4909	4965	5022	5078	5135	5192	5248	5305	
768	5361	5418	5474	5531	5587	5644	5700	5757	5813	5870	
769	5926	5983	6039	6096	6152	6209	6265	6321	6378	6434	56
770	6491	6547	6604	6660	6716	6773	6829	6885	6942	6998	
771	7054	7111	7167	7223	7280	7336	7392	7449	7505	7561	
772	7617	7674	7730	7786	7842	7898	7955	8011	8067	8123	
773	8179	8236	8292	8348	8404	8460	8516	8573	8629	8685	
774	8741	8797	8853	8909	8965	9021	9077	9134	9190	9246	
N.	0	1	2	3	4	5	6	7	8	9	D.

Differ.	Pro. Parts.	1	2	3	4	5	6	7	8	9
60		6	12	18	24	30	36	42	48	54
59		6	12	18	24	30	35	41	47	53
58		6	12	17	23	29	35	41	46	52
57		6	11	17	23	29	34	40	46	51
56		6	11	17	22	28	34	39	45	50

N.	0	1	2	3	4	5	6	7	8	9	D.
775	889302	9358	9414	9470	9526	9582	9638	9694	9750	9806	56
776	9862	9918	9974	..30	..86	.141	.197	.253	.309	.365	
777	890421	0477	0533	0589	0645	0700	0756	0812	0868	0924	
778	0980	1035	1091	1147	1203	1259	1314	1370	1426	1482	
779	1537	1593	1649	1705	1760	1816	1872	1928	1983	2039	
780	2095	2150	2206	2262	2317	2373	2429	2484	2540	2595	
781	2651	2707	2762	2818	2873	2929	2985	3040	3096	3151	
782	3207	3262	3318	3373	3429	3484	3540	3595	3651	3706	
783	3762	3817	3873	3928	3984	4039	4094	4150	4205	4261	55
784	4316	4371	4427	4482	4538	4593	4648	4704	4759	4814	
785	4870	4925	4980	5036	5091	5146	5201	5257	5312	5367	
786	5423	5478	5533	5588	5644	5699	5754	5809	5864	5920	
787	5975	6030	6085	6140	6195	6251	6306	6361	6416	6471	
788	6526	6581	6636	6692	6747	6802	6857	6912	6967	7022	
789	7077	7132	7187	7242	7297	7352	7407	7462	7517	7572	
790	7627	7682	7737	7792	7847	7902	7957	8012	8067	8122	
791	8176	8231	8286	8341	8396	8451	8506	8561	8615	8670	
792	8725	8780	8835	8890	8944	8999	9054	9109	9164	9218	
793	9273	9328	9383	9437	9492	9547	9602	9656	9711	9766	
794	9821	9875	9930	9985	..39	..94	.149	.203	.258	.312	
795	900367	0422	0476	0531	0586	0640	0695	0749	0804	0859	
796	0913	0968	1022	1077	1131	1186	1240	1295	1349	1404	
797	1458	1513	1567	1622	1676	1731	1785	1840	1894	1948	54
798	2003	2057	2112	2166	2221	2275	2329	2384	2438	2492	
799	2547	2601	2655	2710	2764	2818	2873	2927	2981	3036	
800	3090	3144	3199	3253	3307	3361	3416	3470	3524	3578	
801	3633	3687	3741	3795	3849	3904	3958	4012	4066	4120	
802	4174	4229	4283	4337	4391	4445	4499	4553	4607	4661	
803	4716	4770	4824	4878	4932	4986	5040	5094	5148	5202	
804	5256	5310	5364	5418	5472	5526	5580	5634	5688	5742	
805	5796	5850	5904	5958	6012	6066	6119	6173	6227	6281	
806	6335	6389	6443	6497	6551	6604	6658	6712	6766	6820	
807	6874	6927	6981	7035	7089	7143	7196	7250	7304	7358	
808	7411	7465	7519	7573	7626	7680	7734	7787	7841	7895	
809	7949	8002	8056	8110	8163	8217	8270	8324	8378	8431	
810	8485	8539	8592	8646	8699	8753	8807	8860	8914	8967	
811	9021	9074	9128	9181	9235	9289	9342	9396	9449	9503	
812	9556	9610	9663	9716	9770	9823	9877	9930	9984	..37	53
813	910091	0144	0197	0251	0304	0358	0411	0464	0518	0571	
814	0624	0678	0731	0784	0838	0891	0944	0998	1051	1104	
815	1158	1211	1264	1317	1371	1424	1477	1530	1584	1637	
816	1690	1743	1797	1850	1903	1956	2009	2063	2116	2169	
817	2222	2275	2328	2381	2435	2488	2541	2594	2647	2700	
818	2753	2806	2859	2913	2966	3019	3072	3125	3178	3231	
819	3284	3337	3390	3443	3496	3549	3602	3655	3708	3761	
820	3814	3867	3920	3973	4026	4079	4132	4184	4237	4290	
821	4343	4396	4449	4502	4555	4608	4660	4713	4766	4819	
822	4872	4925	4977	5030	5083	5136	5189	5241	5294	5347	
823	5400	5453	5505	5558	5611	5664	5716	5769	5822	5875	
824	5927	5980	6033	6085	6138	6191	6243	6296	6349	6401	
825	6454	6507	6559	6612	6664	6717	6770	6822	6875	6927	
826	6980	7033	7085	7138	7190	7243	7295	7348	7400	7453	
827	7506	7558	7611	7663	7716	7768	7820	7873	7925	7978	52
828	8030	8083	8135	8188	8240	8293	8345	8397	8450	8502	
829	8555	8607	8659	8712	8764	8816	8869	8921	8973	9026	
830	9078	9130	9183	9235	9287	9340	9392	9444	9496	9549	
N.	0	1	2	3	4	5	6	7	8	9	D.
Differ.	55 P. Parts	6	11	17	22	28	33	39	44	50	
	54	5	11	16	22	27	32	38	43	49	
	53	5	11	16	21	27	32	37	42	48	
	52	5	10	16	21	26	31	36	42	47	

N.	0	1	2	3	4	5	6	7	8	9	D.
831	919601	9653	9706	9758	9810	9862	9914	9967	..19	..71	52
832	920123	0176	0228	0280	0332	0384	0436	0489	0541	0593	
833	0645	0697	0749	0801	0853	0906	0958	1010	1062	1114	
834	1166	1218	1270	1322	1374	1426	1478	1530	1582	1634	
835	1686	1738	1790	1842	1894	1946	1998	2050	2102	2154	
836	2206	2258	2310	2362	2414	2466	2518	2570	2622	2674	
837	2725	2777	2829	2881	2933	2985	3037	3089	3140	3192	
838	3244	3296	3348	3399	3451	3503	3555	3607	3658	3710	
839	3762	3814	3865	3917	3969	4021	4072	4124	4176	4228	
840	4279	4331	4383	4434	4486	4538	4589	4641	4693	4744	
841	4796	4848	4899	4951	5003	5054	5106	5157	5209	5261	
842	5312	5364	5415	5467	5518	5570	5621	5673	5725	5776	
843	5828	5879	5931	5982	6034	6085	6137	6188	6240	6291	51
844	6342	6394	6445	6497	6548	6600	6651	6702	6754	6805	
845	6857	6908	6959	7011	7062	7114	7165	7216	7268	7319	
846	7370	7422	7473	7524	7576	7627	7678	7730	7781	7832	
847	7883	7935	7986	8037	8088	8140	8191	8242	8293	8345	
848	8396	8447	8498	8549	8601	8652	8703	8754	8805	8857	
849	8908	8959	9010	9061	9112	9163	9215	9266	9317	9368	
850	9419	9470	9521	9572	9623	9674	9725	9776	9827	9879	
851	9930	9981	..32	..83	.134	.185	.236	.287	.338	.389	
852	930440	0491	0542	0592	0643	0694	0745	0796	0847	0898	
853	0949	1000	1051	1102	1153	1204	1254	1305	1356	1407	
854	1458	1509	1560	1610	1661	1712	1763	1814	1865	1915	
855	1966	2017	2068	2118	2169	2220	2271	2322	2372	2423	
856	2474	2524	2575	2626	2677	2727	2778	2829	2879	2930	
857	2981	3031	3082	3133	3183	3234	3285	3335	3386	3437	
858	3487	3538	3589	3639	3690	3740	3791	3841	3892	3943	
859	3993	4044	4094	4145	4195	4246	4296	4347	4397	4448	
860	4498	4549	4599	4650	4700	4751	4801	4852	4902	4953	50
861	5003	5054	5104	5154	5205	5255	5306	5356	5406	5457	
862	5507	5558	5608	5658	5709	5759	5809	5860	5910	5960	
863	6011	6061	6111	6162	6212	6262	6313	6363	6413	6463	
864	6514	6564	6614	6665	6715	6765	6815	6865	6916	6966	
865	7016	7066	7117	7167	7217	7267	7317	7367	7418	7468	
866	7518	7568	7618	7668	7718	7769	7819	7869	7919	7969	
867	8019	8069	8119	8169	8219	8269	8320	8370	8420	8470	
868	8520	8570	8620	8670	8720	8770	8820	8870	8920	8970	
869	9020	9070	9120	9170	9220	9270	9320	9369	9419	9469	
870	9519	9569	9619	9669	9719	9769	9819	9869	9918	9968	
871	940018	0068	0118	0168	0218	0267	0317	0367	0417	0467	
872	0516	0566	0616	0666	0716	0765	0815	0865	0915	0964	
873	1014	1064	1114	1163	1213	1263	1313	1362	1412	1462	
874	1511	1561	1611	1660	1710	1760	1809	1859	1909	1958	
875	2008	2058	2107	2157	2207	2256	2306	2355	2405	2455	
876	2504	2554	2603	2653	2702	2752	2801	2851	2901	2950	
877	3000	3049	3099	3148	3198	3247	3297	3346	3396	3445	49
878	3495	3544	3593	3643	3692	3742	3791	3841	3890	3939	
879	3989	4038	4088	4137	4186	4236	4285	4335	4384	4433	
880	4483	4532	4581	4631	4680	4729	4779	4828	4877	4927	
881	4976	5025	5074	5124	5173	5222	5272	5321	5370	5419	
882	5469	5518	5567	5616	5665	5715	5764	5813	5862	5912	
883	5961	6010	6059	6108	6157	6207	6256	6305	6354	6403	
884	6452	6501	6551	6600	6649	6698	6747	6796	6845	6894	
885	6943	6992	7041	7090	7140	7189	7238	7287	7336	7385	
886	7434	7483	7532	7581	7630	7679	7728	7777	7826	7875	
N.	**0**	**1**	**2**	**3**	**4**	**5**	**6**	**7**	**8**	**9**	**D.**
Differ.	52 P. Parts.	5	10	16	21	26	31	36	42	47	
	51	5	10	15	20	26	31	36	41	46	
	50	5	10	15	20	25	30	35	40	45	
	49	5	10	15	20	25	29	34	39	44	

N.	0	1	2	3	4	5	6	7	8	9	D.
887	947924	7973	8022	8070	8119	8168	8217	8266	8315	8364	49
888	8413	8462	8511	8560	8609	8657	8706	8755	8804	8853	
889	8902	8951	8999	9048	9097	9146	9195	9244	9292	9341	
890	9390	9439	9488	9536	9585	9634	9683	9731	9780	9829	
891	9878	9926	9975	..24	..73	.121	.170	.219	.267	.316	
892	950365	0414	0462	0511	0560	0608	0657	0706	0754	0803	
893	0851	0900	0949	0997	1046	1095	1143	1192	1240	1289	
894	1338	1386	1435	1483	1532	1580	1629	1677	1726	1775	
895	1823	1872	1920	1969	2017	2066	2114	2163	2211	2260	48
896	2308	2356	2405	2453	2502	2550	2599	2647	2696	2744	
897	2792	2841	2889	2938	2986	3034	3083	3131	3180	3228	
898	3276	3325	3373	3421	3470	3518	3566	3615	3663	3711	
899	3760	3808	3856	3905	3953	4001	4049	4098	4146	4194	
900	4243	4291	4339	4387	4435	4484	4532	4580	4628	4677	
901	4725	4773	4821	4869	4918	4966	5014	5062	5110	5158	
902	5207	5255	5303	5351	5399	5447	5495	5543	5592	5640	
903	5688	5736	5784	5832	5880	5928	5976	6024	6072	6120	
904	6168	6216	6265	6313	6361	6409	6457	6505	6553	6601	
905	6649	6697	6745	6793	6840	6888	6936	6984	7032	7080	
906	7128	7176	7224	7272	7320	7368	7416	7464	7512	7559	
907	7607	7655	7703	7751	7799	7847	7894	7942	7990	8038	
908	8086	8134	8181	8229	8277	8325	8373	8421	8468	8516	
909	8564	8612	8659	8707	8755	8803	8850	8898	8946	8994	
910	9041	9089	9137	9185	9232	9280	9328	9375	9423	9471	
911	9518	9566	9614	9661	9709	9757	9804	9852	9900	9947	
912	9995	..42	..90	.138	.185	.233	.280	.328	.376	.423	
913	960471	0518	0566	0613	0661	0709	0756	0804	0851	0899	
914	0946	0994	1041	1089	1136	1184	1231	1279	1326	1374	47
915	1421	1469	1516	1563	1611	1658	1706	1753	1801	1848	
916	1895	1943	1990	2038	2085	2132	2180	2227	2275	2322	
917	2369	2417	2464	2511	2559	2606	2653	2701	2748	2795	
918	2843	2890	2937	2985	3032	3079	3126	3174	3221	3268	
919	3316	3363	3410	3457	3504	3552	3599	3646	3693	3741	
920	3788	3835	3882	3929	3977	4024	4071	4118	4165	4212	
921	4260	4307	4354	4401	4448	4495	4542	4590	4637	4684	
922	4731	4778	4825	4872	4919	4966	5013	5061	5108	5155	
923	5202	5249	5296	5343	5390	5437	5484	5531	5578	5625	
924	5672	5719	5766	5813	5860	5907	5954	6001	6048	6095	
925	6142	6189	6236	6283	6329	6376	6423	6470	6517	6564	
926	6611	6658	6705	6752	6799	6845	6892	6939	6986	7033	
927	7080	7127	7173	7220	7267	7314	7361	7408	7454	7501	
928	7548	7595	7642	7688	7735	7782	7829	7875	7922	7969	
929	8016	8062	8109	8156	8203	8249	8296	8343	8390	8436	
930	8483	8530	8576	8623	8670	8716	8763	8810	8856	8903	
931	8950	8996	9043	9090	9136	9183	9229	9276	9323	9369	
932	9416	9463	9509	9556	9602	9649	9695	9742	9789	9835	
933	9882	9928	9975	..21	..68	.114	.161	.207	.254	.300	
934	970347	0393	0440	0486	0533	0579	0626	0672	0719	0765	
935	0812	0858	0904	0951	0997	1044	1090	1137	1183	1229	
936	1276	1322	1369	1415	1461	1508	1554	1601	1647	1693	
937	1740	1786	1832	1879	1925	1971	2018	2064	2110	2157	
938	2203	2249	2295	2342	2388	2434	2481	2527	2573	2619	
939	2666	2712	2758	2804	2851	2897	2943	2989	3035	3082	
940	3128	3174	3220	3266	3313	3359	3405	3451	3497	3543	
941	3590	3636	3682	3728	3774	3820	3866	3913	3959	4005	
942	4051	4097	4143	4189	4235	4281	4327	4374	4420	4466	
943	4512	4558	4604	4650	4696	4742	4788	4834	4880	4926	

N.	0	1	2	3	4	5	6	7	8	9
Differ.	48	P. Pts. 5	10	14	19	24	29	34	38	43
	47	5	9	14	19	24	28	33	38	42
	46	5	9	14	18	23	28	32	37	[illegible]

N.	0	1	2	3	4	5	6	7	8	9	D
944	974972	5018	5064	5110	5156	5202	5248	5294	5340	5386	46
945	5432	5478	5524	5570	5616	5662	5707	5753	5799	5845	
946	5891	5937	5983	6029	6075	6121	6167	6212	6258	6304	
947	6350	6396	6442	6488	6533	6579	6625	6671	6717	6763	
948	6808	6854	6900	6946	6992	7037	7083	7129	7175	7220	
949	7266	7312	7358	7403	7449	7495	7541	7586	7632	7678	
950	7724	7769	7815	7861	7906	7952	7998	8043	8089	8135	
951	8181	8226	8272	8317	8363	8409	8454	8500	8546	8591	
952	8637	8683	8728	8774	8819	8865	8911	8956	9002	9047	
953	9093	9138	9184	9230	9275	9321	9366	9412	9457	9503	
954	9548	9594	9639	9685	9730	9776	9821	9867	9912	9958	
955	980003	0049	0094	0140	0185	0231	0276	0322	0367	0412	45
956	0458	0503	0549	0594	0640	0685	0730	0776	0821	0867	
957	0912	0957	1003	1048	1093	1139	1184	1229	1275	1320	
958	1366	1411	1456	1501	1547	1592	1637	1683	1728	1773	
959	1819	1864	1909	1954	2000	2045	2090	2135	2181	2226	
960	2271	2316	2362	2407	2452	2497	2543	2588	2633	2678	
961	2723	2769	2814	2859	2904	2949	2994	3040	3085	3130	
962	3175	3220	3265	3310	3356	3401	3446	3491	3536	3581	
963	3626	3671	3716	3762	3807	3852	3897	3942	3987	4032	
964	4077	4122	4167	4212	4257	4302	4347	4392	4437	4482	
965	4527	4572	4617	4662	4707	4752	4797	4842	4887	4932	
966	4977	5022	5067	5112	5157	5202	5247	5292	5337	5382	
967	5426	5471	5516	5561	5606	5651	5696	5741	5786	5830	
968	5875	5920	5965	6010	6055	6100	6144	6189	6234	6279	
969	6324	6369	6413	6458	6503	6548	6593	6637	6682	6727	
970	6772	6817	6861	6906	6951	6996	7040	7085	7130	7175	
971	7219	7264	7309	7353	7398	7443	7488	7532	7577	7622	
972	7666	7711	7756	7800	7845	7890	7934	7979	8024	8068	
973	8113	8157	8202	8247	8291	8336	8381	8425	8470	8514	
974	8559	8604	8648	8693	8737	8782	8826	8871	8916	8960	
975	9005	9049	9094	9138	9183	9227	9272	9316	9361	9405	
976	9450	9494	9539	9583	9628	9672	9717	9761	9806	9850	44
977	9895	9939	9983	..28	..72	.117	.161	.206	.250	.294	
978	990339	0383	0428	0472	0516	0561	0605	0650	0694	0738	
979	0783	0827	0871	0916	0960	1004	1049	1093	1137	1182	
980	1226	1270	1315	1359	1403	1448	1492	1536	1580	1625	
981	1669	1713	1758	1802	1846	1890	1935	1979	2023	2067	
982	2111	2156	2200	2244	2288	2333	2377	2421	2465	2509	
983	2554	2598	2642	2686	2730	2774	2819	2863	2907	2951	
984	2995	3039	3083	3127	3172	3216	3260	3304	3348	3392	
985	3436	3480	3524	3568	3613	3657	3701	3745	3789	3833	
986	3877	3921	3965	4009	4053	4097	4141	4185	4229	4273	
987	4317	4361	4405	4449	4493	4537	4581	4625	4669	4713	
988	4757	4801	4845	4889	4933	4977	5021	5065	5108	5152	
989	5196	5240	5284	5328	5372	5416	5460	5504	5547	5591	
990	5635	5679	5723	5767	5811	5854	5898	5942	5986	6030	
991	6074	6117	6161	6205	6249	6293	6337	6380	6424	6468	
992	6512	6555	6599	6643	6687	6731	6774	6818	6862	6906	
993	6949	6993	7037	7080	7124	7168	7212	7255	7299	7343	
994	7386	7430	7474	7517	7561	7605	7648	7692	7736	7779	
995	7823	7867	7910	7954	7998	8041	8085	8129	8172	8216	
996	8259	8303	8347	8390	8434	8477	8521	8564	8608	8652	
997	8695	8739	8782	8826	8869	8913	8956	9000	9043	9087	
998	9131	9174	9218	9261	9305	9348	9392	9435	9479	9522	
999	9565	9609	9652	9696	9739	9783	9826	9870	9913	9957	43
N.	0	1	2	3	4	5	6	7	8	9	D.
Differ.	46 P. Parts.	5	9	14	18	23	28	32	37	41	
	45	5	9	14	18	23	27	32	36	41	
	44	4	9	13	18	22	26	31	35	40	
	43	4	9	13	17	22	26	30	34	39	

TABLE

OF

LOGARITHMIC SINES AND TANGENTS

FOR EVERY

TEN SECONDS OF THE QUADRANT.

Min.	Sine of 0 Degree. 0″	10″	20″	30″	40″	50″		P. Part to 1″.
0	Inf. Neg.	5.685575	5.986605	6.162696	6.287635	6.384545	59	
1	6.463726	6.530673	6.588665	639817	685575	726968	58	
2	764756	799518	831703	861666	889695	916024	57	
3	940847	964328	986605	7.007794	7.027997	7.047303	56	
4	7.065786	7.083515	7.100548	116939	132733	147973	55	
5	162696	176936	190725	204089	217054	229643	54	
6	241877	253776	265358	276639	287635	298358	53	
7	308824	319043	329027	338787	348332	357672	52	
8	366816	375771	384544	393145	401578	409850	51	
9	417968	425937	433762	441449	449002	456426	50	
10	463726	470904	477966	484915	491754	498488	49	689.4
11	505118	511649	518083	524423	530672	536832	48	629.4
12	542906	548897	554806	560635	566387	572065	47	579.1
13	577668	583201	588664	594059	599388	604652	46	536.2
14	609853	614993	620072	625093	630056	634964	45	499.2
15	639816	644615	649361	654056	658701	663297	44	467.0
16	667845	672345	676799	681208	685573	689895	43	438.7
17	694173	698410	702606	706762	710879	714957	42	413.6
18	718997	722999	726965	730896	734791	738651	41	391.3
19	742478	746270	750031	753758	757455	761119	40	371.2
20	764754	768358	771932	775477	778994	782482	39	353.1
21	785943	789376	792782	796162	799515	802843	38	336.7
22	806146	809423	812677	815906	819111	822292	37	321.7
23	825451	828586	831700	834791	837860	840907	36	308.0
24	843934	846939	849924	852889	855833	858757	35	295.4
25	861662	864548	867415	870262	873092	875902	34	283.8
26	878695	881470	884228	886968	889690	892396	33	273.1
27	895085	897758	900414	903054	905678	908287	32	263.2
28	910879	913457	916019	918566	921098	923616	31	254.0
29	926119	928608	931082	933543	935989	938422	30	245.4
30	940842	943248	945641	948020	950387	952741	29	237.3
31	955082	957411	959727	962031	964322	966602	28	229.8
32	968870	971126	973370	975603	977824	980034	27	222.7
33	982233	984421	986598	988764	990919	993064	26	2[illegible]6.1
34	995198	997322	999435	8.001538	8.003631	8.005714	25	209.8
35	8.007787	8.009850	8.011903	013947	015981	018005	24	203.9
36	020021	022027	024023	026011	027989	029959	23	198.3
37	031919	033871	035814	037749	039675	041592	22	193.0
38	043501	045401	047294	049178	051054	052922	21	188.0
39	054781	056633	058477	060314	062142	063963	20	183.2
40	065776	067582	069380	071171	072955	074731	19	178.7
41	076500	078261	080016	081764	083504	085238	18	174.4
42	086965	088684	090398	092104	093804	095497	17	170.3
43	097183	098863	100537	102204	103864	105519	16	166.4
44	107167	108809	110444	112074	113697	115315	15	162.6
45	116926	118532	120131	121725	123313	124895	14	159.1
46	126471	128042	129607	131166	132720	134268	13	155.6
47	135810	137348	138879	140406	141927	143443	12	152.4
48	144953	146458	147959	149453	150943	152428	11	149.2
49	153907	155382	156852	158316	159776	161231	10	146.2
50	162681	164126	165566	167002	168433	169859	9	143.3
51	171280	172697	174109	175517	176920	178319	8	140.5
52	179713	181103	182488	183869	185245	186617	7	137.9
53	187985	189348	190707	192062	193413	194760	6	135.3
54	196102	197440	198774	200104	201430	202752	5	132.8
55	204070	205384	206694	208000	209302	210601	4	130.4
56	211895	213185	214472	215755	217034	218309	3	128.1
57	219581	220849	222113	223374	224631	225884	2	125.9
58	227134	228380	229622	230861	232096	233328	1	123.7
[illegible]	[illegible]34557	235782	237003	238221	2[illegible]9436	240647	0	121.6

Min.	Tangent of 0 Degree.							P. Part to 1″.
	0″	10″	20″	30″	40″	50″		
0	Inf. Neg.	5.685575	5.986605	6.162696	6.287635	6.384545	59	
1	6.463726	6.530673	6.588665	639817	685575	726968	58	
2	764756	799518	831703	861666	889695	916024	57	
3	940847	964329	986605	7.007794	7.027998	7.047303	56	
4	7.065786	7.083515	7.100548	116939	132733	147973	55	
5	162696	176937	190725	204089	217054	229643	54	
6	241878	253777	265359	276640	287635	298359	53	
7	308825	319044	329028	338788	348333	357673	52	
8	366817	375772	384546	393146	401579	409852	51	
9	417970	425939	433764	441451	449004	456428	50	
10	463727	470906	477968	484917	491756	498490	49	689.4
11	505120	511651	518085	524426	530675	536835	48	629.4
12	542909	548900	554808	560638	566390	572068	47	579.1
13	577672	583204	588667	594062	599391	604655	46	536.2
14	609857	614996	620076	625097	630060	634968	45	499.2
15	639820	644619	649366	654061	658706	663302	44	467.0
16	667849	672350	676804	681213	685578	689900	43	438.7
17	694179	698416	702612	706768	710885	714963	42	413.6
18	719003	723005	726972	730902	734797	738658	41	391.3
19	742484	746277	750037	753765	757462	761127	40	371.2
20	764761	768365	771940	775485	779002	782490	39	353.1
21	785951	789384	792790	796170	799524	802852	38	336.7
22	806155	809433	812686	815915	819120	822302	37	321.7
23	825460	828596	831710	834801	837870	840918	36	308.0
24	843944	846950	849935	852900	855844	858769	35	295.4
25	861674	864560	867426	870274	873104	875915	34	283.9
26	878708	881483	884240	886981	889704	892410	33	273.2
27	895099	897772	900428	903068	905692	908301	32	263.2
28	910894	913471	916034	918581	921113	923631	31	254.0
29	926134	928623	931098	933559	936006	938439	30	245.4
30	940858	943265	945658	948037	950404	952758	29	237.3
31	955100	957428	959745	962049	964341	966621	28	229.8
32	968889	971145	973389	975622	977844	980054	27	222.7
33	982253	984441	986618	988785	990940	993085	26	216.2
34	995219	997343	999457	8.001560	8.003653	8.005736	25	209.8
35	8 007809	8.009872	8.011926	013970	016004	018029	24	203.9
36	020044	022051	024048	026035	028014	029984	23	198.3
37	031945	033897	035840	037775	039701	041618	22	193.0
38	043527	045428	047321	049205	051081	052949	21	188.0
39	054809	056662	058506	060342	062171	063992	20	183.3
40	065806	067612	069410	071201	072985	074761	19	178.7
41	076531	078293	080047	081795	083536	085270	18	174.4
42	086997	088717	090431	092137	093837	095530	17	170.3
43	097217	098897	100571	102239	103900	105554	16	166.4
44	107203	108845	110481	112110	113734	115352	15	162.7
45	116963	118569	120169	121763	123351	124933	14	159.1
46	126510	128081	129646	131206	132760	134308	13	155.7
47	135851	137389	138921	140447	141969	143485	12	152.4
48	144996	146501	148001	149497	150987	152472	11	149.3
49	153952	155426	156896	158361	159821	161276	10	146.2
50	162727	164172	165613	167049	168480	169906	9	143.4
51	171328	172745	174158	175566	176969	178368	8	140.6
52	179763	181153	182538	183919	185296	186668	7	137.[illegible]
53	188036	189400	190760	192115	193466	194813	6	135.3
54	196156	197494	198829	200159	201485	202808	5	132.8
55	204126	205440	206750	208057	209359	210658	4	130.[illegible]
56	211953	213243	214530	215814	217093	218369	3	128[illegible]
57	219641	220909	222174	223434	224692	225945	2	12[illegible]
58	227195	228442	229685	230924	232160	233392	1	12[illegible]
59	234621	235846	237068	238286	339502	2407[illegible]		[illegible]

Min.	Sine of 1 Degree. 0″	10″	20″	30″	40″	50″		P. Part to 1″.
0	8.241855	8.243060	8.244261	8.245459	8.246654	8.247845	59	119.6
1	249033	250218	251400	252578	253753	254925	58	117.7
2	256094	257260	258423	259582	260739	261892	57	115.8
3	263042	264190	265334	266475	267613	268749	56	114.0
4	269881	271010	272137	273260	274381	275499	55	112.2
5	276614	277726	278835	279941	281045	282145	54	110.5
6	283243	284339	285431	286521	287608	288692	53	108.8
7	289773	290852	291928	293002	294073	295141	52	107.2
8	296207	297270	298330	299388	300443	301496	51	105.7
9	302546	303594	304639	305681	306721	307759	50	104.1
10	8.308794	8.309827	8.310857	8.311885	8.312910	8.313933	49	102.6
11	314954	315972	316987	318001	319012	320021	48	101.2
12	321027	322031	323033	324032	325029	326024	47	99.8
13	327016	328007	328995	329980	330964	331945	46	98.5
14	332924	333901	334876	335848	336819	337787	45	97.1
15	338753	339717	340679	341638	342596	343551	44	95.8
16	344504	345456	346405	347352	348297	349240	43	94.6
17	350181	351119	352056	352991	353924	354855	42	93.4
18	355783	356710	357635	358558	359479	360398	41	92.2
19	361315	362230	363143	364055	364964	365871	40	91.0
20	8.366777	8.367681	8.368582	8.369482	8.370380	8.371277	39	89.9
21	372171	373063	373954	374843	375730	376615	38	88.8
22	377499	378380	379260	380138	381015	381889	37	87.7
23	382762	383633	384502	385370	386236	387100	36	86.7
24	387962	388823	389682	390539	391395	392249	35	85.6
25	393101	393951	394800	395647	396493	397337	34	84.6
26	398179	399020	399859	400696	401532	402366	33	83.7
27	403199	404030	404859	405687	406514	407338	32	82.7
28	408161	408983	409803	410621	411438	412254	31	81.8
29	413068	413880	414691	415500	416308	417114	30	80.8
30	8.417919	8.418722	8.419524	8.420325	8.421123	8.421921	29	80.0
31	422717	423511	424304	425096	425886	426675	28	79.1
32	427462	428248	429032	429815	430597	431377	27	78.2
33	432156	432934	433710	434484	435257	436029	26	77.4
34	436800	437569	438337	439103	439868	440632	25	76.6
35	441394	442156	442915	443674	444431	445186	24	75.8
36	445941	446694	447446	448196	448946	449694	23	75.0
37	450440	451186	451930	452673	453414	454154	22	74.2
38	454893	455631	456368	457103	457837	458570	21	73.5
39	459301	460032	460761	461489	462215	462941	20	72.7
40	8.463665	8.464388	8.465110	8.465830	8.466550	8.467268	19	72.0
41	467985	468701	469416	470129	470841	471553	18	71.3
42	472263	472971	473679	474386	475091	475795	17	70.6
43	476498	477200	477901	478601	479299	479997	16	69.9
44	480693	481388	482083	482776	483467	484158	15	69.2
45	484848	485536	486224	486910	487596	488280	14	68.6
46	488963	489645	490326	491006	491685	492363	13	67.9
47	493040	493715	494390	495064	495736	496408	12	67.3
48	497078	497748	498416	499084	499750	500416	11	66.7
49	501080	501743	502405	503067	503727	504386	10	66.1
50	8.505045	8.505702	8.506358	8.507014	8.507668	8.508321	9	65.5
51	508974	509625	510275	510925	511573	512221	8	64.9
52	512867	513513	514157	514801	515444	516086	7	64.3
53	516726	517366	518005	518643	519280	519916	6	63.7
54	520551	521186	521819	522451	523083	523713	5	63.2
55	524343	524972	525599	526226	526852	527477	4	62.6
56	528102	528725	529347	529969	530590	531209	3	62.1
57	531828	532446	533063	533679	534295	534909	2	61.6
58	535523	536136	536747	537358	537969	538578	1	61.1
59	539186	539794	540401	541007	541612	542216	0	60.5
	60″	50″	40″	30″	20″	10″		

Min.	Tangent of 1 Degree. 0″	10″	20″	30″	40″	50″		P. Part to 1″.
0	8.241921	8.243126	8.244328	8.245526	8.246721	8.247913	59	119.7
1	249102	250287	251469	252648	253823	254996	58	117.7
2	256165	257331	258494	259654	260811	261965	57	115.8
3	263115	264263	265408	266549	267688	268824	56	114.0
4	269956	271086	272213	273337	274458	275576	55	112.2
5	276691	277804	278913	280020	281124	282225	54	110.5
6	283323	284419	285512	286602	287689	288774	53	108.9
7	289856	290935	292012	293086	294157	295226	52	107.2
8	296292	297355	298416	299474	300530	301583	51	105.7
9	302634	303682	304727	305770	306811	307849	50	104.2
10	8.308884	8.309917	8.310948	8.311976	8.313002	8.314025	49	102.7
11	315046	316065	317081	318095	319106	320115	48	101.3
12	321122	322127	323129	324129	325126	326121	47	99.8
13	327114	328105	329093	330080	331064	332045	46	98.5
14	333025	334002	334977	335950	336921	337890	45	97.2
15	338856	339821	340783	341743	342701	343657	44	95.9
16	344610	345562	346512	347459	348405	349348	43	94.6
17	350289	351229	352166	353101	354035	354966	42	93.4
18	355895	356823	357748	358671	359593	360512	41	92.2
19	361430	362345	363259	364171	365081	365988	40	91.1
20	8.366895	8.367799	8.368701	8.369601	8.370500	8.371397	39	89.9
21	372292	373185	374076	374965	375853	376738	38	88.8
22	377622	378504	379385	380263	381140	382015	37	87.8
23	382889	383760	384630	385498	386364	387229	36	86.7
24	388092	388953	389813	390670	391526	392381	35	85.7
25	393234	394085	394934	395782	396628	397472	34	84.7
26	398315	399156	399996	400834	401670	402505	33	83.7
27	403338	404170	405000	405828	406655	407480	32	82.8
28	408304	409126	409946	410765	411583	412399	31	81.8
29	413213	414026	414837	415647	416456	417263	30	80.9
30	8.418068	8.418872	8.419674	8.420475	8.421274	8.422072	29	80.0
31	422869	423664	424458	425250	426041	426830	28	79.1
32	427618	428404	429189	429973	430755	431536	27	78.3
33	432315	433093	433870	434645	435419	436191	26	77.5
34	436962	437732	438500	439267	440033	440797	25	76.6
35	441560	442322	443082	443841	444599	445355	24	75.8
36	446110	446864	447616	448368	449117	449866	23	75.0
37	450613	451359	452104	452847	453589	454330	22	74.3
38	455070	455808	456545	457281	458016	458749	21	73.5
39	459481	460212	460942	461670	462398	463124	20	72.8
40	8.463849	8.464572	8.465295	8.466016	8.466736	8.467455	19	72.1
41	468172	468889	469604	470318	471031	471743	18	71.3
42	472454	473163	473872	474579	475285	475990	17	70.7
43	476693	477396	478097	478798	479497	480195	16	70.0
44	480892	481588	482283	482976	483669	484360	15	69.3
45	485050	485740	486428	487115	487801	488486	14	68.6
46	489170	489852	490534	491215	491894	492573	13	68.0
47	493250	493927	494602	495276	495949	496622	12	67.4
48	497293	497963	498632	499300	499967	500633	11	66.8
49	501298	501962	502625	503287	503948	504608	10	66.1
50	8.505267	8.505925	8.506582	8.507238	8.507893	8.508547	9	65.5
51	509200	509852	510503	511153	511802	512451	8	65.0
52	513098	513744	514389	515034	515677	516320	7	64.4
53	516961	517602	518241	518880	519518	520154	6	63.8
54	520790	521425	522059	522692	523324	523956	5	63.3
55	524586	525215	525844	526472	527098	527724	4	62.7
56	528349	528973	529596	530218	530840	531460	3	62.2
57	532080	532698	533316	533933	534549	535164	2	61.6
58	535779	536392	537005	537617	538227	538837	1	61.1
59	539447	540055	540662	541269	541875	542480	0	60.6

Sine of 2 Degrees.							Min.	Sine of 3 I		
0"	10"	20"	30"	40"	50"			0"	10"	20"
8.542819	3422	4023	4624	5224	5823	59	0	8.718800	9202	9603
6422	7019	7616	8212	8807	9401	58	1	8.721204	1603	2002
9995	.587	1179	1770	2361	2950	57	2	3595	3992	4389
8 553539	4126	4713	5300	5885	6470	56	3	5972	6367	6762
7054	7637	8219	8801	9381	9961	55	4	8337	8729	9122
8.560540	1119	1696	2273	2849	3425	54	5	8.730688	1079	1469
3999	4573	5146	5719	6290	6861	53	6	3027	3416	3804
7431	8000	8569	9137	9704	.270	52	7	5354	5740	6126
8.570836	1401	1965	2528	3091	3653	51	8	7667	8052	8436
4214	4774	5334	5893	6451	7009	50	9	9969	.352	.734
7566	8122	8678	9232	9786	.340	49	10	8.742259	2639	3019
8.580892	1444	1995	2546	3096	3645	48	11	4536	4914	5293
4193	4741	5288	5834	6380	6925	47	12	6802	7178	7554
7469	8013	8556	9098	9640	.181	46	13	9055	9430	9804
8.590721	1260	1799	2338	2875	3412	45	14	8.751297	1670	2042
3948	4484	5019	5553	6087	6619	44	15	3528	3898	4269
7152	7683	8214	8745	9274	9803	43	16	5747	6116	6484
8.600332	0859	1387	1913	2439	2964	42	17	7955	8321	8688
3489	4012	4536	5058	5580	6102	41	18	8.760151	0516	0881
6623	7143	7662	8181	8699	9217	40	19	2337	2700	3063
9734	.251	.766	1282	1796	2310	39	20	4511	4872	5234
8.612823	3336	3848	4360	4871	5381	38	21	6675	7034	7394
5891	6400	6909	7417	7924	8431	37	22	8828	9185	9543
8937	9442	9947	.452	.956	1459	36	23	8.770970	1326	1681
8.621962	2464	2965	3466	3966	4466	35	24	3101	3456	3810
4965	5464	5962	6459	6956	7453	34	25	5223	5575	5927
7948	8444	8938	9432	9926	.419	33	26	7333	7684	8035
8.630911	1403	1894	2385	2875	3365	32	27	9434	9783	.132
3854	4342	4830	5317	5804	6291	31	28	8.781524	1872	2219
6776	7262	7746	8230	8714	9197	30	29	3605	3951	4296
9680	.162	.643	1124	1604	2084	29	30	5675	6019	6363
8.642563	3042	3520	3998	4475	4952	28	31	7736	8078	8421
5428	5904	6379	6854	7328	7801	27	32	9787	.128	.468
8274	8747	9219	9690	.161	.632	26	33	8.791828	2167	2506
8.651102	1571	2040	2508	2976	3444	25	34	3859	4197	4534
3911	4377	4843	5308	5773	6238	24	35	5881	6218	6553
6702	7165	7628	8090	8552	9014	23	36	7894	8229	8563
9475	9935	.395	.855	1314	1772	22	37	9897	.230	.563
8.662230	2688	3145	3602	4058	4513	21	38	8.801892	2223	2554
4968	5423	5877	6331	6784	7237	20	39	3876	4206	4536
7689	8141	8592	9043	9494	9944	19	40	5852	6181	6509
8.670393	0842	1291	1739	2187	2634	18	41	7819	8146	8473
3080	3527	3972	4418	4863	5307	17	42	9777	.103	.428
5751	6194	6638	7080	7522	7964	16	43	8.811726	2050	2374
8405	8846	9286	9726	.166	.605	15	44	3667	3989	4312
8.681043	1481	1919	2356	2793	3230	14	45	5599	5920	6241
3665	4101	4536	4971	5405	5838	13	46	7522	7841	8161
6272	6705	7137	7569	8001	8432	12	47	9436	9755	..73
8863	9293	9723	.152	.581	1010	11	48	8.821343	1659	1976
8.691438	1866	2293	2720	3146	3572	10	49	3240	3556	3871
3998	4423	4848	5272	5696	6120	9	50	5130	5444	5758
6543	6966	7388	7810	8232	8653	8	51	7011	7324	7637
9073	9494	9913	.333	.752	1171	7	52	8884	9196	9507
8.701589	2007	2424	2841	3258	3674	6	53	8.830749	1060	1369
4090	4505	4920	5335	5749	6163	5	54	2607	2915	3224
6577	6990	7402	7815	8226	8638	4	55	4456	4763	5070
9049	9460	9870	.280	.690	1099	3	56	6297	6603	6909
8.711507	1916	2324	2731	3139	3546	2	57	8130	8435	8740
3952	4358	4764	5169	5574	5979	1	58	9956	.260	.563
6383	6787	7190	7593	7996	8398	0	59	8.841774	2076	2378

of 2 Degrees.					Min.	Tangent of 3 Deg			
20″	30″	40	50″			0″	10″	20″	30″
4289	4891	5492	6092	59	0	8.719396	9798	.201	.603
7887	8483	9079	9674	58	1	8.721806	2207	2607	3007
1454	2046	2637	3227	57	2	4204	4602	5000	5397
4993	5580	6166	6752	56	3	6588	6984	7380	7775
8503	9085	9667	.248	55	4	8959	9353	9746	.140
1985	2563	3140	3716	54	5	8.731317	1709	2101	2492
5440	6013	6585	7157	53	6	3663	4053	4442	4831
8867	9435	...3	.570	52	7	5996	6384	6771	7158
2267	2832	3395	3958	51	8	8317	8703	9088	9473
5642	6201	6760	7319	50	9	8.740626	1009	1393	1776
8990	9545	.100	.654	49	10	2922	3304	3685	4066
2312	2864	3414	3964	48	11	5207	5586	5966	6344
5610	6157	6704	7249	47	12	7479	7857	8234	8611
8883	9426	9968	.510	46	13	9740	.116	.491	.866
2131	2670	3208	3746	45	14	8.751989	2363	2736	3109
5355	5890	6425	6959	44	15	4227	4599	4970	5341
8556	9087	9618	.147	43	16	6453	6823	7193	7562
1733	2260	2787	3313	42	17	8668	9036	9404	9771
4887	5411	5934	6456	41	18	8.760872	1238	1604	1970
8019	8539	9058	9576	40	19	3065	3429	3793	4157
1128	1644	2160	2675	39	20	5246	5609	5971	6333
4215	4728	5240	5751	38	21	7417	7778	8139	8499
7281	7790	8298	8806	37	22	9578	9937	.295	.654
.325	.830	1335	1839	36	23	8.771727	2085	2442	2798
3348	3850	4351	4852	35	24	3866	4222	4577	4932
6350	6849	7346	7844	34	25	5995	6349	6702	7056
9332	9827	.321	.815	33	26	8114	8466	8817	9169
2293	2785	3276	3766	32	27	8.780222	0572	0922	1272
5235	5723	6211	6698	31	28	2320	2669	3017	3365
8156	8641	9126	9610	30	29	4408	4755	5102	5448
1058	1540	2021	2502	29	30	6486	6831	7177	7521
3941	4420	4898	5376	28	31	8554	8898	9242	9585
6805	7281	7756	8230	27	32	8.790613	0955	1297	1639
9651	.123	.595	1067	26	33	2662	3003	3343	3683
2478	2947	3416	3884	25	34	4701	5040	5379	5718
5286	5753	6219	6684	24	35	6731	7069	7406	7743
8077	8541	9004	9466	23	36	8752	9088	9423	9759
.850	1311	1771	2230	22	37	8.800763	1098	1432	1765
3606	4063	4520	4977	21	38	2765	3098	3431	3763
6344	6799	7253	7707	20	39	4758	5090	5421	5751
9065	9517	9968	.419	19	40	6742	7072	7402	7731
1769	2218	2667	3115	18	41	8717	9046	9374	9701
4457	4903	5349	5794	17	42	8.810683	1010	1337	1663
7128	7572	8015	8457	16	43	2641	2966	3291	3616
9783	.224	.664	1104	15	44	4589	4913	5237	5560
2422	2860	3298	3735	14	45	6529	6852	7174	7496
5044	5480	5915	6350	13	46	8461	8782	9103	9423
7652	8085	8517	8950	12	47	8.820384	0703	1023	1342
.244	.674	1104	1534	11	48	2298	2617	2935	3253
2820	3248	3675	4103	10	49	4205	4522	4838	5155
5381	5807	6232	6656	9	50	6103	6418	6733	7049
7928	8351	8773	9195	8	51	7992	8307	8621	8934
.459	.880	1300	1720	7	52	9874	.187	.500	.812
2976	3395	3812	4230	6	53	8.831748	2059	2371	2682
5479	5895	6310	6725	5	54	3613	3924	4234	4543
7967	8381	8794	9206	4	55	5471	5780	6089	6397
.442	.853	1263	1674	3	56	7321	7629	7936	8243
2902	3311	3719	4127	2	57	9163	9470	9776	..81
5348	5755	6161	6567	1	58	8.840998	1303	1607	1912
7781	8186	8589	8993	0	59	2825	3128	3432	3735

Min.	Sine of 4 Degrees.						
	0′	10″	20″	30″	40″	50″	
0	8.843585	3886	4186	4487	4787	5087	59
1	5387	5687	5987	6286	6585	6884	58
2	7183	7481	7780	8078	8376	8673	57
3	8971	9268	9565	9862	.159	.455	56
4	8.850751	1047	1343	1639	1934	2229	55
5	2525	2819	3114	3408	3703	3997	54
6	4291	4584	4878	5171	5464	5757	53
7	6049	6342	6634	6926	7218	7510	52
8	7801	8092	8383	8674	8965	9255	51
9	9546	9836	.126	.415	.705	.994	50
10	8.861283	1572	1861	2149	2438	2726	49
11	3014	3302	3589	3877	4164	4451	48
12	4738	5024	5311	5597	5883	6169	47
13	6455	6740	7025	7310	7595	7880	46
14	8165	8449	8733	9017	9301	9585	45
15	9868	.151	.434	.717	1000	1282	44
16	8.871565	1847	2129	2410	2692	2973	43
17	3255	3536	3817	4097	4378	4658	42
18	4938	5218	5498	5777	6057	6336	41
19	6615	6894	7172	7451	7729	8007	40
20	8285	8563	8841	9118	9395	9672	39
21	9949	.226	.503	.779	1055	1331	38
22	8.881607	1883	2158	2433	2708	2983	37
23	3258	3533	3807	4081	4355	4629	36
24	4903	5177	5450	5723	5996	6269	35
25	6542	6814	7087	7359	7631	7903	34
26	8174	8446	8717	8988	9259	9530	33
27	9801	..71	.341	.612	.882	1151	32
28	8.891421	1690	1960	2229	2498	2767	31
29	3035	3304	3572	3840	4108	4376	30
30	4643	4911	5178	5445	5712	5979	29
31	6246	6512	6778	7044	7310	7576	28
32	7842	8107	8373	8638	8903	9168	27
33	9432	9697	9961	.225	.489	.753	26
34	8.901017	1280	1544	1807	2070	2333	25
35	2596	2858	3121	3383	3645	3907	24
36	4169	4430	4692	4953	5214	5475	23
37	5736	5997	6257	6517	6778	7038	22
38	7297	7557	7817	8076	8335	8595	21
39	8853	9112	9371	9629	9888	.146	20
40	8.910404	0662	0919	1177	1434	1692	19
41	1949	2206	2462	2719	2976	3232	18
42	3488	3744	4000	4256	4511	4767	17
43	5022	5277	5532	5787	6041	6296	16
44	6550	6805	7059	7313	7566	7820	15
45	8073	8327	8580	8833	9086	9338	14
46	9591	9843	..96	.348	.600	.852	13
47	8.921103	1355	1606	1858	2109	2360	12
48	2610	2861	3112	3362	3612	3862	11
49	4112	4362	4612	4861	5111	5360	10
50	5609	5858	6107	6355	6604	6852	9
51	7100	7348	7596	7844	8092	8339	8
52	8587	8834	9081	9328	9575	9821	7
53	8.930068	0314	0560	0806	1052	1298	6
54	1544	1789	2035	2280	2525	2770	5
55	3015	3260	3504	3749	3993	4237	4
56	4481	4725	4969	5212	5456	5699	3
57	5942	6185	6428	6671	6914	7156	2
58	7398	7641	7883	8125	8366	8608	1
59	8850	9091	9332	9573	9814	..55	0

Min.	Sine of 5 Degre[es]			
	0″	10″	20″	30″
0	8.940296	0537	0777	1017
1	1738	1977	2217	2457
2	3174	3413	3652	3891
3	4606	4844	5083	5321
4	6034	6271	6508	6745
5	7456	7693	7929	8166
6	8874	9110	9345	9581
7	8.950287	0522	0757	0992
8	1696	1930	2164	2398
9	3100	3333	3567	3800
10	4499	4732	4965	5197
11	5894	6126	6358	6590
12	7284	7516	7747	7978
13	8670	8901	9131	9362
14	8.960052	0282	0511	0741
15	1429	1658	1887	2116
16	2801	3030	3258	3486
17	4170	4397	4625	4852
18	5534	5761	5987	6214
19	6893	7120	7346	7572
20	8249	8474	8700	8925
21	9600	9825	..49	.274
22	8.970947	1171	1395	1619
23	2289	2513	2736	2959
24	3628	3851	4073	4296
25	4962	5184	5406	5628
26	6293	6514	6735	6956
27	7619	7839	8060	8280
28	8941	9161	9381	9600
29	8.980259	0478	0697	0916
30	1573	1791	2010	2228
31	2883	3101	3319	3536
32	4189	4406	4623	4840
33	5491	5708	5924	6141
34	6789	7005	7221	7437
35	8083	8299	8514	8729
36	9374	9588	9803	..17
37	8.990660	0874	1088	1302
38	1943	2156	2370	2583
39	3222	3435	3647	3860
40	4497	4709	4921	5133
41	5768	5980	6191	6402
42	7036	7247	7457	7668
43	8299	8510	8720	8930
44	9560	9769	9979	.188
45	9.000816	1025	1234	1443
46	2069	2277	2486	2694
47	3318	3526	3733	3941
48	4563	4771	4978	5185
49	5805	6012	6218	6425
50	7044	7250	7456	7661
51	8278	8484	8689	8894
52	9510	9715	9919	.124
53	9.010737	0942	1146	1350
54	1962	2165	2369	2572
55	3182	3385	3588	3791
56	4400	4602	4805	5007
57	5613	5815	6017	6219
58	6824	7025	7227	7428
59	8031	8232	8433	8633

Min.	Tangent of 4 Degrees. 0″	10″	20″	30″	40″	50″	
0	8.844644	4946	5248	5551	5852	6154	59
1	6455	6757	7058	7358	7659	7959	58
2	8260	8560	8859	9159	9458	9758	57
3	8.850057	0355	0654	0952	1250	1548	56
4	1846	2144	2441	2738	3035	3332	55
5	3628	3925	4221	4517	4813	5108	54
6	5403	5699	5993	6288	6583	6877	53
7	7171	7465	7759	8053	8346	8639	52
8	8932	9225	9517	9810	.102	.394	51
9	8.860686	0977	1269	1560	1851	2142	50
10	2433	2723	3013	3303	3593	3883	49
11	4173	4462	4751	5040	5329	5617	48
12	5906	6194	6482	6769	7057	7344	47
13	7632	7919	8206	8492	8779	9065	46
14	9351	9637	9923	.208	.494	.779	45
15	8.871064	1349	1633	1918	2202	2486	44
16	2770	3054	3337	3620	3904	4187	43
17	4469	4752	5034	5317	5599	5881	42
18	6162	6444	6725	7006	7287	7568	41
19	7849	8129	8409	8689	8969	9249	40
20	9529	9808	..87	.366	.645	.924	39
21	8.881202	1480	1759	2037	2314	2592	38
22	2869	3147	3424	3701	3977	4254	37
23	4530	4807	5083	5358	5634	5910	36
24	6185	6460	6735	7010	7285	7559	35
25	7833	8108	8382	8655	8929	9202	34
26	9476	9749	..22	.295	.567	.840	33
27	8 891112	1384	1656	1928	2199	2471	32
28	2742	3013	3284	3555	3825	4096	31
29	4366	4636	4906	5176	5445	5715	30
30	5984	6253	6522	6791	7060	7328	29
31	7596	7864	8132	8400	8668	8935	28
32	9203	9470	9737	...4	.270	.537	27
33	8.900803	1069	1335	1601	1867	2132	26
34	2398	2663	2928	3193	3458	3722	25
35	3987	4251	4515	4779	5043	5306	24
36	5570	5833	6096	6359	6622	6885	23
37	7147	7410	7672	7934	8196	8457	22
38	8719	8980	9242	9503	9764	..25	21
39	8.910285	0546	0806	1066	1326	1586	20
40	1846	2106	2365	2624	2883	3142	19
41	3401	3660	3918	4177	4435	4693	18
42	4951	5209	5466	5724	5981	6238	17
43	6495	6752	7009	7265	7522	7778	16
44	8034	8290	8546	8801	9057	9312	15
45	9568	9823	..78	.332	.587	.841	14
46	8.921096	1350	1604	1858	2112	2365	13
47	2619	2872	3125	3378	3631	3884	12
48	4136	4389	4641	4893	5145	5397	11
49	5649	5900	6152	6403	6654	6905	10
50	7156	7407	7657	7908	8158	8408	9
51	8658	8908	9158	9407	9657	9906	8
52	8.930155	0404	0653	0902	1150	1399	7
53	1647	1895	2143	2391	2639	2887	6
54	3134	3381	3629	3876	4123	4369	5
55	4616	4862	5109	5355	5601	5847	4
56	6093	6339	6584	6830	7075	7320	3
57	7565	7810	8055	8299	8544	8788	2
58	9032	9276	9520	9764	...7	.251	1
59	8.940494	0738	0981	1224	1467	1709	0

Min.	Tangent of 5 Degrees. 0″	10″	20″	30″	40″	50″	
0	8.941952	2194	2437	2679	2921	3163	59
1	3404	3646	3888	4129	4370	4611	58
2	4852	5093	5334	5574	5815	6055	57
3	6295	6535	6775	7015	7255	7494	56
4	7734	7973	8212	8451	8690	8929	55
5	9168	9406	9644	9883	.121	.359	54
6	8.950597	0834	1072	1309	1547	1784	53
7	2021	2258	2495	2732	2968	3205	52
8	3441	3677	3913	4149	4385	4621	51
9	4856	5092	5327	5562	5797	6032	50
10	6267	6502	6736	6971	7205	7439	49
11	7674	7908	8141	8375	8609	8842	48
12	9075	9309	9542	9775	...8	.240	47
13	8.960473	0705	0938	1170	1402	1634	46
14	1866	2098	2329	2561	2792	3023	45
15	3255	3486	3716	3947	4178	4408	44
16	4639	4869	5099	5329	5559	5789	43
17	6019	6248	6478	6707	6936	7165	42
18	7394	7623	7852	8081	8309	8538	41
19	8766	8994	9222	9450	9678	9905	40
20	8.970133	0360	0588	0815	1042	1269	39
21	1496	1723	1949	2176	2402	2628	38
22	2855	3081	3307	3532	3758	3984	37
23	4209	4435	4660	4885	5110	5335	36
24	5560	5784	6009	6233	6458	6682	35
25	6906	7130	7354	7578	7801	8025	34
26	8248	8472	8695	8918	9141	9364	33
27	9586	9809	..32	.254	.476	.699	32
28	8.980921	1143	1364	1586	1808	2029	31
29	2251	2472	2693	2914	3135	3356	30
30	3577	3798	4018	4238	4459	4679	29
31	4899	5119	5339	5559	5778	5998	28
32	6217	6437	6656	6875	7094	7313	27
33	7532	7750	7969	8187	8406	8624	26
34	8842	9060	9278	9496	9714	9931	25
35	8.990149	0366	0583	0801	1018	1235	24
36	1451	1668	1885	2101	2318	2534	23
37	2750	2966	3182	3398	3614	3830	22
38	4045	4261	4476	4692	4907	5122	21
39	5337	5552	5766	5981	6196	6410	20
40	6624	6839	7053	7267	7481	7694	19
41	7908	8122	8335	8549	8762	8975	18
42	9188	9401	9614	9827	..40	.252	17
43	9.000465	0677	0889	1102	1314	1526	16
44	1738	1949	2161	2373	2584	2795	15
45	3007	3218	3429	3640	3851	4061	14
46	4272	4483	4693	4904	5114	5324	13
47	5534	5744	5954	6164	6373	6583	12
48	6792	7002	7211	7420	7629	7838	11
49	8047	8256	8465	8673	8882	9090	10
50	9298	9507	9715	9923	.131	.338	9
51	9.010546	0754	0961	1169	1376	1583	8
52	1790	1997	2204	2411	2618	2824	7
53	3031	3237	3444	3650	3856	4062	6
54	4268	4474	4680	4886	5091	5297	5
55	5502	5707	5913	6118	6323	6528	4
56	6732	6937	7142	7346	7551	7755	3
57	7959	8164	8368	8572	8776	8979	2
58	9183	9387	9590	9794	9997	.200	1
59	9.020403	0606	0809	1012	1215	1418	0

Sine of 6 Degrees.							Min.	Sine of 7 Degre			
0″	10″	20″	30″	40″	50″			0″	10″	20″	30″
9.019235	9435	9635	9835	..35	.235	59	0	9.085894	6066	6237	6409
9.020435	0635	0834	1034	1233	1433	58	1	6922	7093	7264	7435
:632	1831	2030	2229	2428	2627	57	2	7947	8118	8288	8459
2825	3024	3223	3421	3619	3818	56	3	8970	9140	9310	9480
4016	4214	4412	4610	4807	5005	55	4	9990	.160	.330	.500
5203	5400	5598	5795	5992	6189	54	5	9.091008	1178	1347	1516
6386	6583	6780	6977	7174	7370	53	6	2024	2193	2362	2530
7567	7763	7960	8156	8352	8548	52	7	3037	3205	3374	3542
8744	8940	9136	9332	9527	9723	51	8	4047	4216	4384	4552
9918	.114	.309	.504	.699	.894	50	9	5056	5223	5391	5559
9.031089	1284	1479	1673	1868	2062	49	10	6062	6229	6396	6564
2257	2451	2645	2839	3033	3227	48	11	7065	7232	7399	7566
3421	3615	3809	4002	4196	4389	47	12	8066	8233	8399	8566
4582	4776	4969	5162	5355	5548	46	13	9065	9231	9398	9564
5741	5933	6126	6319	6511	6703	45	14	9.100062	0227	0393	0559
6896	7088	7280	7472	7664	7856	44	15	1056	1221	1387	1552
8048	8239	8431	8623	8814	9005	43	16	2048	2213	2378	2543
9197	9388	9579	9770	9961	.152	42	17	3037	3202	3367	3531
9.040342	0533	0724	0914	1105	1295	41	18	4025	4189	4353	4517
1485	1675	1865	2055	2245	2435	40	19	5010	5174	5337	5501
2625	2815	3004	3194	3383	3572	39	20	5992	6156	6319	6483
3762	3951	4140	4329	4518	4707	38	21	6973	7136	7299	7462
4895	5084	5273	5461	5650	5838	37	22	7951	8114	8277	8439
6026	6214	6402	6590	6778	6966	36	23	8927	9090	9252	9414
7154	7342	7529	7717	7904	8091	35	24	9901	..63	.225	.387
8279	8466	8653	8840	9027	9214	34	25	9.110873	1034	1196	1358
9400	9587	9774	9960	.147	.333	33	26	1842	2003	2165	2326
9.050519	0706	0892	1078	1264	1450	32	27	2809	2970	3131	3292
1635	1821	2007	2192	2378	2563	31	28	3774	3935	4095	4256
2749	2934	3119	3304	3489	3674	30	29	4737	4897	5057	5218
3859	4044	4228	4413	4597	4782	29	30	5698	5858	6017	6177
4966	5150	5335	5519	5703	5887	28	31	6656	6816	6975	7135
6071	6254	6438	6622	6805	6989	27	32	7613	7772	7931	8090
7172	7356	7539	7722	7905	8088	26	33	8567	8726	8884	9043
8271	8454	8637	8820	9002	9185	25	34	9519	9677	9836	9994
9367	9550	9732	9914	..96	.278	24	35	9.120469	0627	0785	0943
9.060460	0642	0824	1006	1188	1369	23	36	1417	1574	1732	1890
1551	1732	1914	2095	2276	2457	22	37	2362	2520	2677	2835
2639	2820	3001	3181	3362	3543	21	38	3306	3463	3620	3777
3724	3904	4085	4265	4445	4626	20	39	4248	4404	4561	4718
4806	4986	5166	5346	5526	5705	19	40	5187	5344	5500	5656
5885	6065	6244	6424	6603	6783	18	41	6125	6281	6437	6593
6962	7141	7320	7499	7678	7857	17	42	7060	7216	7371	7527
8036	8215	8393	8572	8751	8929	16	43	7993	8149	8304	8459
9107	9286	9464	9642	9820	9998	15	44	8925	9080	9235	9390
9.070176	0354	0532	0709	0887	1065	14	45	9854	...9	.163	.318
1242	1420	1597	1774	1951	2128	13	46	9.130781	0936	1090	1244
2306	2482	2659	2836	3013	3190	12	47	1706	1860	2014	2168
3366	3543	3719	3896	4072	4248	11	48	2630	2783	2937	3091
4424	4600	4777	4952	5128	5304	10	49	3551	3704	3858	4011
5480	5656	5831	6007	6182	6358	9	50	4470	4623	4776	4929
6533	6708	6883	7058	7233	7408	8	51	5387	5540	5693	5845
7583	7758	7933	8107	8282	8457	7	52	6303	6455	6607	676[illegible]
8631	8805	8980	9154	9328	9502	6	53	7216	7368	7520	7672
9676	9850	..24	.198	.372	.545	5	54	8128	8279	8431	8582
9.080719	0892	1066	1239	1413	1586	4	55	9037	9188	9340	9491
1759	1932	2105	2278	2451	2624	3	56	9944	..96	.247	.398
2797	2969	3142	3314	3487	3659	2	57	9.140850	1001	1151	1302
3832	4004	4176	4348	4520	4692	1	58	1754	1904	2055	2205
4864	5036	5208	5380	5551	5723	0	59	2655	2806	2956	3106

of 6 Degrees.						Tangent of 7 Deg			
20″	30″	40″	50″		Min.	0″	10″	20″	30″
2025	2227	2430	2632	59	0	9.089144	9318	9492	9666
3238	3439	3641	3843	58	1	9.090187	0361	0534	0708
4447	4648	4849	5050	57	2	1228	1401	1574	1747
5653	5853	6054	6254	56	3	2266	2439	2612	2784
6855	7055	7255	7455	55	4	3302	3474	3647	3819
8055	8254	8454	8653	54	5	4336	4508	4680	4851
9251	9450	9649	9848	53	6	5367	5538	5710	5881
0444	0642	0841	1039	52	7	6395	6567	6738	6909
1633	1831	2029	2227	51	8	7422	7593	7764	7934
2820	3017	3215	3412	50	9	8446	8616	8787	8957
4003	4200	4397	4594	49	10	9468	9638	9808	9978
5184	5380	5576	5773	48	11	9.100487	0657	0827	0996
6361	6557	6753	6948	47	12	1504	1674	1843	2012
7535	7730	7926	8121	46	13	2519	2688	2857	3026
8706	8901	9095	9290	45	14	3532	3700	3869	4037
9874	..68	.262	.456	44	15	4542	4710	4878	5046
1039	1232	1426	1620	43	16	5550	5718	5885	6053
2200	2394	2587	2780	42	17	6556	6723	6890	7058
3359	3552	3745	3937	41	18	7559	7726	7893	8060
4515	4707	4899	5092	40	19	8560	8727	8894	9060
5668	5859	6051	6243	39	20	9559	9726	9892	..58
6817	7009	7200	7391	38	21	9.110556	0722	0888	1054
7964	8155	8346	8536	37	22	1551	1716	1882	2047
9108	9298	9489	9679	36	23	2543	2708	2873	3039
.249	.439	.629	.818	35	24	3533	3698	3863	4028
1387	1576	1766	1955	34	25	4521	4686	4850	5015
2522	2711	2900	3088	33	26	5507	5671	5835	5999
3654	3843	4031	4219	32	27	6491	6655	6818	6982
4784	4972	5159	5347	31	28	7472	7636	7799	7962
5910	6098	6285	6472	30	29	8452	8615	8778	8941
7034	7221	7408	7594	29	30	9429	9592	9754	9917
8155	8341	8528	8714	28	31	9.120404	0567	0729	0891
9273	9459	9645	9831	27	32	1377	1539	1701	1863
0388	0573	0759	0944	26	33	2348	2510	2671	2833
1500	1685	1870	2055	25	34	3317	3478	3640	3801
2610	2795	2979	3164	24	35	4284	4445	4606	4766
3717	3901	4085	4269	23	36	5249	5409	5570	5730
4821	5005	5188	5372	22	37	6211	6371	6532	6692
5922	6106	6289	6472	21	38	7172	7332	7492	7651
7021	7204	7387	7570	20	39	8130	8290	8449	8609
8117	8300	8482	8664	19	40	9087	9246	9405	9564
9211	9393	9575	9756	18	41	9.130041	0200	0359	0518
.301	.483	.664	.846	17	42	0994	1152	1311	1469
1389	1570	1751	1932	16	43	1944	2102	2261	2419
2475	2655	2836	3016	15	44	2893	3050	3208	3366
3558	3738	3918	4098	14	45	3839	3997	4154	4312
4638	4817	4997	5177	13	46	4784	4941	5098	5255
5715	5895	6074	6253	12	47	5726	5883	6040	6197
6790	6969	7148	7327	11	48	6667	6823	6980	7136
7862	8041	8219	8398	10	49	7605	7761	7918	8074
8932	9110	9288	9466	9	50	8542	8698	8854	9009
....	.177	.355	.532	8	51	9476	9632	9788	9943
1064	1241	1419	1596	7	52	9.140409	0564	0720	0875
2126	2303	2480	2657	6	53	1340	1495	1650	1805
3186	3362	3539	3715	5	54	2269	2424	2578	2733
4243	4419	4595	4771	4	55	3196	3350	3504	3659
5298	5473	5649	5824	3	56	4121	4275	4429	4583
6350	6525	6700	6875	2	57	5044	5198	5351	5505
7400	7574	7749	7924	1	58	5966	6119	6272	6425
8447	8621	8795	8970	0	59	6885	7038	7191	7344

Min	Sine of 8 Degrees. 0″	10″	20″	30″	40″	50″	
0	9.143555	3705	3855	4005	4154	4304	59
1	4453	4603	4752	4902	5051	5200	58
2	5349	5498	5648	5797	5946	6095	57
3	6243	6392	6541	6690	6839	6987	56
4	7136	7284	7433	7581	7730	7878	55
5	8026	8174	8323	8471	8619	8767	54
6	8915	9063	9211	9358	9506	9654	53
7	9802	9949	..97	.244	.392	.539	52
8	9.150686	0834	0981	1128	1275	1422	51
9	1569	1716	1863	2010	2157	2304	50
10	2451	2597	2744	2891	3037	3184	49
11	3330	3476	3623	3769	3915	4061	48
12	4208	4354	4500	4646	4792	4938	47
13	5083	5229	5375	5521	5666	5812	46
14	5957	6103	6248	6394	6539	6684	45
15	6830	6975	7120	7265	7410	7555	44
16	7700	7845	7990	8135	8279	8424	43
17	8569	8713	8858	9002	9147	9291	42
18	9435	9580	9724	9868	..12	.156	41
19	9.160301	0445	0589	0732	0876	1020	40
20	1164	1308	1451	1595	1738	1882	39
21	2025	2169	2312	2456	2599	2742	38
22	2885	3028	3172	3315	3458	3600	37
23	3743	3886	4029	4172	4314	4457	36
24	4600	4742	4885	5027	5170	5312	35
25	5454	5597	5739	5881	6023	6165	34
26	6307	6449	6591	6733	6875	7017	33
27	7159	7300	7442	7584	7725	7867	32
28	8008	8150	8291	8432	8574	8715	31
29	6856	8997	9138	9279	9420	9561	30
30	9702	9843	9984	.125	.265	.406	29
31	9.170547	0687	0828	0968	1109	1249	28
32	1389	1530	1670	1810	1950	2090	27
33	2230	2370	2510	2650	2790	2930	26
34	3070	3210	3349	3489	3629	3768	25
35	3908	4047	4187	4326	4465	4605	24
36	4744	4883	5022	5161	5300	5439	23
37	5578	5717	5856	5995	6134	6273	22
38	6412	6550	6688	6827	6966	7104	21
39	7242	7381	7519	7657	7796	7934	20
40	8072	8210	8348	8486	8624	8762	19
41	8900	9038	9176	9313	9451	9589	18
42	9726	9864	...2	.139	.276	.414	17
43	9.180551	0689	0826	0963	1100	1237	16
44	1374	1511	1648	1785	1922	2059	15
45	2196	2333	2469	2606	2743	2879	14
46	3016	3152	3289	3425	3562	3698	13
47	3834	3971	4107	4243	4379	4515	12
48	4651	4787	4923	5059	5195	5331	11
49	5466	5602	5738	5874	6009	6145	10
50	6280	6416	6551	6686	6822	6957	9
51	7092	7228	7363	7498	7633	7768	8
52	7903	8038	8173	8308	8442	8577	7
53	[illegible]12	8847	8981	9116	9250	9385	6
54	95[illegible]9	9654	9788	9923	..57	.191	5
55	9.1903[illegible]5	0460	0594	0728	0862	0996	4
56	1130	1264	1398	1532	1665	1799	3
57	1933	2066	2200	2334	2467	2601	2
58	2734	2868	3001	3134	3268	3401	1
59	3534	3667	3800	3933	4066	4199	0

Min.	Sine of 9 Degrees. 0″	10″	20″	30″	40″	50″	
0	9.194332	4465	4598	4731	4864	4997	59
1	5129	5262	5395	5527	5660	5792	58
2	5925	6057	6189	6322	6454	6586	57
3	6719	6851	6983	7115	7247	7379	56
4	7511	7643	7775	7907	8038	8170	55
5	8302	8434	8565	8697	8828	8960	54
6	9091	9223	9354	9486	9617	9748	53
7	9879	..11	.142	.273	.404	.535	52
8	9.200666	0797	0928	1059	1189	1320	51
9	1451	1582	1712	1843	1973	2104	50
10	2234	2365	2495	2626	2756	2886	49
11	3017	3147	3277	3407	3537	3667	48
12	3797	3927	4057	4187	4317	4447	47
13	4577	4706	4836	4966	5095	5225	46
14	5354	5484	5613	5743	5872	6002	45
15	6131	6260	6389	6519	6648	6777	44
16	6906	7035	7164	7293	7422	7551	43
17	7679	7808	7937	8066	8194	8323	42
18	8452	8580	8709	8837	8966	9094	41
19	9222	9351	9479	9607	9735	9864	40
20	9992	.120	.248	.376	.504	.632	39
21	9.210760	0888	1015	1143	1271	1399	38
22	1526	1654	1781	1909	2037	2164	37
23	2291	2419	2546	2674	2801	2928	36
24	3055	3182	3310	3437	3564	3691	35
25	3818	3945	4071	4198	4325	4452	34
26	4579	4705	4832	4959	5085	5212	33
27	5338	5465	5591	5718	5844	5970	32
28	6097	6223	6349	6475	6601	6728	31
29	6854	6980	7106	7232	7358	7483	30
30	7609	7735	7861	7987	8112	8238	29
31	8363	8489	8615	8740	8866	8991	28
32	9116	9242	9367	9492	9618	9743	27
33	9868	9993	.118	.243	.368	.493	26
34	9.220618	0743	0868	0993	1118	1242	25
35	1367	1492	1616	1741	1866	1990	24
36	2115	2239	2364	2488	2612	2737	23
37	2861	2985	3109	3234	3358	3482	22
38	3606	3730	3854	3978	4102	4226	21
39	4349	4473	4597	4721	4845	4968	20
40	5092	5215	5339	5462	5586	5709	19
41	5833	5956	6080	6203	6326	6449	18
42	6573	6696	6819	6942	7065	7188	17
43	7311	7434	7557	7680	7803	7925	16
44	8048	8171	8294	8416	8539	8661	15
45	8784	8906	9029	9151	9274	9396	14
46	9518	9641	9763	9885	...7	.130	13
47	9.230252	0374	0496	0618	0740	0862	12
48	0984	1106	1228	1349	1471	1593	11
49	1715	1836	1958	2079	2201	2323	10
50	2444	2565	2687	2808	2930	3051	9
51	3172	3293	3415	3536	3657	3778	8
52	3899	4020	4141	4262	4383	4504	7
53	4625	4746	4867	4987	5108	5229	6
54	5349	5470	5591	5711	5832	5952	5
55	6073	6193	6313	6434	6554	6674	4
56	6795	6915	7035	7155	7275	7395	3
57	7515	7635	7755	7875	7995	8115	2
58	8235	8355	8474	8594	8714	8834	1
59	8953	9073	9192	9312	9431	9551	0

of 8 Degrees.					Min.	Tangent of 9 Deg			
20″	30″	40″	50″			0″	10″	20″	30″
8108	8261	8413	8566	59	0	9.199713	9849	9985	.121
9023	9175	9328	9480	58	1	9.200529	0665	0801	0937
9936	.88	.240	.392	57	2	1345	1481	1616	1752
0848	0999	1151	1303	56	3	2159	2294	2430	2565
1757	1909	2060	2211	55	4	2971	3107	3242	3377
2665	2816	2967	3118	54	5	3782	3918	4053	4188
3571	3722	3873	4023	53	6	4592	4727	4862	4996
4475	4626	4776	4926	52	7	5400	5535	5669	5804
5377	5528	5678	5828	51	8	6207	6342	6476	6610
6278	6428	6578	6728	50	9	7013	7147	7281	7415
7177	7326	7476	7625	49	10	7817	7950	8084	8218
8074	8223	8372	8521	48	11	8619	8753	8886	9020
8969	9118	9267	9416	47	12	9420	9554	9687	9820
9862	..11	.160	.308	46	13	9.210220	0353	0486	0619
0754	0902	1051	1199	45	14	1018	1151	1284	1417
1644	1792	1940	2088	44	15	1815	1948	2081	2213
2532	2680	2828	2975	43	16	2611	2743	2876	3008
3418	3566	3713	3861	42	17	3405	3537	3670	3802
4303	4450	4598	4745	41	18	4198	4330	4462	4594
5186	5333	5480	5627	40	19	4989	5121	5253	5385
6067	6214	6361	6507	39	20	5780	5911	6043	6174
6947	7093	7240	7386	38	21	6568	6700	6831	6962
7825	7971	8117	8263	37	22	7356	7487	7618	7749
8701	8847	8992	9138	36	23	8142	8273	8403	8534
9575	9721	9866	..12	35	24	8926	9057	9188	9318
0448	0593	0739	0884	34	25	9710	9840	9971	.101
1319	1464	1609	1754	33	26	9.220492	0622	0752	0882
2188	2333	2478	2623	32	27	1272	1402	1532	1662
3056	3201	3345	3489	31	28	2052	2182	2311	2441
3922	4067	4211	4355	30	29	2830	2959	3089	3218
4787	4931	5075	5218	29	30	3607	3736	3865	3994
5650	5793	5937	6080	28	31	4382	4511	4640	4769
6511	6654	6797	6941	27	32	5156	5285	5414	5543
7370	7513	7656	7800	26	33	5929	6058	6186	6315
8228	8371	8514	8657	25	34	6700	6829	6957	7086
9085	9227	9370	9512	24	35	7471	7599	7727	7855
9939	..82	.224	.366	23	36	8239	8368	8496	8623
0792	0934	1076	1218	22	37	9007	9135	9263	9390
1644	1786	1927	2069	21	38	9773	9901	..29	.156
2494	2635	2777	2918	20	39	9.230539	0666	0793	0921
3342	3483	3625	3766	19	40	1302	1430	1557	1684
4189	4330	4471	4612	18	41	2065	2192	2319	2446
5034	5175	5315	5456	17	42	2826	2953	3080	3206
5878	6018	6158	6299	16	43	3586	3713	3839	3966
6720	6860	7000	7140	15	44	4345	4471	4598	4724
7560	7700	7840	7980	14	45	5103	5229	5355	5481
8399	8539	8678	8818	13	46	5859	5985	6111	6237
9236	9376	9515	9655	12	47	6614	6740	6865	6991
..72	.212	.351	.490	11	48	7368	7493	7619	7744
0907	1046	1184	1323	10	49	8120	8246	8371	8496
1739	1878	2017	2155	9	50	8872	8997	9122	9247
2571	2709	2848	2986	8	51	9622	9747	9872	9996
3401	3539	3677	3815	7	52	9.240371	0495	0620	0745
4229	4367	4505	4642	6	53	1118	1243	1367	1492
5056	5193	5331	5468	5	54	1865	1989	2114	2238
5881	6018	6156	6293	4	55	2610	2734	2858	2982
6705	6842	6979	7116	3	56	3354	3478	3602	3726
7527	7664	7801	7938	2	57	4097	4221	4345	4468
8348	8484	8621	8758	1	58	4839	4962	5086	5209
9167	9304	9440	9576	0	59	5579	5703	5826	5949

Sine of 10 Degrees.							Min.	Sine of 11]		
0″	10″	20″	30″	40″	50″			0″	10″	20″
9.239670	9790	9909	..28	.148	.267	59	0	9.280599	0707	0815
9.240386	0505	0624	0744	0863	0982	58	1	1248	1356	1465
1101	1220	1339	1458	1576	1695	57	2	1897	2005	2113
1814	1933	2052	2170	2289	2408	56	3	2544	2652	2760
2526	2645	2763	2882	3001	3119	55	4	3190	3298	3406
3237	3356	3474	3592	3711	3829	54	5	3836	3943	4051
3947	4065	4184	4302	4420	4538	53	6	4480	4588	4695
4656	4774	4892	5010	5128	5245	52	7	5124	5231	5338
5363	5481	5599	5717	5834	5952	51	8	5766	5873	5980
6069	6187	6305	6422	6540	6657	50	9	6408	6514	6621
6775	6892	7009	7127	7244	7361	49	10	7048	7155	7261
7478	7596	7713	7830	7947	8064	48	11	7688	7794	7900
8181	8298	8415	8532	8649	8766	47	12	8326	8432	8539
8883	8999	9116	9233	9350	9466	46	13	8964	9070	9176
9583	9700	9816	9933	..49	.166	45	14	9600	9706	9812
9.250282	0399	0515	0631	0748	0864	44	15	9.290236	0342	0447
0980	1097	1213	1329	1445	1561	43	16	0870	0976	1082
1677	1793	1909	2025	2141	2257	42	17	1504	1610	1715
2373	2489	2605	2720	2836	2952	41	18	2137	2242	2347
3067	3183	3299	3414	3530	3645	40	19	2768	2874	2979
3761	3876	3992	4107	4223	4338	39	20	3399	3504	3609
4453	4568	4684	4799	4914	5029	38	21	4029	4134	4239
5144	5259	5374	5490	5604	5719	37	22	4658	4763	4867
5834	5949	6064	6179	6294	6409	36	23	5286	5391	5495
6523	6638	6753	6867	6982	7096	35	24	5913	6017	6122
7211	7326	7440	7554	7669	7783	34	25	6539	6643	6747
7898	8012	8126	8241	8355	8469	33	26	7164	7268	7372
8583	8697	8811	8926	9040	9154	32	27	7788	7892	7996
9268	9382	9495	9609	9723	9837	31	28	8412	8515	8619
9951	..65	.178	.292	.406	.519	30	29	9034	9138	9241
9.260633	0747	0860	0974	1087	1201	29	30	9655	9759	9862
1314	1428	1541	1654	1768	1881	28	31	9.300276	0379	0482
1994	2107	2220	2334	2447	2560	27	32	0895	0999	1102
2673	2786	2899	3012	3125	3238	26	33	1514	1617	1720
3351	3464	3576	3689	3802	3915	25	34	2132	2235	2337
4027	4140	4253	4365	4478	4590	24	35	2748	2851	2954
4703	4815	4928	5040	5153	5265	23	36	3364	3467	3569
5377	5490	5602	5714	5827	5939	22	37	3979	4082	4184
6051	6163	6275	6387	6499	6611	21	38	4593	4696	4798
6723	6835	6947	7059	7171	7283	20	39	5207	5309	5411
7395	7506	7618	7730	7841	7953	19	40	5819	5921	6023
8065	8176	8288	8399	8511	8622	18	41	6430	6532	6634
8734	8845	8957	9068	9179	9291	17	42	7041	7142	7244
9402	9513	9624	9736	9847	9958	16	43	7650	7752	7853
9.270069	0180	0291	0402	0513	0624	15	44	8259	8360	8462
0735	0846	0957	1067	1178	1289	14	45	8867	8968	9069
1400	1510	1621	1732	1842	1953	13	46	9474	9575	9676
2064	2174	2285	2395	2505	2616	12	47	9.310080	0181	0282
2726	2837	2947	3057	3168	3278	11	48	0685	0786	0886
3388	3498	3608	3718	3829	3939	10	49	1289	1390	1490
4049	4159	4269	4379	4489	4598	9	50	1893	1993	2094
4708	4818	4928	5038	5148	5257	8	51	2495	2595	2696
5367	5477	5586	5696	5805	5915	7	52	3097	3197	3297
6025	6134	6243	6353	6462	6572	6	53	3698	3798	3898
6681	6790	6900	7009	7118	7227	5	54	4297	4397	4497
7337	7446	7555	7664	7773	7882	4	55	4897	4996	5096
7991	8100	8209	8318	8427	8536	3	56	5495	5594	5694
8645	8753	8862	8971	9080	9188	2	57	6092	6192	6291
9297	9406	9514	9623	9731	9840	1	58	6689	6788	6887
9948	..57	.165	.274	.382	.491	0	59	7284	7383	7482

[gent of 10 Degrees.]						Min.	Tangent of 11 Degrees.						
10″	20″	30″	40″	50″	[Min.]		0″	10″	20″	30″	40″	50″	[Min.]
6442	6565	6688	6811	6934	59	0	9.288652	8765	8877	8989	9102	9214	59
7180	7303	7426	7548	7671	58	1	9326	9438	9551	9663	9775	9887	58
7917	8039	8162	8285	8407	57	2	9999	.111	.223	.335	.447	.559	57
8652	8775	8897	9020	9142	56	3	9.290671	0783	0895	1007	1119	1231	56
9387	9509	9631	9753	9876	55	4	1342	1454	1566	1678	1789	1901	55
.120	.242	.364	.486	.608	54	5	2013	2124	2236	2347	2459	2570	54
0852	0974	1096	1218	1339	53	6	2682	2793	2905	3016	3127	3239	53
1583	1705	1826	1948	2070	52	7	3350	3461	3572	3684	3795	3906	52
2313	2434	2556	2677	2799	51	8	4017	4128	4239	4351	4462	4573	51
3041	3163	3284	3405	3527	50	9	4684	4795	4905	5016	5127	5238	50
3769	3890	4011	4132	4253	49	10	5349	5460	5571	5681	5792	5903	49
4495	4616	4737	4858	4979	48	11	6013	6124	6235	6345	6456	6566	48
5221	5341	5462	5583	5703	47	12	6677	6787	6898	7008	7119	7229	47
5945	6065	6186	6306	6427	46	13	7339	7450	7560	7670	7781	7891	46
6668	6788	6908	7029	7149	45	14	8001	8111	8221	8332	8442	8552	45
7389	7510	7630	7750	7870	44	15	8662	8772	8882	8992	9102	9212	44
8110	8230	8350	8470	8590	43	16	9322	9431	9541	9651	9761	9871	43
8830	8950	9069	9189	9309	42	17	9980	..90	.200	.309	.419	.529	42
9548	9668	9787	9907	..27	41	18	9.300638	0748	0857	0967	1076	1186	41
0266	0385	0504	0624	0743	40	19	1295	1405	1514	1624	1733	1842	40
0982	1101	1220	1340	1459	39	20	1951	2061	2170	2279	2388	2497	39
1697	1816	1935	2054	2173	38	21	2607	2716	2825	2934	3043	3152	38
2411	2530	2649	2768	2887	37	22	3261	3370	3479	3588	3697	3805	37
3124	3243	3361	3480	3599	36	23	3914	4023	4132	4241	4349	4458	36
3836	3954	4073	4191	4310	35	24	4567	4675	4784	4893	5001	5110	35
4547	4665	4783	4902	5020	34	25	5218	5327	5435	5544	5652	5761	34
5256	5375	5493	5611	5729	33	26	5869	5977	6086	6194	6302	6410	33
5965	6083	6201	6319	6437	32	27	6519	6627	6735	6843	6951	7059	32
6673	6790	6908	7026	7144	31	28	7168	7276	7384	7492	7600	7708	31
7379	7497	7614	7732	7849	30	29	7816	7923	8031	8139	8247	8355	30
8084	8202	8319	8437	8554	29	30	8463	8570	8678	8786	8893	9001	29
8789	8906	9023	9140	9258	28	31	9109	9216	9324	9432	9539	9647	28
9492	9609	9726	9843	9960	27	32	9754	9862	9969	..76	.184	.291	27
0194	0311	0428	0545	0662	26	33	9.310399	0506	0613	0720	0828	0935	26
0895	1012	1129	1246	1362	25	34	1042	1149	1256	1364	1471	1578	25
1595	1712	1829	1945	2062	24	35	1685	1792	1899	2006	2113	2220	24
2294	2411	2527	2644	2760	23	36	2327	2433	2540	2647	2754	2861	23
2992	3109	3225	3341	3457	22	37	2968	3074	3181	3288	3394	3501	22
3689	3805	3921	4037	4153	21	38	3608	3714	3821	3927	4034	4140	21
4385	4501	4617	4733	4849	20	39	4247	4353	4460	4566	4672	4779	20
5080	5196	5312	5427	5543	19	40	4885	4991	5098	5204	5310	5416	19
5774	5890	6005	6121	6236	18	41	5523	5629	5735	5841	5947	6053	18
6467	6582	6698	6813	6928	17	42	6159	6265	6371	6477	6583	6689	17
7159	7274	7389	7504	7619	16	43	6795	6901	7007	7113	7218	7324	16
7849	7964	8079	8194	8309	15	44	7430	7536	7641	7747	7853	7958	15
8539	8654	8769	8884	8998	14	45	8064	8170	8275	8381	8486	8592	14
9228	9342	9457	9572	9686	13	46	8697	8803	8908	9013	9119	9224	13
9915	..30	.144	.259	.373	12	47	9330	9435	9540	9645	9751	9856	12
0602	0717	0831	0945	1059	11	48	9961	..66	.171	.277	.382	.487	11
1288	1402	1516	1630	1744	10	49	9.320592	0697	0802	0907	1012	1117	10
1973	2087	2201	2314	2428	9	50	1222	1326	1431	1536	1641	1746	9
2656	2770	2884	2998	3111	8	51	1851	1955	2060	2165	2269	2374	8
3339	3453	3566	3680	3793	7	52	2479	2583	2688	2793	2897	3002	7
4021	4134	4248	4361	4474	6	53	3106	3211	3315	3420	3524	3628	6
4701	4815	4928	5041	5154	5	54	3733	3837	3941	4046	4150	4254	5
5381	5494	5607	5720	5833	4	55	4358	4463	4567	4671	4775	4879	4
6060	6173	6286	6399	6512	3	56	4983	5087	5191	5295	5399	5503	3
6737	6850	6963	7076	7189	2	57	5607	5711	5815	5919	6023	6127	2
7414	7527	7639	7752	7865	1	58	6231	6334	6438	6542	6646	6749	1
8090	8202	8315	8427	8540	0	59	6853	6957	7060	7164	7267	7371	0

Sine of 12 Degrees.							Min.	Sine of 13		
0″	10″	20″	30″	40″	50″			0″	10″	20″
317879	7978	8077	8176	8275	8374	59	0	9.352088	2179	2270
8473	8572	8671	8769	8868	8967	58	1	2635	2726	2817
9066	9165	9263	9362	9461	9559	57	2	3181	3272	3363
9658	9757	9855	9954	..52	.151	56	3	3726	3817	3908
320249	0348	0446	0545	0643	0742	55	4	4271	4362	4452
0840	0938	1037	1135	1233	1332	54	5	4815	4906	4996
1430	1528	1626	1724	1822	1921	53	6	5358	5449	5539
2019	2117	2215	2313	2411	2509	52	7	5901	5991	6081
2607	2705	2802	2900	2998	3096	51	8	6443	6533	6623
3194	3292	3389	3487	3585	3683	50	9	6984	7074	7164
323780	3878	3975	4073	4171	4268	49	10	9.357524	7614	7704
4366	4463	4561	4658	4756	4853	48	11	8064	8154	8243
4950	5048	5145	5243	5340	5437	47	12	8603	8692	8782
5534	5632	5729	5826	5923	6020	46	13	9141	9231	9320
6117	6215	6312	6409	6506	6603	45	14	9678	9768	9858
6700	6797	6894	6991	7087	7184	44	15	9.360215	0305	0394
7281	7378	7475	7572	7668	7765	43	16	0752	0841	0930
7862	7958	8055	8152	8248	8345	42	17	1287	1376	1465
8442	8538	8635	8731	8828	8924	41	18	1822	1911	2000
9021	9117	9213	9310	9406	9502	40	19	2356	2445	2534
329599	9695	9791	9888	9984	..80	39	20	9.362889	2978	3067
330176	0272	0368	0465	0561	0657	38	21	3422	3511	3599
0753	0849	0945	1041	1137	1233	37	22	3954	4042	4131
1329	1424	1520	1616	1712	1808	36	23	4485	4574	4662
1903	1999	2095	2191	2286	2382	35	24	5016	5104	5193
2478	2573	2669	2764	2860	2956	34	25	5546	5634	5722
3051	3147	3242	3337	3433	3528	33	26	6075	6163	6251
3624	3719	3814	3910	4005	4100	32	27	6604	6692	6780
4195	4291	4386	4481	4576	4671	31	28	7131	7219	7307
4767	4862	4957	5052	5147	5242	30	29	7659	7747	7834
335337	5432	5527	5622	5716	5811	29	30	9.368185	8273	8361
5906	6001	6096	6191	6285	6380	28	31	8711	8799	8886
6475	6570	6664	6759	6854	6948	27	32	9236	9324	9411
7043	7137	7232	7326	7421	7515	26	33	9761	9848	9936
7610	7704	7799	7893	7988	8082	25	34	9.370285	0372	0459
8176	8271	8365	8459	8553	8648	24	35	0808	0895	0982
8742	8836	8930	9024	9118	9212	23	36	1330	1417	1504
9307	9401	9495	9589	9683	9777	22	37	1852	1939	2026
9871	9964	..58	.152	.246	.340	21	38	2373	2460	2547
340434	0528	0621	0715	0809	0903	20	39	2894	2981	3067
340996	1090	1184	1277	1371	1464	19	40	9.373414	3500	3587
1558	1652	1745	1839	1932	2026	18	41	3933	4020	4106
2119	2212	2306	2399	2493	2586	17	42	4452	4538	4624
2679	2772	2866	2959	3052	3145	16	43	4970	5056	5142
3239	3332	3425	3518	3611	3704	15	44	5487	5573	5659
3797	3890	3983	4076	4169	4262	14	45	6003	6089	6175
4355	4448	4541	4634	4727	4820	13	46	6519	6605	6691
4912	5005	5098	5191	5283	5376	12	47	7035	7120	7206
5469	5561	5654	5747	5839	5932	11	48	7549	7635	7721
6024	6117	6210	6302	6395	6487	10	49	8063	8149	8235
346579	6672	6764	6857	6949	7041	9	50	9.378577	8662	8748
7134	7226	7318	7410	7503	7595	8	51	9089	9175	9260
7687	7779	7871	7963	8056	8148	7	52	9601	9687	9772
8240	8332	8424	8516	8608	8700	6	53	9.380113	0198	0283
8792	8884	8976	9067	9159	9251	5	54	0624	0709	0794
9343	9435	9526	9618	9710	9802	4	55	1134	1219	1304
9893	9985	..77	.168	.260	.352	3	56	1643	1728	1813
350443	0535	0626	0718	0809	0901	2	57	2152	2237	2322
0992	1084	1175	1266	1358	1449	1	58	2661	2745	2830
1540	1632	1723	1814	1906	1997	0	59	3168	3253	3337

Tangent of 12 Degrees. 0″	10″	20″	30″	40″	50″		Min.
9.327475	7578	7682	7785	7888	7992	59	0
8095	8199	8302	8405	8509	8612	58	1
8715	8819	8922	9025	9128	9231	57	2
9334	9438	9541	9644	9747	9850	56	3
9953	..56	.159	.262	.365	.468	55	4
9.330570	0673	0776	0879	0982	1084	54	5
1187	1290	1393	1495	1598	1701	53	6
1803	1906	2008	2111	2213	2316	52	7
2418	2521	2623	2726	2828	2930	51	8
3033	3135	3237	3340	3442	3544	50	9
9.333646	3748	3851	3953	4055	4157	49	10
4259	4361	4463	4565	4667	4769	48	11
4871	4973	5075	5177	5279	5380	47	12
5482	5584	5686	5788	5889	5991	46	13
6093	6194	6296	6398	6499	6601	45	14
6702	6804	6905	7007	7108	7210	44	15
7311	7413	7514	7615	7717	7818	43	16
7919	8021	8122	8223	8324	8426	42	17
8527	8628	8729	8830	8931	9032	41	18
9133	9234	9335	9436	9537	9638	40	19
9.339739	9840	9941	..42	.143	.243	39	20
9.340344	0445	0546	0646	0747	0848	38	21
0948	1049	1150	1250	1351	1451	37	22
1552	1652	1753	1853	1954	2054	36	23
2155	2255	2355	2456	2556	2656	35	24
2757	2857	2957	3057	3158	3258	34	25
3358	3458	3558	3658	3758	3858	33	26
3958	4058	4158	4258	4358	4458	32	27
4558	4658	4758	4858	4957	5057	31	28
5157	5257	5357	5456	5556	5656	30	29
9.345755	5855	5954	6054	6154	6253	29	30
6353	6452	6552	6651	6751	6850	28	31
6949	7049	7148	7248	7347	7446	27	32
7545	7645	7744	7843	7942	8042	26	33
8141	8240	8339	8438	8537	8636	25	34
8735	8834	8933	9032	9131	9230	24	35
9329	9428	9527	9626	9724	9823	23	36
9922	..21	.120	.218	.317	.416	22	37
9.350514	0613	0712	0810	0909	1007	21	38
1106	1204	1303	1401	1500	1598	20	39
9.351697	1795	1894	1992	2090	2189	19	40
2287	2385	2483	2582	2680	2778	18	41
2876	2974	3073	3171	3269	3367	17	42
3465	3563	3661	3759	3857	3955	16	43
4053	4151	4249	4347	4445	4542	15	44
4640	4738	4836	4934	5031	5129	14	45
5227	5324	5422	5520	5617	5715	13	46
5813	5910	6008	6105	6203	6300	12	47
6398	6495	6593	6690	6787	6885	11	48
6982	7079	7177	7274	7371	7469	10	49
9.357566	7663	7760	7857	7954	8052	9	50
8149	8246	8343	8440	8537	8634	8	51
8731	8828	8925	9022	9119	9216	7	52
9313	9409	9506	9603	9700	9797	6	53
9893	9990	..87	.184	.280	.377	5	54
9.360474	0570	0667	0763	0860	0957	4	55
1053	1150	1246	1343	1439	1535	3	56
1632	1728	1825	1921	2017	2114	2	57
2210	2306	2403	2499	2595	2691	1	58
2787	2884	2980	3076	3172	3268	0	59

Min.	Tangent of 13 Degrees. 0″	10″	20″	30″	40″	50″	
0	9.363364	3460	3556	3652	3748	3844	59
1	3940	4036	4132	4228	4324	4420	58
2	4515	4611	4707	4803	4899	4994	57
3	5090	5186	5282	5377	5473	5568	56
4	5664	5760	5855	5951	6046	6142	55
5	6237	6333	6428	6524	6619	6715	54
6	6810	6905	7001	7096	7191	7287	53
7	7382	7477	7572	7668	7763	7858	52
8	7953	8048	8143	8239	8334	8429	51
9	8524	8619	8714	8809	8904	8999	50
10	9.369094	9189	9284	9378	9473	9568	49
11	9663	9758	9853	9947	..42	.137	48
12	9.370232	0326	0421	0516	0610	0705	47
13	0799	0894	0989	1083	1178	1272	46
14	1367	1461	1556	1650	1744	1839	45
15	1933	2028	2122	2216	2311	2405	44
16	2499	2593	2688	2782	2876	2970	43
17	3064	3159	3253	3347	3441	3535	42
18	3629	3723	3817	3911	4005	4099	41
19	4193	4287	4381	4475	4569	4662	40
20	9.374756	4850	4944	5038	5131	5225	39
21	5319	5413	5506	5600	5694	5787	38
22	5881	5975	6068	6162	6255	6349	37
23	6442	6536	6629	6723	6816	6910	36
24	7003	7096	7190	7283	7376	7470	35
25	7563	7656	7750	7843	7936	8029	34
26	8122	8216	8309	8402	8495	8588	33
27	8681	8774	8867	8960	9053	9146	32
28	9239	9332	9425	9518	9611	9704	31
29	9797	9890	9983	..75	.168	.261	30
30	9.380354	0446	0539	0632	0725	0817	29
31	0910	1003	1095	1188	1280	1373	28
32	1466	1558	1651	1743	1836	1928	27
33	2020	2113	2205	2298	2390	2482	26
34	2575	2667	2759	2852	2944	3036	25
35	3129	3221	3313	3405	3497	3589	24
36	3682	3774	3866	3958	4050	4142	23
37	4234	4326	4418	4510	4602	4694	22
38	4786	4878	4970	5062	5153	5245	21
39	5337	5429	5521	5612	5704	5796	20
40	9 385888	5979	6071	6163	6254	6346	19
41	6438	6529	6621	6712	6804	6895	18
42	6987	7078	7170	7261	7353	7444	17
43	7536	7627	7718	7810	7901	7992	16
44	8084	8175	8266	8358	8449	8540	15
45	8631	8722	8814	8905	8996	9087	14
46	9178	9269	9360	9451	9542	9633	13
47	9724	9815	9906	9997	..88	.179	12
48	9.390270	0361	0452	0543	0633	0724	11
49	0815	0906	0997	1087	1178	1269	10
50	9.391360	1450	1541	1632	1722	1813	9
51	1903	1994	2085	2175	2266	2356	8
52	2447	2537	2628	2718	2808	2899	7
53	2989	3080	3170	3260	3351	3441	6
54	3531	3622	3712	3802	3892	3983	5
55	4073	4163	4253	4343	4433	4523	4
56	4614	4704	4794	4884	4974	5064	3
57	5154	5244	5334	5424	5514	5604	2
58	5694	5783	5873	5963	6053	6143	1
59	6233	6322	6412	6502	6592	6681	0

Sine of 14 Degrees.							Min.	Sine of 15]		
0″	10″	20″	30″	40″	50″			0″	10″	20″
383675	3760	3844	3928	4013	4097	59	0	9.412996	3075	3153
4182	4266	4350	4435	4519	4603	58	1	3467	3546	3624
4687	4772	4856	4940	5024	5108	57	2	3938	4016	4095
5192	5277	5361	5445	5529	5613	56	3	4408	4486	4565
5697	5781	5865	5949	6033	6117	55	4	4878	4956	5034
6201	6285	6369	6452	6536	6620	54	5	5347	5425	5503
6704	6788	6872	6955	7039	7123	53	6	5815	5893	5971
7207	7290	7374	7458	7541	7625	52	7	6283	6361	6439
7709	7792	7876	7959	8043	8127	51	8	6751	6828	6906
8210	8294	8377	8461	8544	8627	50	9	7217	7295	7373
388711	8794	8878	8961	9044	9128	49	10	9.417684	7761	7839
9211	9294	9378	9461	9544	9627	48	11	8150	8227	8305
9711	9794	9877	9960	..43	.126	47	12	8615	8692	8770
390210	0293	0376	0459	0542	0625	46	13	9079	9157	9234
0708	0791	0874	0957	1040	1123	45	14	9544	9621	9698
1206	1289	1371	1454	1537	1620	44	15	9.420007	0085	0162
1703	1786	1868	1951	2034	2117	43	16	0470	0548	0625
2199	2282	2365	2447	2530	2613	42	17	0933	1010	1087
2695	2778	2860	2943	3025	3108	41	18	1395	1472	1549
3191	3273	3356	3438	3520	3603	40	19	1857	1933	2010
393685	3768	3850	3932	4015	4097	39	20	9.422318	2394	2471
4179	4262	4344	4426	4508	4591	38	21	2778	2855	2931
4673	4755	4837	4919	5002	5084	37	22	3238	3315	3391
5166	5248	5330	5412	5494	5576	36	23	3697	3774	3850
5658	5740	5822	5904	5986	6068	35	24	4156	4233	4309
6150	6232	6314	6395	6477	6559	34	25	4615	4691	4767
6641	6723	6805	6886	6968	7050	33	26	5073	5149	5225
7132	7213	7295	7377	7458	7540	32	27	5530	5606	5682
7621	7703	7785	7866	7948	8029	31	28	5987	6063	6139
8111	8192	8274	8355	8437	8518	30	29	6443	6519	6595
398600	8681	8762	8844	8925	9007	29	30	9.426899	6975	7051
9088	9169	9250	9332	9413	9494	28	31	7354	7430	7506
9575	9657	9738	9819	9900	9981	27	32	7809	7885	7960
400062	0144	0225	0306	0387	0468	26	33	8263	8339	8414
0549	0630	0711	0792	0873	0954	25	34	8717	8792	8868
1035	1116	1197	1277	1358	1439	24	35	9170	9246	9321
1520	1601	1682	1762	1843	1924	23	36	9623	9698	9774
2005	2085	2166	2247	2328	2408	22	37	9.430075	0150	0226
2489	2570	2650	2731	2811	2892	21	38	0527	0602	0677
2972	3053	3133	3214	3294	3375	20	39	0978	1053	1128
403455	3536	3616	3697	3777	3857	19	40	9.431429	1504	1579
3938	4018	4098	4179	4259	4339	18	41	1879	1954	2029
4420	4500	4580	4660	4741	4821	17	42	2329	2403	2478
4901	4981	5061	5141	5221	5302	16	43	2778	2853	2927
5382	5462	5542	5622	5702	5782	15	44	3226	3301	3376
5862	5942	6022	6102	6181	6261	14	45	3675	3749	3824
6341	6421	6501	6581	6661	6740	13	46	4122	4197	4271
6820	6900	6980	7060	7139	7219	12	47	4569	4644	4718
7299	7378	7458	7538	7617	7697	11	48	5016	5091	5165
7777	7856	7936	8015	8095	8174	10	49	5462	5537	5611
408254	8333	8413	8492	8572	8651	9	50	9.435908	5982	6056
8731	8810	8889	8969	9048	9127	8	51	6353	6427	6502
9207	9286	9365	9445	9524	9603	7	52	6798	6872	6946
9682	9762	9841	9920	9999	..78	6	53	7242	7316	7390
410157	0237	0316	0395	0474	0553	5	54	7686	7760	7834
0632	0711	0790	0869	0948	1027	4	55	8129	8203	8277
1106	1185	1264	1343	1422	1500	3	56	8572	8646	8719
1579	1658	1737	1816	1895	1973	2	57	9014	9088	9162
2052	2131	2210	2288	2367	2446	1	58	9456	9530	9603
2524	2603	2682	2760	2839	2918	0	59	9897	9971	..44

gent of 14 Degrees.						Min.	Tangent of 15 Degrees.					
10″	20″	30″	40″	50″			0″	10″	20″	30″	40″	50″
6861	6950	7040	7130	7219	59	0	9.428052	8137	8221	8305	8389	8473
7399	7488	7578	7667	7757	58	1	8558	8642	8726	8810	8894	8978
7936	8025	8115	8204	8294	57	2	9062	9146	9230	9314	9398	9482
8472	8562	8651	8740	8830	56	3	9566	9650	9734	9818	9902	9986
9008	9098	9187	9276	9365	55	4	9.430070	0154	0237	0321	0405	0489
9544	9633	9722	9811	9900	54	5	0573	0657	0740	0824	0908	0992
..79	.168	.257	.346	.435	53	6	1075	1159	1243	1326	1410	1494
0613	0702	0791	0880	0969	52	7	1577	1661	1745	1828	1912	1995
1147	1236	1325	1413	1502	51	8	2079	2162	2246	2329	2413	2496
1680	1769	1857	1946	2035	50	9	2580	2663	2747	2830	2914	2997
2212	2301	2390	2478	2567	49	10	9.433080	3164	3247	3331	3414	3497
2744	2833	2922	3010	3099	48	11	3580	3664	3747	3830	3914	3997
3276	3364	3453	3541	3630	47	12	4080	4163	4246	4330	4413	4496
3807	3895	3983	4072	4160	46	13	4579	4662	4745	4828	4912	4995
4337	4425	4514	4602	4690	45	14	5078	5161	5244	5327	5410	5493
4867	4955	5043	5131	5219	44	15	5576	5659	5742	5825	5907	5990
5396	5484	5572	5660	5748	43	16	6073	6156	6239	6322	6405	6488
5924	6012	6100	6188	6276	42	17	6570	6653	6736	6819	6901	6984
6452	6540	6628	6716	6804	41	18	7067	7150	7232	7315	7398	7480
6980	7068	7155	7243	7331	40	19	7563	7646	7728	7811	7894	7976
7507	7594	7682	7770	7858	39	20	9.438059	8141	8224	8306	8389	8471
8033	8121	8208	8296	8384	38	21	8554	8636	8719	8801	8884	8966
8559	8646	8734	8821	8909	37	22	9048	9131	9213	9296	9378	9460
9084	9171	9259	9346	9434	36	23	9543	9625	9707	9790	9872	9954
9609	9696	9783	9871	9958	35	24	9.440036	0119	0201	0283	0365	0447
0133	0220	0307	0395	0482	34	25	0529	0612	0694	0776	0858	0940
0656	0743	0831	0918	1005	33	26	1022	1104	1186	1268	1350	1432
1179	1266	1353	1441	1528	32	27	1514	1596	1678	1760	1842	1924
1702	1789	1876	1963	2050	31	28	2006	2088	2170	2252	2334	2416
2224	2310	2397	2484	2571	30	29	2497	2579	2661	2743	2825	2907
2745	2832	2919	3005	3092	29	30	9.442988	3070	3152	3234	3315	3397
3266	3352	3439	3526	3613	28	31	3479	3560	3642	3724	3805	3887
3786	3873	3959	4046	4132	27	32	3968	4050	4132	4213	4295	4376
4306	4392	4479	4565	4652	26	33	4458	4539	4621	4702	4784	4865
4825	4911	4998	5084	5171	25	34	4947	5028	5110	5191	5272	5354
5343	5430	5516	5603	5689	24	35	5435	5517	5598	5679	5761	5842
5862	5948	6034	6120	6207	23	36	5923	6005	6086	6167	6248	6330
6379	6465	6551	6638	6724	22	37	6411	6492	6573	6654	6735	6817
6896	6982	7068	7154	7240	21	38	6898	6979	7060	7141	7222	7303
7413	7499	7585	7671	7757	20	39	7384	7465	7546	7627	7708	7789
7928	8014	8100	8186	8272	19	40	9.447870	7951	8032	8113	8194	8275
8444	8530	8616	8701	8787	18	41	8356	8437	8518	8599	8680	8760
8959	9044	9130	9216	9302	17	42	8841	8922	9003	9084	9164	9245
9473	9559	9644	9730	9816	16	43	9326	9407	9487	9568	9649	9730
9987	..72	.158	.244	.329	15	44	9810	9891	9972	..52	.133	.213
0500	0586	0671	0757	0842	14	45	9.450294	0375	0455	0536	0616	0697
1013	1098	1184	1269	1354	13	46	0777	0858	0938	1019	1099	1180
1525	1610	1696	1781	1866	12	47	1260	1341	1421	1502	1582	1662
2037	2122	2207	2292	2378	11	48	1743	1823	1903	1984	2064	2144
2548	2633	2718	2803	2888	10	49	2225	2305	2385	2465	2546	2626
3059	3144	3229	3314	3399	9	50	9.452706	2786	2867	2947	3027	3107
3569	3654	3739	3824	3909	8	51	3187	3267	3347	3428	3508	3588
4078	4163	4248	4333	4418	7	52	3668	3748	3828	3908	3988	4068
4587	4672	4757	4842	4927	6	53	4148	4228	4308	4388	4468	4548
5096	5181	5265	5350	5435	5	54	4628	4708	4787	4867	4947	5027
5604	5689	5773	5858	5942	4	55	5107	5187	5267	5346	5426	5506
6112	6196	6281	6365	6450	3	56	5586	5655	5745	5825	5905	5984
6619	6703	6788	6872	6956	2	57	6064	6144	6223	6303	6383	6462
7125	7210	7294	7378	7463	1	58	6542	6622	6701	6781	6860	6940
7631	7715	7800	7884	7968	0	59	7019	7099	7178	7258	7337	7417

Sine of 16 Degrees.							Min.	Sine of 17		
0″	10″	20″	30″	40″	50″			0″	10″	20″
9.440338	0411	0485	0558	0632	0705	59	0	9.465935	6004	60[illegible]3
0778	0852	0925	0998	1072	1145	58	1	6348	6417	6486
1218	1292	1365	1438	1511	1584	57	2	6761	6830	6898
1658	1731	1804	1877	1950	2023	56	3	7173	7242	7310
2096	2170	2243	2316	238[illegible]	2462	55	4	7585	7653	7722
2535	2608	2681	2754	2827	2900	54	5	7996	8065	8133
2973	3046	3119	3192	3265	3337	53	6	8407	8475	8544
3410	3483	3556	3629	3702	3774	52	7	8817	8886	8954
3847	3920	3993	4066	4138	4211	51	8	9227	9296	9364
4284	4356	4429	4502	4574	4647	50	9	9637	9705	9773
9.444720	4792	4865	4938	5010	5083	49	10	9.470046	0114	0182
5155	5228	5300	5373	5445	5518	48	11	0455	0523	0591
5590	5663	5735	5808	5880	5953	47	12	0863	0931	0999
6025	6097	6170	6242	6314	6387	46	13	1271	1339	1407
6459	6531	6604	6676	6748	6820	45	14	1679	1746	1814
6893	6965	7037	7109	7182	7254	44	15	2086	2153	2221
7326	7398	7470	7542	7614	7687	43	16	2492	2560	2628
7759	7831	7903	7975	8047	8119	42	17	2898	2966	3034
8191	8263	8335	8407	8479	8551	41	18	3304	3372	3439
8623	8695	8767	8838	8910	8982	40	19	3710	3777	3845
9.449054	9126	9198	9269	9341	9413	39	20	9.474115	4182	4250
9485	9557	9628	9700	9772	9844	38	21	4519	4587	4654
9915	9987	..59	.130	.202	.274	37	22	4923	4991	5058
9.450345	0417	0488	0560	0632	0703	36	23	5327	5394	5462
0775	0846	0918	0989	1061	1132	35	24	5730	5798	5865
1204	1275	1347	1418	1489	1561	34	25	6133	6200	6268
1632	1704	1775	1846	1918	1989	33	26	6536	6603	6670
2060	2132	2203	2274	2345	2417	32	27	6938	7005	7072
2488	2559	2630	2702	2773	2844	31	28	7340	7407	7473
2915	2986	3057	3129	3200	3271	30	29	7741	7808	7875
9.453342	3413	3484	3555	3626	3697	29	30	9.478142	8209	8275
3768	3839	3910	3981	4052	4123	28	31	8542	8609	8676
4194	4265	4336	4407	4477	4548	27	32	8942	9009	9076
4619	4690	4761	4832	4903	4973	26	33	9342	9409	9475
5044	5115	5186	5256	5327	5398	25	34	9741	9808	9874
5469	5539	5610	5681	5751	5822	24	35	9.480140	0207	0273
5893	5963	6034	6104	6175	6246	23	36	0539	0605	0671
6316	6387	6457	6528	6598	6669	22	37	0937	1003	1069
6739	6810	6880	6951	7021	7091	21	38	1334	1400	1467
7162	7232	7303	7373	7443	7514	20	39	1731	1798	1864
9.457584	7654	7725	7795	7865	7936	19	40	9.482128	2194	2261
8006	8076	8146	8217	8287	8357	18	41	2525	2591	2657
8427	8497	8567	8638	8708	8778	17	42	2921	2987	3053
8848	8918	8988	9058	9128	9198	16	43	3316	3382	3448
9268	9338	9408	9478	9548	9618	15	44	3712	3778	3843
9688	9758	9828	9898	9968	..38	14	45	4107	4172	4238
9.460108	0178	0248	0317	0387	0457	13	46	4501	4567	4632
0527	0597	0667	0736	0806	0876	12	47	4895	4961	5026
0946	1015	1085	1155	1224	1294	11	48	5289	5354	5420
1364	1433	1503	1573	1642	1712	10	49	5682	5748	5813
9.461782	1851	1921	1990	2060	2129	9	50	9.486075	6140	6206
2199	2268	2338	2407	2477	2546	8	51	6467	6533	6598
2616	2685	2755	2824	2894	2963	7	52	6860	6925	6990
3032	3102	3171	3240	3310	3379	6	53	7251	7316	7382
3448	3518	3587	3656	3725	3795	5	54	7643	7708	7773
3864	3933	4002	4072	4141	4210	4	55	8034	8099	8164
4279	4348	4417	4486	4556	4625	3	56	8424	8489	8554
4694	4763	4832	4901	4970	5039	2	57	8814	8879	8944
5108	5177	5246	5315	5384	5453	1	58	9204	9269	9334
5522	5591	5660	5729	5798	5866	0	59	9593	9658	9723

of 16 Degrees.					Min.	Tangent of 17 Degrees.				
20″	30″	40″	50″			0″	10″	20″	30″	40″
7655	7735	7814	7894	59	0	9.485339	5414	5490	5565	5640
8132	8211	8290	8370	58	1	5791	5866	5941	6016	6092
8608	8687	8766	8846	57	2	6242	6317	6392	6467	6543
9083	9163	9242	9321	56	3	6693	6768	6843	6918	6993
9558	9638	9717	9796	55	4	7143	7218	7293	7368	7443
..33	.112	.191	.270	54	5	7593	7668	7743	7818	7893
0507	0586	0665	0744	53	6	8043	8118	8193	8268	8343
0981	1060	1139	1218	52	7	8492	8567	8642	8717	8792
1454	1533	1612	1691	51	8	8941	9016	9091	9166	9240
1927	2006	2085	2164	50	9	9390	9465	9539	9614	9689
2400	2478	2557	2636	49	10	9.489838	9913	9987	..62	.137
2872	2950	3029	3108	48	11	9.490286	0360	0435	0510	0584
3343	3422	3501	3579	47	12	0733	0808	0882	0957	1031
3815	3893	3972	4050	46	13	1180	1255	1329	1404	1478
4285	4364	4442	4521	45	14	1627	1701	1776	1850	1924
4756	4834	4912	4991	44	15	2073	2147	2222	2296	2370
5226	5304	5382	5460	43	16	2519	2593	2668	2742	2816
5695	5773	5851	5930	42	17	2965	3039	3113	3187	3261
6164	6242	6320	6398	41	18	3410	3484	3558	3632	3706
6633	6711	6789	6867	40	19	3854	3929	4003	4077	4151
7101	7179	7257	7335	39	20	9.494299	4373	4447	4521	4595
7569	7647	7724	7802	38	21	4743	4817	4891	4965	5039
8036	8114	8192	8269	37	22	5186	5260	5334	5408	5482
8503	8581	8658	8736	36	23	5630	5704	5777	5851	5925
8969	9047	9125	9202	35	24	6073	6146	6220	6294	6368
9435	9513	9591	9668	34	25	6515	6589	6663	6736	6810
9901	9979	..56	.134	33	26	6957	7031	7105	7178	7252
0366	0444	0521	0599	32	27	7399	7473	7546	7620	7693
0831	0909	0986	1063	31	28	7841	7914	7988	8061	8135
1295	1373	1450	1528	30	29	8282	8355	8429	8502	8575
1759	1837	1914	1991	29	30	9.498722	8796	8869	8943	9016
2223	2300	2377	2455	28	31	9163	9236	9309	9383	9456
2686	2763	2840	2918	27	32	9603	9676	9749	9822	9896
3149	3226	3303	3380	26	33	9.500042	0115	0189	0262	0335
3611	3688	3765	3842	25	34	0481	0555	0628	0701	0774
4073	4150	4227	4304	24	35	0920	0993	1066	1140	1213
4535	4612	4688	4765	23	36	1359	1432	1505	1578	1651
4996	5073	5149	5226	22	37	1797	1870	1943	2016	2089
5456	5533	5610	5687	21	38	2235	2308	2381	2453	2526
5917	5993	6070	6147	20	39	2672	2745	2818	2891	2964
6377	6453	6530	6606	19	40	9.503109	3182	3255	3328	3400
6836	6913	6989	7066	18	41	3546	3619	3691	3764	3837
7295	7372	7448	7524	17	42	3982	4055	4128	4200	4273
7754	7830	7906	7983	16	43	4418	4491	4563	4636	4709
8212	8288	8365	8441	15	44	4854	4926	4999	5072	5144
8670	8746	8822	8899	14	45	5289	5362	5434	5507	5579
9127	9203	9280	9356	13	46	5724	5796	5869	5941	6014
9584	9660	9737	9813	12	47	6159	6231	6303	6376	6448
..41	.117	.193	.269	11	48	6593	6665	6737	6810	6882
0497	0573	0649	0725	10	49	7027	7099	7171	7243	7316
0953	1029	1105	1181	9	50	9.507460	7532	7605	7677	7749
1408	1484	1560	1636	8	51	7893	7965	8038	8110	8182
1863	1939	2015	2091	7	52	8326	8398	8470	8542	8614
2318	2394	2470	2545	6	53	8759	8831	8903	8975	9047
2772	2848	2924	2999	5	54	9191	9263	9335	9407	9479
3226	3302	3377	3453	4	55	9622	9694	9766	9838	9910
3680	3755	3831	3906	3	56	9.510054	0126	0198	0269	0341
4133	4208	4284	4359	2	57	0485	0557	0629	0700	0772
4585	4661	4736	4812	1	58	0916	0987	1059	1131	1203
5038	5113	5188	5264	0	59	1346	1418	1489	1561	1633

Sine of 18 Degrees.							Min.	Sine of 19 Degre			
0″	10″	20″	30″	40″	50″			0″	10″	20″	30″
9.489982	..47	.112	.177	.241	.306	59	0	9.512642	2703	2764	2825
9.490371	0436	0500	0565	0630	0695	58	1	3009	3070	3131	3192
0759	0824	0889	0953	1018	1082	57	2	3375	3436	3497	3558
1147	1212	1276	1341	1405	1470	56	3	3741	3802	3863	3924
1535	1599	1664	1728	1793	1857	55	4	4107	4168	4229	4289
1922	1986	2051	2115	2179	2244	54	5	4472	4533	4594	4655
2308	2373	2437	2502	2566	2630	53	6	4837	4898	4959	5019
2695	2759	2823	2888	2952	3016	52	7	5202	5262	5323	5384
3081	3145	3209	3273	3338	3402	51	8	5566	5627	5687	5748
3466	3530	3595	3659	3723	3787	50	9	5930	5991	6051	6112
9.493851	3915	3980	4044	4108	4172	49	10	9 516294	6354	6415	6475
4236	4300	4364	4428	4492	4556	48	11	6657	6717	6778	6838
4621	4685	4749	4813	4877	4941	47	12	7020	7080	7141	7201
5005	5069	5133	5196	5260	5324	46	13	7382	7443	7503	7564
5388	5452	5516	5580	5644	5708	45	14	7745	7805	7865	7926
5772	5835	5899	5963	6027	6091	44	15	8107	8167	8227	8287
6154	6218	6282	6346	6410	6473	43	16	8468	8528	8589	8649
6537	6601	6664	6728	6792	6856	42	17	8829	8890	8950	9010
6919	6983	7047	7110	7174	7237	41	18	9190	9250	9311	9371
7301	7365	7428	7492	7555	7619	40	19	9551	9611	9671	9731
9.497682	7746	7810	7873	7937	8000	39	20	9.519911	9971	..31	..91
8064	8127	8190	8254	8317	8381	38	21	9.520271	0331	0391	0451
8444	8508	8571	8634	8698	8761	37	22	0631	0691	0750	0810
8825	8888	8951	9015	9078	9141	36	23	0990	1050	1110	1169
9204	9268	9331	9394	9458	9521	35	24	1349	1409	1468	1528
9584	9647	9710	9774	9837	9900	34	25	1707	1767	1827	1887
9963	..26	..90	.153	.216	.279	33	26	2066	2125	2185	2245
9.500342	0405	0468	0531	0594	0658	32	27	2424	2483	2543	2602
0721	0784	0847	0910	0973	1036	31	28	2781	2841	2900	2960
1099	1162	1225	1288	1351	1414	30	29	3138	3198	3257	3317
9.501476	1539	1602	1665	1728	1791	29	30	9.523495	3555	3614	3674
1854	1917	1980	2042	2105	2168	28	31	3852	3911	3971	4030
2231	2294	2356	2419	2482	2545	27	32	4208	4267	4327	4386
2607	2670	2733	2796	2858	2921	26	33	4564	4623	4683	4742
2984	3046	3109	3172	3234	3297	25	34	4920	4979	5038	5097
3360	3422	3485	3548	3610	3673	24	35	5275	5334	5393	5452
3735	3798	3860	3923	3985	4048	23	36	5630	5689	5748	5807
4110	4173	4235	4298	4360	4423	22	37	5984	6044	6103	6162
4485	4548	4610	4673	4735	4797	21	38	6339	6398	6457	6516
4860	4922	4985	5047	5109	5172	20	39	6693	6752	6811	6870
9.505234	5296	5359	5421	5483	5545	19	40	9.527046	7105	7164	7223
5608	5670	5732	5794	5857	5919	18	41	7400	7459	7517	7576
5981	6043	6106	6168	6230	6292	17	42	7753	7811	7870	7929
6354	6416	6478	6541	6603	6665	16	43	8105	8164	8223	8282
6727	6789	6851	6913	6975	7037	15	44	8458	8516	8575	8634
7099	7161	7223	7285	7347	7409	14	45	8810	8868	8927	8986
7471	7533	7595	7657	7719	7781	13	46	9161	9220	9279	9337
7843	7905	7967	8028	8090	8152	12	47	9513	9571	9630	9688
8214	8276	8338	8400	8461	8523	11	48	9864	9922	9981	..39
8585	8647	8709	8770	8832	8894	10	49	9.530215	0273	0331	0390
9.508956	9017	9079	9141	9202	9264	9	50	9.530565	0623	0682	0740
9326	9387	9449	9511	9572	9634	8	51	0915	0973	1032	1090
9696	9757	9819	9880	9942	...4	7	52	1265	1323	1381	1440
9.510065	0127	0188	0250	0311	0373	6	53	1614	1673	1731	1789
0434	0496	0557	0619	0680	0742	5	54	1963	2022	2080	2138
0803	0865	0926	0987	1049	1110	4	55	2312	2370	2428	2487
1172	1233	1294	1356	1417	1478	3	56	2661	2719	2777	2835
1540	1601	1662	1724	1785	1846	2	57	3009	3067	3125	3183
1907	1969	2030	2091	2152	2214	1	58	3357	3415	3473	3531
2275	2336	2397	2458	2520	2581	0	59	3704	3762	3820	3878

...gent of 18 Degrees.					
10″	20″	30″	40″	50″	
1848	1919	1991	2063	2134	59
2277	2349	2420	2492	2564	58
2707	2778	2850	2921	2993	57
3136	3207	3278	3350	3421	56
3564	3636	3707	3778	3850	55
3992	4064	4135	4206	4278	54
4420	4492	4563	4634	4705	53
4848	4919	4990	5062	5133	52
5275	5346	5417	5489	5560	51
5702	5773	5844	5915	5986	50
6129	6200	6271	6342	6413	49
6555	6626	6697	6768	6839	48
6981	7052	7123	7193	7264	47
7406	7477	7548	7619	7690	46
7831	7902	7973	8044	8115	45
8256	8327	8398	8469	8539	44
8681	8752	8822	8893	8964	43
9105	9176	9246	9317	9388	42
9529	9600	9670	9741	9811	41
9953	..23	..94	.164	.235	40
0376	0446	0517	0587	0658	39
0799	0869	0939	1010	1080	38
1221	1292	1362	1432	1503	37
1643	1714	1784	1854	1925	36
2065	2136	2206	2276	2346	35
2487	2557	2627	2698	2768	34
2908	2978	3048	3119	3189	33
3329	3399	3469	3539	3609	32
3750	3820	3890	3960	4030	31
4170	4240	4310	4380	4450	30
4590	4660	4730	4800	4870	29
5009	5079	5149	5219	5289	28
5429	5499	5568	5638	5708	27
5848	5918	5987	6057	6127	26
6266	6336	6406	6476	6545	25
6685	6754	6824	6894	6963	24
7103	7172	7242	7312	7381	23
7520	7590	7660	7729	7799	22
7938	8007	8077	8146	8216	21
8355	8424	8494	8563	8633	20
8772	8841	8910	8980	9049	19
9188	9257	9327	9396	9465	18
9604	9673	9743	9812	9881	17
..20	..89	.158	.228	.297	16
0435	0504	0574	0643	0712	15
0850	0920	0989	1058	1127	14
1265	1334	1403	1473	1542	13
1680	1749	1818	1887	1956	12
2094	2163	2232	2301	2370	11
2508	2577	2646	2715	2784	10
2921	2990	3059	3128	3197	9
3335	3404	3472	3541	3610	8
3748	3816	3885	3954	4023	7
4160	4229	4298	4367	4435	6
4573	4641	4710	4779	4847	5
4985	5053	5122	5191	5259	4
5396	5465	5534	5602	5671	3
5808	5876	5945	6013	6082	2
6219	6287	6356	6424	6493	1
6630	6698	6767	6835	6903	0
50″	40″	30″	20″	10″	Min.
...ngent of 71 Degrees.					

Min.	Tangent of 19 Degrees.						
	0″	10″	20″	30″	40″	50″	
0	9.536972	7040	7109	7177	7245	7314	59
1	7382	7450	7519	7587	7655	7724	58
2	7792	7860	7929	7997	8065	8133	57
3	8202	8270	8338	8406	8475	8543	56
4	8611	8679	8747	8816	8884	8952	55
5	9020	9088	9156	9224	9293	9361	54
6	9429	9497	9565	9633	9701	9769	53
7	9837	9905	9973	..41	.109	.177	52
8	9.540245	0313	0381	0449	0517	0585	51
9	0653	0721	0789	0857	0925	0993	50
10	9.541061	1128	1196	1264	1332	1400	49
11	1468	1536	1603	1671	1739	1807	48
12	1875	1943	2010	2078	2146	2214	47
13	2281	2349	2417	2485	2552	2620	46
14	2688	2755	2823	2891	2958	3026	45
15	3094	3161	3229	3297	3364	3432	44
16	3499	3567	3635	3702	3770	3837	43
17	3905	3972	4040	4107	4175	4242	42
18	4310	4377	4445	4512	4580	4647	41
19	4715	4782	4850	4917	4985	5052	40
20	9.545119	5187	5254	5322	5389	5456	39
21	5524	5591	5658	5726	5793	5860	38
22	5928	5995	6062	6129	6197	6264	37
23	6331	6398	6466	6533	6600	6667	36
24	6735	6802	6869	6936	7003	7071	35
25	7138	7205	7272	7339	7406	7473	34
26	7540	7608	7675	7742	7809	7876	33
27	7943	8010	8077	8144	8211	8278	32
28	8345	8412	8479	8546	8613	8680	31
29	8747	8814	8881	8948	9015	9082	30
30	9.549149	9216	9283	9349	9416	9483	29
31	9550	9617	9684	9751	9817	9884	28
32	9951	..18	..85	.152	.218	.285	27
33	9.550352	0419	0485	0552	0619	0686	26
34	0752	0819	0886	0952	1019	1086	25
35	1153	1219	1286	1353	1419	1486	24
36	1552	1619	1686	1752	1819	1885	23
37	1952	2019	2085	2152	2218	2285	22
38	2351	2418	2484	2551	2617	2684	21
39	2750	2817	2883	2950	3016	3083	20
40	9.553149	3216	3282	3348	3415	3481	19
41	3548	3614	3680	3747	3813	3880	18
42	3946	4012	4079	4145	4211	4278	17
43	4344	4410	4476	4543	4609	4675	16
44	4741	4808	4874	4940	5006	5073	15
45	5139	5205	5271	5337	5404	5470	14
46	5536	5602	5668	5734	5800	5867	13
47	5933	5999	6065	6131	6197	6263	12
48	6329	6395	6461	6527	6593	6659	11
49	6725	6791	6857	6923	6989	7055	10
50	9.557121	7187	7253	7319	7385	7451	9
51	7517	7583	7649	7715	7781	7847	8
52	7913	7978	8044	8110	8176	8242	7
53	8308	8373	8439	8505	8571	8637	6
54	8703	8768	8834	8900	8966	9031	5
55	9097	9163	9229	9294	9360	9426	4
56	9491	9557	9623	9688	9754	9820	3
57	9885	9951	..17	..82	.148	.214	2
58	9.560279	0345	0410	0476	0542	0607	1
59	0673	0738	0804	0869	0935	1000	0
	60″	50″	40″	30″	20″	10″	Min.
	Co-tangent of 70 Degrees.						

Sine of 20 Degrees.							Min.	Sine of 21		
″	10″	20″	30″	40″	50″			0″	10″	20″
4052	4110	4167	4225	4283	4341	59	0	9.554329	4384	4439
4399	4456	4514	4572	4630	4687	58	1	4658	4713	4768
4745	4803	4861	4918	4976	5034	57	2	4987	5042	5096
5092	5149	5207	5265	5322	5380	56	3	5315	5370	5425
5438	5495	5553	5610	5668	5726	55	4	5643	5698	5753
5783	5841	5898	5956	6014	6071	54	5	5971	6026	6080
6129	6186	6244	6301	6359	6416	53	6	6299	6353	6408
6474	6531	6589	6646	6704	6761	52	7	6626	6680	6735
6818	6876	6933	6991	7048	7105	51	8	6953	7007	7062
7163	7220	7278	7335	7392	7450	50	9	7280	7334	7388
7507	7564	7622	7679	7736	7794	49	10	9.557606	7660	7715
7851	7908	7965	8023	8080	8137	48	11	7932	7986	8041
8194	8252	8309	8366	8423	8480	47	12	8258	8312	8366
8538	8595	8652	8709	8766	8823	46	13	8583	8638	8692
8880	8938	8995	9052	9109	9166	45	14	8909	8963	9017
9223	9280	9337	9394	9451	9508	44	15	9234	9288	9342
9565	9622	9679	9736	9793	9850	43	16	9558	9613	9667
9907	9964	..21	..78	.135	.192	42	17	9883	9937	9991
0249	0306	0363	0420	0477	0533	41	18	9.560207	0261	0315
0590	0647	0704	0761	0818	0875	40	19	0531	0585	0639
0931	0988	1045	1102	1159	1215	39	20	9.560855	0908	0962
1272	1329	1386	1442	1499	1556	38	21	1178	1232	1286
1613	1669	1726	1783	1839	1896	37	22	1501	1555	1609
1953	2009	2066	2123	2179	2236	36	23	1824	1878	1931
2293	2349	2406	2462	2519	2576	35	24	2146	2200	2254
2632	2689	2745	2802	2858	2915	34	25	2468	2522	2576
2971	3028	3084	3141	3197	3254	33	26	2790	2844	2898
3310	3367	3423	3480	3536	3593	32	27	3112	3166	3219
3649	3705	3762	3818	3875	3931	31	28	3433	3487	3541
3987	4044	4100	4156	4213	4269	30	29	3755	3808	3862
4325	4382	4438	4494	4550	4607	29	30	9.564075	4129	4182
4663	4719	4776	4832	4888	4944	28	31	4396	4449	4503
5000	5057	5113	5169	5225	5281	27	32	4716	4770	4823
5338	5394	5450	5506	5562	5618	26	33	5036	5090	5143
5674	5731	5787	5843	5899	5955	25	34	5356	5409	5463
6011	6067	6123	6179	6235	6291	24	35	5676	5729	5782
6347	6403	6459	6515	6571	6627	23	36	5995	6048	6101
6683	6739	6795	6851	6907	6963	22	37	6314	6367	6420
7019	7075	7131	7187	7242	7298	21	38	6632	6685	6739
7354	7410	7466	7522	7578	7633	20	39	6951	7004	7057
7689	7745	7801	7857	7912	7968	19	40	9.567269	7322	7375
8024	8080	8136	8191	8247	8303	18	41	7587	7640	7693
8359	8414	8470	8526	8581	8637	17	42	7904	7957	8010
8693	8748	8804	8860	8915	8971	16	43	8222	8275	8327
9027	9082	9138	9193	9249	9305	15	44	8539	8592	8644
9360	9416	9471	9527	9582	9638	14	45	8856	8908	8961
9693	9749	9805	9860	9916	9971	13	46	9172	9225	9277
0026	0082	0137	0193	0248	0304	12	47	9488	9541	9594
0359	0415	0470	0525	0581	0636	11	48	9804	9857	9910
0692	0747	0802	0858	0913	0968	10	49	9.570120	0173	0225
1024	1079	1134	1190	1245	1300	9	50	9.570435	0488	0541
1356	1411	1466	1521	1577	1632	8	51	0751	0803	0856
1687	1742	1798	1853	1908	1963	7	52	1066	1118	1170
2018	2074	2129	2184	2239	2294	6	53	1380	1433	1485
2349	2405	2460	2515	2570	2625	5	54	1695	1747	1799
2680	2735	2790	2845	2900	2955	4	55	2009	2061	2113
3010	3065	3121	3176	3231	3286	3	56	2323	2375	2427
3341	3396	3451	3506	3560	3615	2	57	2636	2688	2741
3670	3725	3780	3835	3890	3945	1	58	2950	3002	3054
4000	4055	4110	4165	4219	4274	0	59	3263	3315	3367

of 20 Degrees.					Min.	Tangent of 21 Degrees.				
20″	30″	40″	50″			0″	10″	20″	30″	40″
1197	1262	1328	1393	59	0	9.584177	4240	4303	4366	4429
1590	1655	1721	1786	58	1	4555	4618	4681	4744	4806
1982	2048	2113	2178	57	2	4932	4995	5058	5121	5183
2375	2440	2505	2571	56	3	5309	5372	5435	5498	5560
2767	2832	2897	2963	55	4	5686	5749	5811	5874	5937
3158	3224	3289	3354	54	5	6062	6125	6188	6251	6313
3550	3615	3680	3746	53	6	6439	6501	6564	6627	6689
3941	4006	4071	4137	52	7	6815	6877	6940	7003	7065
4332	4397	4462	4527	51	8	7190	7253	7316	7378	7441
4723	4788	4853	4918	50	9	7566	7629	7691	7754	7816
5113	5178	5243	5308	49	10	9.587941	8004	8066	8129	8191
5503	5568	5633	5698	48	11	8316	8379	8441	8504	8566
5893	5958	6023	6088	47	12	8691	8754	8816	8878	8941
6283	6348	6413	6478	46	13	9066	9128	9191	9253	9315
6672	6737	6802	6867	45	14	9440	9502	9565	9627	9690
7061	7126	7191	7256	44	15	9814	9877	9939	...1	..63
7450	7515	7580	7644	43	16	9.590188	0250	0313	0375	0437
7839	7903	7968	8033	42	17	0562	0624	0686	0748	0811
8227	8292	8356	8421	41	18	0935	0997	1060	1122	1184
8615	8680	8744	8809	40	19	1308	1370	1433	1495	1557
9003	9067	9132	9197	39	20	9.591681	1743	1805	1868	1930
9390	9455	9519	9584	38	21	2054	2116	2178	2240	2302
9777	9842	9906	9971	37	22	2426	2488	2550	2612	2674
0164	0229	0293	0358	36	23	2799	2861	2923	2985	3047
0551	0616	0680	0744	35	24	3171	3232	3294	3356	3418
0938	1002	1066	1131	34	25	3542	3604	3666	3728	3790
1324	1388	1452	1517	33	26	3914	3976	4038	4099	4161
1710	1774	1838	1903	32	27	4285	4347	4409	4471	4532
2095	2160	2224	2288	31	28	4656	4718	4780	4842	4903
2481	2545	2609	2673	30	29	5027	5089	5150	5212	5274
2866	2930	2994	3059	29	30	9.595398	5459	5521	5583	5644
3251	3315	3379	3443	28	31	5768	5830	5891	5953	6015
3636	3700	3764	3828	27	32	6138	6200	6261	6323	6385
4020	4084	4148	4212	26	33	6508	6570	6631	6693	6754
4404	4468	4532	4596	25	34	6878	6939	7001	7062	7124
4788	4852	4916	4980	24	35	7247	7309	7370	7432	7493
5172	5236	5299	5363	23	36	7616	7678	7739	7801	7862
5555	5619	5683	5747	22	37	7985	8047	8108	8170	8231
5938	6002	6066	6130	21	38	8354	8415	8477	8538	8600
6321	6385	6449	6512	20	39	8722	8784	8845	8907	8968
6704	6767	6831	6895	19	40	9.599091	9152	9213	9275	9336
7086	7150	7213	7277	18	41	9459	9520	9581	9643	9704
7468	7532	7595	7659	17	42	9827	9888	9949	..11	..72
7850	7914	7977	8041	16	43	9.600194	0256	0317	0378	0439
8231	8295	8359	8422	15	44	0562	0623	0684	0745	0806
8613	8676	8740	8803	14	45	0929	0990	1051	1112	1174
8994	9057	9121	9184	13	46	1296	1357	1418	1479	1540
5375	9438	9502	9565	12	47	1663	1724	1785	1846	1907
9755	9819	9882	9946	11	48	2029	2090	2151	2212	2273
0136	0199	0262	0326	10	49	2395	2456	2517	2578	2639
0516	0579	0642	0706	9	50	9.602761	2822	2883	2944	3005
0896	0959	1022	1086	8	51	3127	3188	3249	3310	3371
1275	1339	1402	1465	7	52	3493	3554	3615	3675	3736
1655	1718	1781	1844	6	53	3858	3919	3980	4041	4102
2034	2097	2160	2223	5	54	4223	4284	4345	4406	4467
2413	2476	2539	2602	4	55	4588	4649	4710	4771	4831
2791	2854	2917	2980	3	56	4953	5014	5074	5135	5196
3170	3233	3296	3359	2	57	5317	5378	5439	5500	5560
3548	3611	3674	3737	1	58	5682	5742	5803	5864	5924
3926	3989	4052	4114	0	59	6046	6106	6167	6228	6288

Sine of 22 Degrees.							Min.	Sine of 23 Degre			
0″	10″	20″	30″	40″	50″			0″	10″	20″	30″
73575	3628	3680	3732	3784	3836	59	0	9.591878	1928	1977	2027
3888	3940	3992	4044	4096	4148	58	1	2176	2225	2275	2324
4200	4252	4304	4356	4408	4460	57	2	2473	2522	2572	2621
4512	4564	4616	4668	4720	4772	56	3	2770	2819	2869	2918
4824	4876	4928	4980	5032	5084	55	4	3067	3116	3165	3215
5136	5187	5239	5291	5343	5395	54	5	3363	3412	3462	3511
5447	5499	5550	5602	5654	5706	53	6	3659	3709	3758	3807
5758	5810	5861	5913	5965	6017	52	7	3955	4005	4054	4103
6069	6120	6172	6224	6276	6327	51	8	4251	4301	4350	4399
6379	6431	6482	6534	6586	6638	50	9	4547	4596	4645	4695
76689	6741	6793	6844	6896	6948	49	10	9.594842	4891	4941	4990
6999	7051	7102	7154	7206	7257	48	11	5137	5186	5236	5285
7309	7360	7412	7464	7515	7567	47	12	5432	5481	5530	5580
7618	7670	7721	7773	7824	7876	46	13	5727	5776	5825	5874
7927	7979	8030	8082	8133	8185	45	14	6021	6070	6119	6168
8236	8288	8339	8391	8442	8494	44	15	6315	6364	6413	6462
8545	8596	8648	8699	8751	8802	43	16	6609	6658	6707	6756
8853	8905	8956	9008	9059	9110	42	17	6903	6952	7001	7050
9162	9213	9264	9316	9367	9418	41	18	7196	7245	7294	7343
9470	9521	9572	9623	9675	9726	40	19	7490	7539	7587	7636
79777	9828	9880	9931	9982	..33	39	20	9.597783	7831	7880	7929
80085	0136	0187	0238	0289	0341	38	21	8075	8124	8173	8222
0392	0443	0494	0545	0596	0647	37	22	8368	8417	8465	8514
0699	0750	0801	0852	0903	0954	36	23	8660	8709	8758	8806
1005	1056	1107	1158	1209	1261	35	24	8952	9001	9050	9098
1312	1363	1414	1465	1516	1567	34	25	9244	9293	9341	9390
1618	1669	1720	1771	1822	1873	33	26	9536	9584	9633	9681
1924	1975	2025	2076	2127	2178	32	27	9827	9876	9924	9973
2229	2280	2331	2382	2433	2484	31	28	9.600118	0167	0215	0264
2535	2585	2636	2687	2738	2789	30	29	0409	0457	0506	0554
82840	2890	2941	2992	3043	3094	29	30	9.600700	0748	0797	0845
3145	3195	3246	3297	3348	3398	28	31	0990	1038	1087	1135
3449	3500	3551	3601	3652	3703	27	32	1280	1329	1377	1425
3754	3804	3855	3906	3956	4007	26	33	1570	1619	1667	1715
4058	4108	4159	4210	4260	4311	25	34	1860	1908	1957	2005
4361	4412	4463	4513	4564	4615	24	35	2150	2198	2246	2294
4665	4716	4766	4817	4867	4918	23	36	2439	2487	2535	2583
4968	5019	5070	5120	5171	5221	22	37	2728	2776	2824	2872
5272	5322	5373	5423	5474	5524	21	38	3017	3065	3113	3161
5574	5625	5675	5726	5776	5827	20	39	3305	3353	3401	3449
85877	5927	5978	6028	6079	6129	19	40	9.603594	3642	3690	3738
6179	6230	6280	6331	6381	6431	18	41	3882	3930	3978	4026
6482	6532	6582	6633	6683	6733	17	42	4170	4218	4266	4313
6783	6834	6884	6934	6985	7035	16	43	4457	4505	4553	4601
7085	7135	7186	7236	7286	7336	15	44	4745	4793	4841	4888
7386	7437	7487	7537	7587	7637	14	45	5032	5080	5128	5176
7688	7738	7788	7838	7888	7938	13	46	5319	5367	5415	5462
7989	8039	8089	8139	8189	8239	12	47	5606	5654	5701	5749
8289	8339	8389	8439	8489	8540	11	48	5892	5940	5988	6035
8590	8640	8690	8740	8790	8840	10	49	6179	6226	6274	6322
88890	8940	8990	9040	9090	9140	9	50	9.606465	6512	6560	6608
9190	9240	9290	9340	9389	9439	8	51	6751	6798	6846	6893
9489	9539	9589	9639	9689	9739	7	52	7036	7084	7131	7179
9789	9839	9889	9938	9988	..38	6	53	7322	7369	7417	7464
20088	0138	0188	0237	0287	0337	5	54	7607	7654	7702	7749
20387	0437	0487	0536	0586	0636	4	55	7892	7939	7987	8034
3016	0735	0785	0835	0885	0934	3	56	8177	8224	8271	8319
3341	3.034	1084	1133	1183	1233	2	57	8461	8508	8556	8603
3670	3782	1382	1431	1481	1531	1	58	8745	8793	8840	8887
4000	4055	1680	1729	1779	1828	0	59	9029	9077	9124	9171

Min.	Tangent of 22 Degrees. 0″	10″	20″	30″	40″	50″	
0	9.606410	6470	6531	6591	6652	6713	59
1	6773	6834	6894	6955	7015	7076	58
2	7137	7197	7258	7318	7379	7439	57
3	7500	7560	7621	7681	7742	7802	56
4	7863	7923	7984	8044	8105	8165	55
5	8225	8286	8346	8407	8467	8528	54
6	8588	8648	8709	8769	8830	8890	53
7	8950	9011	9071	9131	9192	9252	52
8	9312	9373	9433	9493	9554	9614	51
9	9674	9735	9795	9855	9915	9976	50
10	9.610036	0096	0156	0217	0277	0337	49
11	0397	0458	0518	0578	0638	0698	48
12	0759	0819	0879	0939	0999	1059	47
13	1120	1180	1240	1300	1360	1420	46
14	1480	1540	1601	1661	1721	1781	45
15	1841	1901	1961	2021	2081	2141	44
16	2201	2261	2321	2381	2441	2501	43
17	2561	2621	2681	2741	2801	2861	42
18	2921	2981	3041	3101	3161	3221	41
19	3281	3341	3401	3461	3521	3581	40
20	9.613641	3701	3760	3820	3880	3940	39
21	4000	4060	4120	4180	4239	4299	38
22	4359	4419	4479	4539	4598	4658	37
23	4718	4778	4838	4897	4957	5017	36
24	5077	5136	5196	5256	5316	5375	35
25	5435	5495	5555	5614	5674	5734	34
26	5793	5853	5913	5972	6032	6092	33
27	6151	6211	6271	6330	6390	6450	32
28	6509	6569	6628	6688	6748	6807	31
29	6867	6926	6986	7046	7105	7165	30
30	9 617224	7284	7343	7403	7462	7522	29
31	7582	7641	7701	7760	7820	7879	28
32	7939	7998	8057	8117	8176	8236	27
33	8295	8355	8414	8474	8533	8593	26
34	8652	8711	8771	8830	8890	8949	25
35	9008	9068	9127	9186	9246	9305	24
36	9364	9424	9483	9543	9602	9661	23
37	9720	9780	9839	9898	9958	..17	22
38	9.620076	0136	0195	0254	0313	0373	21
39	0432	0491	0550	0610	0669	0728	20
40	9.620787	0846	0906	0965	1024	1083	19
41	1142	1201	1261	1320	1379	1438	18
42	1497	1556	1616	1675	1734	1793	17
43	1852	1911	1970	2029	2088	2147	16
44	2207	2266	2325	2384	2443	2502	15
45	2561	2620	2679	2738	2797	2856	14
46	2915	2974	3033	3092	3151	3210	13
47	3269	3328	3387	3446	3505	3564	12
48	3623	3682	3741	3800	3858	3917	11
49	3976	4035	4094	4153	4212	4271	10
50	9.624330	4388	4447	4506	4565	4624	9
51	4683	4742	4800	4859	4918	4977	8
52	5036	5094	5153	5212	5271	5330	7
53	5388	5447	5506	5565	5623	5682	6
54	5741	5800	5858	5917	5976	6035	5
55	6093	6152	6211	6269	6328	6387	4
56	6445	6504	6563	6621	6680	6739	3
57	6797	6856	6915	6973	7032	7090	2
58	7149	7208	7266	7325	7383	7442	1
59	7501	7559	7618	7676	7735	7793	0
	60″	50″	40″	30″	20″	10″	Min.
	Co-tangent of 67 Degrees.						

Min.	Tangent of 23 Degrees. 0″	10″	20″	30″	40″	50″	
0	9.627852	7910	7969	8028	8086	8145	59
1	8203	8262	8320	8379	8437	8496	58
2	8554	8612	8671	8729	8788	8846	57
3	8905	8963	9022	9080	9138	9197	56
4	9255	9314	9372	9431	9489	9547	55
5	9606	9664	9722	9781	9839	9897	54
6	9956	..14	..73	.131	.189	.247	53
7	9.630306	0364	0422	0481	0539	0597	52
8	0656	0714	0772	0830	0889	0947	51
9	1005	1063	1122	1180	1238	1296	50
10	9.631355	1413	1471	1529	1587	1646	49
11	1704	1762	1820	1878	1936	1995	48
12	2053	2111	2169	2227	2285	2343	47
13	2402	2460	2518	2576	2634	2692	46
14	2750	2808	2866	2924	2982	3040	45
15	3099	3157	3215	3273	3331	3389	44
16	3447	3505	3563	3621	3679	3737	43
17	3795	3853	3911	3969	4027	4085	42
18	4143	4201	4259	4316	4374	4432	41
19	4490	4548	4606	4664	4722	4780	40
20	9.634838	4896	4954	5011	5069	5127	39
21	5185	5243	5301	5359	5416	5474	38
22	5532	5590	5648	5706	5763	5821	37
23	5879	5937	5995	6052	6110	6168	36
24	6226	6283	6341	6399	6457	6514	35
25	6572	6630	6688	6745	6803	6861	34
26	6919	6976	7034	7092	7149	7207	33
27	7265	7322	7380	7438	7495	7553	32
28	7611	7668	7726	7783	7841	7899	31
29	7956	8014	8072	8129	8187	8244	30
30	9.638302	8359	8417	8475	8532	8590	29
31	8647	8705	8762	8820	8877	8935	28
32	8992	9050	9107	9165	9222	9280	27
33	9337	9395	9452	9510	9567	9625	26
34	9682	9740	9797	9855	9912	9969	25
35	9.640027	0084	0142	0199	0257	0314	24
36	0371	0429	0486	0544	0601	0658	23
37	0716	0773	0830	0888	0945	1002	22
38	1060	1117	1174	1232	1289	1346	21
39	1404	1461	1518	1575	1633	1690	20
40	9.641747	1805	1862	1919	1976	2034	19
41	2091	2148	2205	2263	2320	2377	18
42	2434	2491	2549	2606	2663	2720	17
43	2777	2834	2892	2949	3006	3063	16
44	3120	3177	3235	3292	3349	3406	15
45	3463	3520	3577	3634	3691	3749	14
46	3806	3863	3920	3977	4034	4091	13
47	4148	4205	4262	4319	4376	4433	12
48	4490	4547	4604	4661	4718	4775	11
49	4832	4889	4946	5003	5060	5117	10
50	9.645174	5231	5288	5345	5402	5459	9
51	5516	5573	5630	5687	5744	5801	8
52	5857	5914	5971	6028	6085	6142	7
53	6199	6256	6313	6369	6426	6483	6
54	6540	6597	6654	6710	6767	6824	5
55	6881	6938	6995	7051	7108	7165	4
56	7222	7279	7335	7392	7449	7506	3
57	7562	7619	7676	7733	7789	7846	2
58	7903	7960	8016	8073	8130	8186	1
59	8243	8300	8356	8413	8470	8526	0
	60″	50″	40″	30″	20″	10″	Min.
	Co-tangent of 66 Degrees.						

Min.	Sine of 24 Degrees.							Min.	Sine of 25		
	0"	10"	20"	30"	40"	50"			0"	10"	20"
0	9 609313	9361	9408	9455	9502	9550	59	0	9.625948	5993	6039
1	9597	9644	9691	9739	9786	9833	58	1	6219	6264	6309
2	9880	9928	9975	..22	..69	.116	57	2	6490	6535	6580
3	9.610164	0211	0258	0305	0352	0399	56	3	6760	6805	6850
4	0447	0494	0541	0588	0635	0682	55	4	7030	7075	7120
5	0729	0776	0823	0871	0918	0965	54	5	7300	7345	7390
6	1012	1059	1106	1153	1200	1247	53	6	7570	7615	7660
7	1294	1341	1388	1435	1482	1529	52	7	7840	7885	7929
8	1576	1623	1670	1717	1764	1811	51	8	8109	8154	8199
9	1858	1905	1952	1999	2046	2093	50	9	8378	8423	8468
10	9.612140	2187	2234	2280	2327	2374	49	10	9.628647	8692	8737
11	2421	2468	2515	2562	2609	2655	48	11	8916	8961	9006
12	2702	2749	2796	2843	2890	2936	47	12	9185	9229	9274
13	2983	3030	3077	3124	3171	3217	46	13	9453	9498	9542
14	3264	3311	3358	3404	3451	3498	45	14	9721	9766	9810
15	3545	3591	3638	3685	3732	3778	44	15	9989	..34	..78
16	3825	3872	3918	3965	4012	4058	43	16	9.630257	0301	0346
17	4105	4152	4198	4245	4292	4338	42	17	0524	0569	0613
18	4385	4432	4478	4525	4571	4618	41	18	0792	0836	0881
19	4665	4711	4758	4804	4851	4898	40	19	1059	1103	1148
20	9.614944	4991	5037	5084	5130	5177	39	20	9.631326	1370	1415
21	5223	5270	5316	5363	5409	5456	38	21	1593	1637	1681
22	5502	5549	5595	5642	5688	5735	37	22	1859	1904	1948
23	5781	5828	5874	5921	5967	6013	36	23	2125	2170	2214
24	6060	6106	6153	6199	6245	6292	35	24	2392	2436	2480
25	6338	6385	6431	6477	6524	6570	34	25	2658	2702	2746
26	6616	6663	6709	6755	6802	6848	33	26	2923	2968	3012
27	6894	6941	6987	7033	7080	7126	32	27	3189	3233	3277
28	7172	7218	7265	7311	7357	7403	31	28	3454	3498	3543
29	7450	7496	7542	7588	7635	7681	30	29	3719	3764	3808
30	9.617727	7773	7819	7866	7912	7958	29	30	9.633984	4028	4073
31	8004	8050	8096	8143	8189	8235	28	31	4249	4293	4337
32	8281	8327	8373	8419	8465	8512	27	32	4514	4558	4602
33	8558	8604	8650	8696	8742	8788	26	33	4778	4822	4866
34	8834	8880	8926	8972	9018	9064	25	34	5042	5086	5130
35	9110	9156	9202	9248	9294	9340	24	35	5306	5350	5394
36	9386	9432	9478	9524	9570	9616	23	36	5570	5614	5658
37	9662	9708	9754	9800	9846	9892	22	37	5834	5877	5921
38	9938	9984	..30	..76	.121	.167	21	38	6097	6141	6185
39	9.620213	0259	0305	0351	0397	0443	20	39	6360	6404	6448
40	9.620488	0534	0580	0626	0672	0718	19	40	9.636623	6667	6711
41	0763	0809	0855	0901	0947	0992	18	41	6886	6930	6973
42	1038	1084	1130	1175	1221	1267	17	42	7148	7192	7236
43	1313	1358	1404	1450	1496	1541	16	43	7411	7455	7498
44	1587	1633	1678	1724	1770	1816	15	44	7673	7717	7760
45	1861	1907	1953	1998	2044	2089	14	45	7935	7979	8022
46	2135	2181	2226	2272	2318	2363	13	46	8197	8240	8284
47	2409	2454	2500	2546	2591	2637	12	47	8458	8502	8546
48	2682	2728	2773	2819	2865	2910	11	48	8720	8763	8807
49	2956	3001	3047	3092	3138	3183	10	49	8981	9025	9068
50	9.623229	3274	3320	3365	3411	3456	9	50	9.639242	9286	9329
51	3502	3547	3593	3638	3683	3729	8	51	9503	9546	9590
52	3774	3820	3865	3911	3956	4001	7	52	9764	9807	9851
53	4047	4092	4138	4183	4228	4274	6	53	9.640024	0068	0111
54	4319	4364	4410	4455	4500	4546	5	54	0284	0328	0371
55	4591	4636	4682	4727	4772	4818	4	55	0544	0588	0631
56	4863	4908	4954	4999	5044	5089	3	56	0804	0848	0891
57	5135	5180	5225	5270	5316	5361	2	57	1064	1107	1151

Min.	Tangent of 24 Degrees. 0″	10″	20″	30″	40″	50″	
0	9.648583	8640	8696	8753	8810	8866	59
1	8923	8980	9036	9093	9150	9206	58
2	9263	9319	9376	9433	9489	9546	57
3	9602	9659	9715	9772	9829	9885	56
4	9942	9998	..55	.111	.168	.224	55
5	9.650281	0337	0394	0450	0507	0563	54
6	0620	0676	0733	0789	0846	0902	53
7	0959	1015	1072	1128	1185	1241	52
8	1297	1354	1410	1467	1523	1579	51
9	1636	1692	1749	1805	1861	1918	50
10	9.651974	2031	2087	2143	2200	2256	49
11	2312	2369	2425	2481	2538	2594	48
12	2650	2707	2763	2819	2875	2932	47
13	2988	3044	3101	3157	3213	3269	46
14	3326	3382	3438	3494	3551	3607	45
15	3663	3719	3776	3832	3888	3944	44
16	4000	4057	4113	4169	4225	4281	43
17	4337	4394	4450	4506	4562	4618	42
18	4674	4731	4787	4843	4899	4955	41
19	5011	5067	5123	5179	5236	5292	40
20	9.655348	5404	5460	5516	5572	5628	39
21	5684	5740	5796	5852	5908	5964	38
22	6020	6076	6132	6188	6244	6300	37
23	6356	6412	6468	6524	6580	6636	36
24	6692	6748	6804	6860	6916	6972	35
25	7028	7084	7140	7196	7252	7308	34
26	7364	7419	7475	7531	7587	7643	33
27	7699	7755	7811	7867	7922	7978	32
28	8034	8090	8146	8202	8258	8313	31
29	8369	8425	8481	8537	8592	8648	30
30	9.658704	8760	8816	8871	8927	8983	29
31	9039	9095	9150	9206	9262	9318	28
32	9373	9429	9485	9540	9596	9652	27
33	9708	9763	9819	9875	9930	9986	26
34	9.660042	0098	0153	0209	0265	0320	25
35	0376	0431	0487	0543	0598	0654	24
36	0710	0765	0821	0877	0932	0988	23
37	1043	1099	1155	1210	1266	1321	22
38	1377	1432	1488	1544	1599	1655	21
39	1710	1766	1821	1877	1932	1988	20
40	9.662043	2099	2154	2210	2265	2321	19
41	2376	2432	2487	2543	2598	2654	18
42	2709	2765	2820	2876	2931	2987	17
43	3042	3097	3153	3208	3264	3319	16
44	3375	3430	3485	3541	3596	3651	15
45	3707	3762	3818	3873	3928	3984	14
46	4039	4094	4150	4205	4260	4316	13
47	4371	4426	4482	4537	4592	4648	12
48	4703	4758	4814	4869	4924	4979	11
49	5035	5090	5145	5200	5256	5311	10
50	9.665366	5421	5477	5532	5587	5642	9
51	5698	5753	5808	5863	5918	5974	8
52	6029	6084	6139	6194	6249	6305	7
53	6360	6415	6470	6525	6580	6636	6
54	6691	6746	6801	6856	6911	6966	5
55	7021	7076	7132	7187	7242	7297	4
56	7352	7407	7462	7517	7572	7627	3
57	7682	7737	7792	7847	7903	7958	2
58	8013	8068	8123	8178	8233	8288	1
59	8343	8398	8453	8508	8563	8618	0
	60″	50″	40″	30″	20″	10″	Min.
	Co-tangent of 65 Degrees.						

P. Part { 1″ 2″ 3″ 4″ 5″ 6″ 7″ 8″ 9″

Min.	Tangent of 25 Degrees. 0″	10″	20″	30″	40″	50″	
0	9.668673	8728	8782	8837	8892	8947	59
1	9002	9057	9112	9167	9222	9277	58
2	9332	9387	9442	9497	9552	9606	57
3	9661	9716	9771	9826	9881	9936	56
4	9991	..45	.100	.155	.210	.265	55
5	9.670320	0375	0429	0484	0539	0594	54
6	0649	0703	0758	0813	0868	0923	53
7	0977	1032	1087	1142	1197	1251	52
8	1306	1361	1416	1470	1525	1580	51
9	1635	1689	1744	1799	1853	1908	50
10	9.671963	2018	2072	2127	2182	2236	49
11	2291	2346	2400	2455	2510	2564	48
12	2619	2674	2728	2783	2838	2892	47
13	2947	3001	3056	3111	3165	3220	46
14	3274	3329	3384	3438	3493	3547	45
15	3602	3657	3711	3766	3820	3875	44
16	3929	3984	4038	4093	4148	4202	43
17	4257	4311	4366	4420	4475	4529	42
18	4584	4638	4693	4747	4802	4856	41
19	4911	4965	5019	5074	5128	5183	40
20	9.675237	5292	5346	5401	5455	5509	39
21	5564	5618	5673	5727	5781	5836	38
22	5890	5945	5999	6053	6108	6162	37
23	6217	6271	6325	6380	6434	6488	36
24	6543	6597	6651	6706	6760	6814	35
25	6869	6923	6977	7032	7086	7140	34
26	7194	7249	7303	7357	7412	7466	33
27	7520	7574	7629	7683	7737	7791	32
28	7846	7900	7954	8008	8062	8117	31
29	8171	8225	8279	8334	8388	8442	30
30	9.678496	8550	8604	8659	8713	8767	29
31	8821	8875	8929	8984	9038	9092	28
32	9146	9200	9254	9308	9363	9417	27
33	9471	9525	9579	9633	9687	9741	26
34	9795	9849	9904	9958	..12	..66	25
35	9.680120	0174	0228	0282	0336	0390	24
36	0444	0498	0552	0606	0660	0714	23
37	0768	0822	0876	0930	0984	1038	22
38	1092	1146	1200	1254	1308	1362	21
39	1416	1470	1524	1578	1632	1686	20
40	9.681740	1794	1847	1901	1955	2009	19
41	2063	2117	2171	2225	2279	2333	18
42	2387	2440	2494	2548	2602	2656	17
43	2710	2764	2817	2871	2925	2979	16
44	3033	3087	3140	3194	3248	3302	15
45	3356	3410	3463	3517	3571	3625	14
46	3679	3732	3786	3840	3894	3947	13
47	4001	4055	4109	4162	4216	4270	12
48	4324	4377	4431	4485	4539	4592	11
49	4646	4700	4753	4807	4861	4914	10
50	9.684968	5022	5075	5129	5183	5236	9
51	5290	5344	5397	5451	5505	5558	8
52	5612	5666	5719	5773	5827	5880	7
53	5934	5987	6041	6095	6148	6202	6
54	6255	6309	6363	6416	[illegible]	6523	5
55	6577	6630	6684	6737	6791	6845	4
56	6898	6952	7005	7059	7112	7166	3
57	7219	7273	7326	7380	7433	7487	2
58	7540	7594	7647	7701	7754	7808	1
59	7861	7915	7968	8021	8075	8128	0
	60″	50″	40″	30″	20″	10″	Min.
	Co-tangent of 64 Degrees.						

P. Part { 1″ 2″ 3″ 4″ 5″ 6″ 7″ 8″

Min.	Sine of 26 Degrees. 0″	10″	20″	30″	40″	50″	
0	9.641842	1885	1928	1971	2015	20[illegible]8	59
1	2101	2144	2187	2230	2273	23[illegible]7	58
2	2360	2403	2446	2489	2532	2575	57
3	2618	2661	2704	2747	2790	28[illegible]	56
4	2877	2920	2963	3006	3049	30[illegible]	55
5	3135	3178	3221	3264	3307	3350	54
6	3393	3436	3479	3522	3565	3607	53
7	3650	3693	3736	3779	3822	3865	52
8	3908	3951	3994	4037	4080	4123	51
9	4165	4208	4251	4294	4337	4380	50
10	9.644423	4465	4508	4551	4594	4637	49
11	4680	4722	4765	4808	4851	4894	48
12	4936	4979	5022	5065	5108	5150	47
13	5193	5236	5279	5321	5364	5407	46
14	5450	5492	5535	5578	5620	5663	45
15	5706	5749	5791	5834	5877	5919	44
16	5962	6005	6047	6090	6133	6175	43
17	6218	6260	6303	6346	6388	6431	42
18	6474	6516	6559	6601	6644	6686	41
19	6729	6772	6814	6857	6899	6942	40
20	9.646984	7027	7069	7112	7154	7197	39
21	7240	7282	7325	7367	7409	7452	38
22	7494	7537	7579	7622	7664	7707	37
23	7749	7792	7834	7877	7919	7961	36
24	8004	8046	8089	8131	8173	8216	35
25	8258	8301	8343	8385	8428	8470	34
26	8512	8555	8597	8639	8682	8724	33
27	8766	8809	8851	8893	8936	8978	32
28	9020	9063	9105	9147	9189	9232	31
29	9274	9316	9358	9401	9443	9485	30
30	9.649527	9570	9612	9654	9696	9739	29
31	9781	9823	9865	9907	9949	9992	28
32	9.650034	0076	0118	0160	0202	0245	27
33	0287	0329	0371	0413	0455	0497	26
34	0539	0582	0624	0666	0708	0750	25
35	0792	0834	0876	0918	0960	1002	24
36	1044	1086	1128	1171	1213	1255	23
37	1297	1339	1381	1423	1465	1507	22
38	1549	1591	1633	1675	1716	1758	21
39	1800	1842	1884	1926	1968	2010	20
40	9.652052	2094	2136	2178	2220	2262	19
41	2304	2345	2387	2429	2471	2513	18
42	2555	2597	2638	2680	2722	2764	17
43	2806	2848	2890	2931	2973	3015	16
44	3057	3099	3140	3182	3224	3266	15
45	3308	3349	3391	3433	3475	3516	14
46	3558	3600	3642	3683	3725	3767	13
47	3808	3850	3892	3934	3975	4017	12
48	4059	4100	4142	4184	4225	4267	11
49	4309	4350	4392	4434	4475	4517	10
50	9.654558	4600	4642	4683	4725	4766	9
51	4808	4850	4891	4933	4974	5016	8
52	5058	5099	5141	5182	5224	5265	7
53	5307	5348	5390	5431	5473	5514	6
54	5[illegible]6	5597	5639	5680	5722	5763	5
55	5805	5846	5888	5929	5971	6012	4
56	6054	6095	6136	6178	6219	6261	3
57	6302	6344	6385	6426	6468	6509	2
58	6551	6592	6633	6675	6716	6757	1
59	6799	6840	6881	6923	6964	7005	0

Min.	Sine of 27 Degrees. 0″	10″	20″	30″	40″	50″	
0	9.657047	7088	7129	7171	7212	7253	59
1	7295	7336	7377	7418	7460	7501	58
2	7542	7584	7625	7666	7707	7749	57
3	7790	7831	7872	7913	7955	7996	56
4	8037	8078	8119	8161	8202	8243	55
5	8284	8325	8367	8408	8449	8490	54
6	8531	8572	8613	8655	8696	8737	53
7	8778	8819	8860	8901	8942	8983	52
8	9025	9066	9107	9148	9189	9230	51
9	9271	9312	9353	9394	9435	9476	50
10	9.659517	9558	9599	9640	9681	9722	49
11	9763	9804	9845	9886	9927	9968	48
12	9.660009	0050	0091	0132	0173	0214	47
13	0255	0296	0337	0378	0419	0460	46
14	0501	0541	0582	0623	0664	0705	45
15	0746	0787	0828	0869	0909	0950	44
16	0991	1032	1073	1114	1154	1195	43
17	1236	1277	1318	1359	1399	1440	42
18	1481	1522	1563	1603	1644	1685	41
19	1726	1766	1807	1848	1889	1929	40
20	9.661970	2011	2052	2092	2133	2174	39
21	2214	2255	2296	2337	2377	2418	38
22	2459	2499	2540	2581	2621	2662	37
23	2703	2743	2784	2825	2865	2906	36
24	2946	2987	3028	3068	3109	3149	35
25	3190	3231	3271	3312	3352	3393	34
26	3433	3474	3515	3555	3596	3636	33
27	3677	3717	3758	3798	3839	3879	32
28	3920	3960	4001	4041	4082	4122	31
29	4163	4203	4244	4284	4325	4365	30
30	9.664406	4446	4486	4527	4567	4608	29
31	4648	4689	4729	4769	4810	4850	28
32	4891	4931	4971	5012	5052	5093	27
33	5133	5173	5214	5254	5294	5335	26
34	5375	5415	5456	5496	5536	5577	25
35	5617	5657	5697	5738	5778	5818	24
36	5859	5899	5939	5979	6020	6060	23
37	6100	6140	6181	6221	6261	6301	22
38	6342	6382	6422	6462	6502	6543	21
39	6583	6623	6663	6703	6743	6784	20
40	9.666824	6864	6904	6944	6984	7025	19
41	7065	7105	7145	7185	7225	7265	18
42	7305	7346	7386	7426	7466	7506	17
43	7546	7586	7626	7666	7706	7746	16
44	7786	7826	7866	7906	7946	7986	15
45	8027	8067	8107	8147	8187	8227	14
46	8267	8307	8347	8386	8426	8466	13
47	8506	8546	8586	8626	8666	8706	12
48	8746	8786	8826	8866	8906	8946	11
49	8986	9026	9065	9105	9145	9185	10
50	9.669225	9265	9305	9345	9384	9424	9
51	9464	9504	9544	9584	9624	9663	8
52	9703	9743	9783	9823	9862	9902	7
53	9942	9982	..22	..61	.101	.141	6
54	9.670181	0220	0260	0300	0340	0379	5
55	0419	0459	0499	0538	0578	0618	4
56	0658	0697	0737	0777	0816	0856	3
57	0896	0935	0975	1015	1054	1094	2
58	1134	1173	1213	1253	1292	1332	1
59	1372	1411	1451	1490	1530	1570	0

of 26 Degrees.					Min.	Tangent of 27 Deg			
20″	30″	40″	50″			0″	10″	20″	30″
8289	8342	8395	8449	59	0	9.707166	7218	7270	7322
8609	8663	8716	8769	58	1	7478	7530	7582	7634
8930	8983	9036	9090	57	2	7790	7842	7894	7946
9250	9303	9356	9410	56	3	8102	8154	8206	8258
9570	9623	9676	9730	55	4	8414	8466	8518	8570
9890	9943	9996	..50	54	5	8726	8778	8830	8882
0210	0263	0316	0369	53	6	9037	9089	9141	9193
0529	0582	0636	0689	52	7	9349	9401	9453	9504
0849	0902	0955	1008	51	8	9660	9712	9764	9816
1168	1221	1274	1328	50	9	9971	..23	..75	.127
1487	1540	1594	1647	49	10	9.710282	0334	0386	0438
1806	1859	1913	1966	48	11	0593	0645	0697	0749
2125	2178	2232	2285	47	12	0904	0956	1008	1059
2444	2497	2550	2603	46	13	1215	1267	1318	1370
2763	2816	2869	2922	45	14	1525	1577	1629	1681
3081	3134	3187	3240	44	15	1836	1887	1939	1991
3400	3453	3506	3559	43	16	2146	2198	2249	2301
3718	3771	3824	3877	42	17	2456	2508	2560	2611
4036	4089	4142	4195	41	18	2766	2818	2870	2921
4354	4407	4460	4513	40	19	3076	3128	3179	3231
4672	4724	4777	4830	39	20	9.713386	3438	3489	3541
4989	5042	5095	5148	38	21	3696	3747	3799	3850
5307	5360	5412	5465	37	22	4005	4057	4108	4160
5624	5677	5730	5783	36	23	4314	4366	4418	4469
5941	5994	6047	6100	35	24	4624	4675	4727	4778
6258	6311	6364	6417	34	25	4933	4984	5036	5087
6575	6628	6681	6734	33	26	5242	5293	5345	5396
6892	6945	6998	7050	32	27	5551	5602	5654	5705
7209	7262	7314	7367	31	28	5860	5911	5962	6014
7525	7578	7631	7684	30	29	6168	6220	6271	6322
7842	7894	7947	8000	29	30	9.716477	6528	6579	6631
8158	8211	8263	8316	28	31	6785	6836	6888	6939
8474	8527	8579	8632	27	32	7093	7145	7196	7247
8790	8843	8895	8948	26	33	7401	7453	7504	7555
9106	9159	9211	9264	25	34	7709	7761	7812	7863
9422	9474	9527	9579	24	35	8017	8069	8120	8171
9737	9790	9842	9895	23	36	8325	8376	8428	8479
..53	.105	.158	.210	22	37	8633	8684	8735	8786
0368	0420	0473	0525	21	38	8940	8991	9043	9094
0683	0736	0788	0841	20	39	9248	9299	9350	9401
0998	1051	1103	1155	19	40	9.719555	9606	9657	9708
1313	1365	1418	1470	18	41	9862	9913	9964	..16
1628	1680	1733	1785	17	42	9.720169	0220	0271	0322
1942	1995	2047	2100	16	43	0476	0527	0578	0629
2257	2309	2362	2414	15	44	0783	0834	0885	0936
2571	2623	2676	2728	14	45	1089	1140	1191	1243
2885	2938	2990	3042	13	46	1396	1447	1498	1549
3199	3252	3304	3356	12	47	1702	1753	1804	1855
3513	3566	3618	3670	11	48	2009	2060	2111	2162
3827	3879	3932	3984	10	49	2315	2366	2417	2468
4141	4193	4245	4297	9	50	9.722621	2672	2723	2774
4454	4506	4559	4611	8	51	2927	2978	3029	3080
4768	4820	4872	4924	7	52	3232	3283	3334	3385
5081	5133	5185	5237	6	53	3538	3589	3640	3691
5394	5446	5498	5551	5	54	3844	3895	3945	3996
5707	5759	5811	5863	4	55	4149	4200	4251	4302
6020	6072	6124	6176	3	56	4454	4505	4556	4607
6333	6385	6437	6489	2	57	4760	4810	4861	4912
6645	6697	6749	6801	1	58	5065	5115	5166	5217
6958	7010	7062	7114	0	59	5370	5420	5471	5522

Min.	Sine of 28 Degrees. 0″	10″	20″	30″	40″	50″	
0	9.671609	1649	1688	1728	1768	1807	59
1	1847	1886	1926	1965	2005	2045	58
2	2084	2124	2163	2203	2242	2282	57
3	2321	2361	2400	2440	2479	2519	56
4	2558	2598	2637	2677	2716	2756	55
5	2795	2835	2874	2914	2953	2992	54
6	3032	3071	3111	3150	3190	3229	53
7	3268	3308	3347	3387	3426	3465	52
8	3505	3544	3583	3623	3662	3702	51
9	3741	3780	3820	3859	3898	3938	50
10	9.673977	4016	4056	4095	4134	4173	49
11	4213	4252	4291	4331	4370	4409	48
12	4448	4488	4527	4566	4606	4645	47
13	4684	4723	4762	4802	4841	4880	46
14	4919	4959	4998	5037	5076	5115	45
15	5155	5194	5233	5272	5311	5350	44
16	5390	5429	5468	5507	5546	5585	43
17	5624	5664	5703	5742	5781	5820	42
18	5859	5898	5937	5976	6016	6055	41
19	6094	6133	6172	6211	6250	6289	40
20	9.676328	6367	6406	6445	6484	6523	39
21	6562	6601	6640	6679	6718	6757	38
22	6796	6835	6874	6913	6952	6991	37
23	7030	7069	7108	7147	7186	7225	36
24	7264	7303	7342	7381	7420	7459	35
25	7498	7536	7575	7614	7653	7692	34
26	7731	7770	7809	7848	7886	7925	33
27	7964	8003	8042	8081	8120	8158	32
28	8197	8236	8275	8314	8353	8391	31
29	8430	8469	8508	8547	8585	8624	30
30	9.678663	8702	8740	8779	8818	8857	29
31	8895	8934	8973	9012	9050	9089	28
32	9128	9167	9205	9244	9283	9321	27
33	9360	9399	9438	9476	9515	9554	26
34	9592	9631	9670	9708	9747	9786	25
35	9824	9863	9902	9940	9979	..17	24
36	9.680056	0095	0133	0172	0210	0249	23
37	0288	0326	0365	0403	0442	0481	22
38	0519	0558	0596	0635	0673	0712	21
39	0750	0789	0828	0866	0905	0943	20
40	9.680982	1020	1059	1097	1136	1174	19
41	1213	1251	1290	1328	1366	1405	18
42	1443	1482	1520	1559	1597	1636	17
43	1674	1713	1751	1789	1828	1866	16
44	1905	1943	1981	2020	2058	2097	15
45	2135	2173	2212	2250	2288	2327	14
46	2365	2403	2442	2480	2519	2557	13
47	2595	2633	2672	2710	2748	2787	12
48	2825	2863	2902	2940	2978	3016	11
49	3055	3093	3131	3170	3208	3246	10
50	9.683284	3323	3361	3399	3437	3475	9
51	3514	3552	3590	3628	3667	3705	8
52	3743	3781	3819	3858	3896	3934	7
53	3972	4010	4048	4087	4125	4163	6
54	4201	4239	4277	4315	4353	4392	5
55	4430	4468	4506	4544	4582	4620	4
56	4658	4696	4735	4773	4811	4849	3
57	4887	4925	4963	5001	5039	5077	2
58	5115	5153	5191	5229	5267	5305	1
59	5343	5381	5419	5457	5495	5533	0

Min.	Sine of 29 Degrees. 0″	10″	20″	30″	40″	50″	
0	9.685571	5609	5647	5685	5723	5761	59
1	5799	5837	5875	5913	5951	5989	58
2	6027	6065	6103	6141	6178	6216	57
3	6254	6292	6330	6368	6406	6444	56
4	6482	6519	6557	6595	6633	6671	55
5	6709	6747	6785	6822	6860	6898	54
6	6936	6974	7012	7049	7087	7125	53
7	7163	7201	7238	7276	7314	7352	52
8	7389	7427	7465	7503	7541	7578	51
9	7616	7654	7692	7729	7767	7805	50
10	9.687843	7880	7918	7956	7993	8031	49
11	8069	8106	8144	8182	8220	8257	48
12	8295	8333	8370	8408	8446	8483	47
13	8521	8559	8596	8634	8671	8709	46
14	8747	8784	8822	8860	8897	8935	45
15	8972	9010	9048	9085	9123	9160	44
16	9198	9235	9273	9311	9348	9386	43
17	9423	9461	9498	9536	9573	9611	42
18	9648	9686	9723	9761	9798	9836	41
19	9873	9911	9948	9986	..23	..61	40
20	9.690098	0136	0173	0211	0248	0286	39
21	0323	0361	0398	0435	0473	0510	38
22	0548	0585	0622	0660	0697	0735	37
23	0772	0809	0847	0884	0922	0959	36
24	0996	1034	1071	1108	1146	1183	35
25	1220	1258	1295	1332	1370	1407	34
26	1444	1482	1519	1556	1594	1631	33
27	1668	1706	1743	1780	1817	1855	32
28	1892	1929	1966	2004	2041	2078	31
29	2115	2153	2190	2227	2264	2302	30
30	9.692339	2376	2413	2450	2488	2525	29
31	2562	2599	2636	2674	2711	2748	28
32	2785	2822	2859	2897	2934	2971	27
33	3008	3045	3082	3119	3157	3194	26
34	3231	3268	3305	3342	3379	3416	25
35	3453	3490	3528	3565	3602	3639	24
36	3676	3713	3750	3787	3824	3861	23
37	3898	3935	3972	4009	4046	4083	22
38	4120	4157	4194	4231	4268	4305	21
39	4342	4379	4416	4453	4490	4527	20
40	9.694564	4601	4638	4675	47.2	4749	19
41	4786	4823	4860	4897	4934	4971	18
42	5007	5044	5081	5118	5155	5192	17
43	5229	5266	5303	5339	5376	5413	16
44	5450	5487	5524	5561	5598	5634	15
45	5671	5708	5745	5782	5819	5855	14
46	5892	5929	5966	6003	6039	6076	13
47	6113	6150	6187	6223	6260	6297	12
48	6334	6370	6407	6444	6481	6517	11
49	6554	6591	6628	6664	6701	6738	10
50	9.696775	6811	6848	6885	6921	6958	9
51	6995	7031	7068	7105	7141	7178	8
52	7215	7251	7288	7325	7361	7398	7
53	7435	7471	7508	7545	7581	7618	6
54	7654	7691	7728	7764	7801	7838	5
55	7874	7911	7947	7984	8020	8057	4
56	8094	8130	8167	8203	8240	8276	3
57	8313	8349	8386	8423	8459	8496	2
58	8532	8569	8605	8642	8678	8715	1
59	8751	8788	8824	8861	8897	8934	0

of 28 Degrees.					Min.	Tangent of 29 Degrees.						
20″	30″	40″	50″			0″	10″	20″	30″	40″	50″	
5776	5827	5878	5928	59	0	9.743752	3802	3851	3901	3951	4000	59
6081	6131	6182	6233	58	1	4050	4099	4149	4199	4248	4298	58
6385	6436	6487	6537	57	2	4348	4397	4447	4496	4546	4596	57
6690	6740	679[illegible]	6842	56	3	4645	4695	4744	4794	4844	4893	56
6994	7045	7095	7146	55	4	4943	4992	5042	5092	5141	5191	55
7298	7349	7399	7450	54	5	5240	5290	5339	5389	5439	5488	54
7602	7653	7703	7754	53	6	5538	5587	5637	5686	5736	5785	53
7906	7957	8007	8058	52	7	5835	5884	5934	5983	6033	6082	52
8210	8261	8311	8362	51	8	6132	6182	6231	6281	6330	6380	51
8514	8564	8615	8665	50	9	6429	6479	6528	6577	6627	6676	50
8817	8868	8918	8969	49	10	9.746726	6775	6825	6874	6924	6973	49
9121	9171	9222	9272	48	11	7023	7072	7122	7171	7221	7270	48
9424	9475	9525	9576	47	12	7319	7369	7418	7468	7517	7567	47
9727	9778	9828	9879	46	13	7616	7665	7715	7764	7814	7863	46
..30	..81	.132	.182	45	14	7913	7962	8011	8061	8110	8160	45
0333	0384	0434	0485	44	15	8209	8258	8308	8357	8406	8456	44
0636	0687	0737	0788	43	16	8505	8555	8604	8653	8703	8752	43
0939	0990	1040	1091	42	17	8801	8851	8900	8949	8999	9048	42
1242	1292	1343	1393	41	18	9097	9147	9196	9245	9295	9344	41
1544	1595	1645	1696	40	19	9393	9443	9492	9541	9591	9640	40
1847	1897	1948	1998	39	20	9.749689	9739	9788	9837	9886	9936	39
2149	2200	2250	2300	38	21	9985	..34	..84	.133	.182	.231	38
2451	2502	2552	2602	37	22	9.750281	0330	0379	0428	0478	0527	37
2753	2804	2854	2904	36	23	0576	0625	0675	0724	0773	0822	36
3055	3106	3156	3206	35	24	0872	0921	0970	1019	1069	1118	35
3357	3408	3458	3508	34	25	1167	1216	1265	1315	1364	1413	34
3659	3709	3760	3810	33	26	1462	1511	1561	1610	1659	1708	33
3961	4011	4061	4111	32	27	1757	1806	1856	1905	1954	2003	32
4262	4312	4363	4413	31	28	2052	2101	2151	2200	2249	2298	31
4564	4614	4664	4714	30	29	2347	2396	2446	2495	2544	2593	30
4865	4915	4965	5015	29	30	9.752642	2691	2740	2789	2839	2888	29
5166	5216	5266	5317	28	31	2937	2986	3035	3084	3133	3182	28
5467	5517	5567	5618	27	32	3231	3280	3330	3379	3428	3477	27
5768	5818	5868	5918	26	33	3526	3575	3624	3673	3722	3771	26
6069	6119	6169	6219	25	34	3820	3869	3918	3967	4016	4066	25
6370	6420	6470	6520	24	35	4115	4164	4213	4262	4311	4360	24
6670	6720	6770	6820	23	36	4409	4458	4507	4556	4605	4654	23
6971	7021	7071	7121	22	37	4703	4752	4801	4850	4899	4948	22
7271	7321	7371	7421	21	38	4997	5046	5095	5144	5193	5242	21
7571	7621	7671	7721	20	39	5291	5340	5389	5438	5487	5536	20
7871	7921	7971	8021	19	40	9.755585	5634	5682	5731	5780	5829	19
8171	8221	8271	8321	18	41	5878	5927	5976	6025	6074	6123	18
8471	8521	8571	8621	17	42	6172	6221	6270	6319	6368	6416	17
8771	8821	8871	8921	16	43	6465	6514	6563	6612	6661	6710	16
9071	9121	9171	9221	15	44	6759	6808	6857	6905	6954	7003	15
9371	9420	9470	9520	14	45	7052	7101	7150	7199	7247	7296	14
9670	9720	9770	9820	13	46	7345	7394	7443	7492	7541	7589	13
9969	..19	..69	.119	12	47	7638	7687	7736	7785	7834	7882	12
0269	0319	0368	0418	11	48	7931	7980	8029	8078	8127	8175	11
0568	0618	0668	0717	10	49	8224	8273	8322	8371	8419	8468	10
0867	0917	0967	1016	9	50	9.758517	8566	8615	8663	8712	8761	9
1166	1216	1265	1315	8	51	8810	8858	8907	8956	9005	9053	8
1465	1514	1564	1614	7	52	9102	9151	9200	9248	9297	9346	7
1763	1813	1863	1913	6	53	9395	9443	9492	9541	9590	9638	6
2062	2112	2161	2211	5	54	9687	9736	9785	9833	9882	9931	5
2360	2410	2460	2510	4	55	9979	..28	..77	.126	.174	.223	4
2659	2709	2758	2808	3	56	9.760272	0320	0369	0418	0466	0515	3
2957	3007	3056	3106	2	57	0564	0612	0661	0710	0758	0807	2
3255	3305	3355	3404	1	58	0856	0904	0953	1002	1050	1099	1
3553	3603	3653	3702	0	59	1148	1196	1245	1293	1342	1391	0

Min.	Sine of 30 Degrees.						
	0″	10″	20″	30″	40″	50″	
0	9.698970	9006	9043	9079	9116	9152	59
1	9189	9225	9262	9298	9334	9371	58
2	9407	9444	9480	9517	9553	9589	57
3	9626	9662	9699	9735	9771	9808	56
4	9844	9880	9917	9953	9990	..26	55
5	9.700062	0099	0135	0171	0208	0244	54
6	0280	0317	0353	0389	0425	0462	53
7	0498	0534	0571	0607	0643	0680	52
8	0716	0752	0788	0825	0861	0897	51
9	0933	0970	1006	1042	1078	1115	50
10	9.701151	1187	1223	1259	1296	1332	49
11	1368	1404	1440	1477	1513	1549	48
12	1585	1621	1658	1694	1730	1766	47
13	1802	1838	1874	1911	1947	1983	46
14	2019	2055	2091	2127	2164	2200	45
15	2236	2272	2308	2344	2380	2416	44
16	2452	2488	2524	2561	2597	2633	43
17	2669	2705	2741	2777	2813	2849	42
18	2885	2921	2957	2993	3029	3065	41
19	3101	3137	3173	3209	3245	3281	40
20	9.703317	3353	3389	3425	3461	3497	39
21	3533	3569	3605	3641	3677	3713	38
22	3749	3784	3820	3856	3892	3928	37
23	3964	4000	4036	4072	4108	4144	36
24	4179	4215	4251	4287	4323	4359	35
25	4395	4431	4466	4502	4538	4574	34
26	4610	4646	4682	4717	4753	4789	33
27	4825	4861	4896	4932	4968	5004	32
28	5040	5075	5111	5147	5183	5219	31
29	5254	5290	5326	5362	5397	5433	30
30	9.705469	5505	5540	5576	5612	5648	29
31	5683	5719	5755	5790	5826	5862	28
32	5898	5933	5969	6005	6040	6076	27
33	6112	6147	6183	6219	6254	6290	26
34	6326	6361	6397	6433	6468	6504	25
35	6539	6575	6611	6646	6682	6718	24
36	6753	6789	6824	6860	6895	6931	23
37	6967	7002	7038	7073	7109	7145	22
38	7180	7216	7251	7287	7322	7358	21
39	7393	7429	7464	7500	7535	7571	20
40	9.707606	7642	7677	7713	7748	7784	19
41	7819	7855	7890	7926	7961	7997	18
42	8032	8068	8103	8139	8174	8210	17
43	8245	8280	8316	8351	9387	8422	16
44	8458	8493	8528	8564	8599	8635	15
45	8670	8705	8741	8776	8811	8847	14
46	8882	8918	8953	8988	9024	9059	13
47	9094	9130	9165	9200	9236	9271	12
48	9306	9342	9377	9412	9448	9483	11
49	9518	9553	9589	9624	9659	9695	10
50	9.709730	9765	9800	9836	9871	9906	9
51	9941	9977	..12	..47	..82	.118	8
52	9.710153	0188	0223	0259	0294	0329	7
53	0364	0399	0435	0470	0505	0540	6
54	0575	0611	0646	0681	0716	0751	5
55	0786	0822	0857	0892	0927	0962	4
56	0997	1032	1067	1103	1138	1173	3
57	1208	1243	1278	1313	1348	1383	2
58	1419	1454	1489	1524	1559	1594	1
59	1629	1664	1699	1734	1769	1804	0

Min.	Sine of 31 Degrees.						
	0″	10″	20″	30″	40″	50″	
0	9.711839	1874	1909	1944	1979	2014	59
1	2050	2085	2120	2155	2190	2225	58
2	2260	2295	2330	2365	2400	2434	57
3	[illegible]469	2504	2539	2574	2609	2644	56
4	2679	2714	2749	2784	2819	2854	55
5	2889	2924	2959	2994	3029	3063	54
6	3098	3133	3168	3203	3238	3273	53
7	3308	3343	3377	3412	3447	3482	52
8	3517	3552	3587	3621	3656	3691	51
9	3726	3761	3796	3830	3865	3900	50
10	9.713935	3970	4005	4039	4074	4109	49
11	4144	4179	4213	4248	4283	4318	48
12	4352	4387	4422	4457	4491	4526	47
13	4561	4596	4630	4665	4700	4735	46
14	4769	4804	4839	4873	4908	4943	45
15	4978	5012	5047	5082	5116	5151	44
16	5186	5220	5255	5290	5324	5359	43
17	5394	5428	5463	5498	5532	5567	42
18	5602	5636	5671	5705	5740	5775	41
19	5809	5844	5878	5913	5948	5982	40
20	9.716017	6051	6086	6121	6155	6190	39
21	6224	6259	6293	6328	6362	6397	38
22	6432	6466	6501	6535	6570	6604	37
23	6639	6673	6708	6742	6777	6811	36
24	6846	6880	6915	6949	6984	7018	35
25	7053	7087	7122	7156	7191	7225	34
26	7259	7294	7328	7363	7397	7432	33
27	7466	7500	7535	7569	7604	7638	32
28	7673	7707	7741	7776	7810	7844	31
29	7879	7913	7948	7982	8016	8051	30
30	9.718085	8119	8154	8188	8223	8257	29
31	8291	8326	8360	8394	8429	8463	28
32	8497	8531	8566	8600	8634	8669	27
33	8703	8737	8772	8806	8840	8874	26
34	8909	8943	8977	9011	9046	9080	25
35	9114	9148	9183	9217	9251	9285	24
36	9320	9354	9388	9422	9456	9491	23
37	9525	9559	9593	9627	9662	9696	22
38	9730	9764	9798	9833	9867	9901	21
39	9935	9969	...3	..38	..72	.106	20
40	9.720140	0174	0208	0242	0276	0311	19
41	0345	0379	0413	0447	0481	0515	18
42	0549	0583	0617	0652	0686	0720	17
43	0754	0788	0822	0856	0890	0924	16
44	0958	0992	1026	1060	1094	1128	15
45	1162	1196	1230	1264	1298	1332	14
46	1366	1400	1434	1468	1502	1536	13
47	1570	1604	1638	1672	1706	1740	12
48	1774	1808	1842	1876	1910	1944	11
49	1978	2012	2046	2080	2114	2148	10
50	9.722181	2215	2249	2283	2317	2351	9
51	2385	2419	2453	2487	2520	2554	8
52	2588	2622	2656	2690	2724	2757	7
53	2791	2825	2859	2893	2927	2960	6
54	2994	3028	3062	3096	3130	3163	5
55	3197	3231	3265	3299	3332	3366	4
56	3400	3434	3468	3501	3535	3569	3
57	3603	3636	3670	3704	3738	3771	2
58	3805	3839	3873	3906	3940	3974	1
59	4007	4041	4075	4109	4142	4176	0

of 30 Degrees.					Min.	Tangent of 31 Degrees.						
20″	30″	40″	50″			0″	10″	20″	30″	40″	50″	
1537	1585	1634	1682	59	0	9.778774	8821	8869	8917	8964	9012	59
1828	1877	1925	1974	58	1	9060	9108	9155	9203	9251	9298	58
2120	2168	2217	2266	57	2	9346	9394	9441	9489	9537	9584	57
2411	2460	2508	2557	56	3	9632	9679	9727	9775	9822	9870	56
2703	2751	2800	2848	55	4	9918	9965	..13	..61	.108	.156	55
2994	3043	3091	3140	54	5	9.780203	0251	0299	0346	0394	0441	54
3285	3334	3382	3431	53	6	0489	0537	0584	0632	0679	0727	53
3576	3625	3673	3722	52	7	0775	0822	0870	0917	0965	1013	52
3867	3916	3964	4013	51	8	1060	1108	1155	1203	1250	1298	51
4158	4207	4255	4304	50	9	1346	1393	1441	1488	1536	1583	50
4449	4497	4546	4594	49	10	9.781631	1678	1726	1774	1821	1869	49
4740	4788	4836	4885	48	11	1916	1964	2011	2059	2106	2154	48
5030	5079	5127	5175	47	12	2201	2249	2296	2344	2391	2439	47
5321	5369	5418	5466	46	13	2486	2534	2581	2629	2676	2724	46
5611	5660	5708	5756	45	14	2771	2819	2866	2914	2961	3009	45
5901	5950	5998	6047	44	15	3056	3104	3151	3199	3246	3294	44
6192	6240	6288	6337	43	16	3341	3388	3436	3483	3531	3578	43
6482	6530	6578	6627	42	17	3626	3673	3721	3768	3816	3863	42
6772	6820	6868	6917	41	18	3910	3958	4005	4053	4100	4148	41
7062	7110	7158	7207	40	19	4195	4242	4290	4337	4385	4432	40
7352	7400	7448	7496	39	20	9.784479	4527	4574	4622	4669	4716	39
7641	7690	7738	7786	38	21	4764	4811	4859	4906	4953	5001	38
7931	7979	8027	8076	37	22	5048	5095	5143	5190	5238	5285	37
8221	8269	8317	8365	36	23	5332	5380	5427	5474	5522	5569	36
8510	8558	8606	8655	35	24	5616	5664	5711	5758	5806	5853	35
8799	8848	8896	8944	34	25	5900	5948	5995	6042	6090	6137	34
9089	9137	9185	9233	33	26	6184	6232	6279	6326	6374	6421	33
9378	9426	9474	9522	32	27	6468	6516	6563	6610	6657	6705	32
9667	9715	9763	9811	31	28	6752	6799	6847	6894	6941	6988	31
9956	...4	..52	.100	30	29	7036	7083	7130	7178	7225	7272	30
0245	0293	0341	0389	29	30	9.787319	7367	7414	7461	7508	7556	29
0534	0582	0630	0678	28	31	7603	7650	7697	7745	7792	7839	28
0822	0870	0919	0967	27	32	7886	7934	7981	8028	8075	8122	27
1111	1159	1207	1255	26	33	8170	8217	8264	8311	8359	8406	26
1399	1448	1496	1544	25	34	8453	8500	8547	8595	8642	8689	25
1688	1736	1784	1832	24	35	8736	8783	8830	8878	8925	8972	24
1976	2024	2072	2120	23	36	9019	9066	9114	9161	9208	9255	23
2264	2312	2361	2409	22	37	9302	9349	9397	9444	9491	9538	22
2553	2601	2649	2697	21	38	9585	9632	9679	9727	9774	9821	21
2841	2889	2937	2985	20	39	9868	9915	9962	...9	..57	.104	20
3129	3177	3225	3273	19	40	9.790151	0198	0245	0292	0339	0386	19
3417	3465	3512	3560	18	41	0434	0481	0528	0575	0622	0669	18
3704	3752	3800	3848	17	42	0716	0763	0810	0857	0905	0952	17
3992	4040	4088	4136	16	43	0999	1046	1093	1140	1187	1234	16
4280	4328	4375	4423	15	44	1281	1328	1375	1422	1469	1516	15
4567	4615	4663	4711	14	45	1563	1611	1658	1705	1752	1799	14
4855	4902	4950	4998	13	46	1846	1893	1940	1987	2034	2081	13
5142	5190	5238	5286	12	47	2128	2175	2222	2269	2316	2363	12
5429	5477	5525	5573	11	48	2410	2457	2504	2551	2598	2645	11
5716	5764	5812	5860	10	49	2692	2739	2786	2833	2880	2927	10
6003	6051	6099	6147	9	50	9.792974	3021	3068	3115	3162	3209	9
6290	6338	6386	6434	8	51	3256	3303	3350	3397	3444	3491	8
6577	6625	6673	6721	7	52	3538	3585	3632	3679	3726	3773	7
6864	6912	6960	7007	6	53	3819	3866	3913	3960	4007	4054	6
7151	7199	7246	7294	5	54	4101	4148	4195	4242	4289	4336	5
7437	7485	7533	7581	4	55	4383	4430	4476	4523	4570	4617	4
7724	7772	7819	7867	3	56	4664	4711	4758	4805	4852	4899	3
8010	8058	8106	8154	2	57	4946	4992	5039	5086	5133	5180	2
8297	8344	8392	8440	1	58	5227	5274	5321	5367	5414	5461	1
8583	8631	8678	8726	0	59	5508	5555	5602	5649	5696	5742	0

Sine of 32 Degrees.							Min.	Sine of 33]		
0″	10″	20″	30″	40″	50″			0″	10″	20″
724210	4243	4277	4311	4344	4378	59	0	9.736109	6141	6174
4412	4445	4479	4513	4546	4580	58	1	6303	6336	6368
4614	4647	4681	4715	4748	4782	57	2	6498	6530	6562
4816	4849	4883	4917	4950	4984	56	3	6692	6724	6757
5017	5051	5085	5118	5152	5185	55	4	6886	6918	6951
5219	5253	5286	5320	5353	5387	54	5	7080	7112	7145
5420	5454	5488	5521	5555	5588	53	6	7274	7306	7338
5622	5655	5689	5722	5756	5789	52	7	7467	7500	7532
5823	5856	5890	5923	5957	5990	51	8	7661	7693	7726
6024	6057	6091	6124	6158	6191	50	9	7855	7887	7919
726225	6258	6292	6325	6359	6392	49	10	9.738048	8080	8112
6426	6459	6493	6526	6560	6593	48	11	8241	8273	8306
6626	6660	6693	6727	6760	6794	47	12	8434	8466	8499
6827	6860	6894	6927	6961	6994	46	13	8627	8659	8692
7027	7061	7094	7128	7161	7194	45	14	8820	8852	8884
7228	7261	7294	7328	7361	7394	44	15	9013	9045	9077
7428	7461	7494	7528	7561	7594	43	16	9206	9238	9270
7628	7661	7694	7728	7761	7794	42	17	9398	9430	9462
7828	7861	7894	7928	7961	7994	41	18	9590	9622	9654
8027	8061	8094	8127	8161	8194	40	19	9783	9815	9847
728227	8260	8294	8327	8360	8393	39	20	9.739975	...7	..39
8427	8460	8493	8526	8560	8593	38	21	9.740167	0199	0231
8626	8659	8692	8726	8759	8792	37	22	0359	0391	0423
8825	8858	8892	8925	8958	8991	36	23	0550	0582	0614
9024	9058	9091	9124	9157	9190	35	24	0742	0774	0806
9223	9257	9290	9323	9356	9389	34	25	0934	0966	0997
9422	9455	9489	9522	9555	9588	33	26	1125	1157	1189
9621	9654	9687	9720	9753	9787	32	27	1316	1348	1380
9820	9853	9886	9919	9952	9985	31	28	1508	1539	1571
730018	0051	0084	0117	0150	0183	30	29	1699	1730	1762
730217	0250	0283	0316	0349	0382	29	30	9.741889	1921	1953
0415	0448	0481	0514	0547	0580	28	31	2080	2112	2144
0613	0646	0679	0712	0745	0778	27	32	2271	2303	2335
0811	0844	0877	0910	0943	0976	26	33	2462	2493	2525
1009	1042	1075	1108	1141	1173	25	34	2652	2684	2715
1206	1239	1272	1305	1338	1371	24	35	2842	2874	2906
1404	1437	1470	1503	1536	1569	23	36	3033	3064	3096
1602	1634	1667	1700	1733	1766	22	37	3223	3254	3286
1799	1832	1865	1897	1930	1963	21	38	3413	3444	3476
1996	2029	2062	2095	2127	2160	20	39	3602	3634	3666
732193	2226	2259	2292	2325	2357	19	40	9.743792	3824	3855
2390	2423	2456	2489	2521	2554	18	41	3982	4013	4045
2587	2620	2653	2685	2718	2751	17	42	4171	4203	4234
2784	2816	2849	2882	2915	2948	16	43	4361	4392	4424
2980	3013	3046	3079	3111	3144	15	44	4550	4581	4613
3177	3210	3242	3275	3308	3340	14	45	4739	4770	4802
3373	3406	3439	3471	3504	3537	13	46	4928	4959	4991
3569	3602	3635	3667	3700	3733	12	47	5117	5148	5180
3765	3798	3831	3863	3896	3929	11	48	5306	5337	5369
3961	3994	4027	4059	4092	4125	10	49	5494	5526	5557
734157	4190	4222	4255	4288	4320	9	50	9.745683	5714	5746
4353	4386	4418	4451	4483	4516	8	51	5871	5903	5934
4549	4581	4614	4646	4679	4711	7	52	6060	6091	6122
4744	4777	4809	4842	4874	4907	6	53	6248	6279	6310
4939	4972	5004	5037	5069	5102	5	54	6436	6467	6498
5135	5167	5200	5232	5265	5297	4	55	6624	6655	6686
5330	5362	5395	5427	5460	5492	3	56	6812	6843	6874
5525	5557	5590	5622	5655	5687	2	57	6999	7031	7062
5719	5752	5784	5817	5849	5882	1	58	7187	7218	7249
5914	5947	5979	6011	6044	6076	0	59	7374	7406	7437

of 32 Degrees.					Min.	Tangent of 33 Degrees.				
20″	30″	40″	50″			0″	10″	20″	30″	40″
5883	5930	5977	6023	59	0	9.812517	2563	2610	2656	2702
6164	6211	6258	6304	58	1	2794	2840	2886	2932	2978
6445	6492	6539	6585	57	2	3070	3116	3163	3209	3255
6726	6773	6819	6866	56	3	3347	3393	3439	3485	3531
7007	7053	7100	7147	55	4	3623	3669	3715	3761	3807
7287	7334	7381	7428	54	5	3899	3945	3991	4037	4083
7568	7615	7662	7708	53	6	4176	4222	4268	4314	4360
7849	7895	7942	7989	52	7	4452	4498	4544	4590	4636
8129	8176	8223	8269	51	8	4728	4774	4820	4866	4912
8409	8456	8503	8550	50	9	5004	5050	5096	5142	5188
8690	8737	8783	8830	49	10	9.815280	5326	5371	5417	5463
8970	9017	9063	9110	48	11	5555	5601	5647	5693	5739
9250	9297	9344	9390	47	12	5831	5877	5923	5969	6015
9530	9577	9624	9670	46	13	6107	6153	6199	6245	6291
9810	9857	9904	9950	45	14	6382	6428	6474	6520	6566
..90	.137	.184	.230	44	15	6658	6704	6750	6796	6842
0370	0417	0463	0510	43	16	6933	6979	7025	7071	7117
0650	0697	0743	0790	42	17	7209	7255	7301	7347	7392
0930	0976	1023	1070	41	18	7484	7530	7576	7622	7668
1209	1256	1303	1349	40	19	7759	7805	7851	7897	7943
1489	1535	1582	1629	39	20	9.818035	8081	8126	8172	8218
1768	1815	1862	1908	38	21	8310	8356	8402	8447	8493
2048	2094	2141	2187	37	22	8585	8631	8677	8722	8768
2327	2374	2420	2467	36	23	8860	8906	8952	8997	9043
2606	2653	2699	2746	35	24	9135	9181	9226	9272	9318
2886	2932	2979	3025	34	25	9410	9455	9501	9547	9593
3165	3211	3258	3304	33	26	9684	9730	9776	9822	9868
3444	3490	3537	3583	32	27	9959	...5	..51	..96	.142
3723	3769	3816	3862	31	28	9.820234	0280	0325	0371	0417
4001	4048	4094	4141	30	29	0508	0554	0600	0646	0691
4280	4327	4373	4420	29	30	9.820783	0829	0874	0920	0966
4559	4605	4652	4698	28	31	1057	1103	1149	1195	1240
4838	4884	4930	4977	27	32	1332	1377	1423	1469	1515
5116	5163	5209	5255	26	33	1606	1652	1697	1743	1789
5395	5441	5487	5534	25	34	1880	1926	1972	2017	2063
5673	5719	5766	5812	24	35	2154	2200	2246	2292	2337
5951	5998	6044	6091	23	36	2429	2474	2520	2566	2611
6230	6276	6322	6369	22	37	2703	2748	2794	2840	2885
6508	6554	6601	6647	21	38	2977	3022	3068	3114	3159
6786	6832	6879	6925	20	39	3251	3296	3342	3387	3433
7064	7110	7157	7203	19	40	9.823524	3570	3616	3661	3707
7342	7388	7435	7481	18	41	3798	3844	3889	3935	3981
7620	7666	7713	7759	17	42	4072	4117	4163	4209	4254
7898	7944	7990	8037	16	43	4345	4391	4437	4482	4528
8176	8222	8268	8314	15	44	4619	4665	4710	4756	4801
8453	8499	8546	8592	14	45	4893	4938	4984	5029	5075
8731	8777	8823	8870	13	46	5166	5212	5257	5303	5348
9008	9055	9101	9147	12	47	5439	5485	5531	5576	5622
9286	9332	9378	9424	11	48	5713	5758	5804	5849	5895
9563	9609	9656	9702	10	49	5986	6032	6077	6123	6168
9840	9887	9933	9979	9	50	9.826259	6305	6350	6396	6441
0118	0164	0210	0256	8	51	6532	6578	6623	6669	6714
0395	0441	0487	0533	7	52	6805	6851	6896	6942	6987
0672	0718	0764	0810	6	53	7078	7124	7169	7215	7260
0949	0995	1041	1087	5	54	7351	7397	7442	7488	7533
1226	1272	1318	1364	4	55	7624	7670	7715	7761	7806
1503	1549	1595	1641	3	56	7897	7942	7988	8033	8079
1780	1826	1872	1918	2	57	8170	8215	8261	8306	8351
2056	2102	2149	2195	1	58	8442	8488	8533	8579	8624
2333	2379	2425	2471	0	59	8715	8760	8806	8851	8897

)egrees.				Min.	Sine of 35 Degree[illegible].						
30″	40″	50″			0″	10″	20″	30″	40″	50″	
7655	7686	7718	59	0	9.758591	8621	8651	8681	8712	8742	59
7842	7874	7905	58	1	8772	8802	8832	8862	8892	8922	58
8030	8061	8092	57	2	8952	8982	9012	9042	9072	9102	57
8216	8248	8279	56	3	9132	9162	9192	9222	9252	9282	56
8403	8434	8466	55	4	9312	9342	9372	9402	9432	9462	55
8590	8621	8652	54	5	9492	9522	9552	9582	9612	9642	54
8777	8808	8839	53	6	9672	9702	9732	9762	9792	9822	53
8963	8994	9025	52	7	9852	9881	9911	9941	9971	...1	52
9149	9180	9212	51	8	9.760031	0061	0091	0121	0151	0181	51
9336	9367	9398	50	9	0211	0240	0270	0300	0330	0360	50
9522	9553	9584	49	10	9.760390	0420	0450	0480	0509	0539	49
9708	9739	9770	48	11	0569	0599	0629	0659	0689	0718	48
9894	9925	9956	47	12	0748	0778	0808	0838	0868	0898	47
..79	.110	.141	46	13	0927	0957	0987	1017	1047	1076	46
0265	0296	0327	45	14	1106	1136	1166	1196	1225	1255	45
0451	0482	0512	44	15	1285	1315	1345	1374	1404	1434	44
0636	0667	0698	43	16	1464	1494	1523	1553	1583	1613	43
0821	0852	0883	42	17	1642	1672	1702	1732	1761	1791	42
1007	1037	1068	41	18	1821	1851	1880	1910	1940	1969	41
1192	1222	1253	40	19	1999	2029	2059	2088	2118	2148	40
1377	1407	1438	39	20	9.762177	2207	2237	2267	2296	2326	39
1561	1592	1623	38	21	2356	2385	2415	2445	2474	2504	38
1746	1777	1808	37	22	2534	2563	2593	2623	2652	2682	37
1931	1962	1992	36	23	2712	2741	2771	2801	2830	2860	36
2115	2146	2177	35	24	2889	2919	2949	2978	3008	3038	35
2300	2330	2361	34	25	3067	3097	3126	3156	3186	3215	34
2484	2515	2545	33	26	3245	3274	3304	3333	3363	3393	33
2668	2699	2729	32	27	3422	3452	3481	3511	3540	3570	32
2852	2883	2914	31	28	3600	3629	3659	3688	3718	3747	31
3036	3067	3097	30	29	3777	3806	3836	3865	3895	3925	30
3220	3251	3281	29	30	9.763954	3984	4013	4043	4072	4102	29
3404	3434	3465	28	31	4131	4161	4190	4220	4249	4279	28
3587	3618	3648	27	32	4308	4338	4367	4396	4426	4455	27
3771	3801	3832	26	33	4485	4514	4544	4573	4603	4632	26
3954	3985	4015	25	34	4662	4691	4720	4750	4779	4809	25
4137	4168	4198	24	35	4838	4868	4897	492[illegible]	4956	4985	24
4320	4351	4381	23	36	5015	5044	5074	5103	5132	5162	23
4503	4534	4564	22	37	5191	5221	5250	5279	5309	5338	22
4686	4717	4747	21	38	5367	5397	5426	5456	5485	5514	21
4869	4900	4930	20	39	5544	5573	5602	5632	5661	5690	20
5052	5082	5113	19	40	9.765720	5749	5778	5808	5837	5866	19
5234	5265	5295	18	41	5896	5925	5954	5984	6013	6042	18
5417	5447	5478	17	42	6072	6101	6130	6159	6189	6218	17
5599	5629	5660	16	43	6247	6277	6306	6335	6364	6394	16
5781	5812	5842	15	44	6423	6452	6481	6511	6540	6569	15
5963	5994	6024	14	45	6598	6628	6657	6686	6715	6745	14
6145	6176	6206	13	46	6774	6803	6832	6862	6891	6920	13
6327	6358	6388	12	47	6949	6978	7008	7037	7066	7095	12
6509	6539	6570	11	48	7124	7154	7183	7212	7241	7270	11
6691	6721	6751	10	49	7300	7329	7358	7387	7416	7445	10
6872	6903	6933	9	50	9.767475	7504	7533	7562	7591	7620	9
7054	7084	7114	8	51	7649	7679	7708	7737	7766	7795	8
7235	7265	7295	7	52	7824	7853	7882	7912	7941	7970	7
7416	7446	7477	6	53	7999	8028	8057	8086	8115	8144	6
7597	7627	7658	5	54	8173	8203	8232	8261	8290	8319	5
7778	7808	7839	4	55	8348	8377	8406	8435	8464	8493	4
7959	7989	8019	3	56	8522	8551	8580	8609	8638	8668	3
8140	8170	8200	2	57	8697	8726	8755	8784	8813	8842	2
8321	8351	8381	1	58	8871	8900	8929	8958	8987	9016	1
8501	8531	8561	0	59	9045	9074	9103	9132	9161	9190	0

gent of 34 Degrees.						Min.	Tangent of 35 Degrees.				
10″	20″	30″	40″	50″			0″	10″	20″	30″	40″
9033	9078	9124	9169	9215	59	0	9.845227	5272	5316	5361	5406
9305	9351	9396	9442	9487	58	1	5496	5540	5585	5630	5675
9578	9623	9669	9714	9759	57	2	5764	5809	5854	5899	5944
9850	9895	9941	9986	..32	56	3	6033	6078	6123	6168	6212
0122	0168	0213	0258	0304	55	4	6302	6347	6391	6436	6481
0395	0440	0485	0531	0576	54	5	6570	6615	6660	6705	6750
0667	0712	0757	0803	0848	53	6	6839	6884	6929	6973	7018
0939	0984	1029	1075	1120	52	7	7108	7152	7197	7242	7287
1211	1256	1301	1347	1392	51	8	7376	7421	7465	7510	7555
1483	1528	1573	1619	1664	50	9	7644	7689	7734	7779	7823
1755	1800	1845	1891	1936	49	10	9.847913	7957	8002	8047	8092
2026	2072	2117	2162	2208	48	11	8181	8226	8270	8315	8360
2298	2343	2389	2434	2479	47	12	8449	8494	8539	8583	8628
2570	2615	2660	2706	2751	46	13	8717	8762	8807	8851	8896
2842	2887	2932	2977	3023	45	14	8986	9030	9075	9120	9164
3113	3158	3204	3249	3294	44	15	9254	9298	9343	9388	9432
3385	3430	3475	3520	3566	43	16	9522	9566	9611	9656	9700
3656	3701	3747	3792	3837	42	17	9790	9834	9879	9924	9968
3927	3973	4018	4063	4108	41	18	9.850057	0102	0147	0191	0236
4199	4244	4289	4334	4380	40	19	0325	0370	0415	0459	0504
4470	4515	4561	4606	4651	39	20	9.850593	0638	0682	0727	0772
4741	4787	4832	4877	4922	38	21	0861	0905	0950	0995	1039
5012	5058	5103	5148	5193	37	22	1129	1173	1218	1262	1307
5284	5329	5374	5419	5464	36	23	1396	1441	1485	1530	1575
5555	5600	5645	5690	5735	35	24	1664	1708	1753	1797	1842
5826	5871	5916	5961	6006	34	25	1931	1976	2020	2065	2110
6096	6142	6187	6232	6277	33	26	2199	2243	2288	2332	2377
6367	6412	6458	6503	6548	32	27	2466	2511	2555	2600	2644
6638	6683	6728	6773	6819	31	28	2733	2778	2823	2867	2912
6909	6954	6999	7044	7089	30	29	3001	3045	3090	3134	3179
7179	7225	7270	7315	7360	29	30	9.853268	3313	3357	3402	3446
7450	7495	7540	7585	7630	28	31	3535	3580	3624	3669	3713
7721	7766	7811	7856	7901	27	32	3802	3847	3891	3936	3980
7991	8036	8081	8126	8171	26	33	4069	4114	4158	4203	4247
8261	8307	8352	8397	8442	25	34	4336	4381	4425	4470	4514
8532	8577	8622	8667	8712	24	35	4603	4648	4692	4737	4781
8802	8847	8892	8937	8982	23	36	4870	4915	4959	5004	5048
9072	9117	9162	9207	9252	22	37	5137	5182	5226	5271	5315
9343	9388	9433	9478	9523	21	38	5404	5449	5493	5537	5582
9613	9658	9703	9748	9793	20	39	5671	5715	5760	5804	5849
9883	9928	9973	..18	..63	19	40	9.855938	5982	6026	6071	6115
0153	0198	0243	0288	0333	18	41	6204	6249	6293	6338	6382
0423	0468	0513	0558	0603	17	42	6471	6515	6560	6604	6649
0693	0737	0782	0827	0872	16	43	6737	6782	6826	6871	6915
0962	1007	1052	1097	1142	15	44	7004	7048	7093	7137	7182
1232	1277	1322	1367	1412	14	45	7270	7315	7359	7404	7448
1502	1547	1592	1637	1682	13	46	7537	7581	7626	7670	7714
1771	1816	1861	1906	1951	12	47	7803	7848	7892	7936	7981
2041	2086	2131	2176	2221	11	48	8069	8114	8158	8203	8247
2311	2355	2400	2445	2490	10	49	8336	8380	8424	8469	8513
2580	2625	2670	2715	2760	9	50	9.858602	8646	8691	8735	8779
2849	2894	2939	2984	3029	8	51	8868	8912	8957	9001	9045
3119	3164	3209	3253	3298	7	52	9134	9178	9223	9267	9311
3388	3433	3478	3523	3568	6	53	9400	9444	9489	9533	9577
3657	3702	3747	3792	3837	5	54	9666	9710	9755	9799	9843
3927	3971	4016	4061	4106	4	55	9932	9976	..21	..65	.109
4196	4241	4285	4330	4375	3	56	9.860198	0242	0287	0331	0375
4465	4510	4554	4599	4644	2	57	0464	0508	0552	0597	0641
4734	4779	4823	4868	4913	1	58	0730	0774	0818	0862	0907
5003	5048	5092	5137	5182	0	59	0995	1040	1084	1128	1172

Sine of 36 Degrees.							Min.	Sine of 37 Degrees.				
0″	10″	20″	30″	40″	50″			0″	10″	20″	30″	40″
9·769219	9248	9277	9306	9335	9364	59	0	9·779463	9491	9519	9547	9575
9393	9421	9450	9479	9508	9537	58	1	9631	9659	9686	9714	9742
9566	9595	9624	9653	9682	9711	57	2	9798	9826	9854	9882	9910
9740	9769	9798	9827	9856	9884	56	3	9966	9993	..21	..49	..77
9913	9942	9971	....	..29	..58	55	4	9·780133	0161	0189	0216	0244
9·770087	0116	0145	0173	0202	0231	54	5	0300	0328	0356	0384	0411
0260	0289	0318	0347	0376	0404	53	6	0467	0495	0523	0551	0578
0433	0462	0491	0520	0549	0577	52	7	0634	0662	0690	0718	0745
0606	0635	0664	0693	0722	0750	51	8	0801	0829	0857	0884	0912
0779	0808	0837	0866	0895	0923	50	9	0968	0996	1023	1051	1079
9·770952	0981	1010	1039	1067	1096	49	10	9·781134	1162	1190	1218	1246
1125	1154	1183	1211	1240	1269	48	11	1301	1329	1357	1384	1412
1298	1326	1355	1384	1413	1441	47	12	1468	1495	1523	1551	1578
1470	1499	1528	1556	1585	1614	46	13	1634	1662	1689	1717	1745
1643	1671	1700	1729	1758	1786	45	14	1800	1828	1856	1883	1911
1815	1844	1872	1901	1930	1959	44	15	1966	1994	2022	2049	2077
1987	2016	2045	2073	2102	2131	43	16	2132	2160	2188	2215	2243
2159	2188	2217	2245	2274	2303	42	17	2298	2326	2354	2381	2409
2331	2360	2389	2417	2446	2475	41	18	2464	2492	2520	2547	2575
2503	2532	2561	2589	2618	2646	40	19	2630	2658	2685	2713	2741
9·772675	2704	2732	2761	2790	2818	39	20	9·782796	2823	2851	2879	2906
2847	2875	2904	2933	2961	2990	38	21	2961	2989	3017	3044	3072
3018	3047	3076	3104	3133	3161	37	22	3127	3154	3182	3210	3237
3190	3219	3247	3276	3304	3333	36	23	3292	3320	3347	3375	3402
3361	3390	3418	3447	3476	3504	35	24	3458	3485	3513	3540	3568
3533	3561	3590	3618	3647	3675	34	25	3623	3650	3678	3705	3733
3704	3732	3761	3789	3818	3846	33	26	3788	3815	3843	3870	3898
3875	3903	3932	3960	3989	4017	32	27	3953	3980	4008	4035	4063
4046	4074	4103	4131	4160	4188	31	28	4118	4145	4173	4200	4228
4217	4245	4274	4302	4331	4359	30	29	4282	4310	4337	4365	4392
9·774388	4416	4445	4473	4501	4530	29	30	9·784447	4475	4502	4529	4557
4558	4587	4615	4644	4672	4700	28	31	4612	4639	4667	4694	4721
4729	4757	4786	4814	4842	4871	27	32	4776	4804	4831	4858	4886
4899	4928	4956	4985	5013	5041	26	33	4941	4968	4995	5023	5050
5070	5098	5126	5155	5183	5212	25	34	5105	5132	5160	5187	5214
5240	5268	5297	5325	5353	5382	24	35	5269	5296	5324	5351	5378
5410	5438	5467	5495	5523	5552	23	36	5433	5461	5488	5515	5543
5580	5608	5637	5665	5693	5722	22	37	5597	5624	5652	5679	5706
5750	5778	5807	5835	5863	5892	21	38	5761	5788	5816	5843	5870
5920	5948	5977	6005	6033	6061	20	39	5925	5952	5979	6007	6034
9·776090	6118	6146	6175	6203	6231	19	40	9·786089	6116	6143	6170	6198
6259	6288	6316	6344	6372	6401	18	41	6252	6279	6307	6334	6361
6429	6457	6485	6514	6542	6570	17	42	6416	6443	6470	6497	6525
6598	6627	6655	6683	6711	6739	16	43	6579	6606	6634	6661	6688
6768	6796	6824	6852	6880	6909	15	44	6742	6770	6797	6824	6851
6937	6965	6993	7021	7050	7078	14	45	6906	6933	6960	6987	7014
7106	7134	7162	7191	7219	7247	13	46	7069	7096	7123	7150	7177
7275	7303	7331	7359	7388	7416	12	47	7232	7259	7286	7313	7340
7444	7472	7500	7528	7556	7585	11	48	7395	7422	7449	7476	7503
7613	7641	7669	7697	7725	7753	10	49	7557	7585	7612	7639	7666
9·777781	7810	7838	7866	7894	7922	9	50	9·787720	7747	7774	7801	7829
7950	7978	8006	8034	8062	8091	8	51	7883	7910	7937	7964	7991
8119	8147	8175	8203	8231	8259	7	52	8045	8072	8099	8127	8154
8287	8315	8343	8371	8399	8427	6	53	8208	8235	8262	8289	8316
8455	8483	8511	8539	8567	8595	5	54	8370	8397	8424	8451	8478
8624	8652	8680	8708	8736	8764	4	55	8532	8559	8586	8613	8640
8792	8820	8848	8876	8904	8932	3	56	8694	8721	8748	8775	8802
8960	8988	9016	9044	9072	9100	2	57	8856	8883	8910	8937	8964
9128	9156	9183	9211	9239	9267	1	58	9018	9045	9072	9099	9126
9295	9323	9351	9379	9407	9435	0	59	9180	9207	9234	9261	9288

Tangent of 36 Degrees.							Min.	Tangent of 37 Degrees.						
0″	10″	20″	30″	40″	50″			0″	10″	20″	30″	40″	50″	
.861261	1305	1350	1394	1438	1482	59	0	9.877114	7158	7202	7246	7290	7333	59
1527	1571	1615	1659	1704	1748	58	1	7377	7421	7465	7509	7552	7596	58
1792	1837	1881	1925	1969	2014	57	2	7640	7684	7728	7771	7815	7859	57
2058	2102	2146	2191	2235	2279	56	3	7903	7947	7990	8034	8078	8122	56
2323	2368	2412	2456	2500	2545	55	4	8165	8209	8253	8297	8341	8384	55
2589	2633	2677	2721	2766	2810	54	5	8428	8472	8516	8559	8603	8647	54
2854	2898	2943	2987	3031	3075	53	6	8691	8734	8778	8822	8866	8909	53
3119	3164	3208	3252	3296	3341	52	7	8953	8997	9041	9085	9128	9172	52
3385	3429	3473	3517	3562	3606	51	8	9216	9260	9303	9347	9391	9435	51
3650	3694	3738	3783	3827	3871	50	9	9478	9522	9566	9609	9653	9697	50
.863915	3959	4004	4048	4092	4136	49	10	9.879741	9784	9828	9872	9916	9959	49
4180	4225	4269	4313	4357	4401	48	11	9.880003	0047	0091	0134	0178	0222	48
4445	4490	4534	4578	4622	4666	47	12	0265	0309	0353	0397	0440	0484	47
4710	4755	4799	4843	4887	4931	46	13	0528	0571	0615	0659	0703	0746	46
4975	5020	5064	5108	5152	5196	45	14	0790	0834	0877	0921	0965	1008	45
5240	5285	5329	5373	5417	5461	44	15	1052	1096	1140	1183	1227	1271	44
5505	5549	5594	5638	5682	5726	43	16	1314	1358	1402	1445	1489	1533	43
5770	5814	5858	5903	5947	5991	42	17	1577	1620	1664	1708	1751	1795	42
6035	6079	6123	6167	6211	6256	41	18	1839	1882	1926	1970	2013	2057	41
6300	6344	6388	6432	6476	6520	40	19	2101	2144	2188	2232	2275	2319	40
.866564	6609	6653	6697	6741	6785	39	20	9.882363	2406	2450	2494	2537	2581	39
6829	6873	6917	6961	7006	7050	38	21	2625	2668	2712	2756	2799	2843	38
7094	7138	7182	7226	7270	7314	37	22	2887	2930	2974	3018	3061	3105	37
7358	7402	7446	7491	7535	7579	36	23	3148	3192	3236	3279	3323	3367	36
7623	7667	7711	7755	7799	7843	35	24	3410	3454	3498	3541	3585	3628	35
7887	7931	7975	8019	8064	8108	34	25	3672	3716	3759	3803	3847	3890	34
8152	8196	8240	8284	8328	8372	33	26	3934	3977	4021	4065	4108	4152	33
8416	8460	8504	8548	8592	8636	32	27	4196	4239	4283	4326	4370	4414	32
8680	8724	8768	8813	8857	8901	31	28	4457	4501	4544	4588	4632	4675	31
8945	8989	9033	9077	9121	9165	30	29	4719	4762	4806	4850	4893	4937	30
.869209	9253	9297	9341	9385	9429	29	30	9.884980	5024	5068	5111	5155	5198	29
9473	9517	9561	9605	9649	9693	28	31	5242	5286	5329	5373	5416	5460	28
9737	9781	9825	9869	9913	9957	27	32	5504	5547	5591	5634	5678	5721	27
.870001	0045	0089	0133	0177	0221	26	33	5765	5809	5852	5896	5939	5983	26
0265	0309	0353	0397	0441	0485	25	34	6026	6070	6114	6157	6201	6244	25
0529	0573	0617	0661	0705	0749	24	35	6288	6331	6375	6419	6462	6506	24
0793	0837	0881	0925	0969	1013	23	36	6549	6593	6636	6680	6723	6767	23
1057	1101	1145	1189	1233	1277	22	37	6811	6854	6898	6941	6985	7028	22
1321	1365	1409	1453	1497	1541	21	38	7072	7115	7159	7202	7246	7289	21
1585	1629	1673	1717	1761	1805	20	39	7333	7377	7420	7464	7507	7551	20
.871849	1893	1937	1980	2024	2068	19	40	9.887594	7638	7681	7725	7768	7812	19
2112	2156	2200	2244	2288	2332	18	41	7855	7899	7942	7986	8029	8073	18
2376	2420	2464	2508	2552	2596	17	42	8116	8160	8203	8247	8291	8334	17
2640	2684	2727	2771	2815	2859	16	43	8378	8421	8465	8508	8552	8595	16
2903	2947	2991	3035	3079	3123	15	44	8639	8682	8726	8769	8813	8856	15
3167	3211	3255	3299	3342	3386	14	45	8900	8943	8987	9030	9074	9117	14
3430	3474	3518	3562	3606	3650	13	46	9161	9204	9248	9291	9334	9378	13
3694	3738	3781	3825	3869	3913	12	47	9421	9465	9508	9552	9595	9639	12
3957	4001	4045	4089	4133	4177	11	48	9682	9726	9769	9813	9856	9900	11
4220	4264	4308	4352	4396	4440	10	49	9943	9987	..30	..74	.117	.160	10
.874484	4528	4572	4615	4659	4703	9	50	9.890204	0247	0291	0334	0378	0421	9
4747	4791	4835	4879	4923	4966	8	51	0465	0508	0552	0595	0639	0682	8
5010	5054	5098	5142	5186	5230	7	52	0725	0769	0812	0856	0899	0943	7
5273	5317	5361	5405	5449	5493	6	53	0986	1030	1073	1116	1160	1203	6
5537	5580	5624	5668	5712	5756	5	54	1247	1290	1334	1377	1421	1464	5
5800	5843	5887	5931	5975	6019	4	55	1507	1551	1594	1638	1681	1725	4
6063	6107	6150	6194	6238	6282	3	56	1768	1811	1855	1898	1942	1985	3
6326	6370	6413	6457	6501	6545	2	57	2028	2072	2115	2159	2202	2246	2
6589	6632	6676	6720	6764	6808	1	58	2289	2332	2376	2419	2463	2506	1
6852	6895	6939	6983	7027	7071	0	59	2549	2593	2636	2680	2723	2766	0
60″	50″	40″	30″	20″	10″	Min.		60″	50″	40″	30″	20″	10″	Min.
Co-tangent of 53 Degrees.								Co-tangent of 52 Degrees.						

Sine of 38 Degrees.							Min.	Sine of 39]		
0	10″	20″	30″	40″	50″			0″	10″	20″
789342	9369	9396	9423	9450	9477	59	0	9.798872	8898	8924
9504	9531	9557	9584	9611	9638	58	1	9028	9054	9080
9665	9692	9719	9746	9773	9800	57	2	9184	9210	9236
9827	9854	9880	9907	9934	9961	56	3	9339	9365	9391
9988	..15	..42	..69	..96	.122	55	4	9495	9521	9547
790149	0176	0203	0230	0257	0284	54	5	9651	9677	9703
0310	0337	0364	0391	0418	0445	53	6	9806	9832	9858
0471	0498	0525	0552	0579	0606	52	7	9962	9987	..13
0632	0659	0686	0713	0740	0767	51	8	9.800117	0143	0169
0793	0820	0847	0874	0901	0927	50	9	0272	0298	0324
790954	0981	1008	1034	1061	1088	49	10	9.800427	0453	0479
1115	1142	1168	1195	1222	1249	48	11	0582	0608	0634
1275	1302	1329	1356	1382	1409	47	12	0737	0763	0789
1436	1463	1489	1516	1543	1570	46	13	0892	0918	0944
1596	1623	1650	1676	1703	1730	45	14	1047	1073	1098
1757	1783	1810	1837	1863	1890	44	15	1201	1227	1253
1917	1943	1970	1997	2024	2050	43	16	1356	1382	1408
2077	2104	2130	2157	2184	2210	42	17	1511	1536	1562
2237	2264	2290	2317	2344	2370	41	18	1665	1691	1716
2397	2423	2450	2477	2503	2530	40	19	1819	1845	1871
792557	2583	2610	2636	2663	2690	39	20	9.801973	1999	2025
2716	2743	2770	2796	2823	2849	38	21	2128	2153	2179
2876	2903	2929	2956	2982	3009	37	22	2282	2307	2333
3035	3062	3089	3115	3142	3168	36	23	2436	2461	2487
3195	3222	3248	3275	3301	3328	35	24	2589	2615	2641
3354	3381	3407	3434	3460	3487	34	25	2743	2769	2794
3514	3540	3567	3593	3620	3646	33	26	2897	2922	2948
3673	3699	3726	3752	3779	3805	32	27	3050	3076	3102
3832	3858	3885	3911	3938	3964	31	28	3204	3229	3255
3991	4017	4044	4070	4097	4123	30	29	3357	3383	3408
794150	4176	4203	4229	4255	4282	29	30	9.803511	3536	3562
4308	4335	4361	4388	4414	4441	28	31	3664	3689	3715
4467	4493	4520	4546	4573	4599	27	32	3817	3842	3868
4626	4652	4678	4705	4731	4758	26	33	3970	3995	4021
4784	4810	4837	4863	4890	4916	25	34	4123	4148	4174
4942	4969	4995	5022	5048	5074	24	35	4276	4301	4327
5101	5127	5154	5180	5206	5233	23	36	4428	4454	4479
5259	5285	5312	5338	5364	5391	22	37	4581	4607	4632
5417	5443	5470	5496	5522	5549	21	38	4734	4759	4784
5575	5601	5628	5654	5680	5707	20	39	4886	4912	4937
795733	5759	5786	5812	5838	5865	19	40	9.805039	5064	5089
5891	5917	5943	5970	5996	6022	18	41	5191	5216	5242
6049	6075	6101	6127	6154	6180	17	42	5343	5368	5394
6206	6233	6259	6285	6311	6338	16	43	5495	5520	5546
6364	6390	6416	6443	6469	6495	15	44	5647	5673	5698
6521	6547	6574	6600	6626	6652	14	45	5799	5824	5850
6679	6705	6731	6757	6783	6810	13	46	5951	5976	6002
6836	6862	6888	6914	6941	6967	12	47	6103	6128	6153
6993	7019	7045	7072	7098	7124	11	48	6254	6280	6305
7150	7176	7202	7229	7255	7281	10	49	6406	6431	6456
797307	7333	7359	7386	7412	7438	9	50	9.806557	6583	6608
7464	7490	7516	7542	7569	7595	8	51	6709	6734	6759
7621	7647	7673	7699	7725	7751	7	52	6860	6885	6911
7777	7804	7830	7856	7882	7908	6	53	7011	7037	7062
[illegible]34	7960	7986	8012	8038	8065	5	54	7163	7188	7213
[illegible]	8117	8143	8169	8195	8221	4	55	7314	7339	7364
[illegible]	8273	8299	8325	8351	8377	3	56	7465	7490	7515
[illegible]	[illegible]429	8455	8481	8508	8534	2	57	7615	7641	7666
[illegible]	[illegible]86	8612	8638	8664	8690	1	58	7766	7791	7816
[illegible]	[illegible]	8768	8794	8820	8846	0	59	7917	7942	7967

8 828 8455 8624 8792 8960 9128 9295

88 898 9156 9323

of 38 Degrees.					Min.	Tangent of 39 Deg			
20″	30″	40″	50″			0″	10″	20″	30″
2897	2940	2983	3027	59	0	9.908369	8412	8455	8498
3157	3200	3244	3287	58	1	8628	8671	8714	8757
3417	3461	3504	3547	57	2	8886	8929	8972	9015
3678	3721	3764	3808	56	3	9144	9187	9230	9273
3938	3981	4025	4068	55	4	9402	9445	9488	9531
4198	4241	4285	4328	54	5	9660	9703	9746	9789
4458	4502	4545	4588	53	6	9918	9961	...5	..48
4718	4762	4805	4848	52	7	9.910177	0220	0263	0306
4979	5022	5065	5109	51	8	0435	0478	0521	0564
5239	5282	5325	5369	50	9	0693	0736	0779	0822
5499	5542	5585	5629	49	10	9.910951	0994	1037	1080
5759	5802	5845	5889	48	11	1209	1252	1295	1338
6019	6062	6105	6149	47	12	1467	1510	1553	1596
6278	6322	6365	6408	46	13	1725	1768	1810	1853
6538	6582	6625	6668	45	14	1982	2025	2068	2111
6798	6842	6885	6928	44	15	2240	2283	2326	2369
7058	7101	7145	7188	43	16	2498	2541	2584	2627
7318	7361	7404	7448	42	17	2756	2799	2842	2685
7578	7621	7664	7707	41	18	3014	3057	3100	3143
7837	7881	7924	7967	40	19	3271	3314	3357	3400
8097	8140	8183	8227	39	20	9.913529	3572	3615	3658
8357	8400	8443	8486	38	21	3787	3830	3873	3916
8616	8659	8703	8746	37	22	4044	4087	4130	4173
8876	8919	8962	9005	36	23	4302	4345	4388	4431
9135	9178	9222	9265	35	24	4560	4603	4645	4688
9395	9438	9481	9524	34	25	4817	4860	4903	4946
9654	9697	9741	9784	33	26	5075	5118	5161	5203
9914	9957	....	..43	32	27	5332	5375	5418	5461
0173	0216	0259	0303	31	28	5590	5633	5675	5718
0432	0476	0519	0562	30	29	5847	5890	5933	5976
0692	0735	0778	0821	29	30	9.916104	6147	6190	6233
0951	0994	1037	1081	28	31	6362	6405	6448	6491
1210	1253	1297	1340	27	32	6619	6662	6705	6748
1469	1513	1556	1599	26	33	6877	6919	6962	7005
1729	1772	1815	1858	25	34	7134	7177	7220	7262
1988	2031	2074	2117	24	35	7391	7434	7477	7520
2247	2290	2333	2376	23	36	7648	7691	7734	7777
2506	2549	2592	2635	22	37	7906	7948	7991	8034
2765	2808	2851	2894	21	38	8163	8206	8248	8291
3024	3067	3110	3153	20	39	8420	8463	8506	8548
3283	3326	3369	3412	19	40	9.918677	8720	8763	8805
3542	3585	3628	3671	18	41	8934	8977	9020	9063
3801	3844	3887	3930	17	42	9191	9234	9277	9320
4060	4103	4146	4189	16	43	9448	9491	9534	9577
4318	4362	4405	4448	15	44	9705	9748	9791	9834
4577	4620	4663	4707	14	45	9962	...5	. 48	..91
4836	4879	4922	4965	13	46	9.920219	0262	0305	0348
5095	5138	5181	5224	12	47	0476	0519	0562	0604
5353	5397	5440	5483	11	48	0733	0776	0819	0861
5612	5655	5698	5741	10	49	0990	1033	1075	1118
5871	5914	5957	6000	9	50	9.921247	1289	1332	1375
6129	6172	6216	6259	8	51	1503	1546	1589	1632
6388	6431	6474	6517	7	52	1760	1803	1846	1889
6646	6690	6733	6776	6	53	2017	2060	2103	2145
6905	6948	6991	7034	5	54	2274	2316	2359	2402
7163	7207	7250	7293	4	55	2530	2573	2616	2659
7422	7465	7508	7551	3	56	2787	2830	2873	2915
7680	7723	7766	7809	2	57	3044	3087	3129	3172
7939	7982	8025	8068	1	58	3300	3343	3386	3429
8197	8240	8283	8326	0	59	3557	3600	3642	3685

Sine of 40 Degrees.							Min.	Sine of 41 Degrees.				
0″	10″	20″	30″	40″	50″			0″	10″	20″	30″	40″
9.808067	8093	8118	8143	8168	8193	59	0	9.816943	6967	6991	7016	7040
8218	8243	8268	8293	8318	8343	58	1	7088	7112	7137	7161	7185
8368	8393	8419	8444	8469	8494	57	2	7233	7258	7282	7306	7330
8519	8544	8569	8594	8619	8644	56	3	7379	7403	7427	7451	7475
8669	8694	8719	8744	8769	8794	55	4	7524	7548	7572	7596	7620
8819	8844	8869	8894	8919	8944	54	5	7668	7693	7717	7741	7765
8969	8994	9019	9044	9069	9094	53	6	7813	7837	7862	7886	7910
9119	9144	9169	9194	9219	9244	52	7	7958	7982	8006	8030	8055
9269	9294	9319	9344	9369	9394	51	8	8103	8127	8151	8175	8199
9419	9444	9469	9494	9519	9544	50	9	8247	8272	8296	8320	8344
9.809569	9594	9619	9643	9668	9693	49	10	9.818392	8416	8440	8464	8488
9718	9743	9768	9793	9818	9[illegible]3	48	11	8536	8560	8584	8609	8633
9868	9893	9918	9943	9967	9992	47	12	8681	8705	8729	8753	8777
9.810017	0042	0067	0092	0117	0142	46	13	8825	8849	8873	8897	8921
0167	0191	0216	0241	0266	0291	45	14	8969	8993	9017	9041	9065
0316	0341	0366	0390	0415	0440	44	15	9113	9137	9161	9185	9209
0465	0490	0515	0540	0564	0589	43	16	9257	9281	9305	9329	9353
0614	0639	0664	0689	0713	0738	42	17	9401	9425	9449	9473	9497
0763	0788	0813	0838	0862	0887	41	18	9545	9569	9593	9617	9641
0912	0937	0962	0986	1011	1036	40	19	9689	9713	9737	9761	9785
9 811061	1086	1110	1135	1160	1185	39	20	9.819832	9856	9880	9904	9928
1210	1234	1259	1284	1309	1334	38	21	9976	....	..24	..48	..72
1358	1383	1408	1433	1457	1482	37	22	9.820120	0143	0167	0191	0215
1507	1532	1556	1581	1606	1631	36	23	0263	0287	0311	0335	0359
1655	1680	1705	1730	1754	1779	35	24	0406	0430	0454	0478	0502
1804	1828	1853	1878	1903	1927	34	25	0550	0573	0597	0621	0645
1952	1977	2001	2026	2051	2076	33	26	0693	0717	0740	0764	0788
2100	2125	2150	2174	2199	2224	32	27	0836	0860	0883	0907	0931
2248	2273	2298	2322	2347	2372	31	28	0979	1003	1026	1050	1074
2396	2421	2446	2470	2495	2520	30	29	1122	1146	1169	1193	1217
9.812544	2569	2594	2618	2643	2668	29	30	9.821265	1288	1312	1336	1360
2692	2717	2742	2766	2791	2815	28	31	1407	1431	1455	1479	1502
2840	2865	2889	2914	2939	2963	27	32	1550	1574	1598	1621	1645
2988	3012	3037	3062	3086	3111	26	33	1693	1716	1740	1764	1788
3135	3160	3185	3209	3234	3258	25	34	1835	1859	1883	1906	1930
3283	3307	3332	3357	3381	3406	24	35	1977	2001	2025	2049	2072
3430	3455	3479	3504	3529	3553	23	36	2120	2144	2167	2191	2215
3578	3602	3627	3651	3676	3700	22	37	2262	2286	2309	2333	2357
3725	3749	3774	3799	3823	3848	21	38	2404	2428	2452	2475	2499
3872	3897	3921	3946	3970	3995	20	39	2546	2570	2594	2617	2641
9.814019	4044	4068	4093	4117	4142	19	40	9.822688	2712	2736	2759	2783
4166	4191	4215	4240	4264	4289	18	41	2830	2854	2878	2901	2925
4313	4338	4362	4387	4411	4436	17	42	2972	2996	3019	3043	3067
4460	4484	4509	4533	4558	4582	16	43	3114	3137	3161	3185	3208
4607	4631	4656	4680	4704	4729	15	44	3255	3279	3303	3326	3350
4753	4778	4802	4827	4851	4876	14	45	3397	3421	3444	3468	3491
4900	4924	4949	4973	4998	5022	13	46	3539	3562	3586	3609	3633
5046	5071	5095	5120	5144	5168	12	47	3680	3704	3727	3751	3774
5193	5217	5242	5266	5290	5315	11	48	3821	3845	3868	3892	3915
5339	5364	5388	5412	5437	5461	10	49	3963	3986	4010	4033	4057
9.815485	5510	5534	5558	5583	5607	9	50	9.824104	4127	4151	4174	4198
5632	5656	5680	5705	5729	5753	8	51	4245	4268	4292	4315	4339
5778	5802	5826	5851	5875	5899	7	52	4386	4409	4433	4456	4480
5924	5948	5972	5996	6021	6045	6	53	4527	4550	4574	4597	4621
6069	6094	6118	6142	6167	6191	5	54	4668	4691	4715	4738	4761
6215	6240	6264	6288	6312	6337	4	55	4808	4832	4855	4879	4902
6361	6385	6409	6434	6458	6482	3	56	4949	4972	4996	5019	5043
6507	6531	6555	6579	6604	6628	2	57	5090	5113	5136	5160	5183
6652	6676	6701	6725	6749	6773	1	58	5230	5254	5277	5300	5324
6798	6822	6846	6870	6894	6919	0	59	5371	5394	5417	5441	5464

Min.	Tangent of 40 Degrees. 0″	10″	20″	30″	40″	50″	
0	9.923814	3856	3899	3942	3985	4027	59
1	4070	4113	4156	4198	4241	4284	58
2	4327	4369	4412	4455	4498	4540	57
3	4583	4626	4669	4711	4754	4797	56
4	4840	4882	4925	4968	5011	5053	55
5	5096	5139	5181	5224	5267	5310	54
6	5352	5395	5438	5481	5523	5566	53
7	5609	5652	5694	5737	5780	5822	52
8	5865	5908	5951	5993	6036	6079	51
9	6122	6164	6207	6250	6292	6335	50
10	9.926378	6421	6463	6506	6549	6591	49
11	6634	6677	6720	6762	6805	6848	48
12	6890	6933	6976	7019	7061	7104	47
13	7147	7189	7232	7275	7317	7360	46
14	7403	7446	7488	7531	7574	7616	45
15	7659	7702	7744	7787	7830	7872	44
16	7915	7958	8001	8043	8086	8129	43
17	8171	8214	8257	8299	8342	8385	42
18	8427	8470	8513	8555	8598	8641	41
19	8684	8726	8769	8812	8854	8897	40
20	9.928940	8982	9025	9068	9110	9153	39
21	9196	9238	9281	9324	9366	9409	38
22	9452	9494	9537	9580	9622	9665	37
23	9708	9750	9793	9836	9878	9921	36
24	9964	...6	..49	..92	.134	.177	35
25	9.930220	0262	0305	0348	0390	0433	34
26	0475	0518	0561	0603	0646	0689	33
27	0731	0774	0817	0859	0902	0945	32
28	0987	1030	1073	1115	1158	1200	31
29	1243	1286	1328	1371	1414	1456	30
30	9.931499	1542	1584	1627	1669	1712	29
31	1755	1797	1840	1883	1925	1968	28
32	2010	2053	2096	2138	2181	2224	27
33	2266	2309	2351	2394	2437	2479	26
34	2522	2565	2607	2650	2692	2735	25
35	2778	2820	2863	2906	2948	2991	24
36	3033	3076	3119	3161	3204	3246	23
37	3289	3332	3374	3417	3459	3502	22
38	3545	3587	3630	3672	3715	3758	21
39	3800	3843	3885	3928	3971	4013	20
40	9.934056	4098	4141	4184	4226	4269	19
41	4311	4354	4397	4439	4482	4524	18
42	4567	4610	4652	4695	4737	4780	17
43	4822	4865	4908	4950	4993	5035	16
44	5078	5121	5163	5206	5248	5291	15
45	5333	5376	5419	5461	5504	5546	14
46	5589	5632	5674	5717	5759	5802	13
47	5844	5887	5930	5972	6015	6057	12
48	6100	6142	6185	6227	6270	6313	11
49	6355	6398	6440	6483	6525	6568	10
50	9.936611	6653	6696	6738	6781	6823	9
51	6866	6908	6951	6994	7036	7079	8
52	7121	7164	7206	7249	7291	7334	7
53	7377	7419	7462	7504	7547	7589	6
54	7632	7674	7717	7759	7802	7845	5
55	7887	7930	7972	8015	8057	8100	4
56	8142	8185	8227	8270	8312	8355	3
57	8398	8440	8483	8525	8568	8610	2
58	8653	8695	8738	8780	8823	8865	1
59	8908	8950	8993	9035	9078	9121	0

Min.	Tangent of 41 Degrees. 0″	10″	20″	30″	40″	50″	
0	9.939163	9206	9248	9291	9333	9376	59
1	9418	9461	9503	9546	9588	9631	58
2	9673	9716	9758	9801	9843	9886	57
3	9928	9971	..13	..56	..98	.141	56
4	9.940183	0226	0268	0311	0354	0396	55
5	0439	0481	0524	0566	0609	0651	54
6	0694	0736	0779	0821	0864	0906	53
7	0949	0991	1034	1076	1119	1161	52
8	1204	1246	1289	1331	1374	1416	51
9	1459	1501	1544	1586	1628	1671	50
10	9.941713	1756	1798	1841	1883	1926	49
11	1968	2011	2053	2096	2138	2181	48
12	2223	2266	2308	2351	2393	2436	47
13	2478	2521	2563	2606	2648	2691	46
14	2733	2776	2818	2861	2903	2945	45
15	2988	3030	3073	3115	3158	3200	44
16	3243	3285	3328	3370	3413	3455	43
17	3498	3540	3583	3625	3667	3710	42
18	3752	3795	3837	3880	3922	3965	41
19	4007	4050	4092	4135	4177	4219	40
20	9.944262	4304	4347	4389	4432	4474	39
21	4517	4559	4602	4644	4686	4729	38
22	4771	4814	4856	4899	4941	4984	37
23	5026	5069	5111	5153	5196	5238	36
24	5281	5323	5366	5408	5451	5493	35
25	5535	5578	5620	5663	5705	5748	34
26	5790	5832	5875	5917	5960	6002	33
27	6045	6087	6130	6172	6214	6257	32
28	6299	6342	6384	6427	6469	6511	31
29	6554	6596	6639	6681	6724	6766	30
30	9.946808	6851	6893	6936	6978	7021	29
31	7063	7105	7148	7190	7233	7275	28
32	7318	7360	7402	7445	7487	7530	27
33	7572	7614	7657	7699	7742	7784	26
34	7827	7869	7911	7954	7996	8039	25
35	8081	8123	8166	8208	8251	8293	24
36	8335	8378	8420	8463	8505	8548	23
37	8590	8632	8675	8717	8760	8802	22
38	8844	8887	8929	8972	9014	9056	21
39	9099	9141	9184	9226	9268	9311	20
40	9.949353	9396	9438	9480	9523	9565	19
41	9608	9650	9692	9735	9777	9819	18
42	9862	9904	9947	9989	..31	..74	17
43	9.950116	0159	0201	0243	0286	0328	16
44	0371	0413	0455	0498	0540	0582	15
45	0625	0667	0710	0752	0794	0837	14
46	0879	0921	0964	1006	1049	1091	13
47	1133	1176	1218	1261	1303	1345	12
48	1388	1430	1472	1515	1557	1600	11
49	1642	1684	1727	1769	1811	1854	10
50	9.951896	1938	1981	2023	2066	2108	9
51	2150	2193	2235	2277	2320	2362	8
52	2405	2447	2489	2532	2574	2616	7
53	2659	2701	2743	2786	2828	2870	6
54	2913	2955	2998	3040	3082	3125	5
55	3167	3209	3252	3294	3336	3379	4
56	3421	3463	3506	3548	3591	3633	3
57	3675	3718	3760	3802	3845	3887	2
58	3929	3972	4014	4056	4099	4141	1
59	4183	4226	4268	4310	4353	4395	0

Min.	Sine of 42 Degrees. 0″	10″	20″	30″	40″	50″		Min.	Sine of 43 [D] 0″	10″	20″
0	9.825511	5534	5558	5581	5604	5628	59	0	9.833783	3806	3828
1	5651	5675	5698	5721	5745	5768	58	1	3919	3941	3964
2	5791	5815	5838	5861	5885	5908	57	2	4054	4077	4099
3	5931	5955	5978	6001	6025	6048	56	3	4189	4212	4234
4	6071	6095	6118	6141	6165	6188	55	4	4325	4347	4370
5	6211	6235	6258	6281	6305	6328	54	5	4460	4482	4505
6	6351	6375	6398	6421	6444	6468	53	6	4595	4617	4640
7	6491	6514	6538	6561	6584	6607	52	7	4730	4752	4775
8	6631	6654	6677	6701	6724	6747	51	8	4865	4887	4910
9	6770	6794	6817	6840	6863	6887	50	9	4999	5022	5044
10	9.826910	6933	6956	6980	7003	7026	49	10	9.835134	5157	5179
11	7049	7073	7096	7119	7142	7165	48	11	5269	5291	5314
12	7189	7212	7235	7258	7282	7305	47	12	5403	5426	5448
13	7328	7351	7374	7398	7421	7444	46	13	5538	5560	5583
14	7467	7490	7514	7537	7560	7583	45	14	5672	5695	5717
15	7606	7629	7653	7676	7699	7722	44	15	5807	5829	5851
16	7745	7768	7792	7815	7838	7861	43	16	5941	5963	5986
17	7884	7907	7931	7954	7977	8000	42	17	6075	6097	6120
18	8023	8046	8069	8093	8116	8139	41	18	6209	6231	6254
19	8162	8185	8208	8231	8254	8278	40	19	6343	6365	6388
20	9.828301	8324	8347	8370	8393	8416	39	20	9.836477	6499	6522
21	8439	8462	8485	8509	8532	8555	38	21	6611	6633	6656
22	8578	8601	8624	8647	8670	8693	37	22	6745	6767	6789
23	8716	8739	8762	8786	8809	8832	36	23	6878	6901	6923
24	8855	8878	8901	8924	8947	8970	35	24	7012	7034	7057
25	8993	9016	9039	9062	9085	9108	34	25	7146	7168	7190
26	9131	9154	9177	9200	9223	9246	33	26	7279	7301	7324
27	9269	9292	9315	9338	9361	9384	32	27	7412	7435	7457
28	9407	9430	9453	9476	9499	9522	31	28	7546	7568	7590
29	9545	9568	9591	9614	9637	9660	30	29	7679	7701	7723
30	9.829683	9706	9729	9752	9775	9798	29	30	9.837812	7834	7857
31	9821	9844	9867	9890	9913	9936	28	31	7945	7967	7990
32	9959	9982	...5	..28	..51	..74	27	32	8078	8100	8123
33	9.830097	0120	0142	0165	0188	0211	26	33	8211	8233	8256
34	0234	0257	0280	0303	0326	0349	25	34	8344	8366	8388
35	0372	0395	0417	0440	0463	0486	24	35	8477	8499	8521
36	0509	0532	0555	0578	0601	0624	23	36	8610	8632	8654
37	0646	0669	0692	0715	0738	0761	22	37	8742	8764	8786
38	0784	0807	0829	0852	0875	0898	21	38	8875	8897	8919
39	0921	0944	0967	0989	1012	1035	20	39	9007	9029	9051
40	9.831058	1081	1104	1127	1149	1172	19	40	9.839140	9162	9184
41	1195	1218	1241	1263	1286	1309	18	41	9272	9294	9316
42	1332	1355	1378	1400	1423	1446	17	42	9404	9426	9448
43	1469	1492	1514	1537	1560	1583	16	43	9536	9558	9580
44	1606	1628	1651	1674	1697	1720	15	44	9668	9690	9712
45	1742	1765	1788	1811	1833	1856	14	45	9800	9822	9844
46	1879	1902	1924	1947	1970	1993	13	46	9932	9954	9976
47	2015	2038	2061	2084	2106	2129	12	47	9.840064	0086	0108
48	2152	2175	2197	2220	2243	2266	11	48	0196	0218	0240
49	2288	2311	2334	2356	2379	2402	10	49	0328	0350	0372
50	9.832425	2447	2470	2493	2515	2538	9	50	9.840459	0481	0503
51	2561	2584	2606	2629	2652	2674	8	51	0591	0613	0635
52	2697	2720	2742	2765	2788	2810	7	52	0722	0744	0766
53	2833	2856	2878	2901	2924	2946	6	53	0854	0876	0897
54	2969	2992	3014	3037	3060	3082	5	54	0985	1007	1029
55	3105	3128	3150	3173	3196	3218	4	55	1116	1138	1160
56	3241	3263	3286	3309	3331	3354	3	56	1247	1269	1291
57	3377	3399	3422	3444	3467	3490	2	57	1378	1400	1422
58	3512	3535	3557	3580	3603	3625	1	58	1509	1531	1553
59	3648	3670	3693	3716	3738	3761	0	59	1640	1662	1684

of 42 Degrees.					Min.	Tangent of 43 De			
20″	30″	40″	50″			0″	10″	20″	30″
4522	4564	4607	4649	59	0	9.969656	9698	9740	9783
4776	4819	4861	4903	58	1	9909	9951	9994	..36
5030	5073	5115	5157	57	2	9.970162	0205	0247	0289
5284	5327	5369	5411	56	3	0416	0458	0500	0542
5538	5581	5623	5665	55	4	0669	0711	0753	0796
5792	5835	5877	5919	54	5	0922	0964	1007	1049
6046	6088	6131	6173	53	6	1175	1218	1260	1302
6300	6342	6385	6427	52	7	1429	1471	1513	1555
6554	6596	6639	6681	51	8	1682	1724	1766	1808
6808	6850	6893	6935	50	9	1935	1977	2019	2062
7062	7104	7146	7189	49	10	9.972188	2230	2273	2315
7316	7358	7400	7443	48	11	2441	2484	2526	2568
7570	7612	7654	7697	47	12	2695	2737	2779	2821
7823	7866	7908	7950	46	13	2948	2990	3032	3074
8077	8120	8162	8204	45	14	3201	3243	3285	3327
8331	8373	8416	8458	44	15	3454	3496	3538	3581
8585	8627	8670	8712	43	16	3707	3749	3791	3834
8839	8881	8923	8966	42	17	3960	4002	4045	4087
9093	9135	9177	9219	41	18	4213	4255	4298	4340
9346	9389	9431	9473	40	19	4466	4509	4551	4593
9600	9642	9685	9727	39	20	9.974720	4762	4804	4846
9854	9896	9938	9981	38	21	4973	5015	5057	5099
0108	0150	0192	0234	37	22	5226	5268	5310	5352
0361	0404	0446	0488	36	23	5479	5521	5563	5605
0615	0657	0700	0742	35	24	5732	5774	5816	5858
0869	0911	0953	0996	34	25	5985	6027	6069	6111
1122	1165	1207	1249	33	26	6238	6280	6322	6364
1376	1418	1461	1503	32	27	6491	6533	6575	6617
1630	1672	1714	1757	31	28	6744	6786	6828	6870
1883	1926	1968	2010	30	29	6997	7039	7081	7123
2137	2179	2222	2264	29	30	9.977250	7292	7334	7377
2391	2433	2475	2517	28	31	7503	7545	7587	7630
2644	2686	2729	2771	27	32	7756	7798	7840	7882
2898	2940	2982	3025	26	33	8009	8051	8093	8135
3151	3194	3236	3278	25	34	8262	8304	8346	8388
3405	3447	3489	3532	24	35	8515	8557	8599	8641
3659	3701	3743	3785	23	36	8768	8810	8852	8894
3912	3954	3997	4039	22	37	9021	9063	9105	9147
4166	4208	4250	4292	21	38	9274	9316	9358	9400
4419	4461	4504	4546	20	39	9527	9569	9611	9653
4673	4715	4757	4799	19	40	9.979780	9822	9864	9906
4926	4968	5011	5053	18	41	9.980033	0075	0117	0159
5180	5222	5264	5306	17	42	0286	0328	0370	0412
5433	5475	5518	5560	16	43	0538	0581	0623	0665
5687	5729	5771	5813	15	44	0791	0834	0876	0918
5940	5982	6024	6067	14	45	1044	1086	1129	1171
6193	6236	6278	6320	13	46	1297	1339	1382	1424
6447	6489	6531	6574	12	47	1550	1592	1634	1677
6700	6742	6785	6827	11	48	1803	1845	1887	1929
6954	6996	7038	7080	10	49	2056	2098	2140	2182
7207	7249	7291	7334	9	50	9.982309	2351	2393	2435
7460	7503	7545	7587	8	51	2562	2604	2646	2688
7714	7756	7798	7840	7	52	2814	2857	2899	2941
7967	8009	8052	8094	6	53	3067	3109	3152	3194
8220	8263	8305	8347	5	54	3320	3362	3404	3447
8474	8516	8558	8600	4	55	3573	3615	3657	3699
8727	8769	8812	8854	3	56	3826	3868	3910	3952
8980	9023	9065	9107	2	57	4079	4121	4163	4205
9234	9276	9318	9360	1	58	4332	4374	4416	4458
9487	9529	9571	9614	0	59	4584	4627	4669	4711
40″	30″	20″	10″	Min.		60″	50″	40″	30″
t of 47 Degrees.						Co-tangent of 46 De			

Sine of 44 Degrees.							Min.	Sine of 45 Degrees.						
0″	10″	20″	30″	40″	50″			0″	10″	20″	30″	40″	50″	
9.841771	1793	1815	1837	1858	1880	59	0	9.849485	9506	9527	9548	9569	9590	59
1902	1924	1946	1967	1989	2011	58	1	9611	9632	9653	9674	9695	9716	58
2033	2055	2076	2098	2120	2142	57	2	9738	9759	9780	9801	9822	9843	57
2163	2185	2207	2229	2250	2272	56	3	9864	9885	9906	9927	9948	9969	56
2294	2316	2337	2359	2381	2403	55	4	9990	..11	..32	..53	..74	..95	55
2424	2446	2468	2490	2511	2533	54	5	9.850116	0137	0158	0179	0200	0221	54
2555	2577	2598	2620	2642	2663	53	6	0242	0263	0284	0305	0326	0347	53
2685	2707	2729	2750	2772	2794	52	7	0368	0388	0409	0430	0451	0472	52
2815	2837	2859	2880	2902	2924	51	8	0493	0514	0535	0556	0577	0598	51
2946	2967	2989	3011	3032	3054	50	9	0619	0640	0661	0682	0703	0724	50
9.843076	3097	3119	3141	3162	3184	49	10	9.850745	0766	0787	0807	0828	0849	49
3206	3227	3249	3271	3292	3314	48	11	0870	0891	0912	0933	0954	0975	48
3336	3357	3379	3401	3422	3444	47	12	0996	1017	1038	1058	1079	1100	47
3466	3487	3509	3530	3552	3574	46	13	1121	1142	1163	1184	1205	1226	46
3595	3617	3639	3660	3682	3703	45	14	1246	1267	1288	1309	1330	1351	45
3725	3747	3768	3790	3811	3833	44	15	1372	1393	1413	1434	1455	1476	44
3855	3876	3898	3919	3941	3963	43	16	1497	1518	1539	1559	1580	1601	43
3984	4006	4027	4049	4071	4092	42	17	1622	1643	1664	1685	1705	1726	42
4114	4135	4157	4178	4200	4222	41	18	1747	1768	1789	1810	1830	1851	41
4243	4265	4286	4308	4329	4351	40	19	1872	1893	1914	1935	1955	1976	40
9.844372	4394	4416	4437	4459	4480	39	20	9.851997	2018	2039	2059	2080	2101	39
4502	4523	4545	4566	4588	4609	38	21	2122	2143	2163	2184	2205	2226	38
4631	4652	4674	4696	4717	4739	37	22	2247	2267	2288	2309	2330	2350	37
4760	4782	4803	4825	4846	4868	36	23	2371	2392	2413	2434	2454	2475	36
4889	4911	4932	4954	4975	4997	35	24	2496	2517	2537	2558	2579	2600	35
5018	5040	5061	5083	5104	5126	34	25	2620	2641	2662	2683	2703	2724	34
5147	5168	5190	5211	5233	5254	33	26	2745	2766	2786	2807	2828	2849	33
5276	5297	5319	5340	5362	5383	32	27	2869	2890	2911	2931	2952	2973	32
5405	5426	5447	5469	5490	5512	31	28	2994	3014	3035	3056	3076	3097	31
5533	5555	5576	5598	5619	5640	30	29	3118	3139	3159	3180	3201	3221	30
9.845662	5683	5705	5726	5747	5769	29	30	9.853242	3263	3283	3304	3325	3345	29
5790	5812	5833	5855	5876	5897	28	31	3366	3387	3408	3428	3449	3470	28
5919	5940	5962	5983	6004	6026	27	32	3490	3511	3532	3552	3573	3594	27
6047	6069	6090	6111	6133	6154	26	33	3614	3635	3655	3676	3697	3717	26
6175	6197	6218	6240	6261	6282	25	34	3738	3759	3779	3800	3821	3841	25
6304	6325	6346	6368	6389	6410	24	35	3862	3883	3903	3924	3944	3965	24
6432	6453	6474	6496	6517	6539	23	36	3986	4006	4027	4047	4068	4089	23
6560	6581	6603	6624	6645	6667	22	37	4109	4130	4151	4171	4192	4212	22
6688	6709	6731	6752	6773	6794	21	38	4233	4254	4274	4295	4315	4336	21
6816	6837	6858	6880	6901	6922	20	39	4356	4377	4398	4418	4439	4459	20
9.846944	6965	6986	7008	7029	7050	19	40	9.854480	4500	4521	4542	4562	4583	19
7071	7093	7114	7135	7157	7178	18	41	4603	4624	4644	4665	4686	4706	18
7199	7220	7242	7263	7284	7305	17	42	4727	4747	4768	4788	4809	4829	17
7327	7348	7369	7391	7412	7433	16	43	4850	4870	4891	4911	4932	4953	16
7454	7476	7497	7518	7539	7561	15	44	4973	4994	5014	5035	5055	5076	15
7582	7603	7624	7645	7667	7688	14	45	5096	5117	5137	5158	5178	5199	14
7709	7730	7752	7773	7794	7815	13	46	5219	5240	5260	5281	5301	5322	13
7836	7858	7879	7900	7921	7943	12	47	5342	5363	5383	5404	5424	5445	12
7964	7985	8006	8027	8049	8070	11	48	5465	5485	5506	5526	5547	5567	11
8091	8112	8133	8154	8176	8197	10	49	5588	5608	5629	5649	5670	5690	10
9.848218	8239	8260	8282	8303	8324	9	50	9.855711	5731	5751	5772	5792	5813	9
8345	8366	8387	8409	8430	8451	8	51	5833	5854	5874	5895	5915	5935	8
8472	8493	8514	8535	8557	8578	7	52	5956	5976	5997	6017	6038	6058	7
8599	8620	8641	8662	8683	8705	6	53	6078	6099	6119	6140	6160	6180	6
8726	8747	8768	8789	8810	8831	5	54	6201	6221	6242	6262	6282	6303	5
8852	8874	8895	8916	8937	8958	4	55	6323	6344	6364	6384	6405	6425	4
8979	9000	9021	9042	9063	9085	3	56	6446	6466	6486	6507	6527	6547	3
9106	9127	9148	9169	9190	9211	2	57	6568	6588	6609	6629	6649	6670	2
9232	9253	9274	9295	9316	9338	1	58	6690	6710	6731	6751	6771	6792	1
9359	9380	9401	9422	9443	9464	0	59	6812	6832	6853	6873	6893	6914	0

Min.	Tangent of 44 Degrees.						
	0″	10″	20″	30″	40″	50″	
0	9.984837	4879	4921	4964	5006	5048	59
1	5090	5132	5174	5216	5259	5301	58
2	5343	5385	5427	5469	5511	5553	57
3	5596	5638	5680	5722	5764	5806	56
4	5848	5891	5933	5975	6017	6059	55
5	6101	6143	6185	6228	6270	6312	54
6	6354	6396	6438	6480	6523	6565	53
7	6607	6649	6691	6733	6775	6817	52
8	6860	6902	6944	6986	7028	7070	51
9	7112	7154	7197	7239	7281	7323	50
10	9.987365	7407	7449	7491	7534	7576	49
11	7618	7660	7702	7744	7786	7829	48
12	7871	7913	7955	7997	8039	8081	47
13	8123	8166	8208	8250	8292	8334	46
14	8376	8418	8460	8503	8545	8587	45
15	8629	8671	8713	8755	8797	8840	44
16	8882	8924	8966	9008	9050	9092	43
17	9134	9177	9219	9261	9303	9345	42
18	9387	9429	9471	9513	9556	9598	41
19	9640	9682	9724	9766	9808	9850	40
20	9.989893	9935	9977	..19	..61	.103	39
21	9.990145	0187	0230	0272	0314	0356	38
22	0398	0440	0482	0524	0567	0609	37
23	0651	0693	0735	0777	0819	0861	36
24	0903	0946	0988	1030	1072	1114	35
25	1156	1198	1240	1283	1325	1367	34
26	1409	1451	1493	1535	1577	1620	33
27	1662	1704	1746	1788	1830	1872	32
28	1914	1956	1999	2041	2083	2125	31
29	2167	2209	2251	2293	2336	2378	30
30	9.992420	2462	2504	2546	2588	2630	29
31	2672	2715	2757	2799	2841	2883	28
32	2925	2967	3009	3051	3094	3136	27
33	3178	3220	3262	3304	3346	3388	26
34	3431	3473	3515	3557	3599	3641	25
35	3683	3725	3767	3810	3852	3894	24
36	3936	3978	4020	4062	4104	4146	23
37	4189	4231	4273	4315	4357	4399	22
38	4441	4483	4526	4568	4610	4652	21
39	4694	4736	4778	4820	4862	4905	20
40	9.994947	4989	5031	5073	5115	5157	19
41	5199	5241	5284	5326	5368	5410	18
42	5452	5494	5536	5578	5620	5663	17
43	5705	5747	5789	5831	5873	5915	16
44	5957	5999	6042	6084	6126	6168	15
45	6210	6252	6294	6336	6378	6421	14
46	6463	6505	6547	6589	6631	6673	13
47	6715	6757	6800	6842	6884	6926	12
48	6968	7010	7052	7094	7136	7179	11
49	7221	7263	7305	7347	7389	7431	10
50	9.997473	7515	7558	7600	7642	7684	9
51	7726	7768	7810	7852	7894	7937	8

Min.	Tangent of 45 Degrees.						
	0″	10″	20″	30″	40″	50″	
0	10.000000	0042	0084	0126	0168	0211	59
1	0253	0295	0337	0379	0421	0463	58
2	0505	0547	0590	0632	0674	0716	57
3	0758	0800	0842	0884	0926	0969	56
4	1011	1053	1095	1137	1179	1221	55
5	1263	1305	1348	1390	1432	1474	54
6	1516	1558	1600	1642	1684	1727	53
7	1769	1811	1853	1895	1937	1979	52
8	2021	2063	2106	2148	2190	2232	51
9	2274	2316	2358	2400	2442	2485	50
10	10.002527	2569	2611	2653	2695	2737	49
11	2779	2821	2864	2906	2948	2990	48
12	3032	3074	3116	3158	3200	3243	47
13	3285	3327	3369	3411	3453	3495	46
14	3537	3579	3622	3664	3706	3748	45
15	3790	3832	3874	3916	3958	4001	44
16	4043	4085	4127	4169	4211	4253	43
17	4295	4337	4380	4422	4464	4506	42
18	4548	4590	4632	4674	4716	4759	41
19	4801	4843	4885	4927	4969	5011	40
20	10.005053	5095	5138	5180	5222	5264	39
21	5306	5348	5390	5432	5474	5517	38
22	5559	5601	5643	5685	5727	5769	37
23	5811	5854	5896	5938	5980	6022	36
24	6064	6106	6148	6190	6233	6275	35
25	6317	6359	6401	6443	6485	6527	34
26	6569	6612	6654	6696	6738	6780	33
27	6822	6864	6906	6949	6991	7033	32
28	7075	7117	7159	7201	7243	7285	31
29	7328	7370	7412	7454	7496	7538	30
30	10.007580	7622	7664	7707	7749	7791	29
31	7833	7875	7917	7959	8001	8044	28
32	8086	8128	8170	8212	8254	8296	27
33	8338	8380	8423	8465	8507	8549	26
34	8591	8633	8675	8717	8760	8802	25
35	8844	8886	8928	8970	9012	9054	24
36	9097	9139	9181	9223	9265	9307	23
37	9349	9391	9433	9476	9518	9560	22
38	9602	9644	9686	9728	9770	9813	21
39	9855	9897	9939	9981	..23	..65	20
40	10.010107	0150	0192	0234	0276	0318	19
41	0360	0402	0444	0487	0529	0571	18
42	0613	0655	0697	0739	0781	0823	17
43	0866	0908	0950	0992	1034	1076	16
44	1118	1160	1203	1245	1287	1329	15
45	1371	1413	1455	1497	1540	1582	14
46	1624	1666	1708	1750	1792	1834	13
47	1877	1919	1961	2003	2045	2087	12
48	2129	2171	2214	2256	2298	2340	11
49	2382	2424	2466	2509	2551	2593	10
50	10.012635	2677	2719	2761	2803	2846	9
51	2888	2930	2972	3014	3056	3098	8

Sine of 46 Degrees.							Min.	Sine of 47 Degrees.						
0″	10″	20″	30″	40″	50″			0″	10″	20″	30″	40″	50″	
9.856934	6954	6975	6995	7015	7036	59	0	9.864127	4147	4167	4186	4206	4226	59
7056	7076	7097	7117	7137	7158	58	1	4245	4265	4284	4304	4324	4343	58
7178	7198	7219	7239	7259	7279	57	2	4363	4383	4402	4422	4441	4461	57
7300	7320	7340	7361	7381	7401	56	3	4481	4500	4520	4539	4559	4579	56
7422	7442	7462	7482	7503	7523	55	4	4598	4618	4637	4657	4676	4696	55
7543	7563	7584	7604	7624	7645	54	5	4716	4735	4755	4774	4794	4813	54
7665	7685	7705	7726	7746	7766	53	6	4833	4853	4872	4892	4911	4931	53
7786	7807	7827	7847	7867	7888	52	7	4950	4970	4990	5009	5029	5048	52
7908	7928	7948	7968	7989	8009	51	8	5068	5087	5107	5126	5146	5165	51
8029	8049	8070	8090	8110	8130	50	9	5185	5204	5224	5244	5263	5283	50
9.858151	8171	8191	8211	8231	8252	49	10	9.865302	5322	5341	5361	5380	5400	49
8272	8292	8312	8332	8353	8373	48	11	5419	5439	5458	5478	5497	5517	48
8393	8413	8433	8454	8474	8494	47	12	5536	5556	5575	5595	5614	5634	47
8514	8534	8554	8575	8595	8615	46	13	5653	5673	5692	5712	5731	5751	46
8635	8655	8675	8696	8716	8736	45	14	5770	5790	5809	5828	5848	5867	45
8756	8776	8796	8817	8837	8857	44	15	5887	5906	5926	5945	5965	5984	44
8877	8897	8917	8937	8958	8978	43	16	6004	6023	6042	6062	6081	6101	43
8998	9018	9038	9058	9078	9098	42	17	6120	6140	6159	6179	6198	6217	42
9119	9139	9159	9179	9199	9219	41	18	6237	6256	6276	6295	6315	6334	41
9239	9259	9279	9300	9320	9340	40	19	6353	6373	6392	6412	6431	6450	40
9.859360	9380	9400	9420	9440	9460	39	20	9.866470	6489	6509	6528	6547	6567	39
9480	9501	9521	9541	9561	9581	38	21	6586	6606	6625	6644	6664	6683	38
9601	9621	9641	9661	9681	9701	37	22	6703	6722	6741	6761	6780	6800	37
9721	9741	9761	9781	9802	9822	36	23	6819	6838	6858	6877	6896	6916	36
9842	9862	9882	9902	9922	9942	35	24	6935	6954	6974	6993	7013	7032	35
9962	9982	...2	..22	..42	..62	34	25	7051	7071	7090	7109	7129	7148	34
9.860082	0102	0122	0142	0162	0182	33	26	7167	7187	7206	7225	7245	7264	33
0202	0222	0242	0262	0282	0302	32	27	7283	7303	7322	7341	7361	7380	32
0322	0342	0362	0382	0402	0422	31	28	7399	7419	7438	7457	7476	7496	31
0442	0462	0482	0502	0522	0542	30	29	7515	7534	7554	7573	7592	7612	30
9.860562	0582	0602	0622	0642	0662	29	30	9.867631	7650	7669	7689	7708	7727	29
0682	0702	0722	0742	0762	0782	28	31	7747	7766	7785	7804	7824	7843	28
0802	0822	0842	0862	0882	0902	27	32	7862	7882	7901	7920	7939	7959	27
0922	0941	0961	0981	1001	1021	26	33	7978	7997	8016	8036	8055	8074	26
1041	1061	1081	1101	1121	1141	25	34	8093	8113	8132	8151	8170	8190	25
1161	1181	1201	1221	1240	1260	24	35	8209	8228	8247	8267	8286	8305	24
1280	1300	1320	1340	1360	1380	23	36	8324	8343	8363	8382	8401	8420	23
1400	1420	1439	1459	1479	1499	22	37	8440	[illegible]	[illegible]	[illegible]	8516	8536	[illegible]
1519	1539	1559	1579	1599	1618	21	38	[illegible]	[illegible]	[illegible]	[illegible]	[illegible]	[illegible]	[illegible]
1638	1658	1678	1698	1718	1738	20	39	86[illegible]	[illegible]	[illegible]	[illegible]	[illegible]	[illegible]	[illegible]
9.861758	1777	1797	1817	1837	1857	19	40	9.86878[illegible]	[illegible]	[illegible]	[illegible]	[illegible]	[illegible]	[illegible]
1877	1897	1916	1936	1956	1976	18	41	8900	[illegible]	[illegible]	[illegible]	[illegible]	[illegible]	[illegible]
1996	2016	2035	2055	2075	2095	17	42	9015	[illegible]	[illegible]	[illegible]	[illegible]	[illegible]	[illegible]
2115	2135	2154	2174	2194	2214	16	43	9130	9149	[illegible]	[illegible]	[illegible]	[illegible]	[illegible]
2234	2254	2273	2293	2313	2333	15	44	9245	9264	[illegible]	[illegible]	[illegible]	[illegible]	[illegible]
2353	2372	2392	2412	2432	2452	14	45	9360	9379	939[illegible]	[illegible]	[illegible]	[illegible]	[illegible]
2471	2491	2511	2531	2551	2570	13	46	9474	9494	9513	[illegible]	[illegible]	[illegible]	[illegible]
2590	2610	2630	2650	2669	2689	12	47	9589	9608	9627	9646	[illegible]	[illegible]	[illegible]
2709	2729	2748	2768	2788	2808	11	48	9704	9723	9742	9761	97[illegible]	[illegible]	[illegible]
2827	2847	2867	2887	2906	2926	10	49	9818	9837	9856	9875	9894	[illegible]	[illegible]
9.862946	2966	2985	3005	3025	3045	9	50	9.869933	9952	9971	9990	...9	..2[illegible]	[illegible]
3064	3084	3104	3124	3143	3163	8	51	9.870047	0066	0085	0104	0123	0142	[illegible]
3183	3203	3222	3242	3262	3281	7	52	0161	0180	0199	0218	0238	0257	7
3301	3321	3341	3360	3380	3400	6	53	0276	0295	0314	0333	0352	0371	6
[illegible]	3439	3459	3478	3498	3518	5	54	0390	0409	0428	0447	0466	0485	5
[illegible]	3557	3577	3597	3616	3636	4	55	0504	0523	0542	0561	0580	0599	4
[illegible]	[illegible]	3695	3715	3734	3754	3	56	0618	0637	0656	0675	0694	0713	3
[illegible]	[illegible]	3813	3833	3852	3872	2	57	0732	0751	0770	0789	0808	0827	2
[illegible]	[illegible]	3931	3951	3970	3990	1	58	0846	0865	0884	0903	0922	0941	1
[illegible]	[illegible]	[illegible]	4069	4088	4108	0	59	0960	0979	0998	1017	1036	1054	0

Tangent of 46 Degrees.							Min.	Tangent of 47 Deg			
0″	10″	20″	30″	40″	50″			0″	10″	20″	30″
015163	5205	5247	5289	5331	5373	59	0	10.030344	0386	0429	0471
5416	5458	5500	5542	5584	5626	58	1	0597	0640	0682	0724
5668	5711	5753	5795	5837	5879	57	2	0851	0893	0935	0977
5921	5963	6006	6048	6090	6132	56	3	1104	1146	1188	1231
6174	6216	6258	6301	6343	6385	55	4	1357	1400	1442	1484
6427	6469	6511	6553	6596	6638	54	5	1611	1653	1695	1737
6680	6722	6764	6806	6848	6891	53	6	1864	1906	1948	1991
6933	6975	7017	7059	7101	7143	52	7	2117	2160	2202	2244
7186	7228	7270	7312	7354	7396	51	8	2371	2413	2455	2497
7438	7481	7523	7565	7607	7649	50	9	2624	2666	2709	2751
017691	7733	7776	7818	7860	7902	49	10	10.032877	2920	2962	3004
7944	7986	8028	8071	8113	8155	48	11	3131	3173	3215	3258
8197	8239	8281	8323	8366	8408	47	12	3384	3426	3469	3511
8450	8492	8534	8576	8618	8661	46	13	3638	3680	3722	3764
8703	8745	8787	8829	8871	8914	45	14	3891	3933	3976	4018
8956	8998	9040	9082	9124	9166	44	15	4145	4187	4229	4271
9209	9251	9293	9335	9377	9419	43	16	4398	4440	4482	4525
9462	9504	9546	9588	9630	9672	42	17	4651	4694	4736	4778
9714	9757	9799	9841	9883	9925	41	18	4905	4947	4989	5032
9967	..10	..52	..94	.136	.178	40	19	5158	5201	5243	5285
020220	0262	0305	0347	0389	0431	39	20	10.035412	5454	5496	5539
0473	0515	0558	0600	0642	0684	38	21	5665	5708	5750	5792
0726	0768	0810	0853	0895	0937	37	22	5919	5961	6003	6046
0979	1021	1063	1106	1148	1190	36	23	6172	6215	6257	6299
1232	1274	1316	1359	1401	1443	35	24	6426	6468	6511	6553
1485	1527	1569	1612	1654	1696	34	25	6680	6722	6764	6806
1738	1780	1822	1865	1907	1949	33	26	6933	6975	7018	7060
1991	2033	2075	2118	2160	2202	32	27	7187	7229	7271	7314
2244	2286	2328	2370	2413	2455	31	28	7440	7483	7525	7567
2497	2539	2581	2623	2666	2708	30	29	7694	7736	7778	7821
022750	2792	2834	2877	2919	2961	29	30	10.037948	7990	8032	8074
3003	3045	3087	3130	3172	3214	28	31	8201	8243	8286	8328
3256	3298	3340	3383	3425	3467	27	32	8455	8497	8539	8582
3509	3551	3593	3636	3678	3720	26	33	8708	8751	8793	8835
3762	3804	3846	3889	3931	3973	25	34	8962	9004	9047	9089
4015	4057	4099	4142	4184	4226	24	35	9216	9258	9300	9343
4268	4310	4353	4395	4437	4479	23	36	9470	9512	9554	9596
4521	4563	4606	4648	4690	4732	22	37	9723	9766	9808	9850
4774	4817	4859	4901	4943	4985	21	38	9977	..19	..62	.104
5027	5070	5112	5154	5196	5238	20	39	10.040231	0273	0315	0358
025280	5323	5365	5407	5449	5491	19	40	10.040484	0527	0569	0611
5534	5576	5618	5660	5702	5745	18	41	0738	0781	0823	0865
5787	5829	5871	5913	5955	5998	17	42	0992	1034	1077	1119
6040	6082	6124	6166	6209	6251	16	43	1246	1288	1330	1373
6293	6335	6377	6419	6462	6504	15	44	1500	1542	1584	1627
6546	6588	6630	6673	6715	6757	14	45	1753	1796	1833	1880
6799	6841	6884	6926	6968	7010	13	46	2007	2050	2092	2134
7052	7095	7137	7179	7221	7263	12	47	2261	2303	2346	2388
7305	7348	7390	7432	7474	7516	11	48	2515	2557	2600	2642
7559	7601	7643	7685	7727	7770	10	49	2769	2811	2854	2896
027812	7854	7896	7938	7981	8023	9	50	10.043023	3065	3107	3150
8065	8107	8149	8192	8234	8276	8	51	3277	3319	3361	3404
8318	8360	8403	8445	8487	8529	7	52	3531	3573	3615	3658
8571	8614	8656	8698	8740	8782	6	53	3785	3827	3869	3912
8825	8867	8909	8951	8993	9036	5	54	4039	4081	4123	4165
9078	9120	9162	9204	9247	9289	4	55	4292	4335	4377	4419
9331	9373	9416	9458	9500	9542	3	56	4546	4589	4631	4673
9584	9627	9669	9711	9753	9795	2	57	4800	4843	4885	4927
9838	9880	9922	9964	...6	..49	1	58	5054	5097	5139	5181
030091	0133	0175	0217	0260	0302	0	59	5309	5351	5393	5436

Sine of 48 Degrees.							Min.	Sine of 49		
0"	10"	20"	30"	40"	50"			0"	10"	20"
71073	1092	1111	1130	1149	1168	59	0	9.877780	7798	7816
1187	1206	1225	1244	1263	1282	58	1	7890	7908	7926
1301	1320	1339	1358	1377	1395	57	2	7999	8018	8036
1414	1433	1452	1471	1490	1509	56	3	8109	8127	8146
1528	1547	1566	1585	1604	1622	55	4	8219	8237	8255
1641	1660	1679	1698	1717	1736	54	5	8328	8346	8365
1755	1774	1793	1811	1830	1849	53	6	8438	8456	8474
1868	1887	1906	1925	1944	1962	52	7	8547	8565	8583
1981	2000	2019	2038	2057	2076	51	8	8656	8675	8693
2095	2113	2132	2151	2170	2189	50	9	8766	8784	8802
2208	2226	2245	2264	2283	2302	49	10	9.878875	8893	8911
2321	2340	2358	2377	2396	2415	48	11	8984	9002	9020
2434	2452	2471	2490	2509	2528	47	12	9093	9111	9129
2547	2565	2584	2603	2622	2641	46	13	9202	9220	9238
2659	2678	2697	2716	2735	2753	45	14	9311	9329	9347
2772	2791	2810	2829	2847	2866	44	15	9420	9438	9456
2885	2904	2923	2941	2960	2979	43	16	9529	9547	9565
2998	3016	3035	3054	3073	3091	42	17	9637	9656	9674
3110	3129	3148	3166	3185	3204	41	18	9746	9764	9782
3223	3241	3260	3279	3298	3316	40	19	9855	9873	9891
3335	3354	3373	3391	3410	3429	39	20	9.879963	9981	....
3448	3466	3485	3504	3522	3541	38	21	9.880072	0090	0108
3560	3579	3597	3616	3635	3653	37	22	0180	0198	0216
3672	3691	3710	3728	3747	3766	36	23	0289	0307	0325
3784	3803	3822	3840	3859	3878	35	24	0397	0415	0433
3896	3915	3934	3953	3971	3990	34	25	0505	0523	0541
4009	4027	4046	4065	4083	4102	33	26	0613	0631	0649
4121	4139	4158	4177	4195	4214	32	27	0722	0740	0758
4232	4251	4270	4288	4307	4326	31	28	0830	0848	0866
4344	4363	4382	4400	4419	4438	30	29	0938	0956	0974
4456	4475	4493	4512	4531	4549	29	30	9.881046	1063	1081
4568	4586	4605	4624	4642	4661	28	31	1153	1171	1189
4680	4698	4717	4735	4754	4773	27	32	1261	1279	1297
4791	4810	4828	4847	4866	4884	26	33	1369	1387	1405
4903	4921	4940	4958	4977	4996	25	34	1477	1495	1512
5014	5033	5051	5070	5088	5107	24	35	1584	1602	1620
5126	5144	5163	5181	5200	5218	23	36	1692	1710	1728
5237	5255	5274	5293	5311	5330	22	37	1799	1817	1835
5348	5367	5385	5404	5422	5441	21	38	1907	1925	1942
5459	5478	5496	5515	5534	5552	20	39	2014	2032	2050
5571	5589	5608	5626	5645	5663	19	40	9.882121	2139	2157
5682	5700	5719	5737	5756	5774	18	41	2229	2246	2264
5793	5811	5830	5848	5867	5885	17	42	2336	2354	2371
5904	5922	5941	5959	5978	5996	16	43	2443	2461	2479
6014	6033	6051	6070	6088	6107	15	44	2550	2568	2586
6125	6144	6162	6181	6199	6218	14	45	2657	2675	2692
6236	6255	6273	6291	6310	6328	13	46	2764	2782	2799
6347	6365	6384	6402	6421	6439	12	47	2871	2888	2906
6457	6476	6494	6513	6531	6550	11	48	2977	2995	3013
6568	6586	6605	6623	6642	6660	10	49	3084	3102	3120
6678	6697	6715	6734	6752	6770	9	50	9.883191	3209	3226
6789	6807	6826	6844	6862	6881	8	51	3297	3315	3333
6899	6918	6936	6954	6973	6991	7	52	3404	3422	3439
7010	7028	7046	7065	7083	7101	6	53	3510	3528	3546
7120	7138	7157	7175	7193	7212	5	54	3617	3635	3652
7230	7248	7267	7285	7303	7322	4	55	3723	3741	3759
7340	7358	7377	7395	7413	7432	3	56	3829	3847	3865
7450	7468	7487	7505	7523	7542	2	57	3936	3953	3971
7560	7578	7597	7615	7633	7652	1	58	4042	4060	4077
7670	7688	7707	7725	7743	7762	0	59	4148	4166	4183
60"	50"	40"	30"	20"	10"	Min.		60"	50"	40"
Co-sine of 41 Degrees.								Co-sine of 40		

ent of 48 Degrees.						Min.	Tangent of 49 Degrees.				
10″	20″	30″	40″	50″			0″	10″	20″	30″	40″
5605	5647	5690	5732	5774	59	0	10.060837	0879	0922	0965	1007
5859	5901	5944	5986	6028	58	1	1092	1135	1177	1220	1262
6113	6155	6198	6240	6282	57	2	1347	1390	1432	1475	1517
6367	6409	6452	6494	6537	56	3	1602	1645	1688	1730	1773
6621	6664	6706	6748	6791	55	4	1858	1900	1943	1985	2028
6875	6918	6960	7002	7045	54	5	2113	2155	2198	2241	2283
7130	7172	7214	7257	7299	53	6	2368	2411	2453	2496	2538
7384	7426	7468	7511	7553	52	7	2623	2666	2709	2751	2794
7638	7680	7723	7765	7807	51	8	2879	2921	2964	3006	3049
7892	7934	7977	8019	8062	50	9	3134	3177	3219	3262	3304
8146	8189	8231	8273	8316	49	10	10.063389	3432	3475	3517	3560
8400	8443	8485	8528	8570	48	11	3645	3687	3730	3773	3815
8655	8697	8739	8782	8824	47	12	3900	3943	3985	4028	4070
8909	8951	8994	9036	9079	46	13	4156	4198	4241	4283	4326
9163	9206	9248	9290	9333	45	14	4411	4454	4496	4539	4581
9418	9460	9502	9545	9587	44	15	4667	4709	4752	4794	4837
9672	9714	9757	9799	9841	43	16	4922	4965	5007	5050	5092
9926	9969	..11	..53	..96	42	17	5178	5220	5263	5305	5348
0181	0223	0265	0308	0350	41	18	5433	5476	5518	5561	5603
0435	0477	0520	0562	0604	40	19	5689	5731	5774	5816	5859
0689	0732	0774	0816	0859	39	20	10.065944	5987	6029	6072	6115
0944	0986	1028	1071	1113	38	21	6200	6242	6285	6328	6370
1198	1240	1283	1325	1368	37	22	6455	6498	6541	6583	6626
1452	1495	1537	1580	1622	36	23	6711	6754	6796	6839	6881
1707	1749	1792	1834	1877	35	24	6967	7009	7052	7094	7137
1961	2004	2046	2089	2131	34	25	7222	7265	7308	7350	7393
2216	2258	2301	2343	2386	33	26	7478	7521	7563	7606	7649
2470	2513	2555	2598	2640	32	27	7734	7776	7819	7862	7904
2725	2767	2810	2852	2895	31	28	7990	8032	8075	8117	8160
2979	3022	3064	3107	3149	30	29	8245	8288	8331	8373	8416
3234	3276	3319	3361	3404	29	30	10.068501	8544	8586	8629	8672
3489	3531	3573	3616	3658	28	31	8757	8800	8842	8885	8927
3743	3786	3828	3870	3913	27	32	9013	9055	9098	9141	9183
3998	4040	4083	4125	4168	26	33	9269	9311	9354	9397	9439
4252	4295	4337	4380	4422	25	34	9525	9567	9610	9652	9695
4507	4549	4592	4634	4677	24	35	9780	9823	9866	9908	9951
4762	4804	4847	4889	4931	23	36	10.070036	0079	0122	0164	0207
5016	5059	5101	5144	5186	22	37	0292	0335	0378	0420	0463
5271	5314	5356	5398	5441	21	38	0548	0591	0634	0676	0719
5526	5568	5611	5653	5696	20	39	0804	0847	0890	0932	0975
5781	5823	5865	5908	5950	19	40	10.071060	1103	1146	1188	1231
6035	6078	6120	6163	6205	18	41	1316	1359	1402	1445	1487
6290	6333	6375	6417	6460	17	42	1573	1615	1658	1701	1743
6545	6587	6630	6672	6715	16	43	1829	1871	1914	1957	1999
6800	6842	6885	6927	6970	15	44	2085	2128	2170	2213	2256
7055	7097	7139	7182	7224	14	45	2341	2384	2426	2469	2512
7309	7352	7394	7437	7479	13	46	2597	2640	2683	2725	2768
7564	7607	7649	7692	7734	12	47	2853	2896	2939	2981	3024
7819	7862	7904	7947	7989	11	48	3110	3152	3195	3238	3280
8074	8117	8159	8202	8244	10	49	3366	3409	3451	3494	3537
8329	8372	8414	8456	8499	9	50	10.073622	3665	3708	3750	3793
8584	8626	8669	8711	8754	8	51	3878	3921	3964	4007	4049
8839	8881	8924	8966	9009	7	52	4135	4178	4220	4263	4306
9094	9136	9179	9221	9264	6	53	4391	4434	4477	4519	4562
9349	9391	9434	9476	9519	5	54	4648	4590	4733	4776	4819
9604	9646	9689	9732	9774	4	55	4904	4947	4989	5032	5075
9859	9902	9944	9987	..29	3	56	5160	5203	5246	5289	5331
0114	0157	0199	0242	0284	2	57	5417	5460	5502	5545	5588
0369	0412	0454	0497	0539	1	58	5673	5716	5759	5802	5844
0624	0667	0709	0752	0794	0	59	5930	5973	6015	6058	6101

Min.	Sine of 50 Degrees. 0"	10"	20"	30"	40"	50"	
0	9.884254	4272	4289	4307	4325	4342	59
1	4360	4378	4395	4413	4431	4448	58
2	4466	4483	4501	4519	4536	4554	57
3	4572	4589	4607	4625	4642	4660	56
4	4677	4695	4713	4730	4748	4766	55
5	4783	4801	4818	4836	4854	4871	54
6	4889	4906	4924	4942	4959	4977	53
7	4994	5012	5030	5047	5065	5082	52
8	5100	5118	5135	5153	5170	5188	51
9	5205	5223	5241	5258	5276	5293	50
10	9.885311	5328	5346	5364	5381	5399	49
11	5416	5434	5451	5469	5486	5504	48
12	5522	5539	5557	5574	5592	5609	47
13	5627	5644	5662	5679	5697	5714	46
14	5732	5749	5767	5784	5802	5819	45
15	5837	5855	5872	5890	5907	5925	44
16	5942	5960	5977	5995	6012	6030	43
17	6047	6065	6082	6099	6117	6134	42
18	6152	6169	6187	6204	6222	6239	41
19	6257	6274	6292	6309	6327	6344	40
20	9.886362	6379	6396	6414	6431	6449	39
21	6466	6484	6501	6519	6536	6554	38
22	6571	6588	6606	6623	6641	6658	37
23	6676	6693	6710	6728	6745	6763	36
24	6780	6798	6815	6832	6850	6867	35
25	6885	6902	6919	6937	6954	6972	34
26	6989	7006	7024	7041	7059	7076	33
27	7093	7111	7128	7146	7163	7180	32
28	7198	7215	7232	7250	7267	7285	31
29	7302	7319	7337	7354	7371	7389	30
30	9.887406	7423	7441	7458	7475	7493	29
31	7510	7528	7545	7562	7580	7597	28
32	7614	7632	7649	7666	7684	7701	27
33	7718	7736	7753	7770	7787	7805	26
34	7822	7839	7857	7874	7891	7909	25
35	7926	7943	7961	7978	7995	8012	24
36	8030	8047	8064	8082	8099	8116	23
37	8134	8151	8168	8185	8203	8220	22
38	8237	8254	8272	8289	8306	8324	21
39	8341	8358	8375	8393	8410	8427	20
40	9.888444	8462	8479	8496	8513	8531	19
41	8548	8565	8582	8600	8617	8634	18
42	8651	8669	8686	8703	8720	8737	17
43	8755	8772	8789	8806	8824	8841	16
44	8858	8875	8892	8910	8927	8944	15
45	8961	8978	8996	9013	9030	9047	14
46	9064	9082	9099	9116	9133	9150	13
47	9168	9185	9202	9219	9236	9253	12
48	9271	9288	9305	9322	9339	9356	11
49	9374	9391	9408	9425	9442	9459	10
50	9.889477	9494	9511	9528	9545	9562	9
51	9579	9597	9614	9631	9648	9665	8
52	9682	9699	9716	9734	9751	9768	7
53	9785	9802	9819	9836	9853	9871	6
54	9888	9905	9922	9939	9956	9973	5
55	9990	...7	..25	..42	..59	..76	4
56	9.890093	0110	0127	0144	0161	0178	3
57	0195	0212	0230	0247	0264	0281	2
58	0298	0315	0332	0349	0366	0383	1
59	0400	0417	0434	0451	0468	0486	0

Min.	Sine of 51 0"	10"	20"
0	9.890503	0520	0537
1	0605	0622	0639
2	0707	0724	0741
3	0809	0826	0843
4	0911	0928	0945
5	1013	1030	1047
6	1115	1132	1149
7	1217	1234	1251
8	1319	1336	1353
9	1421	1438	1455
10	9.891523	1540	1556
11	1624	1641	1658
12	1726	1743	1760
13	1827	1844	1861
14	1929	1946	1963
15	2030	2047	2064
16	2132	2149	2165
17	2233	2250	2267
18	2334	2351	2368
19	2435	2452	2469
20	9.892536	2553	2570
21	2638	2654	2671
22	2739	2755	2772
23	2839	2856	2873
24	2940	2957	2974
25	3041	3058	3075
26	3142	3159	3176
27	3243	3259	3276
28	3343	3360	3377
29	3444	3461	3477
30	9.893544	3561	3578
31	3645	3662	3678
32	3745	3762	3779
33	3846	3862	3879
34	3946	3963	3979
35	4046	4063	4079
36	4146	4163	4180
37	4246	4263	4280
38	4346	4363	4380
39	4446	4463	4480
40	9.894546	4563	4580
41	4646	4663	4679
42	4746	4763	4779
43	4846	4862	4879
44	4945	4962	4979
45	5045	5062	5078
46	5145	5161	5178
47	5244	5261	5277
48	5343	5360	5377
49	5443	5459	5476
50	9.895542	5559	5575
51	5641	5658	5675
52	5741	5757	5774
53	5840	5856	5873
54	5939	5955	5972
55	6038	6054	6071
56	6137	6153	6170
57	6236	6252	6269
58	6335	6351	6368
59	6433	6450	6466

Min.	Tangent of 50 Degrees. 0″	10″	20″	30″	40″	50″	
0	10.076186	6229	6272	6315	6358	6400	59
1	6443	6486	6529	6571	6614	6657	58
2	6700	6742	6785	7828	6871	6913	57
3	6956	6999	7042	7085	7127	7170	56
4	7213	7256	7298	7341	7384	7427	55
5	7470	7512	7555	7598	7641	7684	54
6	7726	7769	7812	7855	7897	7940	53
7	7983	8026	8069	8111	8154	8197	52
8	8240	8283	8325	8368	8411	8454	51
9	8497	8539	8582	8625	8668	8711	50
10	10.078753	8796	8839	8882	8925	8967	49
11	9010	9053	9096	9139	9181	9224	48
12	9267	9310	9353	9396	9438	9481	47
13	9524	9567	9610	9652	9695	9738	46
14	9781	9824	9867	9909	9952	9995	45
15	10.080038	0081	0124	0166	0209	0252	44
16	0295	0338	0381	0423	0466	0509	43
17	0552	0595	0638	0680	0723	0766	42
18	0809	0852	0895	0937	0980	1023	41
19	1066	1109	1152	1195	1237	1280	40
20	10.081323	1366	1409	1452	1494	1537	39
21	1580	1623	1666	1709	1752	1794	38
22	1837	1880	1923	1966	2009	2052	37
23	2094	2137	2180	2223	2266	2309	36
24	2352	2395	2437	2480	2523	2566	35
25	2609	2652	2695	2738	2780	2823	34
26	2866	2909	2952	2995	3038	3081	33
27	3123	3166	3209	3252	3295	3338	32
28	3381	3424	3467	3509	3552	3595	31
29	3638	3681	3724	3767	3810	3853	30
30	10.083896	3938	3981	4024	4067	4110	29
31	4153	4196	4239	4282	4325	4367	28
32	4410	4453	4496	4539	4582	4625	27
33	4668	4711	4754	4797	4839	4882	26
34	4925	4968	5011	5054	5097	5140	25
35	5183	5226	5269	5312	5355	5397	24
36	5440	5483	55[illegible]	5569	5612	5655	23
37	5698	5741	5784	5827	5870	5913	22
38	5956	5999	6041	6084	6127	6170	21
39	6213	6256	6299	6342	6385	6428	20
40	10.086471	6514	6557	6600	6643	6686	19
41	6729	6772	6815	6857	6900	6943	18
42	6986	7029	7072	7115	7158	7201	17
43	7244	7287	7330	7373	7416	7459	16
44	7502	7545	7588	7631	7674	7717	15
45	7760	7803	7846	7889	7932	7975	14
46	8018	8061	8104	8147	8190	8232	13
47	8275	8318	8361	8404	8447	8490	12
48	8533	8576	8619	8662	8705	8748	11
49	8791	8834	8877	8920	8963	9006	10
50	10.089049	9092	9135	9178	9221	9264	9
51	9307	9350	9393	9436	9479	9522	8
52	0565	0608	0651	0694	0737	0780	7

Min.	Tangent of 51 Degrees. 0″	10″	20″	30″	40″	50″	
0	10.091631	1674	1717	1760	1803	1846	59
1	1889	1932	1975	2018	2061	2104	58
2	2147	2191	2234	2277	2320	2363	57
3	2406	2449	2492	2535	2578	2621	56
4	2664	2707	2750	2793	2837	2880	55
5	2923	2966	3009	3052	3095	3138	54
6	3181	3224	3267	3310	3354	3397	53
7	3440	3483	3526	3569	3612	3655	52
8	3698	3741	3784	3828	3871	3914	51
9	3957	4000	4043	4086	4129	4172	50
10	10.094215	4259	4302	4345	4388	4431	49
11	4474	4517	4560	4603	4647	4690	48
12	4733	4776	4819	4862	4905	4948	47
13	4992	5035	5078	5121	5164	5207	46
14	5250	5293	5337	5380	5423	5466	45
15	5509	5552	5595	5638	5682	5725	44
16	5768	5811	5854	5897	5940	5984	43
17	6027	6070	6113	6156	6199	6242	42
18	6286	6329	6372	6415	6458	6501	41
19	6544	6588	6631	6674	6717	6760	40
20	10.096803	6847	6890	6933	6976	7019	39
21	7062	7106	7149	7192	7235	7278	38
22	7321	7365	7408	7451	7494	7537	37
23	7580	7624	7667	7710	7753	7796	36
24	7840	7883	7926	7969	8012	8056	35
25	8099	8142	8185	8228	8271	8315	34
26	8358	8401	8444	8487	8531	8574	33
27	8617	8660	8703	8747	8790	8833	32
28	8876	8919	8963	9006	9049	9092	31
29	9136	9179	9222	9265	9308	9352	30
30	10.099395	9438	9481	9524	9568	9611	29
31	9654	9697	9741	9784	9827	9870	28
32	9913	9957	....	..43	..86	.130	27
33	10.100173	0216	0259	0303	0346	0389	26
34	0432	0476	0519	0562	0605	0649	25
35	0692	0735	0778	0822	0865	0908	24
36	0951	0995	1038	1081	1124	1168	23
37	1211	1254	1297	1341	1384	1427	22
38	1470	1514	1557	1600	1643	1687	21
39	1730	1773	1817	1860	1903	1946	20
40	10.101990	2033	2076	2119	2163	2206	19
41	2249	2293	2336	2379	2422	2466	18
42	2509	2552	2596	2639	2682	2725	17
43	2769	2812	2855	2899	2942	2985	16
44	3029	3072	3115	3158	3202	3245	15
45	3288	3332	3375	3418	3462	3505	14
46	3548	3592	3635	3678	3722	3765	13
47	3808	3851	3895	3938	3981	4025	12
48	4068	4111	4155	4198	4241	4285	11
49	4328	4371	4415	4458	4501	4545	10
50	10.104588	4631	4675	4718	4761	4805	9
51	4848	4891	4935	4978	5021	5065	8
52	5108	5152	5105	5238	5282	5325	7

Sine of 52 Degrees.							Min.	Sine of 53		
0″	10″	20″	30″	40″	50″			0″	10″	20″
896532	6549	6565	6581	6598	6614	59	0	9.902349	2364	2380
6631	6647	6664	6680	6697	6713	58	1	2444	2460	2475
6729	6746	6762	6779	6795	6812	57	2	2539	2555	2571
6828	6844	6861	6877	6894	6910	56	3	2634	2650	2666
6926	6943	6959	6976	6992	7009	55	4	2729	2745	2761
7025	7041	7058	7074	7090	7107	54	5	2824	2840	2856
7123	7140	7156	7172	7189	7205	53	6	2919	2935	2950
7222	7238	7254	7271	7287	7303	52	7	3014	3029	3045
7320	7336	7353	7369	7385	7402	51	8	3108	3124	3140
7418	7434	7451	7467	7483	7500	50	9	3203	3219	3235
897516	7533	7549	7565	7582	7598	49	10	9.903298	3313	3329
7614	7631	7647	7663	7680	7696	48	11	3392	3408	3424
7712	7729	7745	7761	7778	7794	47	12	3487	3503	3518
7810	7827	7843	7859	7876	7892	46	13	3581	3597	3613
7908	7924	7941	7957	7973	7990	45	14	3676	3691	3707
8006	8022	8039	8055	8071	8088	44	15	3770	3786	3802
8104	8120	8136	8153	8169	8185	43	16	3864	3880	3896
8202	8218	8234	8250	8267	8283	42	17	3959	3974	3990
8299	8315	8332	8348	8364	8381	41	18	4053	4069	4084
8397	8413	8429	8446	8462	8478	40	19	4147	4163	4178
898494	8511	8527	8543	8559	8576	39	20	9.904241	4257	4272
8592	8608	8624	8641	8657	8673	38	21	4335	4351	4366
8689	8706	8722	8738	8754	8770	37	22	4429	4445	4460
8787	8803	8819	8835	8852	8868	36	23	4523	4539	4554
8884	8900	8916	8933	8949	8965	35	24	4617	4632	4648
8981	8997	9014	9030	9046	9062	34	25	4711	4726	4742
9078	9095	9111	9127	9143	9159	33	26	4804	4820	4836
9176	9192	9208	9224	9240	9256	32	27	4898	4914	4929
9273	9289	9305	9321	9337	9354	31	28	4992	5007	5023
9370	9386	9402	9418	9434	9450	30	29	5085	5101	5116
899467	9483	9499	9515	9531	9547	29	30	9.905179	5194	5210
9564	9580	9596	9612	9628	9644	28	31	5272	5288	5303
9660	9677	9693	9709	9725	9741	27	32	5366	5381	5397
9757	9773	9789	9806	9822	9838	26	33	5459	5474	5490
9854	9870	9886	9902	9918	9935	25	34	5552	5568	5583
9951	9967	9983	9999	..15	..31	24	35	5645	5661	5676
900047	0063	0079	0096	0112	0128	23	36	5739	5754	5770
0144	0160	0176	0192	0208	0224	22	37	5832	5847	5863
0240	0256	0272	0289	0305	0321	21	38	5925	5940	5956
0337	0353	0369	0385	0401	0417	20	39	6018	6033	6049
900433	0449	0465	0481	0497	0513	19	40	9.906111	6126	6142
0529	0545	0562	0578	0594	0610	18	41	6204	6219	6235
0626	0642	0658	0674	0690	0706	17	42	6296	6312	6327
0722	0738	0754	0770	0786	0802	16	43	6389	6405	6420
0818	0834	0850	0866	0882	0898	15	44	6482	6497	6513
0914	0930	0946	0962	0978	0994	14	45	6575	6590	6605
1010	1026	1042	1058	1074	1090	13	46	6667	6683	6698
1106	1122	1138	1154	1170	1186	12	47	6760	6775	6791
1202	1218	1234	1250	1266	1282	11	48	6852	6868	6883

Tangent of 52 Degrees.							Min.	Tangent of 53 Degrees.						
0″	10″	20″	30″	40″	50″			0″	10″	20″	30″	40″	50″	
10.107190	7234	7277	7320	7364	7407	59	0	10.122886	2929	2973	3017	3061	3105	59
7451	7494	7537	7581	7624	7668	58	1	3148	3192	3236	3280	3324	3368	58
7711	7754	7798	7841	7885	7928	57	2	3411	3455	3499	3543	3587	3630	57
7972	8015	8058	8102	8145	8189	56	3	3674	3718	3762	3806	3850	3893	56
8232	8275	8319	8362	8406	8449	55	4	3937	3981	4025	4069	4113	4157	55
8493	8536	8579	8623	8666	8710	54	5	4200	4244	4288	4332	4376	4420	54
8753	8797	8840	8884	8927	8970	53	6	4463	4507	4551	4595	4639	4683	53
9014	9057	9101	9144	9188	9231	52	7	4727	4770	4814	4858	4902	4946	52
9275	9318	9361	9405	9448	9492	51	8	4990	5034	5077	5121	5165	5209	51
9535	9579	9622	9666	9709	9753	50	9	5253	5297	5341	5385	5428	5472	50
10.109796	9840	9883	9926	9970	..13	49	10	10.125516	5560	5604	5648	5692	5736	49
10.110057	0100	0144	0187	0231	0274	48	11	5780	5823	5867	5911	5955	5999	48
0318	0361	0405	0448	0492	0535	47	12	6043	6087	6131	6175	6219	6262	47
0579	0622	0666	0709	0752	0796	46	13	6306	6350	6394	6438	6482	6526	46
0839	0883	0926	0970	1013	1057	45	14	6570	6614	6658	6701	6745	6789	45
1100	1144	1187	1231	1274	1318	44	15	6833	6877	6921	6965	7009	7053	44
1361	1405	1448	1492	1535	1579	43	16	7097	7141	7185	7229	7273	7316	43
1622	1666	1709	1753	1797	1840	42	17	7360	7404	7448	7492	7536	7580	42
1884	1927	1971	2014	2058	2101	41	18	7624	7668	7712	7756	7800	7844	41
2145	2188	2232	2275	2319	2362	40	19	7888	7932	7976	8020	8063	8107	40
10.112406	2449	2493	2536	2580	2623	39	20	10.128151	8195	8239	8283	8327	8371	39
2667	2711	2754	2798	2841	2885	38	21	8415	8459	8503	8547	8591	8635	38
2928	2972	3015	3059	3102	3146	37	22	8679	8723	8767	8811	8855	8899	37
3189	3233	3277	3320	3364	3407	36	23	8943	8987	9031	9075	9119	9163	36
3451	3494	3538	3581	3625	3669	35	24	9207	9251	9295	9339	9383	9427	35
3712	3756	3799	3843	3886	3930	34	25	9471	9515	9559	9603	9647	9691	34
3974	4017	4061	4104	4148	4191	33	26	9735	9779	9823	9867	9911	9955	33
4235	4279	4322	4366	4409	4453	32	27	9999	..43	..87	.131	.175	.219	32
4496	4540	4584	4627	4671	4714	31	28	10.130263	0307	0351	0395	0439	0483	31
4758	4802	4845	4889	4932	4976	30	29	0527	0571	0615	0659	0703	0747	30
10.115020	5063	5107	5150	5194	5238	29	30	10.130791	0835	0879	0923	0967	1011	29
5281	5325	5368	5412	5456	5499	28	31	1055	1099	1143	1187	1232	1276	28
5543	5586	5630	5674	5717	5761	27	32	1320	1364	1408	1452	1496	1540	27
5804	5848	5892	5935	5979	6023	26	33	1584	1628	1672	1716	1760	1804	26
6066	6110	6153	6197	6241	6284	25	34	1848	1892	1936	1981	2025	2069	25
6328	6372	6415	6459	6502	6546	24	35	2113	2157	2201	2245	2289	2333	24
6590	6633	6677	6721	6764	6808	23	36	2377	2421	2465	2509	2554	2598	23
6852	6895	6939	6982	7026	7070	22	37	2642	2686	2730	2774	2818	2862	22
7113	7157	7201	7244	7288	7332	21	38	2906	2950	2994	3039	3083	3127	21
7375	7419	7463	7506	7550	7594	20	39	3171	3215	3259	3303	3347	3391	20
10.117637	7681	7725	7768	7812	7856	19	40	10.133436	3480	3524	3568	3612	3656	19
7899	7943	7987	8030	8074	8118	18	41	3700	3744	3789	3833	3877	3921	18
8161	8205	8249	8292	8336	8380	17	42	3965	4009	4053	4097	4142	4186	17
8423	8467	8511	8555	8598	8642	16	43	4230	4274	4318	4362	4406	4451	16
8686	8729	8773	8817	8860	8904	15	44	4495	4539	4583	4627	4671	4715	15
8948	8992	9035	9079	9123	9166	14	45	4760	4804	4848	4892	4936	4980	14
9210	9254	9297	9341	9385	9429	13	46	5025	5069	5113	5157	5201	5245	13
9472	9516	9560	9603	9647	9691	12	47	5290	5334	5378	5422	5466	5510	12
9735	9778	9822	9866	9909	9953	11	48	5555	5599	5643	5687	5731	5775	11
9997	..41	..84	.128	.172	.216	10	49	5820	5864	5908	5952	5996	6041	10
10.120259	0303	0347	0391	0434	0478	9	50	10.136085	6129	6173	6217	6262	6306	9
0522	0565	0609	0653	0697	0740	8	51	6350	6394	6438	6483	6527	6571	8
0784	0828	0872	0915	0959	1003	7	52	6615	6659	6704	6748	6792	6836	7
1047	1091	1134	1178	1222	1266	6	53	6881	6925	6969	7013	7057	7102	6
1309	1353	1397	1441	1484	1528	5	54	7146	7190	7234	7279	7323	7367	5
1572	1616	1659	1703	1747	1791	4	55	7411	7455	7500	7544	7588	7632	4
1835	1878	1922	1966	2010	2053	3	56	7677	7721	7765	7809	7854	7898	3
2097	2141	2185	2229	2272	2316	2	57	7942	7986	8031	8075	8119	8163	2
2360	2404	2448	2491	2535	2579	1	58	8208	8252	8296	8341	8385	8429	1
2623	2667	2710	2754	2798	2842	0	59	8473	8518	8562	8606	8650	8695	0
60″	50″	40″	30″	20″	10″	Min.		60″	50″	40″	30″	20″	10″	Min.
Co-tangent of 37 Degrees.								Co-tangent of 36 Degrees.						

Min.	Sine of 54 Degrees.						
	0″	10″	20″	30″	40″	50″	
0	9.907958	7973	7988	8004	8019	8034	59
1	8049	8065	8080	8095	8111	8126	58
2	8141	8156	8172	8187	8202	8217	57
3	8233	8248	8263	8279	8294	8309	56
4	8324	8340	8355	8370	8385	8401	55
5	8416	8431	8446	8462	8477	8492	54
6	8507	8523	8538	8553	8568	8584	53
7	8599	8614	8629	8644	8660	8675	52
8	8690	8705	8721	8736	8751	8766	51
9	8781	8797	8812	8827	8842	8857	50
10	9.908873	8888	8903	8918	8933	8949	49
11	8964	8979	8994	9009	9025	9040	48
12	9055	9070	9085	9101	9116	9131	47
13	9146	9161	9176	9192	9207	9222	46
14	9237	9252	9267	9283	9298	9313	45
15	9328	9343	9358	9374	9389	9404	44
16	9419	9434	9449	9464	9480	9495	43
17	9510	9525	9540	9555	9570	9586	42
18	9601	9616	9631	9646	9661	9676	41
19	9691	9707	9722	9737	9752	9767	40
20	9.909782	9797	9812	9827	9843	9858	39
21	9873	9888	9903	9918	9933	9948	38
22	9963	9978	9994	...9	..24	..39	37
23	9.910054	0069	0084	0099	0114	0129	36
24	0144	0159	0175	0190	0205	0220	35
25	0235	0250	0265	0280	0295	0310	34
26	0325	0340	0355	0370	0385	0400	33
27	0415	0430	0446	0461	0476	0491	32
28	0506	0521	0536	0551	0566	0581	31
29	0596	0611	0626	0641	0656	0671	30
30	9.910686	0701	0716	0731	0746	0761	29
31	0776	0791	0806	0821	0836	0851	28
32	0866	0881	0896	0911	0926	0941	27
33	0956	0971	0986	1001	1016	1031	26
34	1046	1061	1076	1091	1106	1121	25
35	1136	1151	1166	1181	1196	1211	24
36	1226	1241	1256	1271	1286	1300	23
37	1315	1330	1345	1360	1375	1390	22
38	1405	1420	1435	1450	1465	1480	21
39	1495	1510	1525	1540	1555	1569	20
40	9.911584	1599	1614	1629	1644	1659	19
41	1674	1689	1704	1719	1734	1748	18
42	1763	1778	1793	1808	1823	1838	17
43	1853	1868	1883	1897	1912	1927	16
44	1942	1957	1972	1987	2002	2017	15
45	2031	2046	2061	2076	2091	2106	14
46	2121	2136	2150	2165	2180	2195	13
47	2210	2225	2240	2255	2269	2284	12
48	2299	2314	2329	2344	2358	2373	11
49	2388	2403	2418	2433	2448	2462	10
50	9.912477	2492	2507	2522	2537	2551	9
51	2566	2581	2596	2611	2625	2640	8
52	2655	2670	2685	2700	2714	2729	7
53	2744	2759	2774	2788	2803	2818	6
54	2833	2848	2862	2877	2892	2907	5
55	2922	2936	2951	2966	2981	2995	4
56	3010	3025	3040	3055	3069	3084	3
57	3099	3114	3128	3143	3158	3173	2
58	3187	3202	3217	3232	3247	3261	1

Min.	Sine of 55 Degrees.						
	0″	10″	20″	30″	40″	50″	
0	9.913365	3379	3394	3409	3423	3438	59
1	3453	3468	3482	3497	3512	3527	58
2	3541	3556	3571	3585	3600	3615	57
3	3630	3644	3659	3674	3688	3703	56
4	3718	3733	3747	3762	3777	3791	55
5	3806	3821	3836	3850	3865	3880	54
6	3894	3909	3924	3938	3953	3968	53
7	3982	3997	4012	4026	4041	4056	52
8	4070	4085	4100	4114	4129	4144	51
9	4158	4173	4188	4202	4217	4232	50
10	9.914246	4261	4276	4290	4305	4320	49
11	4334	4349	4364	4378	4393	4407	48
12	4422	4437	4451	4466	4481	4495	47
13	4510	4524	4539	4554	4568	4583	46
14	4598	4612	4627	4641	4656	4671	45
15	4685	4700	4714	4729	4744	4758	44
16	4773	4787	4802	4817	4831	4846	43
17	4860	4875	4890	4904	4919	4933	42
18	4948	4962	4977	4992	5006	5021	41
19	5035	5050	5064	5079	5094	5108	40
20	9.915123	5137	5152	5166	5181	5196	39
21	5210	5225	5239	5254	5268	5283	38
22	5297	5312	5326	5341	5356	5370	37
23	5385	5399	5414	5428	5443	5457	36
24	5472	5486	5501	5515	5530	5544	35
25	5559	5573	5588	5602	5617	5631	34
26	5646	5660	5675	5689	5704	5718	33
27	5733	5747	5762	5776	5791	5805	32
28	5820	5834	5849	5863	5878	5892	31
29	5907	5921	5936	5950	5965	5979	30
30	9.915994	6008	6023	6037	6052	6066	29
31	6081	6095	6109	6124	6138	6153	28
32	6167	6182	6196	6211	6225	6240	27
33	6254	6268	6283	6297	6312	6326	26
34	6341	6355	6369	6384	6398	6413	25
35	6427	6442	6456	6470	6485	6499	24
36	6514	6528	6543	6557	6571	6586	23
37	6600	6615	6629	6643	6658	6672	22
38	6687	6701	6715	6730	6744	6759	21
39	6773	6787	6802	6816	6830	6845	20
40	9.916859	6874	6888	6902	6917	6931	19
41	6946	6960	6974	6989	7003	7017	18
42	7032	7046	7060	7075	7089	7104	17
43	7118	7132	7147	7161	7175	7190	16
44	7204	7218	7233	7247	7261	7276	15
45	7290	7304	7319	7333	7347	7362	14
46	7376	7390	7405	7419	7433	7448	13
47	7462	7476	7491	7505	7519	7534	12
48	7548	7562	7576	7591	7605	7619	11
49	7634	7648	7662	7677	7691	7705	10
50	9.917719	7734	7748	7762	7777	7791	9
51	7805	7819	7834	7848	7862	7877	8
52	7891	7905	7919	7934	7948	7962	7
53	7976	7991	8005	8019	8033	8048	6
54	8062	8076	8090	8105	8119	8133	5
55	8147	8162	8176	8190	8204	8219	4
56	8233	8247	8261	8276	8290	8304	3
57	8318	8333	8347	8361	8375	8389	2
58	8404	8418	8432	8446	8461	8475	1

...ent of 54 Degrees.						Min.	Tangent of 55 Degrees.						
10″	20″	30″	40″	50″			0″	10″	20″	30″	40″	50″	
8783	8828	8872	8916	8960	59	0	10.154773	4818	4863	4908	4952	4997	59
9049	9093	9138	9182	9226	58	1	5042	5087	5132	5177	5221	5266	58
9315	9359	9403	9448	9492	57	2	5311	5356	5401	5446	5490	5535	57
9580	9625	9669	9713	9758	56	3	5580	5625	5670	5715	5759	5804	56
9846	9891	9935	9979	..24	55	4	5849	5894	5939	5984	6029	6073	55
0112	0157	0201	0245	0290	54	5	6118	6163	6208	6253	6298	6343	54
0378	0423	0467	0511	0556	53	6	6388	6432	6477	6522	6567	6612	53
0644	0689	0733	0777	0822	52	7	6657	6702	6747	6791	6836	6881	52
0910	0955	0999	1043	1088	51	8	6926	6971	7016	7061	7106	7151	51
1176	1221	1265	1309	1354	50	9	7195	7240	7285	7330	7375	7420	50
1442	1487	1531	1576	1620	49	10	10.157465	7510	7555	7600	7645	7689	49
1709	1753	1797	1842	1886	48	11	7734	7779	7824	7869	7914	7959	48
1975	2019	2064	2108	2152	47	12	8004	8049	8094	8139	8184	8229	47
2241	2286	2330	2374	2419	46	13	8273	8318	8363	8408	8453	8498	46
2508	2552	2596	2641	2685	45	14	8543	8588	8633	8678	8723	8768	45
2774	2818	2863	2907	2952	44	15	8813	8858	8903	8948	8993	9038	44
3041	3085	3129	3174	3218	43	16	9083	9128	9173	9218	9263	9307	43
3307	3351	3396	3440	3485	42	17	9352	9397	9442	9487	9532	9577	42
3574	3618	3662	3707	3751	41	18	9622	9667	9712	9757	9802	9847	41
3840	3885	3929	3974	4018	40	19	9892	9937	9982	..27	..72	.117	40
4107	4151	4196	4240	4285	39	20	10.160162	0207	0252	0297	0342	0387	39
4374	4418	4463	4507	4551	38	21	0432	0477	0522	0567	0612	0657	38
4640	4685	4729	4774	4818	37	22	0703	0748	0793	0838	0883	0928	37
4907	4952	4996	5041	5085	36	23	0973	1018	1063	1108	1153	1198	36
5174	5219	5263	5308	5352	35	24	1243	1288	1333	1378	1423	1468	35
5441	5486	5530	5575	5619	34	25	1513	1558	1603	1648	1693	1739	34
5708	5753	5797	5842	5886	33	26	1784	1829	1874	1919	1964	2009	33
5975	6020	6064	6109	6153	32	27	2054	2099	2144	2189	2234	2279	32
6242	6287	6331	6376	6420	31	28	2325	2370	2415	2460	2505	2550	31
6509	6554	6598	6643	6687	30	29	2595	2640	2685	2730	2775	2821	30
6777	6821	6866	6910	6955	29	30	10.162866	2911	2956	3001	3046	3091	29
7044	7088	7133	7177	7222	28	31	3136	3181	3227	3272	3317	3362	28
7311	7356	7400	7445	7489	27	32	3407	3452	3497	3542	3588	3633	27
7578	7623	7668	7712	7757	26	33	3678	3723	3768	3813	3858	3904	26
7846	7890	7935	7980	8024	25	34	3949	3994	4039	4084	4129	4174	25
8113	8158	8203	8247	8292	24	35	4220	4265	4310	4355	4400	4445	24
8381	8425	8470	8515	8559	23	36	4491	4536	4581	4626	4671	4716	23
8648	8693	8738	8782	8827	22	37	4762	4807	4852	4897	4942	4988	22
8916	8961	9005	9050	9095	21	38	5033	5078	5123	5168	5213	5259	21
9184	9228	9273	9318	9362	20	39	5304	5349	5394	5439	5485	5530	20
9452	9496	9541	9585	9630	19	40	10.165575	5620	5666	5711	5756	5801	19
9719	9764	9809	9853	9898	18	41	5846	5892	5937	5982	6027	6073	18
9987	..32	..76	.121	.166	17	42	6118	6163	6208	6253	6299	6344	17
0255	0300	0344	0389	0434	16	43	6389	6434	6480	6525	6570	6615	16
0523	0568	0612	0657	0702	15	44	6661	6706	6751	6796	6842	6887	15
0791	0836	0880	0925	0970	14	45	6932	6977	7023	7068	7113	7158	14
1059	1104	1149	1193	1238	13	46	7204	7249	7294	7340	7385	7430	13
1327	1372	1417	1461	1506	12	47	7475	7521	7566	7611	7657	7702	12
1595	1640	1685	1730	1774	11	48	7747	7792	7838	7883	7928	7974	11
1864	1908	1953	1998	2043	10	49	8019	8064	8109	8155	8200	8245	10
2132	2177	2221	2266	2311	9	50	10.168291	8336	8381	8427	8472	8517	9
2400	2445	2490	2535	2579	8	51	8563	8608	8653	8699	8744	8789	8
2669	2713	2758	2803	2848	7	52	8835	8880	8925	8971	9016	9061	7
2937	2982	3027	3071	3116	6	53	9107	9152	9197	9243	9288	9333	6
3206	3250	3295	3340	3385	5	54	9379	9424	9469	9515	9560	9605	5
3474	3519	3564	3609	3653	4	55	9651	9696	9742	9787	9832	9878	4
3743	3788	3832	3877	3922	3	56	9923	9968	..14	..59	.105	.150	3
4012	4056	4101	4146	4191	2	57	10.170195	0241	0286	0331	0377	0422	2
4280	4325	4370	4415	4460	1	58	0468	0513	0558	0604	0649	0695	1
4549	4594	4639	4684	4728	0	59	0740	0785	0831	0876	0922	0967	0
50″	40″	30″	20″	10″	Min.		60″	50″	40″	30″	20″	10″	Min.
...gent of 35 Degrees.							Co-tangent of 34 Degrees.						

Min.	Sine of 56 Degrees. 0″	10″	20″	30″	40″	50″	
0	9.918574	8588	8603	8617	8631	8645	59
1	8659	8674	8688	8702	8716	8730	58
2	8745	8759	8773	8787	8801	8815	57
3	8830	8844	8858	8872	8886	8900	56
4	8915	8929	8943	8957	8971	8985	55
5	9000	9014	9028	9042	9056	9070	54
6	9085	9099	9113	9127	9141	9155	53
7	9169	9184	9198	9212	9226	9240	52
8	9254	9268	9282	9297	9311	9325	51
9	9339	9353	9367	9381	9395	9410	50
10	9.919424	9438	9452	9466	9480	9494	49
11	9508	9522	9537	9551	9565	9579	48
12	9593	9607	9621	9635	9649	9663	47
13	9677	9692	9706	9720	9734	9748	46
14	9762	9776	9790	9804	9818	9832	45
15	9846	9860	9875	9889	9903	9917	44
16	9931	9945	9959	9973	9987	...1	43
17	9.920015	0029	0043	0057	0071	0085	42
18	0099	0113	0127	0141	0156	0170	41
19	0184	0198	0212	0226	0240	0254	40
20	9.920268	0282	0296	0310	0324	0338	39
21	0352	0366	0380	0394	0408	0422	38
22	0436	0450	0464	0478	0492	0506	37
23	0520	0534	0548	0562	0576	0590	36
24	0604	0618	0632	0646	0660	0674	35
25	0688	0702	0716	0730	0744	0758	34
26	0772	0786	0800	0814	0828	0842	33
27	0856	0869	0883	0897	0911	0925	32
28	0939	0953	0967	0981	0995	1009	31
29	1023	1037	1051	1065	1079	1093	30
30	9.921107	1121	1134	1148	1162	1176	29
31	1190	1204	1218	1232	1246	1260	28
32	1274	1288	1302	1315	1329	1343	27
33	1357	1371	1385	1399	1413	1427	26
34	1441	1455	1468	1482	1496	1510	25
35	1524	1538	1552	1566	1580	1593	24
36	1607	1621	1635	1649	1663	1677	23
37	1691	1704	1718	1732	1746	1760	22
38	1774	1788	1802	1815	1829	1843	21
39	1857	1871	1885	1899	1912	1926	20
40	9.921940	1954	1968	1982	1995	2009	19
41	2023	2037	2051	2065	2079	2092	18
42	2106	2120	2134	2148	2162	2175	17
43	2189	2203	2217	2231	2244	2258	16
44	2272	2286	2300	2313	2327	2341	15
45	2355	2369	2383	2396	2410	2424	14
46	2438	2452	2465	2479	2493	2507	13
47	2520	2534	2548	2562	2576	2589	12
48	2603	2617	2631	2644	2658	2672	11
49	2686	2700	2713	2727	2741	2755	10
50	9.922768	2782	2796	2810	2823	2837	9
51	2851	2865	2878	2892	2906	2920	8
52	2933	2947	2961	2975	2988	3002	7
53	3016	3030	3043	3057	3071	3084	6
54	3098	3112	3126	3139	3153	3167	5
55	3181	3194	3208	3222	3235	3249	4
56	3263	3277	3290	3304	3318	3331	3
57	3345	3359	3372	3386	3400	3414	2
58	3427	3441	3455	3468	3482	3496	1
59	3509	3523	3537	3550	3564	3578	0

Min.	Sine of 57 Degr 0″	10″	20″	30″
0	9.923591	3605	3619	3632
1	3673	3687	3701	3714
2	3755	3769	3783	3796
3	3837	3851	3865	3878
4	3919	3933	3946	3960
5	4001	4015	4028	4042
6	4083	4096	4110	4124
7	4164	4178	4192	4205
8	4246	4260	4273	4287
9	4328	4341	4355	4368
10	9.924409	4423	4436	4450
11	4491	4504	4518	4531
12	4572	4586	4599	4613
13	4654	4667	4681	4694
14	4735	4748	4762	4776
15	4816	4830	4843	4857
16	4897	4911	4924	4938
17	4979	4992	5006	5019
18	5060	5073	5087	5100
19	5141	5154	5168	5181
20	9.925222	5235	5249	5262
21	5303	5316	5330	5343
22	5384	5397	5411	5424
23	5465	5478	5491	5505
24	5545	5559	5572	5586
25	5626	5640	5653	5667
26	5707	5720	5734	5747
27	5788	5801	5814	5828
28	5868	5882	5895	5908
29	5949	5962	5976	5989
30	9.926029	6043	6056	6069
31	6110	6123	6136	6150
32	6190	6203	6217	6230
33	6270	6284	6297	6311
34	6351	6364	6377	6391
35	6431	6444	6458	6471
36	6511	6525	6538	6551
37	6591	6605	6618	6631
38	6671	6685	6698	6711
39	6751	6765	6778	6791
40	9.926831	6845	6858	6871
41	6911	6925	6938	6951
42	6991	7005	7018	7031
43	7071	7084	7098	7111
44	7151	7164	7177	7191
45	7231	7244	7257	7270
46	7310	7324	7337	7350
47	7390	7403	7416	7430
48	7470	7483	7496	7509
49	7549	7562	7576	7589
50	9.927629	7642	7655	7668
51	7708	7721	7734	7748
52	7787	7801	7814	7827
53	7867	7880	7893	7906
54	7946	7959	7972	7986
55	8025	8038	8052	8065
56	8104	8118	8131	8144
57	8183	8197	8210	8223
58	8263	8276	8289	8302
59	8342	8355	8368	8381

7529	7575	7621	7667	7713	59
7805	7851	7898	7944	7990	58
8082	8128	8174	8220	8267	57
8359	8405	8451	8497	8543	56
8636	8682	8728	8774	8820	55
8913	8959	9005	9051	9097	54
9190	9236	9282	9328	9374	53
9467	9513	9559	9605	9651	52
9744	9790	9836	9882	9929	51
..21	..67	.113	.160	.206	50
0298	0344	0391	0437	0483	49
0576	0622	0668	0714	0760	48
0853	0899	0945	0992	1038	47
1130	1177	1223	1269	1315	46
1408	1454	1501	1547	1593	45
1686	1732	1778	1824	1871	44
1963	2010	2056	2102	2149	43
2241	2287	2334	2380	2426	42
2519	2565	2612	2658	2704	41
2797	2843	2890	2936	2982	40
3075	3121	3168	3214	3260	39
3353	3399	3446	3492	3538	38
3631	3678	3724	3770	3817	37
3909	3956	4002	4049	4095	36
4188	4234	4281	4327	4373	35
4466	4513	4559	4605	4652	34
4745	4791	4837	4884	4930	33
5023	5070	5116	5162	5209	32
5302	5348	5395	5441	5487	31
5580	5627	5673	5720	5766	30
5859	5906	5952	5999	6045	29
6138	6184	6231	6277	6324	28
6417	6463	6510	6556	6603	27
6696	6742	6789	6835	6882	26
6975	7021	7068	7114	7161	25
7254	7301	7347	7394	7440	24
7533	7580	7626	7673	7719	23
7813	7859	7906	7952	7999	22
8092	8138	8185	8232	8278	21
8371	8418	8465	8511	8558	20
8651	8697	8744	8791	8837	19
8930	8977	9024	9070	9117	18
9210	9257	9303	9350	9397	17
9490	9537	9583	9630	9676	16
9770	9816	9863	9910	9956	15
0050	0096	0143	0190	0236	14
0330	0376	0423	0470	0516	13
0610	0656	0703	0750	0796	12
0890	0937	0983	1030	1077	11
1170	1217	1263	1310	1357	10
1450	1497	1544	1591	1637	9
1731	1777	1824	1871	1918	8
2011	2058	2105	2151	2198	7
2292	2338	2385	2432	2479	6
2572	2619	2666	2713	2759	5
2853	2900	2947	2993	3040	4
3134	3181	3227	3274	3321	3
3415	3461	3508	3555	3602	2
3696	3742	3789	3836	3883	1
3977	4023	4070	4117	4164	0
50″	40″	30″	20″	10″	Min.

ngent of 32 Degrees.

Degrees.				Min.	Sine of 59 Degrees.						
30″	40″	50″			0″	10″	20″	30″	40″	50″	
8460	8473	8486	59	0	9.933066	3078	3091	3104	3116	3129	59
8539	8552	8565	58	1	3141	3154	3167	3179	3192	3205	58
8618	8631	8644	57	2	3217	3230	3243	3255	3268	3280	57
8696	8710	8723	56	3	3293	3306	3318	3331	3344	3356	56
8775	8788	8801	55	4	3369	3381	3394	3407	3419	3432	55
8854	8867	8880	54	5	3445	3457	3470	3482	3495	3508	54
8933	8946	8959	53	6	3520	3533	3545	3558	3571	3583	53
9011	9024	9037	52	7	3596	3608	3621	3633	3646	3659	52
9090	9103	9116	51	8	3671	3684	3696	3709	3722	3734	51
9168	9181	9194	50	9	3747	3759	3772	3784	3797	3810	50
9247	9260	9273	49	10	9.933822	3835	3847	3860	3872	3885	49
9325	9338	9351	48	11	3898	3910	3923	3935	3948	3960	48
9403	9416	9429	47	12	3973	3985	3998	4011	4023	4036	47
9482	9495	9508	46	13	4048	4061	4073	4086	4098	4111	46
9560	9573	9586	45	14	4123	4136	4148	4161	4174	4186	45
9638	9651	9664	44	15	4199	4211	4224	4236	4249	4261	44
9716	9729	9742	43	16	4274	4286	4299	4311	4324	4336	43
9794	9807	9820	42	17	4349	4361	4374	4386	4399	4411	42
9872	9885	9898	41	18	4424	4436	4449	4461	4474	4486	41
9950	9963	9976	40	19	4499	4511	4524	4536	4549	4561	40
..28	..41	..54	39	20	9.934574	4586	4599	4611	4624	4636	39
0106	0119	0132	38	21	4649	4661	4674	4686	4699	4711	38
0184	0197	0210	37	22	4723	4736	4748	4761	4773	4786	37
0262	0274	0287	36	23	4798	4811	4823	4836	4848	4861	36
0339	0352	0365	35	24	4873	4885	4898	4910	4923	4935	35
0417	0430	0443	34	25	4948	4960	4973	4985	4997	5010	34
0495	0507	0520	33	26	5022	5035	5047	5060	5072	5084	33
0572	0585	0598	32	27	5097	5109	5122	5134	5147	5159	32
0650	0663	0675	31	28	5171	5184	5196	5209	5221	5234	31
0727	0740	0753	30	29	5246	5258	5271	5283	5296	5308	30
0804	0817	0830	29	30	9.935320	5333	5345	5358	5370	5382	29
0882	0895	0908	28	31	5395	5407	5420	5432	5444	5457	28
0959	0972	0985	27	32	5469	5482	5494	5506	5519	5531	27
1036	1049	1062	26	33	5543	5556	5568	5581	5593	5605	26
1114	1127	1139	25	34	5618	5630	5642	5655	5667	5679	25
1191	1204	1217	24	35	5692	5704	5717	5729	5741	5754	24
1268	1281	1294	23	36	5766	5778	5791	5803	5815	5828	23
1345	1358	1371	22	37	5840	5852	5865	5877	5889	5902	22
1422	1435	1448	21	38	5914	5926	5939	5951	5963	5976	21
1499	1512	1525	20	39	5988	6000	6013	6025	6037	6050	20
1576	1589	1601	19	40	9.936062	6074	6087	6099	6111	6124	19
1653	1666	1678	18	41	6136	6148	6161	6173	6185	6198	18
1730	1742	1755	17	42	6210	6222	6234	6247	6259	6271	17
1806	1819	1832	16	43	6284	6296	6308	6320	6333	6345	16
1883	1896	1909	15	44	6357	6370	6382	6394	6406	6419	15
1960	1972	1985	14	45	6431	6443	6456	6468	6480	6492	14
2036	2049	2062	13	46	6505	6517	6529	6542	6554	6566	13
2113	2126	2138	12	47	6578	6591	6603	6615	6627	6640	12
2189	2202	2215	11	48	6652	6664	6676	6689	6701	6713	11
2266	2279	2291	10	49	6725	6738	6750	6762	6774	6787	10
2342	2355	2368	9	50	9.936799	6811	6823	6836	6848	6860	9
2419	2431	2444	8	51	6872	6884	6897	6909	6921	6933	8
2495	2508	2520	7	52	6946	6958	6970	6982	6994	7007	7
2571	2584	2597	6	53	7019	7031	7043	7056	7068	7080	6
2647	2660	2673	5	54	7092	7104	7117	7129	7141	7153	5
2724	2736	2749	4	55	7165	7178	7190	7202	7214	7226	4
2800	2812	2825	3	56	7238	7251	7263	7275	7287	7299	3
2876	2888	2901	2	57	7312	7324	7336	7348	7360	7372	2
2952	2964	2977	1	58	7385	7397	7409	7421	7433	7446	1
3028	3040	3053	0	59	7458	7470	7482	7494	7506	7518	0

Min.	Tangent of 58 Degrees. 0″	10″	20″	30″	40″	50″	
0	10.204211	4258	4304	4351	4398	4445	59
1	4492	4539	4586	4633	4679	4726	58
2	4773	4820	4867	4914	4961	5008	57
3	5054	5101	5148	5195	5242	5289	56
4	5336	5383	5430	5477	5524	5570	55
5	5617	5664	5711	5758	5805	5852	54
6	5899	5946	5993	6040	6087	6134	53
7	6181	6227	6274	6321	6368	6415	52
8	6462	6509	6556	6603	6650	6697	51
9	6744	6791	6838	6885	6932	6979	50
10	10.207026	7073	7120	7167	7214	7261	49
11	7308	7355	7402	7449	7496	7543	48
12	7590	7637	7684	7731	7778	7825	47
13	7872	7919	7966	8013	8060	8107	46
14	8154	8201	8248	8295	8342	8389	45
15	8437	8484	8531	8578	8625	8672	44
16	8719	8766	8813	8860	8907	8954	43
17	9001	9048	9095	9143	9190	9237	42
18	9284	9331	9378	9425	9472	9519	41
19	9566	9614	9661	9708	9755	9802	40
20	10.209849	9896	9943	9991	..38	..85	39
21	10.210132	0179	0226	0273	0321	0368	38
22	0415	0462	0509	0556	0603	0651	37
23	0698	0745	0792	0839	0886	0934	36
24	0981	1028	1075	1122	1170	1217	35
25	1264	1311	1358	1405	1453	1500	34
26	1547	1594	1641	1689	1736	1783	33
27	1830	1878	1925	1972	2019	2066	32
28	2114	2161	2208	2255	2303	2350	31
29	2397	2444	2492	2539	2586	2633	30
30	10.212681	2728	2775	2822	2870	2917	29
31	2964	3012	3059	3106	3153	3201	28
32	3248	3295	3343	3390	3437	3484	27
33	3532	3579	3626	3674	3721	3768	26
34	3816	3863	3910	3958	4005	4052	25
35	4100	4147	4194	4242	4289	4336	24
36	4384	4431	4478	4526	4573	4620	23
37	4668	4715	4762	4810	4857	4905	22
38	4952	4999	5047	5094	5141	5189	21
39	5236	5284	5331	5378	5426	5473	20
40	10.215521	5568	5615	5663	5710	5758	19
41	5805	5852	5900	5947	5995	6042	18
42	6090	6137	6184	6232	6279	6327	17
43	6374	6422	6469	6517	6564	6612	16
44	6659	6706	6754	6801	6849	6896	15
45	6944	6991	7039	7086	7134	7181	14
46	7229	7276	7324	7371	7419	7466	13
47	7514	7561	7609	7656	7704	7751	12
48	7799	7846	7894	7941	7989	8036	11
49	8084	8131	8179	8226	8274	8322	10
50	10.218369	8417	8464	8512	8559	8607	9
51	8654	8702	8750	8797	8845	8892	8
52	8940	8987	9035	9083	9130	9178	7
53	9225	9273	9321	9368	9416	9463	6
54	9511	9559	9606	9654	9701	9749	5
55	9797	9844	9892	9939	9987	..35	4
56	10.220082	0130	0178	0225	0273	0321	3
57	0368	0416	0463	0511	0559	0606	2
58	0654	0702	0749	0797	0845	0892	1
59	0940	0988	1036	1083	1131	1179	0
	60″	50″	40″	30″	20″	10″	Min.
	Co-tangent of 31 Degrees.						

Min.	Tangent of 59 Degrees. 0″	10″	20″	30″	40″	50″	
0	10.221226	1274	1322	1369	1417	1465	59
1	1512	1560	1608	1656	1703	1751	58
2	1799	1846	1894	1942	1990	2037	57
3	2085	2133	2181	2228	2276	2324	56
4	2372	2419	2467	2515	2563	2610	55
5	2658	2706	2754	2801	2849	2897	54
6	2945	2993	3040	3088	3136	3184	53
7	3232	3279	3327	3375	3423	3471	52
8	3518	3566	3614	3662	3710	3757	51
9	3805	3853	3901	3949	3997	4044	50
10	10.224092	4140	4188	4236	4284	4332	49
11	4379	4427	4475	4523	4571	4619	48
12	4667	4714	4762	4810	4858	4906	47
13	4954	5002	5050	5098	5145	5193	46
14	5241	5289	5337	5385	5433	5481	45
15	5529	5577	5625	5672	5720	5768	44
16	5816	5864	5912	5960	6008	6056	43
17	6104	6152	6200	6248	6296	6344	42
18	6392	6440	6488	6535	6583	6631	41
19	6679	6727	6775	6823	6871	6919	40
20	10.226967	7015	7063	7111	7159	7207	39
21	7255	7303	7351	7399	7447	7495	38
22	7543	7591	7639	7688	7736	7784	37
23	7832	7880	7928	7976	8024	8072	36
24	8120	8168	8216	8264	8312	8360	35
25	8408	8456	8504	8552	8601	8649	34
26	8697	8745	8793	8841	8889	8937	33
27	8985	9033	9081	9130	9178	9226	32
28	9274	9322	9370	9418	9466	9515	31
29	9563	9611	9659	9707	9755	9803	30
30	10.229852	9900	9948	9996	..44	..92	29
31	10.230140	0189	0237	0285	0333	0381	28
32	0429	0478	0526	0574	0622	0670	27
33	0719	0767	0815	0863	0911	0960	26
34	1008	1056	1104	1152	1201	1249	25
35	1297	1345	1394	1442	1490	1538	24
36	1586	1635	1683	1731	1779	1828	23
37	1876	1924	1973	2021	2069	2117	22
38	2166	2214	2262	2310	2359	2407	21
39	2455	2504	2552	2600	2648	2697	20
40	10.232745	2793	2842	2890	2938	2987	19
41	3035	3083	3132	3180	3228	3277	18
42	3325	3373	3422	3470	3518	3567	17
43	3615	3663	3712	3760	3808	3857	16
44	3905	3953	4002	4050	4099	4147	15
45	4195	4244	4292	4340	4389	4437	14
46	4486	4534	4582	4631	4679	4728	13
47	4776	4825	4873	4921	4970	5018	12
48	5067	5115	5164	5212	5260	5309	11
49	5357	5406	5454	5503	5551	5600	10
50	10.235648	5696	5745	5793	5842	5890	9
51	5939	5987	6036	6084	6133	6181	8
52	6230	6278	6327	6375	6424	6472	7
53	6521	6569	6618	6666	6715	6763	6
54	6812	6860	6909	6957	7006	7055	5
55	7103	7152	7200	7249	7297	7346	4
56	7394	7443	7492	7540	7589	7637	3
57	7686	7734	7783	7832	7880	7929	2
58	7977	8026	8075	8123	8172	8220	1
59	8269	8318	8366	8415	8463	8512	0
	60″	50″	40″	30″	20″	10″	Min.
	Co-tangent of 30 Degrees.						

P. P. { 1″ 2″ 3″ 4″ 5″ 6″ 7″ 8″ 9″ ‖ P. P. { 1″ 2″ 3″ 4″ 5″ 6″ 7″ 8″ 9″

Sine of 60 Degrees.							Min.	Sine of 61 D			
0″	10″	20″	30″	40″	50″			0″	10″	20″	
9.937531	7543	7555	7567	7579	7591	59	0	9.941819	1831	1843	1
7604	7616	7628	7640	7652	7664	58	1	1889	1901	1913	1
7676	7689	7701	7713	7725	7737	57	2	1959	1971	1983	1
7749	7761	7773	7786	7798	7810	56	3	2029	2041	2052	2
7822	7834	7846	7858	7870	7883	55	4	2099	2111	2122	2
7895	7907	7919	7931	7943	7955	54	5	2169	2180	2192	2
7967	7979	7992	8004	8016	8028	53	6	2239	2250	2262	2
8040	8052	8064	8076	8088	8100	52	7	2308	2320	2331	2
8113	8125	8137	8149	8161	8173	51	8	2378	2390	2401	2
8185	8197	8209	8221	8233	8245	50	9	2448	2459	2471	2
9.938258	8270	8282	8294	8306	8318	49	10	9.942517	2529	2540	2
8330	8342	8354	8366	8378	8390	48	11	2587	2598	2610	2
8402	8414	8426	8439	8451	8463	47	12	2656	2668	2679	2
8475	8487	8499	8511	8523	8535	46	13	2726	2737	2749	2
8547	8559	8571	8583	8595	8607	45	14	2795	2806	2818	2
8619	8631	8643	8655	8667	8679	44	15	2864	2876	2887	2
8691	8703	8715	8727	8739	8751	43	16	2934	2945	2957	2
8763	8776	8788	8800	8812	8824	42	17	3003	3014	3026	3
8836	8848	8860	8872	8884	8896	41	18	3072	3083	3095	3
8908	8920	8932	8944	8956	8968	40	19	3141	3153	3164	3
9.938980	8992	9004	9016	9028	9040	39	20	9.943210	3222	3233	3
9052	9064	9076	9087	9099	9111	38	21	3279	3291	3302	3
9123	9135	9147	9159	9171	9183	37	22	3348	3360	3371	3
9195	9207	9219	9231	9243	9255	36	23	3417	3429	3440	3
9267	9279	9291	9303	9315	9327	35	24	3486	3498	3509	3
9339	9351	9363	9375	9387	9399	34	25	3555	3566	3578	3
9410	9422	9434	9446	9458	9470	33	26	3624	3635	3647	3
9482	9494	9506	9518	9530	9542	32	27	3693	3704	3715	3
9554	9566	9578	9590	9601	9613	31	28	3761	3773	3784	3
9625	9637	9649	9661	9673	9685	30	29	3830	3841	3853	3
9.939697	9709	9721	9733	9744	9756	29	30	9.943899	3910	3921	3
9768	9780	9792	9804	9816	9828	28	31	3967	3978	3990	4
9840	9852	9863	9875	9887	9899	27	32	4036	4047	4058	4
9911	9923	9935	9947	9959	9970	26	33	4104	4115	4127	4
9982	9994	...6	..18	..30	..42	25	34	4172	4184	4195	4
9.940054	0065	0077	0089	0101	0113	24	35	4241	4252	4264	4
0125	0137	0148	0160	0172	0184	23	36	4309	4321	4332	4
0196	0208	0220	0231	0243	0255	22	37	4377	4389	4400	4
0267	0279	0291	0303	0314	0326	21	38	4446	4457	4468	4
0338	0350	0362	0374	0385	0397	20	39	4514	4525	4537	4
9.940409	0421	0433	0445	0456	0468	19	40	9.944582	4593	4605	4
0480	0492	0504	0516	0527	0539	18	41	4650	4661	4673	4
0551	0563	0575	0586	0598	0610	17	42	4718	4730	4741	4
0622	0634	0645	0657	0669	0681	16	43	4786	4798	4809	4
0693	0704	0716	0728	0740	0752	15	44	4854	4865	4877	4
0763	0775	0787	0799	0811	0822	14	45	4922	4933	4945	4
0834	0846	0858	0870	0881	0893	13	46	4990	5001	5013	5
0905	0917	0928	0940	0952	0964	12	47	5058	5069	5080	5
0975	0987	0999	1011	1023	1034	11	48	5125	5137	5148	5
1046	1058	1070	1081	1093	1105	10	49	5193	5204	5216	5
9.941117	1128	1140	1152	1164	1175	9	50	9.945261	5272	5283	5
1187	1199	1211	1222	1234	1246	8	51	5328	5340	5351	5
1258	1269	1281	1293	1304	1316	7	52	5396	5407	5419	5
1328	1340	1351	1363	1375	1387	6	53	5464	5475	5486	5
1398	1410	1422	1433	1445	1457	5	54	5531	5542	5554	5
1469	1480	1492	1504	1515	1527	4	55	5598	5610	5621	5
1539	1550	1562	1574	1586	1597	3	56	5666	5677	5688	5
1609	1621	1632	1644	1656	1667	2	57	5733	5744	5756	5
1679	1691	1702	1714	1726	1738	1	58	5800	5812	5823	5
1749	1761	1773	1784	1796	1808	0	59	5868	5879	5890	5

gent of 60 Degrees.						Min.	Tangent of 61		
10″	20″	30″	40″	50″			0″	10″	20″
8609	8658	8707	8755	8804	59	0	10.256248	6298	6347
8901	8950	8998	9047	9096	58	1	6546	6596	6645
9193	9242	9290	9339	9388	57	2	6844	6894	6944
9485	9534	9582	9631	9680	56	3	7142	7192	7242
9777	9826	9874	9923	9972	55	4	7441	7490	7540
0069	0118	0167	0215	0264	54	5	7739	7789	7839
0362	0410	0459	0508	0557	53	6	8038	8087	8137
0654	0703	0752	0800	0849	52	7	8336	8386	8436
0947	0995	1044	1093	1142	51	8	8635	8685	8735
1239	1288	1337	1385	1434	50	9	8934	8984	9033
1532	1581	1629	1678	1727	49	10	10.259233	9283	9332
1825	1873	1922	1971	2020	48	11	9532	9582	9632
2118	2166	2215	2264	2313	47	12	9831	9881	9931
2411	2459	2508	2557	2606	46	13	10.260130	0180	0230
2704	2753	2801	2850	2899	45	14	0430	0480	0530
2997	3046	3095	3143	3192	44	15	0729	0779	0829
3290	3339	3388	3437	3486	43	16	1029	1079	1129
3584	3632	3681	3730	3779	42	17	1329	1379	1429
3877	3926	3975	4024	4073	41	18	1629	1679	1729
4171	4220	4269	4318	4366	40	19	1929	1979	2029
4464	4513	4562	4611	4660	39	20	10.262229	2279	2329
4758	4807	4856	4905	4954	38	21	2529	2579	2629
5052	5101	5150	5199	5248	37	22	2829	2879	2929
5346	5395	5444	5493	5542	36	23	3130	3180	3230
5640	5689	5738	5787	5836	35	24	3430	3480	3530
5934	5984	6033	6082	6131	34	25	3731	3781	3831
6229	6278	6327	6376	6425	33	26	4031	4082	4132
6523	6572	6621	6670	6720	32	27	4332	4382	4433
6818	6867	6916	6965	7014	31	28	4633	4683	4734
7112	7161	7211	7260	7309	30	29	4934	4985	5035
7407	7456	7505	7554	7604	29	30	10.265236	5286	5336
7702	7751	7800	7849	7899	28	31	5537	5587	5637
7997	8046	8095	8144	8194	27	32	5838	5889	5939
8292	8341	8390	8439	8489	26	33	6140	6190	6240
8587	8636	8685	8735	8784	25	34	6442	6492	6542
8882	8931	8981	9030	9079	24	35	6743	6794	6844
9178	9227	9276	9325	9375	23	36	7045	7096	7146
9473	9522	9572	9621	9670	22	37	7347	7398	7448
9769	9818	9867	9916	9966	21	38	7649	7700	7750
0064	0114	0163	0212	0261	20	39	7952	8002	8052
0360	0409	0459	0508	0557	19	40	10.268254	8304	8355
0656	0705	0755	0804	0853	18	41	8556	8607	8657
0952	1001	1051	1100	1149	17	42	8859	8909	8960
1248	1297	1347	1396	1445	16	43	9162	9212	9263
1544	1594	1643	1692	1742	15	44	9465	9515	9566
1840	1890	1939	1989	2038	14	45	9767	9818	9868
2137	2186	2236	2285	2335	13	46	10.270071	0121	0172
2433	2483	2532	2582	2631	12	47	0374	0424	0475

			Min.	Sine of 63 Degre			
0″	50″			0″	10″	20″	30″
80	5991	59	0	9.949881	9892	9902	9913
47	6058	58	1	9945	9956	9967	9977
14	6125	57	2	9.950010	0020	0031	0042
81	6192	56	3	0074	0084	0095	0106
48	6259	55	4	0138	0149	0159	0170
15	6326	54	5	0202	0213	0224	0234
82	6393	53	6	0266	0277	0288	0298
49	6460	52	7	0330	0341	0352	0362
15	6526	51	8	0394	0405	0416	0426
82	6593	50	9	0458	0469	0480	0490
49	6660	49	10	9.950522	0533	0544	0554
15	6726	48	11	0586	0597	0607	0618
82	6793	47	12	0650	0661	0671	0682
49	6860	46	13	0714	0724	0735	0746
15	6926	45	14	0778	0788	0799	0809
82	6993	44	15	0841	0852	0862	0873
48	7059	43	16	0905	0915	0926	0937
14	7125	42	17	0968	0979	0990	1000
81	7192	41	18	1032	1043	1053	1064
47	7258	40	19	1096	1106	1117	1127
13	7324	39	20	9.951159	1170	1180	1191
79	7390	38	21	1222	1233	1244	1254
45	7456	37	22	1286	1296	1307	1317
11	7522	36	23	1349	1360	1370	1381
78	7589	35	24	1412	1423	1434	1444
44	7655	34	25	1476	1486	1497	1507
09	7720	33	26	1539	1549	1560	1570
75	7786	32	27	1602	1613	1623	1634
41	7852	31	28	1665	1676	1686	1697
07	7918	30	29	1728	1739	1749	1760
73	7984	29	30	9.951791	1802	1812	1823
38	8049	28	31	1854	1865	1875	1886
04	8115	27	32	1917	1928	1938	1949
70	8181	26	33	1980	1990	2001	2011
35	8246	25	34	2043	2053	2064	2074
01	8312	24	35	2106	2116	2126	2137
66	8377	23	36	2168	2179	2189	2200
32	8443	22	37	2231	2241	2252	2262

Tangent of 62 Degrees.							Min.	Tangent of 63 Degrees.						
0″	10″	20″	30″	40″	50″			0″	10″	20″	30″	40″	50″	
10.274326	4376	4427	4478	4529	4580	59	0	10.292834	2886	2938	2990	3042	3094	5
4630	4681	4732	4783	4834	4885	58	1	3146	3199	3251	3303	3355	3407	5
4935	4986	5037	5088	5139	5190	57	2	3459	3511	3563	3615	3667	3720	5
5240	5291	5342	5393	5444	5495	56	3	3772	3824	3876	3928	3980	4032	5
5546	5597	5647	5698	5749	5800	55	4	4084	4137	4189	4241	4293	4345	5
5851	5902	5953	6004	6055	6105	54	5	4397	4449	4502	4554	4606	4658	5
6156	6207	6258	6309	6360	6411	53	6	4710	4763	4815	4867	4919	4971	5
6462	6513	6564	6615	6666	6717	52	7	5024	5076	5128	5180	5232	5285	5
6768	6819	6870	6920	6971	7022	51	8	5337	5389	5441	5494	5546	5598	5
7073	7124	7175	7226	7277	7328	50	9	5650	5703	5755	5807	5859	5912	5
10.277379	7430	7481	7532	7583	7634	49	10	10.295964	6016	6068	6121	6173	6225	4
7685	7736	7787	7838	7889	7940	48	11	6278	6330	6382	6434	6487	6539	4
7991	8043	8094	8145	8196	8247	47	12	6591	6644	6696	6748	6801	6853	4
8298	8349	8400	8451	8502	8553	46	13	6905	6958	7010	7062	7115	7167	4
8604	8655	8706	8757	8809	8860	45	14	7219	7272	7324	7377	7429	7481	4
8911	8962	9013	9064	9115	9166	44	15	7534	7586	7638	7691	7743	7796	4
9217	9268	9320	9371	9422	9473	43	16	7848	7900	7953	8005	8058	8110	4
9524	9575	9626	9678	9729	9780	42	17	8163	8215	8267	8320	8372	8425	4
9831	9882	9933	9984	..36	..87	41	18	8477	8530	8582	8635	8687	8740	4
10.280138	0189	0240	0292	0343	0394	40	19	8792	8845	8897	8949	9002	9054	4
10.280445	0496	0548	0599	0650	0701	39	20	10.299107	9159	9212	9264	9317	9370	3
0752	0804	0855	0906	0957	1009	38	21	9422	9475	9527	9580	9632	9685	3
1060	1111	1162	1214	1265	1316	37	22	9737	9790	9842	9895	9947	....	3
1367	1419	1470	1521	1572	1624	36	23	10.300053	0105	0158	0210	0263	0315	3
1675	1726	1777	1829	1880	1931	35	24	0368	0421	0473	0526	0578	0631	3
1983	2034	2085	2137	2188	2239	34	25	0684	0736	0789	0841	0894	0947	3
2291	2342	2393	2445	2496	2547	33	26	0999	1052	1105	1157	1210	1263	3
2599	2650	2701	2753	2804	2855	32	27	1315	1368	1421	1473	1526	1579	3
2907	2958	3009	3061	3112	3164	31	28	1631	1684	1737	1789	1842	1895	3
3215	3266	3318	3369	3421	3472	30	29	1947	2000	2053	2106	2158	2211	3
10.283523	3575	3626	3678	3729	3780	29	30	10.302264	2316	2369	2422	2475	2527	2
3832	3883	3935	3986	4038	4089	28	31	2580	2633	2686	2738	2791	2844	2
4140	4192	4243	4295	4346	4398	27	32	2897	2950	3002	3055	3108	3161	2
4449	4501	4552	4604	4655	4707	26	33	3213	3266	3319	3372	3425	3478	2
4758	4810	4861	4913	4964	5016	25	34	3530	3583	3636	3689	3742	3794	2
5067	5119	5170	5222	5273	5325	24	35	3847	3900	3953	4006	4059	4112	2
5376	5428	5479	5531	5582	5634	23	36	4164	4217	4270	4323	4376	4429	2
5686	5737	5789	5840	5892	5943	22	37	4482	4535	4588	4640	4693	4746	2
5995	6046	6098	6150	6201	6253	21	38	4799	4852	4905	4958	5011	5064	2
6304	6356	6408	6459	6511	6562	20	39	5117	5170	5223	5276	5328	5381	2
10.286614	6666	6717	6769	6821	6872	19	40	10.305434	5487	5540	5593	5646	5699	1
6924	6975	7027	7079	7130	7182	18	41	5752	5805	5858	5911	5964	6017	1
7234	7285	7337	7389	7440	7492	17	42	6070	6123	6176	6229	6282	6335	1
7544	7595	7647	7699	7751	7802	16	43	6388	6441	6494	6547	6600	6654	1
7854	7906	7957	8009	8061	8113	15	44	6707	6760	6813	6866	6919	6972	1
8164	8216	8268	8319	8371	8423	14	45	7025	7078	7131	7184	7237	7290	1
8475	8526	8578	8630	8682	8733	13	46	7344	7397	7450	7503	7556	7609	1
8785	8837	8889	8941	8992	9044	12	47	7662	7715	7768	7822	7875	7928	1
9096	9148	9199	9251	9303	9355	11	48	7981	8034	8087	8141	8194	8247	1
9407	9458	9510	9562	9614	9666	10	49	8300	8353	8406	8460	8513	8566	1
10.289718	9769	9821	9873	9925	9977	9	50	10.308619	8672	8726	8779	8832	8885	
10.290029	0081	0132	0184	0236	0288	8	51	8938	8992	9045	9098	9151	9205	
0340	0392	0444	0496	0547	0599	7	52	9258	9311	9364	9418	9471	9524	
0651	0703	0755	0807	0859	0911	6	53	9577	9631	9684	9737	9790	9844	
0963	1015	1066	1118	1170	1222	5	54	9897	9950	...4	..57	.110	.164	
1274	1326	1378	1430	1482	1534	4	55	10.310217	0270	0324	0377	0430	0484	
1586	1638	1690	1742	1794	1846	3	56	0537	0590	0644	0697	0750	0804	
1898	1950	2002	2054	2106	2158	2	57	0857	0910	0964	1017	1070	1124	
2210	2262	2314	2366	2418	2470	1	58	1177	1231	1284	1337	1391	1444	
2522	2574	2626	2678	2730	2782	0	59	1498	1551	1605	1658	1711	1765	

Sine of 64 Degrees.							Min.	Sine of 65 Degr			
0″	10″	20″	30″	40″	50″			0″	10″	20″	30″
.953660	3670	3681	3691	3701	3712	59	0	9.957276	7286	7295	7305
3722	3732	3742	3753	3763	3773	58	1	7335	7344	7354	7364
3783	3794	3804	3814	3824	3835	57	2	7393	7403	7413	7423
3845	3855	3865	3876	3886	3896	56	3	7452	7462	7472	7482
3906	3917	3927	3937	3947	3957	55	4	7511	7521	7531	7540
3968	3978	3988	3998	4009	4019	54	5	7570	7579	7589	7599
4029	4039	4050	4060	4070	4080	53	6	7628	7638	7648	7658
4090	4101	4111	4121	4131	4141	52	7	7687	7697	7707	7716
4152	4162	4172	4182	4192	4203	51	8	7746	7755	7765	7775
4213	4223	4233	4243	4254	4264	50	9	7804	7814	7824	7833
.954274	4284	4294	4305	4315	4325	49	10	9.957863	7872	7882	7892
4335	4345	4356	4366	4376	4386	48	11	7921	7931	7940	7950
4396	4406	4417	4427	4437	4447	47	12	7979	7989	7999	8009
4457	4468	4478	4488	4498	4508	46	13	8038	8047	8057	8067
4518	4529	4539	4549	4559	4569	45	14	8096	8106	8115	8125
4579	4589	4600	4610	4620	4630	44	15	8154	8164	8174	8183
4640	4650	4661	4671	4681	4691	43	16	8213	8222	8232	8242
4701	4711	4721	4732	4742	4752	42	17	8271	8280	8290	8300
4762	4772	4782	4792	4802	4813	41	18	8329	8339	8348	8358
4823	4833	4843	4853	4863	4873	40	19	8387	8397	8406	8416
.954883	4894	4904	4914	4924	4934	39	20	9.958445	8455	8464	8474
4944	4954	4964	4974	4985	4995	38	21	8503	8513	8522	8532
5005	5015	5025	5035	5045	5055	37	22	8561	8571	8580	8590
5065	5075	5086	5096	5106	5116	36	23	8619	8628	8638	8648
5126	5136	5146	5156	5166	5176	35	24	8677	8686	8696	8706
5186	5196	5207	5217	5227	5237	34	25	8734	8744	8754	8763
5247	5257	5267	5277	5287	5297	33	26	8792	8802	8812	8821
5307	5317	5327	5337	5348	5358	32	27	8850	8860	8869	8879
5368	5378	5388	5398	5408	5418	31	28	8908	8917	8927	8937
5428	5438	5448	5458	5468	5478	30	29	8965	8975	8985	8994
.955488	5498	5508	5518	5528	5538	29	30	9.959023	9033	9042	9052
5548	5559	5569	5579	5589	5599	28	31	9080	9090	9100	9109
5609	5619	5629	5639	5649	5659	27	32	9138	9148	9157	9167
5669	5679	5689	5699	5709	5719	26	33	9195	9205	9215	9224
5729	5739	5749	5759	5769	5779	25	34	9253	9262	9272	9282
5789	5799	5809	5819	5829	5839	24	35	9310	9320	9329	9339
5849	5859	5869	5879	5889	5899	23	36	9368	9377	9387	9396
5909	5919	5929	5939	5949	5959	22	37	9425	9434	9444	9453
5969	5979	5989	5999	6009	6019	21	38	9482	9492	9501	9511
6029	6039	6049	6059	6069	6079	20	39	9539	9549	9558	9568
.956089	6099	6108	6118	6128	6138	19	40	9.959596	9606	9615	9625
6148	6158	6168	6178	6188	6198	18	41	9654	9663	9673	9682
6208	6218	6228	6238	6248	6258	17	42	9711	9720	9730	9739
6268	6278	6288	6298	6308	6317	16	43	9768	9777	9787	9796
6327	6337	6347	6357	6367	6377	15	44	9825	9834	9844	9853
6387	6397	6407	6417	6427	6437	14	45	9882	9891	9900	9910
6447	6457	6466	6476	6486	6496	13	46	9938	9948	9957	9967
6506	6516	6526	6536	6546	6556	12	47	9995	...5	..14	..24
6566	6575	6585	6595	6605	6615	11	48	9.960052	0061	0071	0080
6625	6635	6645	6655	6665	6674	10	49	0109	0118	0128	0137
.956684	6694	6704	6714	6724	6734	9	50	9.960165	0175	0184	0194
6744	6754	6763	6773	6783	6793	8	51	0222	0232	0241	0250
6803	6813	6823	6833	6843	6852	7	52	0279	0288	0298	0307
6862	6872	6882	6892	6902	6912	6	53	0335	0345	0354	0364
6921	6931	6941	6951	6961	6971	5	54	0392	0401	0411	0420
6981	6990	7000	7010	7020	7030	4	55	0448	0458	0467	0477
7040	7050	7059	7069	7079	7089	3	56	0505	0514	0524	0533
7099	7109	7118	7128	7138	7148	2	57	0561	0571	0580	0589
7158	7168	7177	7187	7197	7207	1	58	0618	0627	0636	0646
7217	7227	7236	7246	7256	7266	0	59	0674	0683	0693	0702

Min.	Tangent of 64 Degrees. 0″	10″	20″	30″	40″	50″	
0	10.311818	1872	1925	1979	2032	2085	59
1	2139	2192	2246	2299	2353	2406	58
2	2460	2513	2567	2620	2674	2727	57
3	2781	2834	2888	2941	2995	3048	56
4	3102	3155	3209	3263	3316	3370	55
5	3423	3477	3530	3584	3637	3691	54
6	3745	3798	3852	3905	3959	4013	53
7	4066	4120	4173	4227	4281	4334	52
8	4388	4442	4495	4549	4603	4656	51
9	4710	4764	4817	4871	4925	4978	50
10	10.315032	5086	5139	5193	5247	5300	49
11	5354	5408	5461	5515	5569	5623	48
12	5676	5730	5784	5838	5891	5945	47
13	5999	6053	6106	6160	6214	6268	46
14	6321	6375	6429	6483	6537	6590	45
15	6644	6698	6752	6806	6860	6913	44
16	6967	7021	7075	7129	7183	7236	43
17	7290	7344	7398	7452	7506	7560	42
18	7613	7667	7721	7775	7829	7883	41
19	7937	7991	8045	8099	8153	8206	40
20	10.318260	8314	8368	8422	8476	8530	39
21	8584	8638	8692	8746	8800	8854	38
22	8908	8962	9016	9070	9124	9178	37
23	9232	9286	9340	9394	9448	9502	36
24	9556	9610	9664	9718	9772	9826	35
25	9880	9934	9988	..42	..96	.151	34
26	10.320205	0259	0313	0367	0421	0475	33
27	0529	0583	0637	0692	0746	0800	32
28	0854	0908	0962	1016	1071	1125	31
29	1179	1233	1287	1341	1396	1450	30
30	·o 321504	1558	1612	1666	1721	1775	29
31	1829	1883	1938	1992	2046	2100	28
32	2154	2209	2263	2317	2371	2426	27
33	2480	2534	2588	2643	2697	2751	26
34	2806	2860	2914	2968	3023	3077	25
35	3131	3186	3240	3294	3349	3403	24
36	3457	3512	3566	3620	3675	3729	23
37	3783	3838	3892	3947	4001	4055	22
38	4110	4164	4219	4273	4327	4382	21
39	4436	4491	4545	4599	4654	4708	20
40	10.324763	4817	4872	4926	4981	5035	19
41	5089	5144	5198	5253	5307	5362	18
42	5416	5471	5525	5580	5634	5689	17
43	5743	5798	5852	5907	5962	6016	16
44	6071	6125	6180	6234	6289	6343	15
45	6398	6453	6507	6562	6616	6671	14
46	6726	6780	6835	6889	6944	6999	13
47	7053	7108	7162	7217	7272	7326	12
48	7381	7436	7490	7545	7600	7654	11
49	7709	7764	7818	7873	7928	7982	10
50	10.328037	8092	8147	8201	8256	8311	9
51	8365	8420	8475	8530	8584	8639	8
52	8694	8749	8803	8858	8913	8968	7

Min.	Tangent of 65 Degrees. 0″	10″	20″	30″	40″	50″	
0	10.331327	1382	1437	1492	1547	1602	59
1	1657	1712	1767	1822	1877	1932	58
2	1987	2042	2097	2153	2208	2263	57
3	2318	2373	2428	2483	2538	2593	56
4	2648	2703	2758	2813	2868	2924	55
5	2979	3034	3089	3144	3199	3254	54
6	3309	3364	3420	3475	3530	3585	53
7	3640	3695	3751	3806	3861	3916	52
8	3971	4026	4082	4137	4192	4247	51
9	4302	4358	4413	4468	4523	4579	50
10	10.334634	4689	4744	4800	4855	4910	49
11	4965	5021	5076	5131	5186	5242	48
12	5297	5352	5408	5463	5518	5574	47
13	5629	5684	5740	5795	5850	5906	46
14	5961	6016	6072	6127	6182	6238	45
15	6293	6349	6404	6459	6515	6570	44
16	6625	6681	6736	6792	6847	6903	43
17	6958	7013	7069	7124	7180	7235	42
18	7291	7346	7402	7457	7513	7568	41
19	7624	7679	7735	7790	7846	7901	40
20	10.337957	8012	8068	8123	8179	8234	39
21	8290	8345	8401	8456	8512	8568	38
22	8623	8679	8734	8790	8845	8901	37
23	8957	9012	9068	9123	9179	9235	36
24	9290	9346	9402	9457	9513	9569	35
25	9624	9680	9735	9791	9847	9902	34
26	9958	..14	..70	.125	.181	.237	33
27	10.340292	0348	0404	0460	0515	0571	32
28	0627	0682	0738	0794	0850	0905	31
29	0961	1017	1073	1129	1184	1240	30
30	10.341296	1352	1408	1463	1519	1575	29
31	1631	1687	1742	1798	1854	1910	28
32	1966	2022	2078	2133	2189	2245	27
33	2301	2357	2413	2469	2525	2581	26
34	2636	2692	2748	2804	2860	2916	25
35	2972	3028	3084	3140	3196	3252	24
36	3308	3364	3420	3476	3532	3588	23
37	3644	3700	3756	3812	3868	3924	22
38	3980	4036	4092	4148	4204	4260	21
39	4316	4372	4428	4484	4540	4596	20
40	10.344652	4708	4764	4821	4877	4933	19
41	4989	5045	5101	5157	5213	5269	18
42	5326	5382	5438	5494	5550	5606	17
43	5663	5719	5775	5831	5887	5943	16
44	6000	6056	6112	6168	6224	6281	15
45	6337	6393	6449	6506	6562	6618	14
46	6674	6731	6787	6843	6899	6956	13
47	7012	7068	7125	7181	7237	7293	12
48	7350	7406	7462	7519	7575	7631	11
49	7688	7744	7800	7857	7913	7969	10
50	10.348026	8082	8139	8195	8251	8308	9
51	8364	8421	8477	8533	8590	8646	8
52	8703	8759	8815	8872	8928	8985	7

)egrees.				Min.	Sine of 67]		
30″	40″	50″			0″	10″	20″
0758	0768	0777	59	0	9.964026	4035	4044
0814	0824	0833	58	1	4080	4089	4098
0871	0880	0889	57	2	4133	4142	4151
0927	0936	0945	56	3	4187	4196	4205
0983	0992	1002	55	4	4240	4249	4258
1039	1048	1058	54	5	4294	4303	4311
1095	1104	1113	53	6	4347	4356	4365
1151	1160	1169	52	7	4400	4409	4416
1207	1216	1225	51	8	4454	4463	4471
1263	1272	1281	50	9	4507	4516	4525
1318	1328	1337	49	10	9.964560	4569	4578
1374	1383	1393	48	11	4613	4622	4631
1430	1439	1448	47	12	4666	4675	4684
1485	1495	1504	46	13	4720	4728	4737
1541	1550	1560	45	14	4773	4781	4790
1597	1606	1615	44	15	4826	4834	4843
1652	1661	1671	43	16	4879	4887	4896
1708	1717	1726	42	17	4931	4940	4949
1763	1772	1782	41	18	4984	4993	5002
1819	1828	1837	40	19	5037	5046	5055
1874	1883	1892	39	20	9.965090	5099	5107
1929	1939	1948	38	21	5143	5151	5160
1985	1994	2003	37	22	5195	5204	5213
2040	2049	2058	36	23	5248	5257	5266
2095	2104	2113	35	24	5301	5309	5318
2150	2159	2169	34	25	5353	5362	5371
2205	2214	2224	33	26	5406	5414	5423
2260	2269	2279	32	27	5458	5467	5476
2315	2325	2334	31	28	5511	5519	5528
2370	2379	2389	30	29	5563	5572	5580
2425	2434	2444	29	30	9.965615	5624	5633
2480	2489	2498	28	31	5668	5676	5685
2535	2544	2553	27	32	5720	5729	5737
2590	2599	2608	26	33	5772	5781	5790
2645	2654	2663	25	34	5824	5833	5842
2699	2708	2717	24	35	5876	5885	5894
2754	2763	2772	23	36	5929	5937	5946
2809	2818	2827	22	37	5981	5989	5998
2863	2872	2881	21	38	6033	6041	6050
2918	2927	2936	20	39	6085	6093	6102
2972	2981	2990	19	40	9.966136	6145	6154
3027	3036	3045	18	41	6188	6197	6206
3081	3090	3099	17	42	6240	6249	6257
3135	3144	3153	16	43	6292	6301	6309
3190	3199	3208	15	44	6344	6352	6361
3244	3253	3262	14	45	6395	6404	6413
3298	3307	3316	13	46	6447	6456	6464
3352	3361	3370	12	47	6499	6507	6516
3407	3416	3425	11	48	6550	6559	6567
3461	3470	3479	10	49	6602	6610	6619
3515	3524	3533	9	50	9.966653	6662	6670
3569	3578	3587	8	51	6705	6713	6722
3623	3632	3641	7	52	6756	6765	6773
3677	3686	3695	6	53	6808	6816	6825
3730	3739	3748	5	54	6859	6867	6876
3784	3793	3802	4	55	6910	6919	6927
3838	3847	3856	3	56	6961	6970	6978
3892	3901	3910	2	57	7013	7021	7030
3946	3955	3963	1	58	7064	7072	7081
3999	4008	4017	0	59	7115	7123	7132

10.372148	2207	2265	2324	2382	2441	59
2499	2558	2617	2675	2734	2792	58
2851	2910	2968	3027	3085	3144	57
3203	3261	3320	3379	3437	3496	56
3555	3613	3672	3731	3789	3848	55
3907	3965	4024	4083	4142	4200	54
4259	4318	4377	4435	4494	4553	53
4612	4670	4729	4788	4847	4906	52
4964	5023	5082	5141	5200	5258	51
5317	5376	5435	5494	5553	5612	50
10.375670	5729	5788	5847	5906	5965	49
6024	6083	6142	6200	6259	6318	48
6377	6436	6495	6554	6613	6672	47
6731	6790	6849	6908	6967	7026	46
7085	7144	7203	7262	7321	7380	45
7439	7498	7557	7616	7675	7734	44
7793	7853	7912	7971	8030	8089	43
8148	8207	8266	8325	8384	8444	42
8503	8562	8621	8680	8739	8799	41
8858	8917	8976	9035	9094	9154	40
10.379213	9272	9331	9390	9450	9509	39
9568	9627	9687	9746	9805	9864	38
9924	9983	..42	.102	.161	.220	37
10.380280	0339	0398	0457	0517	0576	36
0636	0695	0754	0814	0873	0932	35
0992	1051	1110	1170	1229	1289	34
1348	1407	1467	1526	1586	1645	33
1705	1764	1824	1883	1943	2002	32
2061	2121	2180	2240	2299	2359	31
2418	2478	2538	2597	2657	2716	30
10.382776	2835	2895	2954	3014	3074	29
3133	3193	3252	3312	3372	3431	28
3491	3550	3610	3670	3729	3789	27
3849	3908	3968	4028	4087	4147	26
4207	4266	4326	4386	4445	4505	25
4565	4625	4684	4744	4804	4864	24
4923	4983	5043	5103	5162	5222	23
5282	5342	5402	5461	5521	5581	22
5641	5701	5761	5820	5880	5940	21
6000	6060	6120	6180	6240	6299	20
10.386359	6419	6479	6539	6599	6659	19
6719	6779	6839	6899	6959	7019	18
7079	7139	7199	7259	7319	7379	17
7439	7499	7559	7619	7679	7739	16
7799	7859	7919	7979	8039	8099	15
8159	8219	8279	8339	8399	8460	14
8520	8580	8640	8700	8760	8820	13
8880	8941	9001	9061	9121	9181	12
9241	9302	9362	9422	9482	9542	11
9603	9663	9723	9783	9844	9904	10
10.389964	..24	..85	.145	.205	.265	9
10.390326	0386	0446	0507	0567	0627	8
0688	0748	0808	0869	0929	0989	7
1050	1110	1170	1231	1291	1352	6
1412	1472	1533	1593	1654	1714	5
1775	1835	1895	1956	2016	[illegible]	4
2137	2198	2258	2319	[illegible]	[illegible]	[illegible]
2500	2561	2621	2682	[illegible]	[illegible]	[illegible]
2863	2924	2985	3045	[illegible]	[illegible]	[illegible]

Min.	Sine of 68 Degrees. 0″	10″	20″	30″	40″	50″	
0	9·967166	7174	7183	7191	7200	7208	59
1	7217	7225	7234	7242	7251	7259	58
2	7268	7276	7285	7293	7302	7310	57
3	7319	7327	7336	7344	7353	7361	56
4	7370	7378	7387	7395	7404	7412	55
5	7421	7429	7437	7446	7454	7463	54
6	7471	7480	7488	7497	7505	7514	53
7	7522	7531	7539	7547	7556	7564	52
8	7573	7581	7590	7598	7607	7615	51
9	7624	7632	7640	7649	7657	7666	50
10	9·967674	7683	7691	7699	7708	7716	49
11	7725	7733	7742	7750	7758	7767	48
12	7775	7784	7792	7801	7809	7817	47
13	7826	7834	7843	7851	7859	7868	46
14	7876	7885	7893	7901	7910	7918	45
15	7927	7935	7943	7952	7960	7969	44
16	7977	7985	7994	8002	8011	8019	43
17	8027	8036	8044	8053	8061	8069	42
18	8078	8086	8094	8103	8111	8120	41
19	8128	8136	8145	8153	8161	8170	40
20	9·968178	8187	8195	8203	8212	8220	39
21	8228	8237	8245	8253	8262	8270	38
22	8278	8287	8295	8303	8312	8320	37
23	8329	8337	8345	8354	8362	8370	36
24	8379	8387	8395	8404	8412	8420	35
25	8429	8437	8445	8454	8462	8470	34
26	8479	8487	8495	8503	8512	8520	33
27	8528	8537	8545	8553	8562	8570	32
28	8578	8587	8595	8603	8612	8620	31
29	8628	8636	8645	8653	8661	8670	30
30	9·968678	8686	8694	8703	8711	8719	29
31	8728	8736	8744	8752	8761	8769	28
32	8777	8786	8794	8802	8810	8819	27
33	8827	8835	8844	8852	8860	8868	26
34	8877	8885	8893	8901	8910	8918	25
35	8926	8934	8943	8951	8959	8967	24
36	8976	8984	8992	9000	9009	9017	23
37	9025	9033	9042	9050	9058	9066	22
38	9075	9083	9091	9099	9108	9116	21
39	9124	9132	9141	9149	9157	9165	20
40	9·969173	9182	9190	9198	9206	9215	19
41	9223	9231	9239	9247	9256	9264	18
42	9272	9280	9288	9297	9305	9313	17
43	9321	9329	9338	9346	9354	9362	16
44	9370	9379	9387	9395	9403	9411	15
45	9420	9428	9436	9444	9452	9461	14
46	9469	9477	9485	9493	9501	9510	13
47	9518	9526	9534	9542	9550	9559	12
48	9567	9575	9583	9591	9599	9608	11
49	9616	9624	9632	9640	9648	9657	10
50	9·969665	9673	9681	9689	9697	9705	9
51	9714	9722	9730	9738	9746	9754	8
52	9762	9771	9779	9787	9795	9803	7
53	9811	9819	9828	9836	9844	9852	6
54	9860	9868	9876	9884	9893	9901	5
55	9909	9917	9925	9933	9941	9949	4
56	9957	9966	9974	9982	9990	9998	3
57	9·970006	0014	0022	0030	0038	0047	2
58	0055	0063	0071	0079	0087	0095	1
59	0103	0111	0119	0127	0136	0144	0
	60″	50″	40″	30″	20″	10″	Min.
	Co-sine of 21 Degrees.						

Min.	Sine of 69 Degrees. 0″	10″	20″	30″	40″	50″	
0	9 970152	0160	0168	0176	0184	0192	59
1	0200	0208	0216	0224	0233	0241	58
2	0249	0257	0265	0273	0281	0289	57
3	0297	0305	0313	0321	0329	0337	56
4	0345	0353	0361	0370	0378	0386	55
5	0394	0402	0410	0418	0426	0434	54
6	0442	0450	0458	0466	0474	0482	53
7	0490	0498	0506	0514	0522	0530	52
8	0538	0546	0554	0562	0570	0578	51
9	0586	0594	0603	0611	0619	0627	50
10	9·970635	0643	0651	0659	0667	0675	49
11	0683	0691	0699	0707	0715	0723	48
12	0731	0739	0747	0755	0763	0771	47
13	0779	0787	0795	0803	0811	0819	46
14	0827	0835	0842	0850	0858	0866	45
15	0874	0882	0890	0898	0906	0914	44
16	0922	0930	0938	0946	0954	0962	43
17	0970	0978	0986	0994	1002	1010	42
18	1018	1026	1034	1042	1050	1058	41
19	1066	1073	1081	1089	1097	1105	40
20	9·971113	1121	1129	1137	1145	1153	39
21	1161	1169	1177	1185	1193	1200	38
22	1208	1216	1224	1232	1240	1248	37
23	1256	1264	1272	1280	1288	1296	36
24	1303	1311	1319	1327	1335	1343	35
25	1351	1359	1367	1375	1383	1390	34
26	1398	1406	1414	1422	1430	1438	33
27	1446	1454	1462	1469	1477	1485	32
28	1493	1501	1509	1517	1525	1532	31
29	1540	1548	1556	1564	1572	1580	30
30	9·971588	1595	1603	1611	1619	1627	29
31	1635	1643	1651	1658	1666	1674	28
32	1682	1690	1698	1706	1713	1721	27
33	1729	1737	1745	1753	1761	1768	26
34	1776	1784	1792	1800	1808	1815	25
35	1823	1831	1839	1847	1855	1862	24
36	1870	1878	1886	1894	1902	1909	23
37	1917	1925	1933	1941	1949	1956	22
38	1964	1972	1980	1988	1995	2003	21
39	2011	2019	2027	2034	2042	2050	20
40	9·972058	2066	2073	2081	2089	2097	19
41	2105	2112	2120	2128	2136	2144	18
42	2151	2159	2167	2175	2183	2190	17
43	2198	2206	2214	2221	2229	2237	16
44	2245	2253	2260	2268	2276	2284	15
45	2291	2299	2307	2315	2322	2330	14
46	2338	2346	2354	2361	2369	2377	13
47	2385	2392	2400	2408	2416	2423	12
48	2431	2439	2447	2454	2462	2470	11
49	2478	2485	2493	2501	2508	2516	10
50	9·972524	2532	2539	2547	2555	2563	9
51	2570	2578	2586	2593	2601	2609	8
52	2617	2624	2632	2640	2648	2655	7
53	2663	2671	2678	2686	2694	2701	6
54	2709	2717	2725	2732	2740	2748	5
55	2755	2763	2771	2778	2786	2794	4
56	2802	2809	2817	2825	2832	2840	3
57	2848	2855	2863	2871	2878	2886	2
58	2894	2901	2909	2917	2924	2932	1
59	2940	2947	2955	2963	2970	2978	0
	60″	50″	40″	30″	20″	10″	Min.
	Co-sine of 20 Degrees.						

gent of 68 Degrees.						Min.	Tangent of 69 Degrees.					
10″	20″	30″	40″	50″			0″	10″	20″	30″	40″	50″
3651	3712	3772	3833	3894	59	0	10.415823	5886	5948	6011	6074	6137
4015	4076	4136	4197	4258	58	1	6200	6263	6326	6389	6452	6515
4379	4440	4500	4561	4622	57	2	6578	6641	6704	6767	6830	6893
4743	4804	4865	4926	4986	56	3	6956	7020	7083	7146	7209	7272
5108	5169	5229	5290	5351	55	4	7335	7398	7461	7524	7587	7650
5473	5533	5594	5655	5716	54	5	7714	7777	7840	7903	7966	8029
5838	5898	5959	6020	6081	53	6	8093	8156	8219	8282	8345	8409
6203	6264	6325	6385	6446	52	7	8472	8535	8598	8661	8725	8788
6568	6629	6690	6751	6812	51	8	8851	8914	8978	9041	9104	9168
6934	6995	7056	7117	7178	50	9	9231	9294	9358	9421	9484	9547
7300	7361	7422	7483	7544	49	10	10.419611	9674	9738	9801	9864	9928
7666	7727	7788	7849	7910	48	11	9991	..54	.118	.181	.245	.308
8032	8093	8154	8215	8276	47	12	10.420371	0435	0498	0562	0625	0689
8399	8460	8521	8582	8643	46	13	0752	0816	0879	0943	1006	1070
8765	8826	8888	8949	9010	45	14	1133	1197	1260	1324	1387	1451
9132	9194	9255	9316	9377	44	15	1514	1578	1641	1705	1769	1832
9500	9561	9622	9683	9744	43	16	1896	1959	2023	2086	2150	2214
9867	9928	9989	..51	.112	42	17	2277	2341	2405	2468	2532	2596
0235	0296	0357	0419	0480	41	18	2659	2723	2787	2850	2914	2978
0602	0664	0725	0787	0848	40	19	3041	3105	3169	3233	3296	3360
0971	1032	1093	1155	1216	39	20	10.423424	3488	3551	3615	3679	3743
1339	1400	1462	1523	1585	38	21	3807	3870	3934	3998	4062	4126
1707	1769	1830	1892	1953	37	22	4190	4253	4317	4381	4445	4509
2076	2138	2199	2261	2322	36	23	4573	4637	4701	4764	4828	4892
2445	2507	2568	2630	2691	35	24	4956	5020	5084	5148	5212	5276
2815	2876	2938	2999	3061	34	25	5340	5404	5468	5532	5596	5660
3184	3246	3307	3369	3430	33	26	5724	5788	5852	5916	5980	6044
3554	3615	3677	3739	3800	32	27	6108	6172	6236	6300	6364	6429
3924	3985	4047	4109	4170	31	28	6493	6557	6621	6685	6749	6813
4294	4356	4417	4479	4541	30	29	6877	6941	7006	7070	7134	7198
4664	4726	4788	4850	4911	29	30	10.427262	7327	7391	7455	7519	7583
5035	5097	5158	5220	5282	28	31	7648	7712	7776	7840	7905	7969
5406	5468	5529	5591	5653	27	32	8033	8097	8162	8226	8290	8355
5777	5839	5901	5962	6024	26	33	8419	8483	8548	8612	8676	8741
6148	6210	6272	6334	6396	25	34	8805	8869	8934	8998	9062	9127
6520	6582	6644	6706	6768	24	35	9191	9256	9320	9384	9449	9513
6891	6953	7015	7077	7139	23	36	9578	9642	9707	9771	9836	9900
7263	7326	7388	7450	7512	22	37	9965	..29	..94	.158	.223	.287
7636	7698	7760	7822	7884	21	38	10.430352	0416	0481	0545	0610	0674
8008	8070	8132	8195	8257	20	39	0739	0803	0868	0933	0997	1062
8381	8443	8505	8567	8630	19	40	10.431127	1191	1256	1320	1385	1450
8754	8816	8878	8940	9003	18	41	1514	1579	1644	1708	1773	1838
9127	9189	9252	9314	9376	17	42	1902	1967	2032	2097	2161	2226
9501	9563	9625	9687	9750	16	43	2291	2356	2420	2485	2550	2615
9874	9937	9999	..61	.123	15	44	2680	2744	2809	2874	2939	3004
0248	0310	0373	0435	0498	14	45	3068	3133	3198	3263	3328	3393
0622	0685	0747	0809	0872	13	46	3458	3522	3587	3652	3717	3782
0997	1059	1122	1184	1246	12	47	3847	3912	3977	4042	4107	4172
1371	1434	1496	1559	1621	11	48	4237	4302	4367	4432	4497	4562
1746	1809	1871	1934	1996	10	49	4627	4692	4757	4822	4887	4952
2121	2184	2246	2309	2371	9	50	10.435017	5082	5147	5212	5277	5342
2497	2559	2622	2684	2747	8	51	5407	5473	5538	5603	5668	5733
2872	2935	2997	3060	3123	7	52	5798	5863	5929	5994	6059	6124
3248	3311	3373	3436	3499	6	53	6189	6254	6320	6385	6450	6515
3624	3687	3749	3812	3875	5	54	6581	6646	6711	6776	6842	6907
4000	4063	4126	4189	4251	4	55	6972	7037	7103	7168	7233	7299
4377	4440	4502	4565	4628	3	56	7364	7429	7495	7560	7625	7691
4754	4817	4879	4942	5005	2	57	7756	7822	7887	7952	8018	8083
5131	5194	5256	5319	5382	1	58	8149	8214	8279	8345	8410	847[illegible]
5508	5571	5634	5697	5760	0	59	8541	8607	8672	8738	8803	88[illegible]

egrees.				Min.	Sine of 71 Degr			
30″	40″	50″			0″	10″	20″	30″
3009	3016	3024	59	0	9.975670	5677	5685	5692
3055	3062	3070	58	1	5714	5721	5728	5735
3101	3108	3116	57	2	5757	5764	5771	5779
3146	3154	3162	56	3	5800	5808	5815	5822
3192	3200	3208	55	4	5844	5851	5858	5865
3238	3246	3253	54	5	5887	5894	5901	5909
3284	3291	3299	53	6	5930	5938	5945	5952
3330	3337	3345	52	7	5974	5981	5988	5995
3375	3383	3390	51	8	6017	6024	6031	6038
3421	3428	3436	50	9	6060	6067	6074	6081
3466	3474	3482	49	10	9.976103	6110	6117	6125
3512	3519	3527	48	11	6146	6153	6160	6168
3557	3565	3572	47	12	6189	6196	6203	6211
3603	3610	3618	46	13	6232	6239	6246	6254
3648	3656	3663	45	14	6275	6282	6289	6296
3694	3701	3709	44	15	6318	6325	6332	6339
3739	3746	3754	43	16	6361	6368	6375	6382
3784	3792	3799	42	17	6404	6411	6418	6425
3829	3837	3844	41	18	6446	6454	6461	6468
3875	3882	3890	40	19	6489	6496	6503	6510
3920	3927	3935	39	20	9.976532	6539	6546	6553
3965	3972	3980	38	21	6574	6582	6589	6596
4010	4017	4025	37	22	6617	6624	6631	6638
4055	4062	4070	36	23	6660	6667	6674	6681
4100	4107	4115	35	24	6702	6709	6716	6723
4145	4152	4160	34	25	6745	6752	6759	6766
4190	4197	4205	33	26	6787	6794	6801	6808
4235	4242	4250	32	27	6830	6837	6844	6851
4279	4287	4294	31	28	6872	6879	6886	6893
4324	4332	4339	30	29	6914	6921	6928	6935
4369	4376	4384	29	30	9.976957	6964	6971	6978
4414	4421	4428	28	31	6999	7006	7013	7020
4458	4466	4473	27	32	7041	7048	7055	7062
4503	4510	4518	26	33	7083	7090	7097	7104
4547	4555	4562	25	34	7125	7132	7139	7146
4592	4599	4607	24	35	7167	7174	7181	7188
4636	4644	4651	23	36	7209	7216	7223	7230
4681	4688	4696	22	37	7251	7258	7265	7272
4725	4733	4740	21	38	7293	7300	7307	7314
4770	4777	4784	20	39	7335	7342	7349	7356
4814	4821	4829	19	40	9.977377	7384	7391	7398
4858	4866	4873	18	41	7419	7426	7433	7440
4903	4910	4917	17	42	7461	7468	7475	7482
4947	4954	4961	16	43	7503	7510	7517	7524
4991	4998	5006	15	44	7544	7551	7558	7565
5035	5042	5050	14	45	7586	7593	7600	7607
5079	5086	5094	13	46	7628	7635	7642	7648
5123	5130	5138	12	47	7669	7676	7683	7690
5167	5174	5182	11	48	7711	7718	7725	7732
5211	5218	5226	10	49	7752	7759	7766	7773
5255	5262	5270	9	50	9.977794	7801	7808	7815
5299	5306	5313	8	51	7835	7842	7849	7856
5343	5350	5357	7	52	7877	7884	7890	7897
5386	5394	5401	6	53	7918	7925	7932	7939
5430	5437	5445	5	54	7959	7966	7973	7980
5474	5481	5488	4	55	8001	8007	8014	8021
5518	5525	5532	3	56	8042	8049	8056	8062
5561	5568	5576	2	57	8083	8090	8097	8104
5605	5612	5619	1	58	8124	8131	8138	8145
5648	5656	5663	0	59	8165	8172	8179	8186

gent of 70 Degrees.						Min.	Tangent of 71		
10″	20″	30″	40″	50″			0″	10″	20″
9000	9065	9131	9196	9262	59	0	10.463028	3097	3165
9393	9458	9524	9590	9655	58	1	3439	3507	3576
9786	9852	9918	9983	..49	57	2	3850	3918	3987
0180	0246	0312	0377	0443	56	3	4261	4329	4398
0574	0640	0706	0771	0837	55	4	4672	4741	4809
0969	1034	1100	1166	1232	54	5	5084	5153	5221
1363	1429	1495	1561	1627	53	6	5496	5565	5633
1758	1824	1890	1956	2022	52	7	5908	5977	6046
2153	2219	2285	2351	2417	51	8	6321	6390	6459
2549	2615	2681	2747	2813	50	9	6734	6803	6872
2945	3011	3077	3143	3209	49	10	10.467147	7216	7285
3341	3407	3473	3539	3605	48	11	7561	7630	7699
3737	3803	3869	3935	4001	47	12	7975	8044	8113
4133	4200	4266	4332	4398	46	13	8389	8458	8527
4530	4596	4663	4729	4795	45	14	8804	8873	8942
4927	4994	5060	5126	5192	44	15	9219	9288	9357
5325	5391	5457	5524	5590	43	16	9634	9703	9772
5722	5789	5855	5921	5988	42	17	10.470049	0119	0188
6120	6187	6253	6320	6386	41	18	0465	0535	0604
6519	6585	6652	6718	6784	40	19	0881	0951	1020
6917	6984	7050	7117	7183	39	20	10.471298	1367	1437
7316	7383	7449	7516	7582	38	21	1715	1784	1854
7715	7782	7848	7915	7981	37	22	2132	2201	2271
8115	8181	8248	8314	8381	36	23	2549	2619	2688
8514	8581	8647	8714	8781	35	24	2967	3037	3106
8914	8981	9048	9114	9181	34	25	3385	3455	3524
9314	9381	9448	9515	9581	33	26	3803	3873	3943
9715	9782	9848	9915	9982	32	27	4222	4292	4362
0116	0183	0249	0316	0383	31	28	4641	4711	4781
0517	0584	0651	0717	0784	30	29	5060	5130	5200
0918	0985	1052	1119	1186	29	30	10.475480	5550	5620
1320	1387	1454	1521	1588	28	31	5900	5970	6040
1722	1789	1856	1923	1990	27	32	6320	6391	6461
2124	2191	2258	2325	2392	26	33	6741	6811	6881
2527	2594	2661	2728	2795	25	34	7162	7232	7302
2929	2997	3064	3131	3198	24	35	7583	7654	7724
3333	3400	3467	3534	3602	23	36	8005	8075	8146
3736	3803	3871	3938	4005	22	37	8427	8497	8568
4140	4207	4274	4342	4409	21	38	8849	8920	8990
4544	4611	4678	4746	4813	20	39	9272	9342	9413
4948	5015	5083	5150	5218	19	40	10.479695	9765	9836
5353	5420	5488	5555	5623	18	41	10.480118	0189	0259
5758	5825	5893	5969	6028	17	42	0542	0612	0683
6163	6230	6298	6365	6433	16	43	0966	1036	1107
6568	6636	6703	6771	6839	15	44	1390	1461	1531
6974	7042	7109	7177	7245	14	45	1814	1885	1956
7380	7448	7515	7583	7651	13	46	2239	2310	2381
7786	7854	7922	7990	8057	12	47	2665	2736	2807
8193	8261	8329	8397	8464	11	48	3090	3161	3232
8600	8668	8736	8804	8872	10	49	3516	3587	3658
9007	9075	9143	9211	9279	9	50	10.483943	4014	4085
9415	9483	9551	9619	9687	8	51	4369	4440	4511
9823	9891	9959	..27	..95	7	52	4796	4867	4938
0231	0299	0367	0435	0503	6	53	5223	5295	5366
0639	0707	0776	0844	0912	5	54	5651	5722	5794
1048	1116	1184	1253	1321	4	55	6079	6150	6222
1457	1525	1594	1662	1730	3	56	6507	6579	6650
1867	1935	2003	2071	2140	2	57	6936	7007	7079
2276	2345	2413	2481	2550	1	58	7365	7436	7508
2686	2755	2823	2891	2960	0	59	7794	7866	7937

60″	50″	40″	30″	20″	10″	Min.
.978206	8213	8220	8227	8234	8241	59
8247	8254	8261	8268	8275	8282	58
8288	8295	8302	8309	8316	8322	57
8329	8336	8343	8350	8357	8363	56
8370	8377	8384	8391	8397	8404	55
8411	8418	8425	8431	8438	8445	54
8452	8459	8465	8472	8479	8486	53
8493	8499	8506	8513	8520	8527	52
8533	8540	8547	8554	8561	8567	51
8574	8581	8588	8594	8601	8608	50
.978615	8622	8628	8635	8642	8649	49
8655	8662	8669	8676	8682	8689	48
8696	8703	8709	8716	8723	8730	47
8737	8743	8750	8757	8764	8770	46
8777	8784	8791	8797	8804	8811	45
8817	8824	8831	8838	8844	8851	44
8858	8865	8871	8878	8885	8892	43
8898	8905	8912	8918	8925	8932	42
8939	8945	8952	8959	8965	8972	41
8979	8986	8992	8999	9006	9012	40
.979019	9026	9033	9039	9046	9053	39
9059	9066	9073	9079	9086	9093	38
9100	9106	9113	9120	9126	9133	37
9140	9146	9153	9160	9166	9173	36
9180	9186	9193	9200	9206	9213	35
9220	9227	9233	9240	9247	9253	34
9260	9267	9273	9280	9287	9293	33
9300	9306	9313	9320	9326	9333	32
9340	9346	9353	9360	9366	9373	31
9380	9386	9393	9400	9406	9413	30
.979420	9426	9433	9439	9446	9453	29
9459	9466	9473	9479	9486	9492	28
9499	9506	9512	9519	9526	9532	27
9539	9545	9552	9559	9565	9572	26
9579	9585	9592	9598	9605	9612	25
9618	9625	9631	9638	9645	9651	24
9658	9664	9671	9678	9684	9691	23
9697	9704	9711	9717	9724	9730	22
9737	9743	9750	9757	9763	9770	21
9776	9783	9790	9796	9803	9809	20
.979816	9822	9829	9836	9842	9849	19
9855	9862	9868	9875	9881	9888	18
9895	9901	9908	9914	9921	9927	17
9934	9940	9947	9954	9960	9967	16
9973	9980	9986	9993	9999	...6	15
.980012	0019	0026	0032	0039	0045	14
0052	0058	0065	0071	0078	0084	13
0091	0097	0104	0110	0117	0123	12
0130	0136	0143	0149	0156	0163	11
0169	0176	0182	0189	0195	0202	10
.980208	0215	0221	0228	0234	0241	9
0247	0254	0260	0267	0273	0280	8
0286	0293	0299	0306	0312	0318	7
0325	0331	0338	0344	0351	0357	6
0364	0370	0377	0383	0390	0396	5
0403	0409	0416	0422	0429	0435	4
0442	0448	0454	0461	0467	0474	3
0480	0487	0493	0500	0506	0513	2
0519	0525	0532	0538	0545	0551	1
0558	0564	0571	0577	0583	0590	0

Co-sine of 17 Degrees.

Min.	60″	50″	40″	30″	20″
0	9.980596	0603	0609	0616	0622
1	0635	0641	0648	0654	0661
2	0673	0680	0686	0693	0699
3	0712	0718	0725	0731	0738
4	0750	0757	0763	0770	0776
5	0789	0795	0802	0808	0815
6	0827	0834	0840	0847	0853
7	0866	0872	0878	0885	0891
8	0904	0910	0917	0923	0930
9	0942	0949	0955	0961	0968
10	9.980981	0987	0993	1000	1006
11	1019	1025	1031	1038	1044
12	1057	1063	1070	1076	1082
13	1095	1101	1108	1114	1120
14	1133	1139	1146	1152	1158
15	1171	1177	1184	1190	1196
16	1209	1215	1222	1228	1234
17	1247	1253	1260	1266	1272
18	1285	1291	1298	1304	1310
19	1323	1329	1336	1342	1348
20	9.981361	1367	1373	1380	1386
21	1399	1405	1411	1417	1424
22	1436	1443	1449	1455	1461
23	1474	1480	1487	1493	1499
24	1512	1518	1524	1531	1537
25	1549	1556	1562	1568	1574
26	1587	1593	1599	1606	1612
27	1625	1631	1637	1643	1650
28	1662	1668	1675	1681	1687
29	1700	1706	1712	1718	1724
30	9 981737	1743	1749	1756	1762
31	1774	1781	1787	1793	1799
32	18:2	1818	1824	1830	1837
33	1849	1855	1861	1868	1874
34	1886	1893	1899	1905	1911
35	1924	1930	1936	1942	1948
36	1961	1967	1973	1979	1986
37	1998	2004	2010	2016	2023
38	2035	2041	2047	2054	2060
39	2072	2078	2084	2091	2097
40	9.982109	2115	2122	2128	2134
41	2146	2152	2159	2165	2171
42	2183	2189	2195	2202	2208
43	2220	2226	2232	2239	2245
44	2257	2263	2269	2275	2282
45	2294	2300	2306	2312	2318
46	2331	2337	2343	2349	2355
47	2367	2373	2380	2386	2392
48	2404	2410	2416	2422	2429
49	2441	2447	2453	2459	2465
50	9.982477	2484	2490	2496	2502
51	2514	2520	2526	2532	2538
52	2551	2557	2563	2569	2575
53	2587	2593	2599	2605	2611
54	2624	2630	2636	2642	2648
55	2660	2666	2672	2678	2684
56	2696	2702	2709	2715	2721
57	2733	2739	2745	2751	2757
58	2769	2775	2781	2787	2793
59	2805	2811	2817	2824	2830

Co-sine of 16 Degrees.

Tangent of 72 Degrees.							Min.	Tangent of 73 Deg			
0″	10″	20″	30″	40″	50″			0″	10″	20″	30″
10.488224	8296	8367	8439	8511	8582	59	0	10.514661	4736	4812	4887
8654	8726	8797	8869	8941	9013	58	1	5113	5188	5264	5339
9084	9156	9228	9300	9371	9443	57	2	5565	5641	5716	5792
9515	9587	9659	9731	9802	9874	56	3	6018	6094	6169	6245
9946	..18	..90	.162	.234	.306	55	4	6471	6547	6623	6698
10.490378	0449	0521	0593	0665	0737	54	5	6925	7001	7076	7152
0809	0881	0953	1025	1097	1169	53	6	7379	7455	7530	7606
1241	1313	1386	1458	1530	1602	52	7	7833	7909	7985	8061
1674	1746	1818	1890	1962	2035	51	8	8288	8364	8440	8516
2107	2179	2251	2323	2395	2468	50	9	8743	8819	8895	8971
10.492540	2612	2684	2757	2829	2901	49	10	10.519199	9275	9351	9427
2973	3046	3118	3190	3263	3335	48	11	9655	9731	9807	9883
3407	3480	3552	3624	3697	3769	47	12	10.520111	0187	0263	0340
3841	3914	3986	4059	4131	4204	46	13	0568	0644	0720	0797
4276	4348	4421	4493	4566	4638	45	14	1025	1101	1178	1254
4711	4783	4856	4928	5001	5074	44	15	1483	1559	1635	1712
5146	5219	5291	5364	5437	5509	43	16	1941	2017	2094	2170
5582	5654	5727	5800	5872	5945	42	17	2399	2476	2552	2628
6018	6090	6163	6236	6309	6381	41	18	2858	2934	3011	3087
6454	6527	6600	6672	6745	6818	40	19	3317	3394	3470	3547
10.496891	6964	7036	7109	7182	7255	39	20	10.523777	3853	3930	4007
7328	7401	7474	7547	7619	6792	38	21	4237	4313	4390	4467
7765	7838	7911	7984	8057	8130	37	22	4697	4774	4851	4927
8203	8276	8349	8422	8495	8568	36	23	5158	5235	5312	5388
8641	8714	8787	8860	8934	9007	35	24	5619	5696	5773	5850
9080	9153	9226	9299	9372	9445	34	25	6081	6158	6235	6312
9519	9592	9665	9738	9811	9885	33	26	6543	6620	6697	6774
9958	..31	.104	.178	.251	.324	32	27	7005	7082	7160	7237
10.500397	0471	0544	0617	0691	0764	31	28	7468	7545	7623	7700
0837	0911	0984	1057	1131	1204	30	29	7931	8009	8086	8163
10.501278	1351	1425	1498	1571	1645	29	30	10.528395	8472	8550	8627
1718	1792	1865	1939	2012	2086	28	31	8859	8937	9014	9091
2159	2233	2307	2380	2454	2527	27	32	9324	9401	9479	9556
2601	2674	2748	2822	2895	2969	26	33	9789	9866	9944	..21
3043	3116	3190	3264	3337	3411	25	34	10.530254	0332	0409	0487
3485	3559	3632	3706	3780	3854	24	35	0720	0798	0875	0953
3927	4001	4075	4149	4223	4296	23	36	1186	1264	1342	1419
4370	4444	4518	4592	4666	4740	22	37	1653	1731	1808	1886
4814	4887	4961	5035	5109	5183	21	38	2120	2198	2276	2353
5257	5331	5405	5479	5553	5627	20	39	2587	2665	2743	2821
10.505701	5775	5849	5923	5997	6071	19	40	10.533055	3133	3211	3289
6146	6220	6294	6368	6442	6516	18	41	3523	3602	3680	3758
6590	6664	6739	6813	6887	6961	17	42	3992	4070	4149	4227
7035	7110	7184	7258	7332	7407	16	43	4461	4540	4618	4696
7481	7555	7630	7704	7778	7853	15	44	4931	5009	5088	5166
7927	8001	8076	8150	8224	8299	14	45	5401	5479	5558	5636
8373	8448	8522	8596	8671	8745	13	46	5872	5950	6028	6107
8820	8894	8969	9043	9118	9192	12	47	6342	6421	6499	6578
9267	9341	9416	9490	9565	9640	11	48	6814	6892	6971	7050
9714	9789	9863	9938	..13	..87	10	49	7285	7364	7443	7522
10.510162	0237	0311	0386	0461	0535	9	50	10.537758	7836	7915	7994
0610	0685	0760	0834	0909	0984	8	51	8230	8309	8388	8467
1059	1134	1208	1283	1358	1433	7	52	8703	8782	8861	8940
1508	1582	1657	1732	1807	1882	6	53	9177	9256	9335	9414
1957	2032	2107	2182	2257	2332	5	54	9651	9730	9809	9888
2407	2482	2557	2632	2707	2782	4	55	10.540125	0204	0283	0362
2857	2932	3007	3082	3157	3232	3	56	0600	0679	0758	0837
3307	3382	3457	3533	3608	3683	2	57	1075	1154	1234	1313
3758	3833	3908	3984	4059	4134	1	58	1551	1630	1710	1789
4209	4285	4360	4435	4510	4586	0	59	2027	2106	2186	2265
60″	50″	40″	30″	20″	10″	Min.		60″	50″	40″	30″
Co-tangent of 17 Degrees.								Co-tangent of 16 De			

Sine of 74 Degrees.							Min.	Sine of 75 Degrees.						
0"	10"	20"	30"	40'	50'			0"	10"	20"	30"	40"	50"	
9.982842	2848	2854	2860	2866	2872	59	0	9.984944	4949	4955	4961	4966	4972	59
2878	2884	2890	2896	2902	2908	58	1	4978	4983	4989	4995	5000	5006	58
2914	2920	2926	2932	2938	2944	57	2	5011	5017	5023	5028	5034	5040	57
2950	2956	2962	2968	2974	2980	56	3	5045	5051	5056	5062	5068	5073	56
2986	2992	2998	3004	3010	3016	55	4	5079	5084	5090	5096	5101	5107	55
3022	3028	3034	3040	3046	3052	54	5	5113	5118	5124	5129	5135	5141	54
3058	3064	3070	3076	3082	3088	53	6	5146	5152	5157	5163	5169	5174	53
3094	3100	3106	3112	3118	3124	52	7	5180	5185	5191	5197	5202	5208	52
3130	3136	3142	3148	3154	3160	51	8	5213	5219	5224	5230	5236	5241	51
3166	3172	3178	3184	3190	3196	50	9	5247	5252	5258	5264	5269	5275	50
9.983202	3208	3214	3220	3226	3232	49	10	9.985280	5286	5291	5297	5303	5308	49
3238	3244	3250	3256	3262	3268	48	11	5314	5319	5325	5330	5336	5342	48
3273	3279	3285	3291	3297	3303	47	12	5347	5353	5358	5364	5369	5375	47
3309	3315	3321	3327	3333	3339	46	13	5381	5386	5392	5397	5403	5408	46
3345	3351	3357	3363	3369	3375	45	14	5414	5419	5425	5430	5436	5442	45
3381	3386	3392	3398	3404	3410	44	15	5447	5453	5458	5464	5469	5475	44
3416	3422	3428	3434	3440	3446	43	16	5480	5486	5491	5497	5502	5508	43
3452	3458	3464	3469	3475	3481	42	17	5514	5519	5525	5530	5536	5541	42
3487	3493	3499	3505	3511	3517	41	18	5547	5552	5558	5563	5569	5574	41
3523	3529	3535	3540	3546	3552	40	19	5580	5585	5591	5596	5602	5607	40
9.983558	3564	3570	3576	3582	3588	39	20	9.985613	5618	5624	5629	5635	5640	39
3594	3599	3605	3611	3617	3623	38	21	5646	5651	5657	5662	5668	5673	38
3629	3635	3641	3647	3653	3658	37	22	5679	5684	5690	5695	5701	5706	37
3664	3670	3676	3682	3688	3694	36	23	5712	5717	5723	5728	5734	5739	36
3700	3705	3711	3717	3723	3729	35	24	5745	5750	5756	5761	5767	5772	35
3735	3741	3747	3752	3758	3764	34	25	5778	5783	5789	5794	5800	5805	34
3770	3776	3782	3788	3794	3799	33	26	5811	5816	5822	5827	5832	5838	33
3805	3811	3817	3823	3829	3835	32	27	5843	5849	5854	5860	5865	5871	32
3840	3846	3852	3858	3864	3870	31	28	5876	5882	5887	5893	5898	5903	31
3875	3881	3887	3893	3899	3905	30	29	5909	5914	5920	5925	5931	5936	30
9.983911	3916	3922	3928	3934	3940	29	30	9.985942	5947	5952	5958	5963	5969	29
3946	3951	3957	3963	3969	3975	28	31	5974	5980	5985	5991	5996	6001	28
3981	3986	3992	3998	4004	4010	27	32	6007	6012	6018	6023	6029	6034	27
4015	4021	4027	4033	4039	4045	26	33	6039	6045	6050	6056	6061	6067	26
4050	4056	4062	4068	4074	4079	25	34	6072	6077	6083	6088	6094	6099	25
4085	4091	4097	4103	4108	4114	24	35	6104	6110	6115	6121	6126	6132	24
4120	4126	4132	4137	4143	4149	23	36	6137	6142	6148	6153	6159	6164	23
4155	4161	4166	4172	4178	4184	22	37	6169	6175	6180	6186	6191	6196	22
4190	4195	4201	4207	4213	4218	21	38	6202	6207	6212	6218	6223	6229	21
4224	4230	4236	4242	4247	4253	20	39	6234	6239	6245	6250	6256	6261	20
9.984259	4265	4270	4276	4282	4288	19	40	9.986266	6272	6277	6282	6288	6293	19
4294	4299	4305	4311	4317	4322	18	41	6299	6304	6309	6315	6320	6325	18
4328	4334	4340	4345	4351	4357	17	42	6331	6336	6342	6347	6352	6358	17
4363	4368	4374	4380	4386	4391	16	43	6363	6368	6374	6379	6384	6390	16
4397	4403	4409	4414	4420	4426	15	44	6395	6401	6406	6411	6417	6422	15
4432	4437	4443	4449	4455	4460	14	45	6427	6433	6438	6443	6449	6454	14
4466	4472	4477	4483	4489	4495	13	46	6459	6465	6470	6475	6481	6486	13
4500	4506	4512	4518	4523	4529	12	47	6491	6497	6502	6507	6513	6518	12
4535	4540	4546	4552	4558	4563	11	48	6523	6529	6534	6539	6545	6550	11
4569	4575	4580	4586	4592	4598	10	49	6555	6561	6566	6571	6577	6582	10
9.984603	4609	4615	4620	4626	4632	9	50	9.986587	6593	6598	6603	6608	6614	9
4638	4643	4649	4655	4660	4666	8	51	6619	6624	6630	6635	6640	6646	8
4672	4677	4683	4689	4694	4700	7	52	6651	6656	6661	6667	6672	6677	7
4706	4712	4717	4723	4729	4734	6	53	6683	6688	6693	6699	6704	6709	6
4740	4746	4751	4757	4763	4768	5	54	6714	6720	6725	6730	6736	6741	5

Min.	Tangent of 74 Degrees.							Min.	Tangent of 75 Degrees.						
	0″	10″	20″	30″	40″	50″			0″	10″	20″	30″	40″	50″	
0	10.542504	2583	2663	2742	2822	2901	59	0	10.571948	2032	2116	2200	2285	2369	59
1	2981	3060	3140	3219	3299	3378	58	1	2453	2537	2622	2706	2790	2875	58
2	3458	3538	3617	3697	3777	3856	57	2	2959	3044	3128	3212	3297	3381	57
3	3936	4016	4095	4175	4255	4335	56	3	3466	3550	3635	3719	3804	3888	56
4	4414	4494	4574	4654	4733	4813	55	4	3973	4058	4142	4227	4311	4396	55
5	4893	4973	5053	5133	5213	5292	54	5	4481	4565	4650	4735	4819	4904	54
6	5372	5452	5532	5612	5692	5772	53	6	4989	5073	5158	5243	5328	5413	53
7	5852	5932	6012	6092	6172	6252	52	7	5497	5582	5667	5752	5837	5922	52
8	6332	6412	6492	6572	6653	6733	51	8	6007	6091	6176	6261	6346	6431	51
9	6813	6893	6973	7053	7133	7214	50	9	6516	6601	6686	6771	6856	6941	50
10	10.547294	7374	7454	7535	7615	7695	49	10	10.577026	7112	7197	7282	7367	7452	49
11	7775	7856	7936	8016	8097	8177	48	11	7537	7622	7708	7793	7878	7963	48
12	8257	8338	8418	8498	8579	8659	47	12	8048	8134	8219	8304	8390	8475	47
13	8740	8820	8901	8981	9062	9142	46	13	8560	8646	8731	8816	8902	8987	46
14	9223	9303	9384	9464	9545	9625	45	14	9073	9158	9243	9329	9414	9500	45
15	9706	9787	9867	9948	..28	.109	44	15	9585	9671	9756	9842	9928	..13	44
16	10.550190	0270	0351	0432	0513	0593	43	16	10.580099	0184	0270	0356	0441	0527	43
17	0674	0755	0836	0916	0997	1078	42	17	0613	0698	0784	0870	0956	1041	42
18	1159	1240	1320	1401	1482	1563	41	18	1127	1213	1299	1384	1470	1556	41
19	1644	1725	1806	1887	1968	2049	40	19	1642	1728	1814	1900	1986	2072	40
20	10.552130	2211	2292	2373	2454	2535	39	20	10.582158	2243	2329	2415	2501	2587	39
21	2616	2697	2778	2859	2940	3021	38	21	2674	2760	2846	2932	3018	3104	38
22	3102	3183	3265	3346	3427	3508	37	22	3190	3276	3362	3449	3535	3621	37
23	3589	3670	3752	3833	3914	3995	36	23	3707	3793	3880	3966	4052	4138	36
24	4077	4158	4239	4321	4402	4483	35	24	4225	4311	4397	4484	4570	4657	35
25	4565	4646	4728	4809	4890	4972	34	25	4743	4829	4916	5002	5089	5175	34
26	5053	5135	5216	5298	5379	5461	33	26	5262	5348	5435	5521	5608	5694	33
27	5542	5624	5705	5787	5868	5950	32	27	5781	5868	5954	6041	6127	6214	32
28	6032	6113	6195	6276	6358	6440	31	28	6301	6387	6474	6561	6648	6734	31
29	6521	6603	6685	6766	6848	6930	30	29	6821	6908	6995	7081	7168	7255	30
30	10.557012	7093	7175	7257	7339	7421	29	30	10.587342	7429	7516	7603	7690	7776	29
31	7503	7584	7666	7748	7830	7912	28	31	7863	7950	8037	8124	8211	8298	28
32	7994	8076	8158	8240	8322	8404	27	32	8385	8472	8559	8647	8734	8821	27
33	8486	8568	8650	8732	8814	8896	26	33	8908	8995	9082	9169	9257	9344	26
34	8978	9060	9142	9224	9306	9388	25	34	9431	9518	9605	9693	9780	9867	25
35	9471	9553	9635	9717	9799	9881	24	35	9955	..42	.129	.217	.304	.391	24
36	9964	..46	.128	.210	.293	.375	23	36	10.590479	0566	0654	0741	0829	0916	23
37	10.560457	0540	0622	0704	0787	0869	22	37	1004	1091	1179	1266	1354	1441	22
38	0952	1034	1116	1199	1281	1364	21	38	1529	1616	1704	1792	1879	1967	21
39	1446	1529	1611	1694	1776	1859	20	39	2055	2142	2230	2318	2406	2493	20
40	10.561941	2024	2106	2189	2272	2354	19	40	10.592581	2669	2757	2845	2932	3020	19
41	2437	2520	2602	2685	2768	2850	18	41	3108	3196	3284	3372	3460	3548	18
42	2933	3016	3099	3181	3264	3347	17	42	3636	3724	3812	3900	3988	4076	17
43	3430	3512	3595	3678	3761	3844	16	43	4164	4252	4340	4428	4516	4604	16
44	3927	4010	4093	4175	4258	4341	15	44	4692	4781	4869	4957	5045	5133	15
45	4424	4507	4590	4673	4756	4839	14	45	5222	5310	5398	5486	5575	5663	14
46	4922	5005	5088	5172	5255	5338	13	46	5751	5840	5928	6017	6105	6193	13
47	5421	5504	5587	5670	5754	5837	12	47	6282	6370	6459	6547	6636	6724	12
48	5920	6003	6086	6170	6253	6336	11	48	6813	6901	6990	7078	7167	7256	11
49	6420	6503	6586	6669	6753	6836	10	49	7344	7433	7522	7610	7699	7788	10
50	10.566920	7003	7086	7170	7253	7337	9	50	10.597876	7965	8054	8143	8231	8320	9
51	7420	7504	7587	7671	7754	7838	8	51	8409	8498	8587	8675	8764	8853	8
52	7921	8005	8088	8172	8255	8339	7	52	8942	9031	9120	9209	9298	9387	7

…Degrees. 30″	40″	50″		Min.	Sine of 77 Degrees. 0″	10″	20″	30″	40″	50″	
6920	6925	6930	59	0	9.988724	8729	8734	8739	8743	8748	59
6951	6957	6962	58	1	8753	8758	8763	8768	8772	8777	58
6983	6988	6993	57	2	8782	8787	8792	8797	8802	8806	57
7014	7019	7025	56	3	8811	8816	8821	8826	8831	8835	56
7045	7051	7056	55	4	8840	8845	8850	8855	8860	8864	55
7077	7082	7087	54	5	8869	8874	8879	8884	8889	8893	54
7108	7113	7118	53	6	8898	8903	8908	8913	8918	8922	53
7139	7144	7150	52	7	8927	8932	8937	8942	8946	8951	52
7170	7176	7181	51	8	8956	8961	8966	8970	8975	8980	51
7202	7207	7212	50	9	8985	8990	8994	8999	9004	9009	50
7233	7238	7243	49	10	9.989014	9018	9023	9028	9033	9038	49
7264	7269	7274	48	11	9042	9047	9052	9057	9062	9066	48
7295	7300	7305	47	12	9071	9076	9081	9085	9090	9095	47
7326	7331	7336	46	13	9100	9105	9109	9114	9119	9124	46
7357	7362	7367	45	14	9128	9133	9138	9143	9148	9152	45
7388	7393	7398	44	15	9157	9162	9167	9171	9176	9181	44
7419	7424	7429	43	16	9186	9190	9195	9200	9205	9209	43
7449	7454	7460	42	17	9214	9219	9224	9228	9233	9238	42
7480	7485	7490	41	18	9243	9247	9252	9257	9262	9266	41
7511	7516	7521	40	19	9271	9276	9281	9285	9290	9295	40
7542	7547	7552	39	20	9.989300	9304	9309	9314	9318	9323	39
7572	7577	7583	38	21	9328	9333	9337	9342	9347	9352	38
7603	7608	7613	37	22	9356	9361	9366	9370	9375	9380	37
7634	7639	7644	36	23	9385	9389	9394	9399	9403	9408	36
7664	7669	7674	35	24	9413	9417	9422	9427	9432	9436	35
7695	7700	7705	34	25	9441	9446	9450	9455	9460	9464	34
7725	7730	7735	33	26	9469	9474	9479	9483	9488	9493	33
7756	7761	7766	32	27	9497	9502	9507	9511	9516	9521	32
7786	7791	7796	31	28	9525	9530	9535	9539	9544	9549	31
7816	7821	7826	30	29	9553	9558	9563	9568	9572	9577	30
7847	7852	7857	29	30	9.989582	9586	9591	9596	9600	9605	29
7877	7882	7887	28	31	9610	9614	9619	9623	9628	9633	28
7907	7912	7917	27	32	9637	9642	9647	9651	9656	9661	27
7937	7942	7947	26	33	9665	9670	9675	9679	9684	9689	26
7968	7973	7978	25	34	9693	9698	9703	9707	9712	9716	25
7998	8003	8008	24	35	9721	9726	9730	9735	9740	9744	24
8028	8033	8038	23	36	9749	9753	9758	9763	9767	9772	23
8058	8063	8068	22	37	9777	9781	9786	9790	9795	9800	22
8088	8093	8098	21	38	9804	9809	9814	9818	9823	9827	21
8118	8123	8128	20	39	9832	9837	9841	9846	9850	9855	20
8148	8153	8158	19	40	9.989860	9864	9869	9873	9878	9883	19
8178	8183	8188	18	41	9887	9892	9896	9901	9906	9910	18
8208	8213	8218	17	42	9915	9919	9924	9929	9933	9938	17
8237	8242	8247	16	43	9942	9947	9952	9956	9961	9965	16
8267	8272	8277	15	44	9970	9974	9979	9984	9988	9993	15
8297	8302	8307	14	45	9997	...2	...6	..11	..16	..20	14
8327	8332	8337	13	46	9.990025	0029	0034	0038	0043	0048	13
8356	8361	8366	12	47	0052	0057	0061	0066	0070	0075	12
8386	8391	8396	11	48	0079	0084	0088	0093	0098	0102	11
8416	8420	8425	10	49	0107	0111	0116	0120	0125	0129	10
8445	8450	8455	9	50	9.990134	0138	0143	0148	0152	0157	9
8475	8480	8484	8	51	0161	0166	0170	0175	0179	0184	8
8504	8509	8514	7	52	0188	0193	0197	0202	0206	0211	7
8534	8538	8543	6	53	0215	0220	0225	0229	0234	0238	6
8563	8568	8573	5	54	0243	0247	0252	0256	0261	0265	5
8592	8597	8602	4	55	0270	0274	0279	0283	0288	0292	4
8622	8626	8631	3	56	0297	0301	0306	0310	0315	0319	3
8651	8656	8661	2	57	0324	0328	0333	0337	0342	0346	2
8680	8685	8690	1	58	0351	0355	0360	0364	0369	0373	1
8709	8714	8719	0	59	0378	0382	0386	0391	0395	0400	0

ɣent of 76 Degrees.						Min.	Tangent of 77 Degrees.						
10″	20″	30″	40″	50″			0″	10″	20″	30″	40″	50″	
3319	3408	3498	3588	3678	59	0	10.636636	6732	6828	6924	7020	7116	59
3857	3947	4037	4127	4217	58	1	7213	7309	7405	7501	7597	7694	58
4396	4486	4576	4666	4756	57	2	7790	7886	7983	8079	8175	8272	57
4936	5026	5116	5206	5296	56	3	8368	8465	8561	8657	8754	8850	56
5477	5567	5657	5747	5837	55	4	8947	9043	9140	9237	9333	9430	55
6017	6108	6198	6288	6378	54	5	9526	9623	9720	9816	9913	..10	54
6559	6649	6740	6830	6920	53	6	10.640107	0203	0300	0397	0494	0591	53
7101	7192	7282	7372	7463	52	7	0687	0784	0881	0978	1075	1172	52
7644	7734	7825	7915	8006	51	8	1269	1366	1463	1560	1657	1754	51
8187	8278	8368	8459	8550	50	9	1851	1948	2046	2143	2240	2337	50
8731	8822	8913	9003	9094	49	10	10.642434	2531	2629	2726	2823	2921	49
9276	9367	9457	9548	9639	48	11	3018	3115	3213	3310	3407	3505	48
9821	9912	...3	..94	.185	47	12	3602	3700	3797	3895	3992	4090	47
0367	0458	0549	0640	0731	46	13	4187	4285	4383	4480	4578	4676	46
0913	1004	1095	1186	1278	45	14	4773	4871	4969	5066	5164	5262	45
1460	1551	1642	1734	1825	44	15	5360	5458	5555	5653	5751	5849	44
2008	2099	2190	2282	2373	43	16	5947	6045	6143	6241	6339	6437	43
2556	2647	2739	2830	2922	42	17	6535	6633	6731	6829	6927	7026	42
3105	3196	3288	3379	3471	41	18	7124	7222	7320	7418	7517	7615	41
3654	3746	3837	3929	4021	40	19	7713	7811	7910	8008	8106	8205	40
4204	4296	4388	4479	4571	39	20	10.648303	8402	8500	8599	8697	8796	39
4755	4847	4938	5030	5122	38	21	8894	8993	9091	9190	9288	9387	38
5306	5398	5490	5582	5674	37	22	9486	9584	9683	9782	9880	9979	37
5858	5950	6042	6134	6226	36	23	10.650078	0177	0276	0374	0473	0572	36
6411	6503	6595	6687	6779	35	24	0671	0770	0869	0968	1067	1166	35
6964	7056	7148	7241	7333	34	25	1265	1364	1463	1562	1661	1760	34
7518	7610	7702	7795	7887	33	26	1859	1958	2058	2157	2256	2355	33
8072	8164	8257	8349	8442	32	27	4455	2554	2653	2752	2852	2951	32
8627	8720	8812	8905	8997	31	28	3051	3150	3249	3349	3448	3548	31
9183	9275	9368	9461	9554	30	29	3647	3747	3846	3946	4046	4145	30
9739	9832	9925	..17	.110	29	30	10.654245	4344	4444	4544	4643	4743	29
0296	0389	0482	0575	0668	28	31	4843	4943	5043	5142	5242	5342	28
0854	0947	1040	1133	1226	27	32	5442	5542	5642	5742	5842	5942	27
1412	1505	1598	1691	1784	26	33	6042	6142	6242	6342	6442	6542	26
1971	2064	2157	2250	2344	25	34	6642	6742	6842	6943	7043	7143	25
2530	2624	2717	2810	2904	24	35	7243	7344	7444	7544	7645	7745	24
3090	3184	3277	3371	3464	23	36	7845	7946	8046	8147	8247	8348	23
3651	3745	3838	3932	4025	22	37	8448	8549	8649	8750	8850	8951	22
4213	4306	4400	4494	4587	21	38	9052	9152	9253	9354	9454	9555	21
4775	4869	4962	5056	5150	20	39	9656	9757	9857	9958	..59	.160	20
5338	5431	5525	5619	5713	19	40	10.660261	0362	0463	0564	0665	0766	19
5901	5995	6089	6183	6277	18	41	0867	0968	1069	1170	1271	1372	18
6465	6559	6653	6747	6841	17	42	1473	1574	1676	1777	1878	1979	17
7030	7124	7218	7312	7407	16	43	2081	2182	2283	2385	2486	2587	16
7595	7689	7784	7878	7972	15	44	2689	2790	2892	2993	3095	3196	15
8161	8256	8350	8444	8539	14	45	3298	3399	3501	3602	3704	3806	14
8728	8822	8917	9011	9106	13	46	3907	4009	4111	4212	4314	4416	13
9295	9390	9484	9579	9674	12	47	4518	4620	4721	4823	4925	5027	12
9863	9958	..53	.147	.242	11	48	5129	5231	5333	5435	5537	5639	11
0432	0527	0622	0716	0811	10	49	5741	5843	5945	6047	6149	6252	10
1001	1096	1191	1286	1381	9	50	10.666354	6456	6558	6660	6763	6865	9
1571	1666	1761	1857	1952	8	51	6967	7070	7172	7274	7377	7479	8
2142	2237	2332	2428	2523	7	52	7582	7684	7787	7889	7992	8094	7
2713	2809	2904	2999	3095	6	53	8197	8299	8402	8505	8607	8710	6
3285	3381	3476	3572	3667	5	54	8813	8916	9018	9121	9224	9327	5
3858	3954	4049	4145	4240	4	55	9430	9532	9635	9738	9841	9944	4
4432	4527	4623	4718	4814	3	56	10.670047	0150	0253	0356	0459	0562	3
5006	5101	5197	5293	5389	2	57	0666	0769	0872	0975	1078	1181	2
5580	5676	5772	5868	5954	1	58	1285	1388	1491	1595	1698	1801	1
6156	6252	6348	6444	6540	0	59	1905	2008	2112	2215	2318	2422	0

Sine of 78 Degrees.							Min.	Sine of 79		
0″	10″	20″	30″	40″	50″			0″	10″	20″
990404	0409	0413	0418	0422	0427	59	0	9.991947	1951	1955
0431	0436	0440	0445	0449	0454	58	1	1971	1975	1979
0458	0462	0467	0471	0476	0480	57	2	1996	2000	2004
0485	0489	0494	0498	0503	0507	56	3	2020	2024	2028
0511	0516	0520	0525	0529	0534	55	4	2044	2049	2053
0538	0543	0547	0552	0556	0560	54	5	2069	2073	2077
0565	0569	0574	0578	0583	0587	53	6	2093	2097	2101
0591	0596	0600	0605	0609	0614	52	7	2118	2122	2126
0618	0622	0627	0631	0636	0640	51	8	2142	2146	2150
0645	0649	0653	0658	0662	0667	50	9	2166	2170	2174
990671	0675	0680	0684	0689	0693	49	10	9.992190	2194	2198
0697	0702	0706	0711	0715	0719	48	11	2214	2218	2222
0724	0728	0733	0737	0741	0746	47	12	2239	2243	2247
0750	0755	0759	0763	0768	0772	46	13	2263	2267	2271
0777	0781	0785	0790	0794	0798	45	14	2287	2291	2295
0803	0807	0812	0816	0820	0825	44	15	2311	2315	2319
0829	0833	0838	0842	0847	0851	43	16	2335	2339	2343
0855	0860	0864	0868	0873	0877	42	17	2359	2363	2366
0882	0886	0890	0895	0899	0903	41	18	2382	2386	2390
0908	0912	0916	0921	0925	0929	40	19	2406	2410	2414
990934	0938	0942	0947	0951	0955	39	20	9.992430	2434	2438
0960	0964	0969	0973	0977	0982	38	21	2454	2458	2462
0986	0990	0995	0999	1003	1008	37	22	2478	2482	2485
1012	1016	1021	1025	1029	1033	36	23	2501	2505	2509
1038	1042	1046	1051	1055	1059	35	24	2525	2529	2533
1064	1068	1072	1077	1081	1085	34	25	2549	2553	2556
1090	1094	1098	1103	1107	1111	33	26	2572	2576	2580
1115	1120	1124	1128	1133	1137	32	27	2596	2600	2604
1141	1146	1150	1154	1158	1163	31	28	2619	2623	2627
1167	1171	1176	1180	1184	1188	30	29	2643	2647	2651
991193	1197	1201	1206	1210	1214	29	30	9.992666	2670	2674
1218	1223	1227	1231	1235	1240	28	31	2690	2693	2697
1244	1248	1253	1257	1261	1265	27	32	2713	2717	2721
1270	1274	1278	1282	1287	1291	26	33	2736	2740	2744
1295	1299	1304	1308	1312	1316	25	34	2759	2763	2767
1321	1325	1329	1333	1338	1342	24	35	2783	2787	2790
1346	1350	1355	1359	1363	1367	23	36	2806	2810	2814
1372	1376	1380	1384	1389	1393	22	37	2829	2833	2837
1397	1401	1406	1410	1414	1418	21	38	2852	2856	2860
1422	1427	1431	1435	1439	1444	20	39	2875	2879	2883
991448	1452	1456	1460	1465	1469	19	40	9.992898	2902	2906
1473	1477	1482	1486	1490	1494	18	41	2921	2925	2929
1498	1503	1507	1511	1515	1519	17	42	2944	2948	2952
1524	1528	1532	1536	1540	1545	16	43	2967	2971	2975
1549	1553	1557	1561	1566	1570	15	44	2990	2994	2998
1574	1578	1582	1586	1591	1595	14	45	3013	3017	3021
1599	1603	1607	1612	1616	1620	13	46	3036	3040	3043
1624	1628	1632	1637	1641	1645	12	47	3059	3062	3066
1649	1653	1657	1662	1666	1670	11	48	3081	3085	3089
1674	1678	1682	1687	1691	1695	10	49	3104	3108	3112
991699	1703	1707	1712	1716	1720	9	50	9.993127	3131	3134
1724	1728	1732	1736	1741	1745	8	51	3149	3153	3157
1749	1753	1757	1761	1765	1770	7	52	3172	3176	3180
1774	1778	1782	1786	1790	1794	6	53	3195	3198	3202
1799	1803	1807	1811	1815	1819	5	54	3217	3221	3225
1823	1827	1832	1836	1840	1844	4	55	3240	3243	3247
1848	1852	1856	1860	1865	1869	3	56	3262	3266	3270
1873	1877	1881	1885	1889	1893	2	57	3284	3288	3292
1897	1901	1906	1910	1914	1918	1	58	3307	3311	3314
1922	1926	1930	1934	1938	1942	0	59	3329	3333	3337

Tangent of 78 Degrees.							Min.	Tangent of 79 Deg			
0″	10″	20″	30″	40″	50″			0″	10″	20″	30″
10.672525	2629	2733	2836	2940	3043	59	0	10.711348	1460	1573	1685
3147	3251	3354	3458	3562	3666	58	1	2023	2135	2248	2361
3769	3873	3977	4081	4185	4289	57	2	2699	2811	2924	3037
4393	4497	4601	4705	4809	49:3	56	3	3376	3488	3601	3714
5017	5121	5225	5329	5433	5537	55	4	4053	4167	4280	4393
5642	5746	5850	5954	6059	6163	54	5	4732	4846	4959	5072
6267	6372	6476	6580	6685	6789	53	6	5412	5526	5639	5752
6894	6998	7103	7207	7312	7417	52	7	6093	6207	6320	6434
7521	7626	7731	7835	7940	8045	51	8	6775	6889	7002	7116
8149	8254	8359	8464	8569	8674	50	9	7458	7572	7686	7799
8778	8883	8988	9093	9198	9303	49	10	8142	8256	8370	8484
9408	9513	9618	9723	9829	9934	48	11	8826	8941	9055	9169
10 680039	0144	0249	0355	0460	0565	47	12	9512	9627	9741	9856
0670	0776	0881	0987	1092	1197	46	13	10.720199	0314	0428	0543
1303	1408	1514	1619	1725	1830	45	14	0887	1002	1116	1231
1936	2042	2147	2253	2359	2464	44	15	1576	1691	1806	1921
2570	2676	2782	2887	2993	3099	43	16	2260	2381	2496	2611
3205	3311	3417	3523	3629	3735	42	17	2957	3072	3187	3302
3841	3947	4053	4159	4265	4371	41	18	3649	3764	3879	3995
4477	4584	4690	4796	4902	5009	40	19	4342	4457	4573	4688
5115	5221	5328	5434	5540	5647	39	20	5036	5151	5267	5383
5753	5860	5966	6073	6179	6286	38	21	5731	5847	5963	6079
6392	6499	6606	6712	6819	6926	37	22	6427	6543	6659	6775
7032	7139	7246	7353	7460	7567	36	23	7124	7240	7356	7473
7673	7780	7887	7994	8101	8208	35	24	7822	7938	8055	8171
8315	8422	8529	8636	8744	8851	34	25	8521	8638	8754	8871
8958	9065	9172	9280	9387	9494	33	26	9221	9338	9455	9572
9601	9709	9816	9924	..31	.138	32	27	9923	..40	.157	.274
10.690246	0353	0461	0568	0676	0784	31	28	10.730625	0742	0860	0977
0891	0999	1107	1214	1322	1430	30	29	1329	1446	1563	1681
1537	1645	1753	1861	1969	2077	29	30	2033	2151	2268	2386
2184	2292	2400	2508	2616	2724	28	31	2739	2856	2974	3092
2832	2941	3049	3157	3265	3373	27	32	3445	3563	3681	3799
3481	3590	3698	3806	3914	4023	26	33	4153	4271	4389	4507
4131	4239	4348	4456	4565	4673	25	34	4862	4980	5098	5217
4782	4890	4999	5107	5216	5325	24	35	5572	5690	5809	5927
5433	5542	5651	5759	5868	5977	23	36	6283	6401	6520	6639
6086	6195	6303	6412	6521	6630	22	37	6995	7113	7232	7351
6739	6848	6957	7066	7175	7284	21	38	7708	7827	7946	8065
7393	7503	7612	7721	7830	7939	20	39	8422	8541	8660	8780
8049	8158	8267	8376	8486	8595	19	40	9137	9257	9376	9496
8705	8814	8924	9033	9143	9252	18	41	9854	9973	..93	.213
9362	9471	9581	9691	9800	9910	17	42	10.740571	0691	0811	0931
10.700020	0129	0239	0349	0459	0569	16	43	1290	1410	1530	1650
0678	0788	0898	1008	1118	1228	15	44	2010	2130	2250	2370
1338	1448	1558	1668	1779	1889	14	45	2731	2851	2971	3092
1999	2109	2219	2330	2440	2550	13	46	3453	3573	3694	3814
2661	2771	2881	2992	3102	3213	12	47	4176	4297	4417	4538
3323	3434	3544	3655	3765	3876	11	48	4900	5021	5142	5263
3987	4097	4208	4319	4429	4540	10	49	5626	5747	5868	5989
4651	4762	4873	4984	5095	5205	9	50	6352	6473	6595	6716
5316	5427	5538	5649	5761	5872	8	51	7080	7201	7323	7444
5983	6094	6205	6316	6428	6539	7	52	7809	7930	8052	8174
6650	6761	6873	6984	7095	7207	6	53	8539	8661	8782	8904
7318	7430	7541	7653	7764	7876	5	54	9270	9392	9514	9636
7987	8099	8211	8322	8434	8546	4	55	10.750002	0124	0247	0369
8658	8769	8881	8993	9105	9217	3	56	0736	0858	0980	1103
9329	9441	9553	9665	9777	9889	2	57	1470	1593	1715	1838
10.710001	0113	0225	0337	0449	0562	1	58	2206	2329	2452	2574
0674	0786	0898	1011	1123	1235	0	59	2943	3066	3189	3312

Sine of 80 Degrees.							Min.	Sine of 81]		
0″	10″	20″	30″	40″	50″			0″	10″	20″
993351	3355	3359	3363	3366	3370	59	0	9·994620	4623	4627
3374	3377	3381	3385	3389	3392	58	1	4640	4643	4647
3396	3400	3403	3407	3411	3414	57	2	4660	4663	4667
3418	3422	3426	3429	3433	3437	56	3	4680	4683	4686
3440	3444	3448	3451	3455	3459	55	4	4700	4703	4706
3462	3466	3470	3473	3477	3481	54	5	4720	4723	4726
3484	3488	3492	3495	3499	3503	53	6	4739	4743	4746
3506	3510	3514	3517	3521	3525	52	7	4759	4762	4766
3528	3532	3536	3539	3543	3547	51	8	4779	4782	4785
3550	3554	3558	3561	3565	3569	50	9	4798	4802	4805
993572	3576	3580	3583	3587	3591	49	10	9·994818	4821	4825
3594	3598	3601	3605	3609	3612	48	11	4838	4841	4844
3616	3620	3623	3627	3631	3634	47	12	4857	4861	4864
3638	3641	3645	3649	3652	3656	46	13	4877	4880	4883
3660	3663	3667	3670	3674	3678	45	14	4896	4900	4903
3681	3685	3689	3692	3696	3699	44	15	49·6	49·9	4922
3703	3707	3710	3714	3717	3721	43	16	4935	4938	4942
3725	3728	3732	3735	3739	3743	42	17	4955	4958	4961
3746	3750	3753	3757	3761	3764	41	18	4974	4977	4980
3768	3771	3775	3779	3782	3786	40	19	4993	4997	5000
993789	3793	3797	3800	3804	3807	39	20	9·995013	5016	5019
3811	3814	3818	3822	3825	3829	38	21	5032	5035	5038
3832	3836	3840	3843	3847	3850	37	22	5051	5054	5057
3854	3857	3861	3864	3868	3872	36	23	5070	5073	5077
3875	3879	3882	3886	3889	3893	35	24	5089	5092	5096
3897	3900	3904	3907	3911	3914	34	25	5108	5112	5115
3918	3921	3925	3928	3932	3936	33	26	5127	5131	5134
3939	3943	3946	3950	3953	3957	32	27	5146	5150	5153
3960	3964	3967	3971	3974	3978	31	28	5165	5169	5172
3982	3985	3989	3992	3996	3999	30	29	5184	5188	5191
994003	4006	4010	4013	4017	4020	29	30	9·995203	5206	5210
4024	4027	4031	4034	4038	4041	28	31	5222	5225	5228
4045	4048	4052	4055	4059	4062	27	32	5241	5244	5247
4066	4069	4073	4076	4080	4083	26	33	5260	5263	5266
4087	4090	4094	4097	4101	4104	25	34	5278	5282	5285
4108	4111	4115	4118	4122	4125	24	35	5297	5300	5303
4129	4132	4136	4139	4143	4146	23	36	5316	5319	5322
4150	4153	4157	4160	4164	4167	22	37	5334	5338	5341
4171	4174	4178	4181	4184	4188	21	38	5353	5356	5359
4191	4195	4198	4202	4205	4209	20	39	5372	5375	5378
994212	4216	4219	4223	4226	4230	19	40	9·995390	5393	5396
4233	4236	4240	4243	4247	4250	18	41	5409	5412	5415
4254	4257	4261	4264	4267	4271	17	42	5427	5430	5433
4274	4278	4281	4285	4288	4292	16	43	5446	5449	5452
4295	4298	4302	4305	4309	4312	15	44	5464	5467	5470
4316	4319	4322	4326	4329	4333	14	45	5482	5485	5488
4336	4340	4343	4346	4350	4353	13	46	5501	5504	5507
4357	4360	4363	4367	4370	4374	12	47	5519	5522	5525
4377	4381	4384	4387	4391	4394	11	48	5537	5540	5543
4398	4401	4404	4408	4411	4415	10	49	5555	5558	5561
994418	4421	4425	4428	4432	4435	9	50	9·995573	5576	5579
4438	4442	4445	4448	4452	4455	8	51	5591	5594	5597
4459	4462	4465	4469	4472	4476	7	52	5610	5613	5616
4479	4482	4486	4489	4492	4496	6	53	5628	5631	5634
4499	4503	4506	4509	4513	4516	5	54	5646	5649	5052
4519	4523	4526	4530	4533	4536	4	55	5664	5667	5670
4540	4543	4546	4550	4553	4556	3	56	5681	5684	5687
4560	4563	4566	4570	4573	4576	2	57	5699	5702	5705
4580	4583	4587	4590	4593	4597	1	58	5717	5720	5723
4600	4603	4607	4610	4613	4617	0	59	5735	5738	5741

0	10.800287	0424	0560	0696	0833	0969
1	1106	1242	1379	1516	1652	1789
2	1926	2062	2199	2336	2473	2610
3	2747	2884	3021	3158	3295	3433
4	3570	3707	3844	3982	4119	4257
5	4394	4532	4669	4807	4944	5082
6	5220	5358	5495	5633	5771	5909
7	6047	6185	6323	6461	6599	6738
8	6876	7014	7152	7291	7429	7568
9	7706	7845	7983	8122	8261	8399
10	8538	8677	8816	8954	9093	9232
11	9371	9510	9649	9788	9928	..67
12	10.810206	0345	0485	0624	0764	0903
13	1042	1182	1322	1461	1601	1741
14	1880	2020	2160	2300	2440	2580
15	2720	2860	3000	3140	3280	3421
16	3561	3701	3842	3982	4122	4263
17	4403	4544	4685	4825	4966	5107
18	5248	5388	5529	5670	5811	5952
19	6093	6234	6375	6517	6658	6799
20	6941	7082	7223	7365	7506	7648
21	7789	7931	8073	8214	8356	8498
22	8640	8782	8924	9066	9208	9350
23	9492	9634	9776	9918	..61	.203
24	10.820345	0488	0630	0773	0915	1058
25	1201	1343	1486	1629	1772	1915
26	2058	2200	2344	2487	2630	2773
27	2916	3059	3203	3346	3489	3633
28	3776	3920	4063	4207	4350	4494
29	4638	4782	4925	5069	5213	5357
30	5501	5645	5789	5933	6078	6222
31	6366	6511	6655	6799	6944	7088
32	7233	7377	7522	7667	7812	7956
33	8101	8246	8391	8536	8681	8826
34	8971	9116	9261	9407	9552	9697
35	9843	9988	.134	.279	.425	.570
36	10.830716	0862	1008	1153	1299	1445
37	1591	1737	1883	2029	2175	2322
38	2468	2614	2760	2907	3053	3200
39	3346	3493	3639	3786	3933	4080
40	4226	4373	4520	4667	4814	4961
41	5168	5255	5402	5550	5697	5844
42	5992	6139	6287	6434	6582	6729
43	6877	7025	7172	7320	7468	7616
44	7764	7912	8060	8208	8356	8504
45	8653	8801	8949	9098	9246	9395
46	9543	9692	9840	9989	.138	.287
47	10.840435	0584	0733	0882	1031	1180
48	1329	1479	1628	1777	1926	2076
49	2225	2375	2524	2674	2823	2973
50	3123	3272	3422	3572	3722	3872
51	4022	4172	4322	4472	4623	4773
52	4923	5074	5224	5374	5525	5675
53	5826	5977	6127	6278	6429	6580
54	6731	6882	7033	7184	7335	7486
55	7637	7789	7940	8091	8243	8394
56	8546	8697	8849	9001	9152	9304
57	9456	9608	9760	9912	..64	.216
58	10.850368	0520	0672	0825	0977	1129
59	1282	1434	1587	1739	1892	2045

Min.	Sine of 82 Degrees. 0″	10″	20″	30″	40″	50″	
0	9.995753	5756	5759	5762	5765	5768	59
1	5771	5773	5776	5779	5782	5785	58
2	5788	5791	5794	5797	5800	5803	57
3	5806	5809	5812	5815	5818	5821	56
4	5823	5826	5829	5832	5835	5838	55
5	5841	5844	5847	5850	5853	5856	54
6	5859	5862	5864	5867	5870	5873	53
7	5876	5879	5882	5885	5888	5891	52
8	5894	5897	5899	5902	5905	5908	51
9	5911	5914	5917	5920	5923	5926	50
10	9.995928	5931	5934	5937	5940	5943	49
11	5946	5949	5952	5954	5957	5960	48
12	5963	5966	5969	5972	5975	5978	47
13	5980	5983	5986	5989	5992	5995	46
14	5998	6001	6003	6006	6009	6012	45
15	6015	6018	6021	6023	6026	6029	44
16	6032	6035	6038	6041	6043	6046	43
17	6049	6052	6055	6058	6061	6063	42
18	6066	6069	6072	6075	6078	6081	41
19	6083	6086	6089	6092	6095	6098	40
20	9.996100	6103	6106	6109	6112	6115	39
21	6117	6120	6123	6126	6129	6131	38
22	6134	6137	6140	6143	6146	6148	37
23	6151	6154	6157	6160	6162	6165	36
24	6168	6171	6174	6177	6179	6182	35
25	6185	6188	6191	6193	6196	6199	34
26	6202	6205	6207	6210	6213	6216	33
27	6219	6221	6224	6227	6230	6232	32
28	6235	6238	6241	6244	6246	6249	31
29	6252	6255	6257	6260	6263	6266	30
30	9.996269	6271	6274	6277	6280	6282	29
31	6285	6288	6291	6293	6296	6299	28
32	6302	6305	6307	6310	6313	6316	27
33	6318	6321	6324	6327	6329	6332	26
34	6335	6338	6340	6343	6346	6349	25
35	6351	6354	6357	6359	6362	6365	24
36	6368	6370	6373	6376	6379	6381	23
37	6384	6387	6390	6392	6395	6398	22
38	6400	6403	6406	6409	6411	6414	21
39	6417	6419	6422	6425	6428	6430	20
40	9.996433	6436	6438	6441	6444	6447	19
41	6449	6452	6455	6457	6460	6463	18
42	6465	6468	6471	6474	6476	6479	17
43	6482	6484	6487	6490	6492	6495	16
44	6498	6500	6503	6506	6508	6511	15
45	6514	6517	6519	6522	6525	6527	14
46	6530	6533	6535	6538	6541	6543	13
47	6546	6549	6551	6554	6557	6559	12
48	6562	6565	6567	6570	6573	6575	11
49	6578	6580	6583	6586	6588	6591	10
50	9.996594	6596	6599	6602	6604	6607	9
51	6610	6612	6615	6618	6620	6623	8
52	6625	6628	6631	6633	6636	6639	7
53	6641	6644	6646	6649	6652	6654	6
54	6657	6660	6662	6665	6667	6670	5
55	6673	6675	6678	6681	6683	6686	4
56	6688	6691	6694	6696	6699	6701	3
57	6704	6707	6709	6712	6714	6717	2
58	6720	6722	6725	6727	6730	6733	1
59	6735	6738	6740	6743	6746	6748	0
	60″	50″	40″	30″	20″	10″	Min.
	Co-sine of 7 Degrees.						

Min.	Sine of 83 Degrees. 0″	10″	20″	30″	40″	50″	
0	9.996751	6753	6756	6758	6761	6764	59
1	6766	6769	6771	6774	6777	6779	58
2	6782	6784	6787	6789	6792	6795	57
3	6797	6800	6802	6805	6807	6810	56
4	6812	6815	6818	6820	6823	6825	55
5	6828	6830	6833	6835	6838	6841	54
6	6843	6846	6848	6851	6853	6856	53
7	6858	6861	6863	6866	6869	6871	52
8	6874	6876	6879	6881	6884	6886	51
9	6889	6891	6894	6896	6899	6901	50
10	9.996904	6906	6909	6912	6914	6917	49
11	6919	6922	6924	6927	6929	6932	48
12	6934	6937	6939	6942	6944	6947	47
13	6949	6952	6954	6957	6959	6962	46
14	6964	6967	6969	6972	6974	6977	45
15	6979	6982	6984	6987	6989	6992	44
16	6994	6997	6999	7002	7004	7007	43
17	7009	7011	7014	7016	7019	7021	42
18	7024	7026	7029	7031	7034	7036	41
19	7039	7041	7044	7046	7049	7051	40
20	9.997053	7056	7058	7061	7063	7066	39
21	7068	7071	7073	7076	7078	7080	38
22	7083	7085	7088	7090	7093	7095	37
23	7098	7100	7102	7105	7107	7110	36
24	7112	7115	7117	7120	7122	7124	35
25	7127	7129	7132	7134	7137	7139	34
26	7141	7144	7146	7149	7151	7154	33
27	7156	7158	7161	7163	7166	7168	32
28	7170	7173	7175	7178	7180	7182	31
29	7185	7187	7190	7192	7194	7197	30
30	9.997199	7202	7204	7206	7209	7211	29
31	7214	7216	7218	7221	7223	7226	28
32	7228	7230	7233	7235	7238	7240	27
33	7242	7245	7247	7249	7252	7254	26
34	7257	7259	7261	7264	7266	7268	25
35	7271	7273	7276	7278	7280	7283	24
36	7285	7287	7290	7292	7294	7297	23
37	7299	7301	7304	7306	7309	7311	22
38	7313	7316	7318	7320	7323	7325	21
39	7327	7330	7332	7334	7337	7339	20
40	9.997341	7344	7346	7348	7351	7353	19
41	7355	7358	7360	7362	7365	7367	18
42	7369	7372	7374	7376	7379	7381	17
43	7383	7386	7388	7390	7393	7395	16
44	7397	7399	7402	7404	7406	7409	15
45	7411	7413	7416	7418	7420	7423	14
46	7425	7427	7429	7432	7434	7436	13
47	7439	7441	7443	7445	7448	7450	12
48	7452	7455	7457	7459	7461	7464	11
49	7466	7468	7471	7473	7475	7477	10
50	9.997480	7482	7484	7487	7489	7491	9
51	7493	7496	7498	7500	7502	7505	8
52	7507	7509	7511	7514	7516	7518	7
53	7520	7523	7525	7527	7530	7532	6
54	7534	7536	7539	7541	7543	7545	5
55	7547	7550	7552	7554	7556	7559	4
56	7561	7563	7565	7568	7570	7572	3
57	7574	7577	7579	7581	7583	7585	2
58	7588	7590	7592	7594	7597	7599	1
59	7601	7603	7605	7608	7610	7612	0
	60″	50″	40″	30″	20″	10″	Min.
	Co-sine of 6 Degrees.						

Min.	Tangent of 82 Degrees.						
	0″	10″	20″	30″	40″	50″	
0	10.852197	2350	2503	2656	2809	2962	59
1	3215	3268	3421	3575	3728	3881	58
2	4034	4188	4341	4495	4649	4802	57
3	4956	5110	5263	5417	5571	5725	56
4	5879	6033	6187	6341	6496	6650	55
5	6804	6958	7113	7267	7422	7576	54
6	7731	7886	8041	8195	8350	8505	53
7	8660	8815	8970	9125	9280	9436	52
8	9591	9746	9902	..57	.212	.368	51
9	10.860524	0679	0835	0991	1146	1302	50
10	1458	1614	1770	1926	2082	2239	49
11	2395	2551	2708	2864	3020	3177	48
12	3333	3490	3647	3803	3960	4117	47
13	4274	4431	4588	4745	4902	5059	46
14	5216	5374	5531	5688	5846	6003	45
15	6161	6319	6476	6634	6792	6950	44
16	7107	7265	7423	7581	7739	7898	43
17	8056	8214	8372	8531	8689	8848	42
18	9006	9165	9324	9482	9641	9800	41
19	9959	.118	.277	.436	.595	.754	40
20	10.870913	1072	1232	1391	1551	1710	39
21	1870	2029	2189	2349	2508	2668	38
22	2828	2988	3148	3308	3468	3629	37
23	3789	3949	4109	4270	4430	4591	36
24	4751	4912	5073	5234	5394	5555	35
25	5716	5877	6038	6199	6360	6522	34
26	6683	6844	7006	7167	7329	7490	33
27	7652	7813	7975	8137	8299	8461	32
28	8623	8785	8947	9109	9271	9433	31
29	9596	9758	9921	..83	.246	.408	30
30	10.880571	0734	0896	1059	1222	1385	29
31	1548	1711	1874	2038	2201	2364	28
32	2528	2691	2855	3018	3182	3345	27
33	3509	3673	3837	4001	4165	4329	26
34	4493	4657	4821	4985	5150	5314	25
35	5479	5643	5808	5972	6137	6302	24
36	6467	6632	6796	6961	7127	7292	23
37	7457	7622	7787	7953	8118	8284	22
38	8449	8615	8781	8946	9112	9278	21
39	9444	9610	9776	9942	.108	.274	20
40	10.890441	0607	0773	0940	1106	1273	19
41	1440	1606	1773	1940	2107	2274	18
42	2441	2608	2775	2942	3110	3277	17
43	3444	3612	3779	3947	4115	4282	16
44	4450	4618	4786	4954	5122	5290	15
45	5458	5626	5795	5963	6131	6300	14
46	6468	6637	6806	6974	7143	7312	13
47	7481	7650	7819	7988	8157	8326	12
48	8496	8665	8834	9004	9173	9343	11
49	9513	9683	9852	..22	.192	.362	10
50	10.900532	0702	0873	1043	1213	1384	9
51	1554	1724	1895	2066	2236	2407	8
52	[illegible]	[illegible]	[illegible]	[illegible]	[illegible]	3433	7

Min.	Tangent of 83 Degrees.						
	0″	10″	20″	30″	40″	50″	
0	10.910856	1030	1205	1379	1553	1727	59
1	1902	2076	2251	2426	2600	2775	58
2	2950	3125	3300	3475	3650	3825	57
3	4000	4176	4351	4527	4702	4878	56
4	5053	5229	5405	5581	5757	5933	55
5	6109	6285	6461	6638	6814	6990	54
6	7167	7343	7520	7697	7874	8050	53
7	8227	8404	8581	8759	8936	9113	52
8	9290	9468	9645	9823	....	.178	51
9	10.920356	0534	0712	0890	1068	1246	50
10	1424	1602	1781	1959	2138	2316	49
11	2495	2673	2852	3031	3210	3389	48
12	3568	3747	3926	4105	4285	4464	47
13	4644	4823	5003	5183	5362	5542	46
14	5722	5902	6082	6262	6442	6623	45
15	6803	6984	7164	7345	7525	7706	44
16	7887	8068	8249	8430	8611	8792	43
17	8973	9154	9336	9517	9699	9880	42
18	10.930062	0244	0425	0607	0789	0971	41
19	1154	1336	1518	1700	1883	2065	40
20	2248	2430	2613	2796	2979	3162	39
21	3345	3528	3711	3894	4078	4261	38
22	4444	4628	4812	4995	5179	5363	37
23	5547	5731	5915	6099	6283	6467	36
24	6652	6836	7021	7205	7390	7575	35
25	7760	7945	8130	8315	8500	8685	34
26	8870	9056	9241	9427	9612	9798	33
27	9984	.169	.355	.541	.727	.914	32
28	10.941100	1286	1472	1659	1845	2032	31
29	2219	2406	2592	2779	2966	3153	30
30	3341	3528	3715	3902	4090	4277	29
31	4465	4653	4841	5028	5216	5404	28
32	5593	5781	5969	6157	6346	6534	27
33	6723	6912	7100	7289	7478	7667	26
34	7856	8045	8234	8424	8613	8803	25
35	8992	9182	9371	9561	9751	9941	24
36	10.950131	0321	0511	0702	0892	1083	23
37	1273	1464	1654	1845	2036	2227	22
38	2418	2509	2800	2991	3183	3374	21
39	3566	3757	3949	4141	4332	4524	20
40	4716	4908	5101	5293	5485	5678	19
41	5870	6063	6255	6448	6641	6834	18
42	7027	7220	7413	7606	7800	7993	17
43	8187	8380	8574	8768	8961	9155	16
44	9349	9544	9738	9932	.126	.321	15
45	10.960515	0710	0905	1099	1294	1489	14
46	1684	1879	2074	2270	2465	2661	13
47	2856	3052	3247	3443	3639	3835	12
48	4031	4227	4424	4620	4816	5013	11
49	5209	5406	5603	5800	5997	6194	10
50	6391	6588	6785	6983	7180	7377	9
51	7575	7773	7971	8169	8367	8565	8
52	8763	[illegible]	[illegible]	[illegible]	[illegible]	[illegible]	7

Min.	Sine of 84 Degrees. 0″	10″	20″	30″	40″	50″	
0	9·997614	7617	7619	7621	7623	7625	59
1	7628	7630	7632	7634	7636	7639	58
2	7641	7643	7645	7647	7650	7652	57
3	7654	7656	7658	7661	7663	7665	56
4	7667	7669	7672	7674	7676	7678	55
5	7680	7682	7685	7687	7689	7691	54
6	7693	7696	7698	7700	7702	7704	53
7	7706	7709	7711	7713	7715	7717	52
8	7719	7722	7724	7726	7728	7730	51
9	7732	7735	7737	7739	7741	7743	50
10	9·997745	7747	7750	7752	7754	7756	49
11	7758	7760	7762	7765	7767	7769	48
12	7771	7773	7775	7777	7780	7782	47
13	7784	7786	7788	7790	7792	7794	46
14	7797	7799	7801	7803	7805	7807	45
15	7809	7811	7814	7816	7818	7820	44
16	7822	7824	7826	7828	7830	7833	43
17	7835	7837	7839	7841	7843	7845	42
18	7847	7849	7852	7854	7856	7858	41
19	7860	7862	7864	7866	7868	7870	40
20	9·997872	7875	7877	7879	7881	7883	39
21	7885	7887	7889	7891	7893	7895	38
22	7897	7900	7902	7904	7906	7908	37
23	7910	7912	7914	7916	7918	7920	36
24	7922	7924	7926	7929	7931	7933	35
25	7935	7937	7939	7941	7943	7945	34
26	7947	7949	7951	7953	7955	7957	33
27	7959	7961	7963	7965	7967	7970	32
28	7972	7974	7976	7978	7980	7982	31
29	7984	7986	7988	7990	7992	7994	30
30	9·997996	7998	8000	8002	8004	8006	29
31	8008	8010	8012	8014	8016	8018	28
32	8020	8022	8024	8026	8028	8030	27
33	8032	8034	8036	8038	8040	8042	26
34	8044	8046	8048	8050	8052	8054	25
35	8056	8058	8060	8062	8064	8066	24
36	8068	8070	8072	8074	8076	8078	23
37	8080	8082	8084	8086	8088	8090	22
38	8092	8094	8096	8098	8100	8102	21
39	8104	8106	8108	8110	8112	8114	20
40	9·998116	8118	8120	8122	8124	8126	19
41	8128	8130	8131	8133	8135	8137	18
42	8139	8141	8143	8145	8147	8149	17
43	8151	8153	8155	8157	8159	8161	16
44	8163	8165	8167	8168	8170	8172	15
45	8174	8176	8178	8180	8182	8184	14
46	8186	8188	8190	8192	8194	8195	13
47	8197	8199	8201	8203	8205	8207	12
48	8209	8211	8213	8215	8217	8218	11
49	8220	8222	8224	8226	8228	8230	10
50	9·998232	8234	8236	8238	8239	8241	9
51	8243	8245	8247	8249	8251	8253	8
52	8255	8257	8258	8260	8262	8264	7
53	8266	8268	8270	8272	8273	8275	6
54	8277	8279	8281	8283	8285	8287	5
55	8289	8290	8292	8294	8296	8298	4
56	8300	8302	8303	8305	8307	8309	3
57	8311	8313	8315	8316	8318	8320	2
58	8322	8324	8326	8328	8329	8331	1
	8333	8335	8337	8339	8341	8342	0
	60″	50″	40″	30″	20″	10″	Min.
	Co-sine of 5 Degrees.						

Min.	Sine of 85 Degrees. 0″	10″	20″	30″	40″	50″	
0	9·998344	8346	8348	8350	8352	8353	59
1	8355	8357	8359	8361	8363	8364	58
2	8366	8368	8370	8372	8374	8375	57
3	8377	8379	8381	8383	8385	8386	56
4	8388	8390	8392	8394	8395	8397	55
5	8399	8401	8403	8404	8406	8408	54
6	8410	8412	8413	8415	8417	8419	53
7	8421	8422	8424	8426	8428	8430	52
8	8431	8433	8435	8437	8439	8440	51
9	8442	8444	8446	8448	8449	8451	50
10	9·998453	8455	8456	8458	8460	8462	49
11	8464	8465	8467	8469	8471	8472	48
12	8474	8476	8478	8479	8481	8483	47
13	8485	8487	8488	8490	8492	8494	46
14	8495	8497	8499	8501	8502	8504	45
15	8506	8508	8509	8511	8513	8515	44
16	8516	8518	8520	8522	8523	8525	43
17	8527	8529	8530	8532	8534	8535	42
18	8537	8539	8541	8542	8544	8546	41
19	8548	8549	8551	8553	8554	8556	40
20	9·998558	8560	8561	8563	8565	8566	39
21	8568	8570	8572	8573	8575	8577	38
22	8578	8580	8582	8584	8585	8587	37
23	8589	8590	8592	8594	8595	8597	36
24	8599	8601	8602	8604	8606	8607	35
25	8609	8611	8612	8614	8616	8617	34
26	8619	8621	8622	8624	8626	8627	33
27	8629	8631	8633	8634	8636	8638	32
28	8639	8641	8643	8644	8646	8648	31
29	8649	8651	8653	8654	8656	8657	30
30	9·998659	8661	8662	8664	8666	8667	29
31	8669	8671	8672	8674	8676	8677	28
32	8679	8681	8682	8684	8686	8687	27
33	8689	8690	8692	8694	8695	8697	26
34	8699	8700	8702	8704	8705	8707	25
35	8708	8710	8712	8713	8715	8717	24
36	8718	8720	8721	8723	8725	8726	23
37	8728	8729	8731	8733	8734	8736	22
38	8738	8739	8741	8742	8744	8746	21
39	8747	8749	8750	8752	8754	8755	20
40	9·998757	8758	8760	8762	8763	8765	19
41	8766	8768	8769	8771	8773	8774	18
42	8776	8777	8779	8781	8782	8784	17
43	8785	8787	8788	8790	8792	8793	16
44	8795	8796	8798	8799	8801	8803	15
45	8804	8806	8807	8809	8810	8812	14
46	8813	8815	8817	8818	8820	8821	13
47	8823	8824	8826	8827	8829	8831	12
48	8832	8834	8835	8837	8838	8840	11
49	8841	8843	8844	8846	8848	8849	10
50	9·998851	8852	8854	8855	8857	8858	9
51	8860	8861	8863	8864	8866	8867	8
52	8869	8870	8872	8873	8875	8877	7
53	8878	8880	8881	8883	8884	8886	6
54	8887	8889	8890	8892	8893	8895	5
55	8896	8898	8899	8901	8902	8904	4
56	8905	8907	8908	8910	8911	8913	3
57	8914	8916	8917	8919	8920	8922	2
58	8923	8925	8926	8927	8929	8930	1
59	8932	8933	8935	8936	8938	8939	0
	60″	50″	40″	30″	20″	10″	Min.
	Co-sine of 4 Degrees.						

Tangent of 84 Degrees.							Min.	Tangent of 85		
0″	10″	20″	30″	40″	50″			0″	10″	20″
10.978380	8582	8785	8988	9191	9394	59	0	11.058048	8291	8533
9597	9800	...3	.206	.410	.613	58	1	9506	9749	9993
10.980817	1021	1224	1428	1632	1836	57	2	11.060968	1212	1456
2041	2245	2449	2654	2858	3063	56	3	2435	2680	2925
3268	3472	3677	3882	4087	4293	55	4	3907	4153	4399
4498	4703	4909	5114	5320	5526	54	5	5384	5631	5877
5732	5938	6144	6350	6556	6763	53	6	6866	7113	7361
6969	7176	7382	7589	7796	8003	52	7	8353	8601	8850
8210	8417	8624	8831	9039	9246	51	8	9845	..94	.343
9454	9662	9869	..77	.285	.493	50	9	11.071342	1592	1842
10.990702	0910	1118	1327	1535	1744	49	10	2844	3095	3346
1953	2162	2371	2580	2789	2998	48	11	4351	4603	4855
3208	3417	3627	3836	4046	4256	47	12	5864	6116	6369
4466	4676	4886	5096	5307	5517	46	13	7381	7635	7888
5728	5939	6149	6360	6571	6782	45	14	8904	9159	9413
6993	7205	7416	7627	7839	8051	44	15	11.080432	0688	0943
8262	8474	8686	8898	9111	9323	43	16	1966	2222	2478
9535	9748	9960	.173	.386	.599	42	17	3505	3762	4019
11.000812	1025	1238	1451	1665	1878	41	18	5049	5307	5565
2092	2306	2519	2733	2947	3161	40	19	6599	6858	7117
3376	3590	3804	4019	4234	4448	39	20	8154	8414	8674
4663	4878	5093	5308	5524	5739	38	21	9715	9975	.236
5955	6170	6386	6602	6818	7034	37	22	11.091281	1543	1804
7250	7466	7682	7899	8115	8332	36	23	2853	3115	3378
8549	8765	8982	9199	9417	9634	35	24	4430	4694	4957
9851	..69	.286	.504	.722	.940	34	25	6013	6278	6542
11.011158	1376	1594	1813	2031	2250	33	26	7602	7868	8133
2468	2687	2906	3125	3344	3563	32	27	9197	9463	9730
3783	4002	4222	4441	4661	4881	31	28	11.100797	1065	1332
5101	5321	5541	5762	5982	6202	30	29	2404	2672	2940
6423	6644	6865	7086	7307	7528	29	30	4016	4285	4555
7749	7971	8192	8414	8636	8857	28	31	5634	5904	6175
9079	9301	9524	9746	9968	.191	27	32	7258	7529	7801
11.020414	0636	0859	1082	1305	1528	26	33	8888	9160	9433
1752	1975	2199	2422	2646	2870	25	34	11.110524	0798	1071
3094	3318	3542	3767	3991	4216	24	35	2167	2441	2715
4440	4665	4890	5115	5340	5565	23	36	3815	4090	4366
5791	6016	6242	6468	6693	6919	22	37	5470	5746	6023
7145	7372	7598	7824	8051	8277	21	38	7131	7408	7686
8504	8731	8958	9185	9412	9640	20	39	8798	9076	9355
9867	..95	.322	.550	.778	1006	19	40	11.120471	0751	1031
11.031234	1462	1691	1919	2148	2377	18	41	2151	2432	2713
2606	2835	3064	3293	3522	3752	17	42	3838	4119	4401
3981	4211	4441	4671	4901	5131	16	43	5531	5813	6096
5361	5592	5822	6053	6284	6514	15	44	7230	7514	7798
6745	6977	7208	7439	7671	7902	14	45	8936	9221	9506
8134	8366	8598	8830	9062	9295	13	46	11.130649	0935	1221
9527	9760	9992	.225	.458	.691	12	47	2368	2656	2943
11.040925	1158	1391	1625	1859	2092	11	48	4094	4383	4671
2326	2561	2795	3029	3264	3498	10	49	5827	6117	6407
3733	3968	4203	4438	4673	4908	9	50	7567	7858	8149
5144	5379	5615	5851	6087	6323	8	51	9314	9606	9898
6559	6795	7032	7268	7505	7742	7	52	11.141068	1361	1654
7979	8216	8453	8691	8928	9166	6	53	2829	3123	3417
9403	9641	9879	.117	.356	.594	5	54	4597	4892	5187
11.050832	1071	1310	1549	1788	2027	4	55	6372	6668	6965
2266	2506	2745	2985	3225	3465	3	56	8154	8452	8750
3705	3945	4185	4426	4666	4907	2	57	9943	.242	.542
5148	5389	5630	5871	6112	6354	1	58	11.151740	2041	2341
6596	6837	7079	7321	7563	7806	0	59	3545	3846	4148

Min.	Sine of 86 Degrees. 0″	10″	20″	30″	40″	50″	
0	9·998941	8942	8944	8945	8947	8948	59
1	8950	8951	8953	8954	8955	8957	58
2	8958	8960	8961	8963	8964	8966	57
3	8967	8969	8970	8971	8973	8974	56
4	8976	8977	8979	8980	8982	8983	55
5	8984	8986	8987	8989	8990	8992	54
6	8993	8995	8996	8997	8999	9000	53
7	9002	9003	9005	9006	9007	9009	52
8	9010	9012	9013	9015	9016	9017	51
9	9019	9020	9022	9023	9024	9026	50
10	9·999027	9029	9030	9032	9033	9034	49
11	9036	9037	9039	9040	9041	9043	48
12	9044	9046	9047	9048	9050	9051	47
13	9053	9054	9055	9057	9058	9059	46
14	9061	9062	9064	9065	9066	9068	45
15	9069	9071	9072	9073	9075	9076	44
16	9077	9079	9080	9082	9083	9084	43
17	9086	9087	9088	9090	9091	9092	42
18	9094	9095	9097	9098	9099	9101	41
19	9102	9103	9105	9106	9107	9109	40
20	9·999110	9111	9113	9114	9115	9117	39
21	9118	9120	9121	9122	9124	9125	38
22	9126	9128	9129	9130	9132	9133	37
23	9134	9136	9137	9138	9140	9141	36
24	9142	9143	9145	9146	9147	9149	35
25	9150	9151	9153	9154	9155	9157	34
26	9158	9159	9161	9162	9163	9165	33
27	9166	9167	9168	9170	9171	9172	32
28	9174	9175	9176	9178	9179	9180	31
29	9181	9183	9184	9185	9187	9188	30
30	9·999189	9190	9192	9193	9194	9196	29
31	9197	9198	9199	9201	9202	9203	28
32	9205	9206	9207	9208	9210	9211	27
33	9212	9213	9215	9216	9217	9219	26
34	9220	9221	9222	9224	9225	9226	25
35	9227	9229	9230	9231	9232	9234	24
36	9235	9236	9237	9239	9240	9241	23
37	9242	9244	9245	9246	9247	9249	22
38	9250	9251	9252	9254	9255	9256	21
39	9257	9258	9260	9261	9262	9263	20
40	9·999265	9266	9267	9268	9270	9271	19
41	9272	9273	9274	9276	9277	9278	18
42	9279	9280	9282	9283	9284	9285	17
43	9287	9288	9289	9290	9291	9293	16
44	9294	9295	9296	9297	9299	9300	15
45	9301	9302	9303	9305	9306	9307	14
46	9308	9309	9310	9312	9313	9314	13
47	9315	9316	9318	9319	9320	9321	12
48	9322	9323	9325	9326	9327	9328	11
49	9329	9331	9332	9333	9334	9335	10
50	9·999336	9338	9339	9340	9341	9342	9
51	9343	9344	9346	9347	9348	9349	8
52	9350	9351	9353	9354	9355	9356	7
53	9357	9358	9359	9361	9362	9363	6
54	9364	9365	9366	9367	9369	9370	5
55	9371	9372	9373	9374	9375	9377	4
56	9378	9379	9380	9381	9382	9383	3
57	9384	9385	9387	9388	9389	9390	2
58	9391	9392	9393	9394	9396	9397	1
59	9398	9399	9400	9401	9402	9403	0

Min.	Sine of 87 Degrees. 0″	10″	20″	30″	40″	50″	
0	9·999404	9406	9407	9408	9409	9410	59
1	9411	9412	9413	9414	9415	9416	58
2	9418	9419	9420	9421	9422	9423	57
3	9424	9425	9426	9427	9428	9430	56
4	9431	9432	9433	9434	9435	9436	55
5	9437	9438	9439	9440	9441	9442	54
6	9443	9445	9446	9447	9448	9449	53
7	9450	9451	9452	9453	9454	9455	52
8	9456	9457	9458	9459	9460	9461	51
9	9463	9464	9465	9466	9467	9468	50
10	9·999469	9470	9471	9472	9473	9474	49
11	9475	9476	9477	9478	9479	9480	48
12	9481	9482	9483	9484	9485	9486	47
13	9487	9488	9489	9490	9491	9492	46
14	9493	9495	9496	9497	9498	9499	45
15	9500	9501	9502	9503	9504	9505	44
16	9506	9507	9508	9509	9510	9511	43
17	9512	9513	9514	9515	9516	9517	42
18	9518	9519	9520	9521	9522	9523	41
19	9524	9525	9526	9527	9527	9528	40
20	9·999529	9530	9531	9532	9533	9534	39
21	9535	9536	9537	9538	9539	9540	38
22	9541	9542	9543	9544	9545	9546	37
23	9547	9548	9549	9550	9551	9552	36
24	9553	9554	9555	9556	9557	9557	35
25	9558	9559	9560	9561	9562	9563	34
26	9564	9565	9566	9567	9568	9569	33
27	9570	9571	9572	9573	9573	9574	32
28	9575	9576	9577	9578	9579	9580	31
29	9581	9582	9583	9584	9585	9586	30
30	9·999586	9587	9588	9589	9590	9591	29
31	9592	9593	9594	9595	9596	9597	28
32	9597	9598	9599	9600	9601	9602	27
33	9603	9604	9605	9606	9606	9607	26
34	9608	9609	9610	9611	9612	9613	25
35	9614	9614	9615	9616	9617	9618	24
36	9619	9620	9621	9622	9622	9623	23
37	9624	9625	9626	9627	9628	9629	22
38	9629	9630	9631	9632	9633	9634	21
39	9635	9635	9636	9637	9638	9639	20
40	9·999640	9641	9641	9642	9643	9644	19
41	9645	9646	9647	9647	9648	9649	18
42	9650	9651	9652	9653	9653	9654	17
43	9655	9656	9657	9658	9658	9659	16
44	9660	9661	9662	9663	9663	9664	15
45	9665	9666	9667	9668	9668	9669	14
46	9670	9671	9672	9672	9673	9674	13
47	9675	9676	9677	9677	9678	9679	12
48	9680	9681	9681	9682	9683	9684	11
49	9685	9685	9686	9687	9688	9689	10
50	9·999689	9690	9691	9692	9693	9693	9
51	9694	9695	9696	9697	9697	9698	8
52	9699	9700	9700	9701	9702	9703	7
53	9704	9704	9705	9706	9707	9707	6
54	9708	9709	9710	9711	9711	9712	5
55	9713	9714	9714	9715	9716	9717	4
56	9717	9718	9719	9720	9720	9721	3
57	9722	9723	9723	9724	9725	9726	2
58	9726	9727	9728	9729	9729	9730	1
59	9731	9732	9732	9733	9734	9735	0

Min.	Tangent of 86 Degrees. 0″	10″	20″	30″	40″	50″		Min.	Tangent of 87 Degrees. 0″	10″	20″	30″	40″	50″	
0	11.155356	5659	5962	6265	6568	6872	59	0	11.280604	1007	1411	1814	2219	2623	59
1	7175	7479	7784	8088	8393	8697	58	1	3028	3433	3839	4245	4652	5058	58
2	9002	9308	9613	9919	.224	.530	57	2	5466	5873	6281	6689	7098	7507	57
3	11.160837	1143	1450	1757	2064	2371	56	3	7917	8326	8737	9147	9558	9970	56
4	2679	2987	3295	3603	3911	4220	55	4	11.290382	0794	1206	1619	2033	2446	55
5	4529	4838	5147	5457	5766	6076	54	5	2860	3275	3690	4105	4521	4937	54
6	6387	6697	7008	7318	7629	7941	53	6	5354	5770	6188	6605	7024	7442	53
7	8252	8564	8876	9188	9500	9813	52	7	7861	8280	8700	9120	9541	9962	52
8	11.170126	0439	0752	1066	1379	1693	51	8	11.300383	0805	1227	1649	2072	2496	51
9	2008	2322	2637	2951	3267	3582	50	9	2919	3344	3768	4193	4619	5044	50
10	3897	4213	4529	4845	5162	5478	49	10	5471	5897	6325	6752	7180	7608	49
11	5795	6112	6430	6747	7065	7383	48	11	8037	8466	8896	9326	9756	.187	48
12	7702	8020	8339	8658	8977	9297	47	12	11.310619	1050	1483	1915	2348	2782	47
13	9616	9936	.256	.577	.897	1218	46	13	3216	3650	4085	4520	4956	5392	46
14	11.181539	1860	2182	2504	2826	3148	45	14	5828	6265	6702	7140	7578	8017	45
15	3471	3793	4116	4440	4763	5087	44	15	8456	8896	9336	9776	.217	.659	44
16	5411	5735	6059	6384	6709	7034	43	16	11.311100	1543	1985	2428	2872	3316	43
17	7359	7685	8011	8337	8663	8990	42	17	3761	4206	4651	5097	5543	5990	42
18	9317	9644	9971	.299	.626	.954	41	18	6437	6885	7333	7782	8231	8680	41
19	11.191283	1611	1940	2269	2598	2928	40	19	9130	9581	..32	.483	.935	1387	40
20	3258	3588	3918	4249	4579	4910	39	20	11.331840	2293	2747	3201	3656	4111	39
21	5242	5573	5905	6237	6569	6902	38	21	4567	5023	5480	5937	6394	6852	38
22	7235	7568	7901	8235	8568	8902	37	22	7311	7770	8229	8689	9150	9611	37
23	9237	9571	9906	.241	.577	.912	36	23	11.340072	0534	0996	1459	1923	2387	36
24	11.201248	1584	1921	2257	2594	2931	35	24	2851	3316	3781	4247	4714	5180	35
25	3269	3606	3944	4282	4621	4960	34	25	5648	6116	6584	7053	7522	7992	34
26	5299	5638	5977	6317	6657	6997	33	26	8463	8933	9405	9877	.349	.822	33
27	7338	7679	8020	8361	8703	9045	32	27	11.351296	1770	2244	2719	3195	3671	32
28	9387	9729	..72	.415	.758	1102	31	28	4147	4624	5102	5580	6059	6538	31
29	11.211446	1790	2134	2479	2823	3169	30	29	7018	7498	7979	8460	8942	9424	30
30	3514	3860	4206	4552	4898	5245	29	30	9907	.390	.874	1359	1844	2329	29
31	5592	5939	6287	6635	6983	7331	28	31	11.362816	3302	3789	4277	4765	5254	28
32	7680	8029	8378	8728	9078	9428	27	32	5744	6234	6724	7215	7707	8199	27
33	9778	.129	.480	.831	1183	1534	26	33	8692	9185	9679	.173	.668	1164	26
34	11.221886	2239	2591	2944	3298	3651	25	34	11.371660	2156	2654	3151	3650	4149	25
35	4005	4359	4713	5068	5423	5778	24	35	4648	5148	5649	6150	6652	7154	24
36	6134	6489	6845	7202	7558	7915	23	36	7657	8161	8665	9170	9675	.181	23
37	8273	8630	8988	9346	9705	..63	22	37	11.380687	1194	1702	2210	2719	3228	22
38	11.230422	0782	1141	1501	1861	2222	21	38	3738	4249	4760	5272	5785	6298	21
39	2583	2944	3305	3667	4029	4391	20	39	6811	7325	7840	8356	8872	9388	20
40	4754	5116	5480	5843	6207	6571	19	40	9906	.424	.942	1461	1981	2501	19
41	6935	7300	7665	8030	8396	8762	18	41	11.393022	3544	4066	4589	5113	5637	18
42	9128	9495	9861	.229	.596	.964	17	42	6161	6687	7213	7740	8267	8795	17
43	11.241332	1700	2069	2438	2807	3177	16	43	9323	9853	.382	.913	1444	1976	16
44	3547	3917	4288	4659	5030	5401	15	44	11.402508	3041	3575	4110	4645	5180	15
45	5773	6145	6518	6891	7264	7637	14	45	5717	6254	6792	7330	7869	8409	14
46	8011	8385	8759	9134	9509	9884	13	46	8949	9490	..32	.574	1117	1661	13
47	11.250260	0636	1012	1389	1766	2143	12	47	11.412205	2751	3296	3843	4390	4938	12
48	2521	2899	3277	3656	4034	4414	11	48	5486	6036	6586	7136	7688	8240	11
49	4793	5173	5553	5934	6315	6696	10	49	8792	9346	9900	.455	1010	1566	10
50	7078	7460	7842	8224	8607	8991	9	50	11.422123	2681	3240	3799	4358	4919	9
51	9374	9758	.142	.527	.912	1297	8	51	5480	6042	6605	7168	7733	8298	8

Sine of 88 Degrees. 0″	10″	20″	30″	40″	50″		Min.	Sine of 89 Degrees. 0″	10″	20″	30″	40″
999735	9736	9737	9738	9738	9739	59	0	9.999934	9934	9935	9935	9935
9740	9740	9741	9742	9743	9743	58	1	9936	9936	9937	9937	9937
9744	9745	9746	9746	9747	9748	57	2	9938	9939	9939	9939	9940
9748	9749	9750	9751	9751	9752	56	3	9940	9941	9941	9941	9942
9753	9753	9754	9755	9756	9756	55	4	9942	9943	9943	9943	9944
9757	9758	9758	9759	9760	9760	54	5	9944	9945	9945	9945	9946
9761	9762	9763	9763	9764	9765	53	6	9946	9947	9947	9947	9948
9765	9766	9767	9767	9768	9769	52	7	9948	9949	9949	9949	9950
9769	9770	9771	9772	9772	9773	51	8	9950	9951	9951	9951	9952
9774	9774	9775	9776	9776	9777	50	9	9952	9953	9953	9953	9953
999778	9778	9779	9780	9780	9781	49	10	9.999954	9954	9955	9955	9955
9782	9782	9783	9784	9784	9785	48	11	9956	9956	9956	9957	9957
9786	9786	9787	9788	9788	9789	47	12	9958	9958	9958	9959	9959
9790	9790	9791	9792	9792	9793	46	13	9959	9960	9960	9960	9961
9794	9794	9795	9795	9796	9797	45	14	9961	9961	9962	9962	9962
9797	9798	9799	9799	9800	9801	44	15	9963	9963	9963	9964	9964
9801	9802	9803	9803	9804	9804	43	16	9964	9965	9965	9965	9965
9805	9806	9806	9807	9808	9808	42	17	9966	9966	9967	9967	9967
9809	9809	9810	9811	9811	9812	41	18	9968	9968	9968	9968	9969
9813	9813	9814	9814	9815	9816	40	19	9969	9969	9970	9970	9970
999816	9817	9817	9818	9819	9819	39	20	9.999971	9971	9971	9971	9972
9820	9820	9821	9822	9822	9823	38	21	9972	9972	9973	9973	9973
9824	9824	9825	9825	9826	9827	37	22	9973	9974	9974	9974	9974
9827	9828	9828	9829	9829	9830	36	23	9975	9975	9975	9976	9976
9831	9831	9832	9832	9833	9834	35	24	9976	9976	9977	9977	9977
9834	9835	9835	9836	9836	9837	34	25	9977	9978	9978	9978	9978
9838	9838	9839	9839	9840	9840	33	26	9979	9979	9979	9979	9980
9841	9842	9842	9843	9843	9844	32	27	9980	9980	9980	9981	9981
9844	9845	9846	9846	9847	9847	31	28	9981	9981	9982	9982	9982
9848	9848	9849	9849	9850	9851	30	29	9982	9983	9983	9983	9983
999851	9852	9852	9853	9853	9854	29	30	9.999983	9984	9984	9984	9984
9854	9855	9856	9856	9857	9857	28	31	9985	9985	9985	9985	9985
9858	9858	9859	9859	9860	9860	27	32	9986	9986	9986	9986	9986
9861	9861	9862	9863	9863	9864	26	33	9987	9987	9987	9987	9987
9864	9865	9865	9866	9866	9867	25	34	9988	9988	9988	9988	9988
9867	9868	9868	9869	9869	9870	24	35	9989	9989	9989	9989	9989
9870	9871	9871	9872	9872	9873	23	36	9989	9990	9990	9990	9990
9873	9874	9874	9875	9875	9876	22	37	9990	9990	9991	9991	9991
9876	9877	9877	9878	9878	9879	21	38	9991	9991	9991	9992	9992
9879	9880	9880	9881	9881	9882	20	39	9992	9992	9992	9992	9992
999882	9883	9883	9884	9884	9885	19	40	9.999993	9993	9993	9993	9993
9885	9886	9886	9887	9887	9888	18	41	9993	9993	9994	9994	9994
9888	9889	9889	9890	9890	9891	17	42	9994	9994	9994	9994	9994
9891	9892	9892	9892	9893	9893	16	43	9995	9995	9995	9995	9995
9894	9894	9895	9895	9896	9896	15	44	9995	9995	9995	9996	9996
9897	9897	9898	9898	9898	9899	14	45	9996	9996	9996	9996	9996
9899	9900	9900	9901	9901	9902	13	46	9996	9996	9997	9997	9997
9902	9903	9903	9903	9904	9904	12	47	9997	9997	9997	9997	9997
9905	9905	9906	9906	9906	9907	11	48	9997	9997	9997	9998	9998
9907	9908	9908	9909	9909	9910	10	49	9998	9998	9998	9998	9998
999910	9910	9911	9911	9912	9912	9	50	9.999998	9998	9998	9998	9998
9913	9913	9913	9914	9914	9915	8	51	9999	9999	9999	9999	9999
9915	9915	9916	9916	9917	9917	7	52	9999	9999	9999	9999	9999
9918	9918	9918	9919	9919	9920	6	53	9999	9999	9999	9999	9999
9920	9920	9921	9921	9922	9922	5	54	9999	9999	9999	9999	9999
9922	9923	9923	9924	9924	9924	4	55	0.000000	0000	0000	0000	0000
9925	9925	9926	9926	9926	9927	3	56	0000	0000	0000	0000	0000
9927	9927	9928	9928	9929	9929	2	57	0000	0000	0000	0000	0000
9929	9930	9930	9931	9931	9931	1	58	0000	0000	0000	0000	0000
9932	9932	9932	9933	9933	9933	0	59	0000	0000	0000	0000	0000
60″	50″	40″	30″	20″	10″	Min.		60″	50″	40″	30″	20″
Co-sine of 1 Degree.								Co-sine of 0 Degree.				

Tangent of 88 Degrees.						P. Part to 1″.
10″	20″	30″	40″	50″		
11.457520	11.458125	11.458731	11.459338	11.459945	59	60.6
461163	461773	462383	462995	463608	58	61.1
464836	465451	466067	466684	467302	57	61.6
468540	469160	469782	470404	471027	56	62.2
472276	472902	473528	474156	474785	55	62.7
476044	476676	477308	477941	478575	54	63.3
479846	480482	481120	481759	482398	53	63.8
483680	484323	484966	485611	486256	52	64.4
487549	488198	488847	489497	490148	51	65.0
491453	492107	492762	493418	494075	50	65.5
11.495392	11.496052	11.496713	11.497375	11.498038	49	66.1
499367	500033	500700	501368	502037	48	66.8
503378	504051	504724	505398	506073	47	67.4
507427	508106	508785	509466	510148	46	68.0
511514	512199	512885	513572	514260	45	68.6
515640	516331	517024	517717	518412	44	69.3
519805	520503	521202	521903	522604	43	70.0
524010	524715	525421	526128	526837	42	70.7
528257	528969	529682	530396	531111	41	71.3
532545	533264	533984	534705	535428	40	72.1
11.536876	11.537602	11.538330	11.539058	11.539788	39	72.8
541251	541984	542719	543455	544192	38	73.5
545670	546411	547153	547896	548641	37	74.3
550134	550883	551632	552384	553136	36	75.0
554645	555401	556159	556918	557678	35	75.8
559203	559967	560733	561500	562268	34	76.6
563809	564581	565355	566130	566907	33	77.5
568464	569245	570027	570811	571596	32	78.3
573170	573959	574750	575542	576336	31	79.1
577928	578726	579525	580326	581128	30	80 0
11.582737	11.583544	11.584353	11.585163	11.585974	29	80.9
587601	588417	589235	590054	590874	28	81.8
592520	593345	594172	595000	595830	27	82.8
597495	598330	599166	600004	600844	26	83.7
602528	603372	604218	605066	605915	25	84.7
607619	608474	609330	610187	611047	24	85.7
612771	613636	614502	615370	616240	23	86.7
617985	618860	619737	620615	621496	22	87.8
623262	624147	625035	625924	626815	21	88.8
628603	629500	630399	631299	632201	20	89.9
11.634012	11.634919	11.635829	11.636741	11.637655	19	91.1
639488	640407	641329	642252	643177	18	92.2
645034	645965	646899	647834	648771	17	93.4
650652	651595	652541	653488	654438	16	94.6
656343	657299	658257	659217	660179	15	95.9
662110	663079	664050	665023	665998	14	97.2
667955	668936	669920	670907	671895	13	98.5
673879	674874	675871	676871	677873	12	99.8
679885	680894	681905	682919	683935	11	101.3
685975	686998	688024	689052	690083	10	102.7
11.692151	11.693189	11.694230	11.695273	11.696318	9	104.2
698417	699470	700526	701584	702645	8	105.7
704774	705843	706914	707988	709065	7	107.2
711226	712311	713398	714488	715581	6	108.9
717775	718876	719980	721087	722196	5	110.5
724424	725542	726663	727787	728914	4	112 2
731176	732312	733451	734592	735737	3	114.0
738035	739189	740346	741506	742669	2	115.8
745004	746177	747352	748531	749713	1	117.[illegible]
752087	753279	754474	755672	756874	0	119

0 Degree.				Seconds.	Minutes.	1	
—	log. tan. A— log. A″.	log. cot. A+ log. A″.				log. sin. A— log. A″.	lo
5	4.685575	15.314425	60	3600	0	4.685553	4
5	575	425	59	3660	1	552	
5	575	425	58	3720	2	551	
5	575	425	57	3780	3	551	
5	575	425	56	3840	4	550	
5	575	425	55	3900	5	549	
5	575	425	54	3960	6	548	
5	575	425	53	4020	7	547	
4	576	424	52	4080	8	547	
4	576	424	51	4140	9	546	
4	4.685576	15.314424	50	4200	10	4.685545	4
4	576	424	49	4260	11	544	
4	577	423	48	4320	12	543	
4	577	423	47	4380	13	542	
4	577	423	46	4440	14	541	
3	578	422	45	4500	15	540	
3	578	422	44	4560	16	539	
3	578	422	43	4620	17	539	
3	579	421	42	4680	18	538	
3	579	421	41	4740	19	537	
2	4.685580	15.314420	40	4800	20	4.685536	4
2	580	420	39	4860	21	535	
2	581	419	38	4920	22	534	
2	581	419	37	4980	23	533	
1	582	418	36	5040	24	532	
1	583	417	35	5100	25	531	
1	583	417	34	5160	26	530	
0	584	416	33	5220	27	529	
0	584	416	32	5280	28	527	

Min.	Tangent of 89 Degrees. 0″	10″	20″	30″	40″	50″		P. Part to 1″.
0	11.758079	11.759287	11.760498	11.761714	11.762932	11.764154	59	121.7
1	765379	766608	767840	769076	770315	771558	58	123.8
2	772805	774055	775308	776566	777826	779091	57	125.9
3	780359	781631	782907	784186	785470	786757	56	128.1
4	788047	789342	790641	791943	793250	794560	55	130.4
5	795874	797192	798515	799841	801171	802506	54	132.8
6	803844	805187	806534	807885	809240	810600	53	135.3
7	811964	813332	814704	816081	817462	818847	52	137.9
8	820237	821632	823031	824434	825842	827255	51	140.6
9	828672	830094	831520	832951	834387	835828	50	143.4
10	11.837273	11.838724	11.840179	11.841639	11.843104	11.844574	49	146.2
11	846048	847528	849013	850503	851999	853499	48	149.3
12	855004	856515	858031	859553	861079	862611	47	152.4
13	864149	865692	867240	868794	870354	871919	46	155.7
14	873490	875067	876649	878237	879831	881431	45	159.1
15	883037	884648	886266	887890	889519	891155	44	162.7
16	892797	894446	896100	897761	899429	901103	43	166.4
17	902783	904470	906163	907863	909569	911283	42	170.3
18	913003	914730	916464	918205	919953	921707	41	174.4
19	923469	925239	927015	928799	930590	932388	40	178.7
20	11.934194	11.936008	11.937829	11.939658	11.941494	11.943338	39	183.3
21	945191	947051	948919	950795	952679	954572	38	188.0
22	956473	958382	960299	962225	964160	966103	37	193.0
23	968055	970016	971986	973965	975952	977949	36	198.3
24	979956	981971	983996	986030	988074	990128	35	203.9
25	992191	994264	996347	998440	12.000543	12.002657	34	209.8
26	12.004781	12.006915	12.009060	12.011215	013382	015559	33	216.2
27	017747	019946	022156	024378	026611	028855	32	222.7
28	031111	033379	035659	037951	040255	042572	31	229.8
29	044900	047242	049596	051963	054342	056735	30	237.3
30	12.059142	12.061561	12.063994	12.066441	12.068902	12.071377	29	245 4
31	073866	076369	078887	081419	083966	086529	28	254.0
32	089106	091699	094308	096932	099572	102228	27	263.2
33	104901	107590	110296	113019	115760	118517	26	273.2
34	121292	124085	126896	129726	132574	135440	25	283.9
35	138326	141231	144156	147100	150065	153050	24	295.4
36	156056	159082	162130	165199	168290	171404	23	308.0
37	174540	177698	180880	184085	187314	190567	22	321.7
38	193845	197148	200476	203830	207210	210616	21	336.7
39	214049	217510	220998	224515	228060	231635	20	353.1
40	12.235239	12.238873	12.242538	12.246235	12.249963	12.253723	19	371.2
41	257516	261342	265203	269098	273028	276995	18	391.3
42	280997	285037	289115	293232	297388	301584	17	413.6
43	305821	310100	314422	318787	323196	327650	16	438.7
44	332151	336698	341294	345939	350634	355381	15	467.0
45	360180	365032	369940	374903	379924	385004	14	499.2
46	390143	395345	400609	405938	411333	416796	13	536.2
47	422328	427932	433610	439362	445192	451100	12	579.1
48	457091	463165	469325	475574	481915	488349	11	629.4
49	494880	501510	508244	515083	522032	529094	10	689.4
50	12.536273	12.543572	12.550996	12.558549	12.566236	12.574061	9	
51	582030	590148	598421	606854	615454	624228	8	
52	633183	642327	651667	661212	670972	680956	7	
53	691175	701641	712365	723360	734641	746223	6	
54	758122	770357	782946	795911	809275	823063	5	
55	837304	852027	867267	883061	899452	916485	4	
56	934214	952697	972002	992206	13.013395	13.035671	3	
57	13.059153	13.083976	13.110305	13.138334	168297	200482	2	
58	235244	273032	314425	360183	411335	469327	1	
59	536274	615455	712365	837304	14.013395	14.314425	0	

Natural Sines.

3°	4°	5°	6°
052336	069756	087156	104528
2626	070047	7446	4818
2917	0337	7735	5107
3207	0627	8025	5396
3498	0917	8315	5686
3788	1207	8605	5975
4079	1497	8894	6264
4369	1788	9184	6553
4660	2078	9474	6843
4950	2368	9763	7132
055241	072658	090053	107421
5531	2948	0343	7710
5822	3238	0633	7999
6112	3528	0922	8289
6402	3818	1212	8578
6693	4108	1502	8867
6983	4399	1791	9156
7274	4689	2081	9445
7564	4979	2371	9734
7854	5269	2660	110023
058145	075559	092950	110313
8435	5849	3239	0602
8726	6139	3529	0891
9016	6429	3819	1180
9306	6719	4108	1469
9597	7009	4398	1758
9887	7299	4687	2047
060177	7589	4977	2336
0468	7879	5267	2625
0758	8169	5556	2914
061049	078459	095846	113203
1339	8749	6135	3492
1629	9039	6425	3781
1920	9329	6714	4070
2210	9619	7004	4359
2500	9909	7293	4648
2791	080199	7583	4937

0°	1°	2°	3°	4°	5°	6°
000000	017455	034921	052408	069927	087489	105104
0291	7746	5212	2699	070219	7782	5398
0582	8037	5503	2991	0511	8075	5692
0873	8328	5795	3283	0804	8368	5987
.1164	8619	6086	3575	1096	8661	6281
1454	8910	6377	3866	1389	8954	6575
1745	9201	6668	4158	1681	9248	6869
2036	9492	6960	4450	1973	9541	7163
2327	9783	7251	4742	2266	9834	7458
2618	020074	7542	5033	2558	090127	7752
002909	020365	037834	055325	072851	090421	108046
3200	0656	8125	5617	3143	0714	8340
3491	0947	8416	5909	3435	1007	8635
3782	1238	8707	6200	3728	1300	8929
4072	1529	8999	6492	4020	1594	9223
4363	1820	9290	6784	4313	1887	9518
4654	2111	9581	7076	4605	2180	9812
4945	2402	9873	7368	4898	2474	110107
5236	2693	040164	7660	5190	2767	0401
5527	2984	0456	7951	5483	3061	0695
005818	023275	040747	058243	075775	093354	110990
6109	3566	1038	8535	6068	3647	1284
6400	3857	1330	8827	6361	3941	1579
6691	4148	1621	9119	6653	4234	1873
6981	4439	1912	9411	6946	4528	2168
7272	4731	2204	9703	7238	4821	2463
7563	5022	2495	9995	7531	5115	2757
7854	5313	2787	060287	7824	5408	3052
8145	5604	3078	0579	8116	5702	3346
8436	5895	3370	0871	8409	5995	3641
008727	026186	043661	061163	078702	096289	113936
9018	6477	3952	1455	8994	6583	4230
9309	6768	4244	1747	9287	6876	4525
9600	7059	4535	2039	9580	7170	4820
9891	7350	4827	2331	9873	7464	5114
010181	7641	5118	2623	080165	7757	5409
0472	7933	5410	2915	0458	8051	5704
0763	8224	5701	3207	0751	8345	5999
1054	8515	5993	3499	1044	8638	6294
1345	8806	6284	3791	1336	8932	6588
011636	029097	046576	064083	081629	099226	116883
1927	9388	6867	4375	1922	9519	7178
2218	9679	7159	4667	2215	9813	7473
2509	9970	7450	4959	2508	100107	7768
2800	030262	7742	5251	2801	0401	8063
3091	0553	8033	5543	3094	0695	8358
3382	0844	8325	5836	3386	0989	8653
3673	1135	8617	6128	3679	1282	8948
3964	1426	8908	6420	3972	1576	9243
4254	1717	9200	6712	4265	1870	9538
014545	032009	049491	067004	084558	102164	119833
4836	2300	9783	7296	4851	2458	120128
5127	2591	050075	7589	5144	2752	0423
5418	2882	0366	7881	5437	3046	0718
5709	3173	0658	8173	5730	3340	1013
6000	3465	0949	8465	6023	3634	1308
6291	3756	1241	8758	6316	3928	1604
6582	4047	1533	9050	6609	4222	1899
6873	4338	1824	9342	6902	4516	2194
7164	4630	2116	9635	7196	4810	2489
89°	88°	87°	86°	85°	84°	83°

Natural Sines.

13°	14°	15°	16°	17°	18°
24951	241922	258819	275637	292372	309017
5234	2204	9100	5917	2650	9294
5518	2486	9381	6197	2928	9570
5801	2769	9662	6476	3206	9847
6085	3051	9943	6756	3484	310123
6368	3333	260224	7035	3762	0400
6651	3615	0505	7315	4040	0676
6935	3897	0785	7594	4318	0953
7218	4179	1066	7874	4596	1229
7501	4461	1347	8153	4874	1506
27784	244743	261628	278432	295152	311782
8068	5025	1908	8712	5430	2059
8351	5307	2189	8991	5708	2335
8634	5589	2470	9270	5986	2611
8917	5871	2751	9550	6264	2888
9200	6153	3031	9829	6542	3164
9484	6435	3312	280108	6819	3440
9767	6717	3592	0388	7097	3716
30050	6999	3873	0667	7375	3992
0333	7281	4154	0946	7653	4269
30616	247563	264434	281225	297930	314545
0899	7845	4715	1504	8208	4821
1182	8126	4995	1783	8486	5097
1465	8408	5276	2062	8763	5373
1748	8690	5556	2341	9041	5649
2031	8972	5837	2620	9318	5925
2314	9253	6117	2900	9596	6201
2597	9535	6397	3179	9873	6477
2880	9817	6678	3457	300151	6753
3163	250098	6958	3736	0428	7029
33445	250380	267238	284015	300706	317305
3728	0662	7519	4294	0983	7580
4011	0943	7799	4573	1261	7856
4294	1225	8079	4852	1538	8132
4577	1506	8359	5131	1815	8408
4859	1788	8640	5410	2093	8684
5142	2069	8920	5688	2370	8959
5425	2351	9200	5967	2647	9235
5708	2632	9480	6246	2924	9511
5990	2914	9760	6525	3202	9786
36273	253195	270040	286803	303479	320062
6556	3477	0320	7082	3756	0337
6838	3758	0600	7361	4033	0613
7121	4039	0880	7639	4310	0889
7403	4321	1160	7918	4587	1164
7686	4602	1440	8196	4864	1439
7968	4883	1720	8475	5141	1715
8251	5165	2000	8753	5418	1990
8533	5446	2280	9032	5695	2266
8816	5727	2560	9310	5972	2541
39098	256008	272840	289589	306249	322816
9381	6289	3120	9867	6526	3092
9663	6571	3400	290145	6803	3367
9946	6852	3679	0424	7080	3642
40228	7133	3959	0702	7357	3917
0510	7414	4239	0981	7633	4193
0793	7695	4519	1259	7910	4468
1075	7976	4798	1537	8187	4743
1357	8257	5078	1815	8464	5018
1640	8538	5358	2094	8740	5293

Natural Tangents.

11°	12°	13°	14°	15°	16°	17°
194380	212557	230868	249328	267949	286745	305731
4682	2861	1175	9637	8261	7060	6049
4984	3165	1481	9946	8573	7375	6367
5286	3469	1788	250255	8885	7690	6685
5588	3773	2094	0564	9197	8005	7003
5890	4077	2401	0873	9509	8320	7322
6192	4381	2707	1183	9821	8635	7640
6494	4686	3014	1492	270133	8950	7959
6796	4990	3321	1801	0445	9266	8277
7099	5294	3627	2111	0757	9581	8596
197401	215599	233934	252420	271069	289896	308914
7703	5903	4241	2729	1382	290211	9233
8005	6208	4548	3039	1694	0527	9552
8308	6512	4855	3348	2006	0842	9871
8610	6817	5162	3658	2319	1158	310189
8912	7121	5469	3968	2631	1473	0508
9215	7426	5776	4277	2944	1789	0827
9517	7731	6083	4587	3256	2105	1146
9820	8035	6390	4897	3569	2420	1465
200122	8340	6697	5207	3882	2736	1784
200425	218645	237004	255516	274194	293052	312104
0727	8950	7312	5826	4507	3368	2423
1030	9254	7619	6136	4820	3684	2742
1333	9559	7926	6446	5133	4000	3062
1635	9864	8234	6756	5446	4316	3381
1938	220169	8541	7066	5759	4632	3700
2241	0474	8848	7377	6072	4948	4020
2544	0779	9156	7687	6385	5265	4340
2847	1084	9464	7997	6698	5581	4659
3149	1389	9771	8307	7011	5897	4979
203452	221695	240079	258618	277325	296213	315299
3755	2000	0386	8928	7638	6530	5619
4058	2305	0694	9238	7951	6846	5939
4361	2610	1002	9549	8265	7163	6258
4664	2916	1310	9859	8578	7480	6578
4967	3221	1618	260170	8891	7796	6899
5271	3526	1925	0480	9205	8113	7219
5574	3832	2233	0791	9519	8430	7539
5877	4137	2541	1102	9832	8747	7859
6180	4443	2849	1413	280146	9063	8179
206483	224748	243157	261723	280460	299380	318500
6787	5054	3466	2034	0773	9697	8820
7090	5360	3774	2345	1087	300014	9141
7393	5665	4082	2656	1401	0331	9461
7697	5971	4390	2967	1715	0649	9782
8000	6277	4698	3278	2029	0966	320103
8304	6583	5007	3589	2343	1283	0423
8607	6889	5315	3900	2657	1600	0744
8911	7194	5624	4211	2971	1918	1065
9214	7500	5932	4523	3286	2235	1386
209518	227806	246241	264834	283600	302553	321707
9822	8112	6549	5145	3914	2870	2028
210126	8418	6858	5457	4229	3188	2349
0429	8724	7166	5768	4543	3506	2670
0733	9031	7475	6079	4857	3823	2991
1037	9337	7784	6391	5172	4141	3312
1341	9643	8092	6702	5487	4459	3634
1645	9949	8401	7014	5801	4777	3955
1949	230255	8710	7326	6116	5095	4277
2253	0562	9019	7637	6431	5413	4598
78°	77°	76°	75°	74°	73°	72°

Natural Co-tangents.

Natural Sines.

23°	24°	25°	26°	27°	28°	29°	
90731	406737	422618	438371	453990	469472	484810	60
0999	7002	2882	8633	4250	9728	5064	59
1267	7268	3145	8894	4509	9985	5318	58
1534	7534	3409	9155	4768	470242	5573	57
1802	7799	3673	9417	5027	0499	5827	56
2070	8065	3936	9678	5286	0755	6081	55
2337	8330	4199	9939	5545	1012	6335	54
2605	8596	4463	440200	5804	1268	6590	53
2872	8861	4726	0462	6063	1525	6844	52
3140	9127	4990	0723	6322	1782	7098	51
93407	409392	425253	440984	456580	472038	487352	50
3675	9658	5516	1245	6839	2294	7606	49
3942	9923	5779	1506	7098	2551	7860	48
4209	410188	6042	1767	7357	2807	8114	47
4477	0454	6306	2028	7615	3063	8367	46
4744	0719	6569	2289	7874	3320	8621	45
5011	0984	6832	2550	8133	3576	8875	44
5278	1249	7095	2810	8391	3832	9129	43
5546	1514	7358	3071	8650	4088	9382	42
5813	1779	7621	3332	8908	4344	9636	41
96080	412045	427884	443593	459166	474600	489890	40
6347	2310	8147	3853	9425	4856	490143	39
6614	2575	8410	4114	9683	5112	0397	38
6881	2840	8672	4375	9942	5368	0650	37
7148	3104	8935	4635	460200	5624	0904	36
7415	3369	9198	4896	0458	5880	1157	35
7682	3634	9461	5156	0716	6136	1411	34
7949	3899	9723	5417	0974	6392	1664	33
8215	4164	9986	5677	1232	6647	1917	32
8482	4429	430249	5937	1491	6903	2170	31
98749	414693	430511	446198	461749	477159	492424	30
9016	4958	0774	6458	2007	7414	2677	29
9283	5223	1036	6718	2265	7670	2930	28
9549	5487	1299	6979	2523	7925	3183	27
9816	5752	1561	7239	2780	8181	3436	26
00082	6016	1823	7499	3038	8436	3689	25
0349	6281	2086	7759	3296	8692	3942	24
0616	6545	2348	8019	3554	8947	4195	23
0882	6810	2610	8279	3812	9203	4448	22
1149	7074	2873	8539	4069	9458	4700	21
01415	417338	433135	448799	464327	479713	494953	20
1681	7603	3397	9059	4584	9968	5206	19
1948	7867	3659	9319	4842	480223	5459	18
2214	8131	3921	9579	5100	0479	5711	17
2480	8396	4183	9839	5357	0734	5964	16
2747	8660	4445	450098	5615	0989	6217	15
3013	8924	4707	0358	5872	1244	6469	14
3279	9188	4969	0618	6129	1499	6722	13
3545	9452	5231	0878	6387	1754	6974	12
3811	9716	5493	1137	6644	2009	7226	11
04078	419980	435755	451397	466901	482263	497479	10
4344	420244	6017	1656	7158	2518	7731	9
4610	0508	6278	1916	7416	2773	7983	8
4876	0772	6540	2175	7673	3028	8236	7
5142	1036	6802	2435	7930	3282	8488	6
5408	1300	7063	2694	8187	3537	8740	5
5673	1563	7325	2953	8444	3792	8992	4
5939	1827	7587	3213	8701	4046	9244	3
6205	2091	7848	3472	8958	4301	9496	2
6471	2355	8110	3731	9215	4555	9748	1
66°	65°	64°	63°	62°	61°	60°	Min.

Natural Co-sines.

Natural Tangents.

21°	22°	23°	24°	25°	26°	27°
83864	404026	424475	445229	466308	487733	509525
4198	4365	4818	5577	6662	8093	9892
4532	4703	5162	5926	7016	8453	510258
4866	5042	5505	6275	7371	8813	0625
5200	5380	5849	6624	7725	9174	0992
5534	5719	6192	6973	8080	9534	1359
5868	6058	6536	7322	8434	9895	1726
6202	6397	6880	7671	8789	490256	2093
6536	6736	7224	8020	9144	0617	2460
6871	7075	7568	8369	9499	0978	2828
87205	407414	427912	448719	469854	491339	513195
7540	7753	8256	9068	470209	1700	3563
7874	8092	8601	9418	0564	2061	3930
8209	8432	8945	9768	0920	2422	4298
8544	8771	9289	450117	1275	2784	4666
8879	9111	9634	0467	1631	3145	5034
9214	9450	9979	0817	1986	3507	5402
9549	9790	430323	1167	2342	3869	5770
9884	410130	0668	1517	2698	4231	6138
90219	0470	1013	1868	3054	4593	6507
90554	410810	431358	452218	473410	494955	516875
0889	1150	1703	2568	3766	5317	7244
1225	1490	2048	2919	4122	5679	7613
1560	1830	2393	3269	4478	6042	7982
1896	2170	2739	3620	4835	6404	8351
2231	2511	3084	3971	5191	6767	8720
2567	2851	3430	4322	5548	7130	9089
2903	3192	3775	4673	5905	7492	9458
3239	3532	4121	5024	6262	7855	9828
3574	3873	4467	5375	6619	8218	520197
93910	414214	434812	455726	476976	498582	520567
4247	4554	5158	6078	7333	8945	0937
4583	4895	5504	6429	7690	9308	1307
4919	5236	5850	6781	8047	9672	1677
5255	5577	6197	7132	8405	500035	2047
5592	5919	6543	7484	8762	0399	2417
5928	6260	6889	7836	9120	0763	2787
6265	6601	7236	8188	9477	1127	3158
6601	6943	7582	8540	9835	1491	3528
6938	7284	7929	8892	480193	1855	3899
97275	417626	438276	459244	480551	502219	524270
7611	7967	8622	9596	0909	2583	4641
7948	8309	8969	9949	1267	2948	5012
8285	8651	9316	460301	1626	3312	5383
8622	8993	9663	0654	1984	3677	5754
8960	9335	440011	1006	2343	4041	6125
9297	9677	0358	1359	2701	4406	6497
9634	420019	0705	1712	3060	4771	6868
9971	0361	1053	2065	3419	5136	7240
00309	0704	1400	2418	3778	5502	7612
00646	421046	441748	462771	484137	505867	527984
0984	1389	2095	3124	4496	6232	8356
1322	1731	2443	3478	4855	6598	8728
1660	2074	2791	3831	5214	6963	9100
1997	2417	3139	4185	5574	7329	9473
2335	2759	3487	4538	5933	7695	9845
2673	3102	3835	4892	6293	8061	530218
3011	3445	4183	5246	6653	8427	0591
3350	3788	4532	5600	7013	8793	0963
3688	4132	4880	5954	7373	9159	1336
68°	67°	66°	65°	64°	63°	62°

Natural Co-tangents.

Natural Sines.

33°	34°	35°	36°
544639	559193	573576	587785
4883	9434	3815	8021
5127	9675	4053	8256
5371	9916	4291	8491
5615	560157	4529	8726
5858	0398	4767	8961
6102	0639	5005	9196
6346	0880	5243	9431
6589	1121	5481	9666
6833	1361	5719	9901
547076	561602	575957	590136
7320	.843	6195	0371
7563	2083	6432	0606
7807	2324	6670	0840
8050	2564	6908	1075
8293	2805	7145	1310
8536	3045	7383	1544
8780	3286	7620	1779
9023	3526	7853	2013
9266	3766	8095	2248
549509	564007	578332	592482
9752	4247	8570	2716
9995	4487	8807	2951
550238	4727	9044	3185
0481	4967	9281	3419
0724	5207	9518	3653
0966	5447	9755	3887
1209	5687	9992	4121
1452	5927	580229	4355
1694	6166	0466	4589
551937	566406	580703	594823
2180	6646	0940	5057
2422	6886	1176	5290
2664	7125	1413	5524
2907	7365	1650	5758
3149	7604	1886	5991
3392	7844	2123	6225
3634	8083	2359	6458
3876	8323	2596	6692
4118	8562	2832	6925
554360	568801	583069	597159
4602	9040	3305	7392
4844	9280	3541	7625
5086	9519	3777	7858
5328	9758	4014	8092
5570	9997	4250	8325
5812	570236	4486	8558
6054	0475	4722	8791
6296	0714	4958	9024
6537	0952	5194	9256
556779	571191	585429	599489
7021	1430	5665	9722
7262	1669	5901	9955
7504	1907	6137	600188
7745	2146	6372	0420
7987	2384	6608	0653
8228	2623	6844	0885
8469	2861	7079	1118
8710	3100	7314	1350
8952	3338	7550	1583

Natural Tangents.

31°	32°	33°	34°	35°	36°
600861	624869	649408	674509	700208	726543
1257	5274	9821	4932	0641	6987
1653	5679	650235	5355	1075	7432
2049	6083	0649	5779	1509	7877
2445	6488	1063	6203	1943	8322
2842	6894	1477	6627	2377	8767
3239	7299	1892	7051	2812	9213
3635	7704	2306	7475	3246	9658
4032	8110	2721	7900	3681	730104
4429	8516	3136	8324	4116	0550
604827	628921	653551	678749	704551	730996
5224	9327	3966	9174	4987	1443
5622	9734	4382	9599	5422	1889
6019	630140	4797	680025	5858	2336
6417	0546	5213	0450	6294	2783
6815	0953	5629	0876	6730	3230
7213	1360	6045	1302	7166	3678
7611	1767	6461	1728	7603	4125
8010	2174	6877	2154	8039	4573
8408	2581	7294	2580	8476	5021
608807	632988	657710	683007	708913	735469
9205	3396	8127	3433	9350	5917
9604	3804	8544	3860	9788	6366
610003	4211	8961	4287	710225	6815
0403	4619	9379	4714	0663	7264
0802	5027	9796	5142	1101	7713
1201	5436	660214	5569	1539	8162
1601	5844	0631	5997	1977	8611
2001	6253	1049	6425	2416	9061
2401	6661	1467	6853	2854	9511
612801	637070	661886	687281	713293	739961
3201	7479	2304	7709	3732	740411
3601	7888	2723	8138	4171	0862
4002	8298	3141	8567	4611	1312
4402	8707	3560	8995	5050	1763
4803	9117	3979	9425	5490	2214
5204	9527	4398	9854	5930	2666
5605	9937	4818	690283	6370	3117
6006	640347	5237	0713	6810	3569
6408	0757	5657	1143	7250	4020
616809	641167	666077	691572	717691	744472
7211	1578	6497	2003	8132	4925
7613	1989	6917	2433	8573	5377
8015	2399	7337	2863	9014	5830
8417	2810	7758	3294	9455	6282
8819	3222	8179	3725	9897	6735
9221	3633	8599	4156	720339	7189
9624	4044	9020	4587	0781	7642
620026	4456	9442	5018	1223	8096
0429	4868	9863	5450	1665	8549
620832	645280	670284	695881	722108	749003
1235	5692	0706	6313	2550	9458
1638	6104	1128	6745	2993	9912
2042	6516	1550	7177	3436	750366
2445	6929	1972	7610	3879	0821
2849	7342	2394	8042	4323	1276
3253	7755	2817	8475	4766	1731
3657	8168	3240	8908	5210	2187
4061	8581	3662	9341	5654	2642
4465	8994	4085	9774	6098	3098
58°	57°	56°	55°	54°	53°

Natural Co-tangents.

Natural Sines.

43°	44°	45°	46°
681998	694658	707107	719340
2211	4868	7312	9542
2424	5077	7518	9744
2636	5286	7724	9946
2849	5495	7929	720148
3061	5704	8134	0349
3274	5913	8340	0551
3486	6122	8545	0753
3698	6330	8750	0954
3911	6539	8956	1156
684123	696748	709161	721357
4335	6957	9366	1559
4547	7165	9571	1760
4759	7374	9776	1962
4971	7582	9981	2163
5183	7790	710185	2364
5395	7999	0390	2565
5607	8207	0595	2766
5818	8415	0799	2967
6030	8623	1004	3168
686242	698832	711209	723369
6453	9040	1413	3570
6665	9248	1617	3771
6876	9455	1822	3971
7088	9663	2026	4172
7299	9871	2230	4372
7510	700079	2434	4573
7721	0287	2639	4773
7932	0494	2843	4974
8144	0702	3047	5174
688355	700909	713250	725374
8566	1117	3454	5575
8776	1324	3658	5775
8987	1531	3862	5975
9198	1739	4066	6175
9409	1946	4269	6375
9620	2153	4473	6575
9830	2360	4676	6775
690041	2567	4880	6974
0251	2774	5083	7174
690462	702981	715286	727374
0672	3188	5490	7573
0882	3395	5693	7773
1093	3601	5896	7972
1303	3808	6099	8172
1513	4015	6302	8371
1723	4221	6505	8570
1933	4428	6708	8769
2143	4634	6911	8969
2353	4841	7113	9168
692563	705047	717316	729367
2773	5253	7519	9566
2983	5459	7721	9765
3192	5665	7924	9963
3402	5872	8126	730162
3611	6078	8329	0361
3821	6284	8531	0560
4030	6489	8733	0758
4240	6695	8936	0957
4449	6901	9138	1155
46°	45°	44°	43°

Natural Tangents.

42°	43°	44°	45°	46°
900404	932515	965689	1.00000	1.03553
0931	3059	6251	0058	3613
1458	3603	6814	0116	3674
1985	4148	7377	0175	3734
2513	4693	7940	0233	3794
3041	5238	8504	0291	3855
3569	5783	9067	0350	3915
4098	6329	9632	0408	3976
4627	6875	970196	0467	4036
5156	7422	0761	0525	4097
905685	937968	971326	1.00583	1.04158
6215	8515	1892	0642	4218
6745	9063	2458	0701	4279
7275	9610	3024	0759	4340
7805	940158	3590	0818	4401
8336	0706	4157	0876	4461
8867	1255	4724	0935	4522
9398	1803	5291	0994	4583
9930	2352	5859	1053	4644
910462	2902	6427	1112	4705
910994	943451	976996	1.01170	1.04766
1526	4001	7564	1229	4827
2059	4552	8133	1288	4888
2592	5102	8703	1347	4949
3125	5653	9272	1406	5010
3659	5204	9842	1465	5072
4193	6756	980413	1524	5133
4727	7307	0983	1583	5194
5261	7859	1554	1642	5255
5796	8412	2126	1702	5317
916331	948965	982697	1.01761	1.05378
6866	9518	3269	1820	5439
7402	950071	3842	1879	5501
7938	0624	4414	1939	5562
8474	1178	4987	1998	5624
9010	1733	5560	2057	5685
9547	2287	6134	2117	5747
920084	2842	6708	2176	5809
0621	3397	7282	2236	5870
1159	3953	7857	2295	5932
921697	954508	988432	1.02355	1.05994
2235	5064	9007	2414	6056
2773	5621	9582	2474	6117
3312	6177	990158	2533	6179
3851	6734	0735	2593	6241
4390	7292	1311	2653	6303
4930	7849	1888	2713	6365
5470	8407	2465	2772	6427
6010	8966	3043	2832	6489
6551	9524	3621	2892	6551
927091	960083	994199	1.02952	1.06613
7632	0642	4778	3012	6676
8174	1202	5357	3072	6738
8715	1761	5936	3132	6800
9257	2322	6515	3192	6862
9800	2882	7095	3252	6925
930342	3443	7676	3312	6987
0885	4004	8256	3372	7049
1428	4565	8837	3433	7112
1971	5127	9418	3493	7174
47°	46°	45°	44°	43°

Natural Co-tangents.

Natural Sines.

53°	54°	55°	56°	57°	58°
798636	809017	819152	829038	838671	848048
8811	9188	9319	9200	8829	8202
8985	9359	9486	9363	8987	8356
9160	9530	9652	9525	9146	8510
9335	9700	9819	9688	9304	8664
9510	9871	9985	9850	9462	8818
9685	810042	820152	830012	9620	8972
9859	0212	0318	0174	9778	9125
800034	0383	0485	0337	9936	9279
0208	0553	0651	0499	840094	9433
800383	810723	820817	830661	840251	849586
0557	0894	0983	0823	0409	9739
0731	1064	1149	0984	0567	9893
0906	1234	1315	1146	0724	850046
1080	1404	1481	1308	0882	0199
1254	1574	1647	1470	1039	0352
1428	1744	1813	1631	1196	0505
1602	1914	1978	1793	1354	0658
1776	2084	2144	1954	1511	0811
1949	2253	2310	2115	1668	0964
802123	812423	822475	832277	841825	851117
2297	2592	2641	2438	1982	1269
2470	2762	2806	2599	2139	1422
2644	2931	2971	2760	2296	1575
2817	3101	3136	2921	2452	1727
2991	3270	3302	3082	2609	1879
3164	3439	3467	3243	2766	2032
3337	3608	3632	3404	2922	2184
3511	3778	3797	3565	3079	2336
3684	3947	3961	3725	3235	2488
803857	814116	824126	833886	843391	852640
4030	4284	4291	4046	3548	2792
4203	4453	4456	4207	3704	2944
4376	4622	4620	4367	3860	3096
4548	4791	4785	4527	4016	3248
4721	4959	4949	4688	4172	3399
4894	5128	5113	4848	4328	3551
5066	5296	5278	5008	4484	3702
5239	5465	5442	5168	4640	3854
5411	5633	5606	5328	4795	4005
805584	815801	825770	835488	844951	854156
5756	5969	5934	5648	5106	4308
5928	6138	6098	5807	5262	4459
6100	6306	6262	5967	5417	4610
6273	6474	6426	6127	5573	4761
6445	6642	6590	6286	5728	4912
6617	6809	6753	6446	5883	5063
6788	6977	6917	6605	6038	5214
6960	7145	7081	6764	6193	5364
7132	7313	7244	6924	6348	5515
807304	817480	827407	837083	846503	855665
7475	7648	7571	7242	6658	5816
7647	7815	7734	7401	6813	5966
7818	7982	7897	7560	6967	6117
7990	8150	8060	7719	7122	6267
8161	8317	8223	7878	7277	6417
8333	8484	8386	8036	7431	6567
8504	8651	8549	8195	7585	6718
8675	8818	8712	8354	7740	6868
8846	8985	8875	8512	7894	7017
36°	35°	34°	33°	32°	31°

Natural Co-sines.

Natural Tangents.

53°	54°	55°	56°	57°
1.32704	1.37638	1.42815	1.48256	1.53986
2785	7722	2903	8349	4085
2865	7807	2992	8442	4183
2946	7891	3080	8536	4281
3026	7976	3169	8629	4379
3107	8060	3258	8722	4478
3187	8145	3347	8816	4576
3268	8229	3436	8909	4675
3349	8314	3525	9003	4774
3430	8399	3614	9097	4873
1.33511	1.38484	1.43703	1.49190	1.54972
3592	8568	3792	9284	5071
3673	8653	3881	9378	5170
3754	8738	3970	9472	5269
3835	8824	4060	9566	5368
3916	8909	4149	9661	5467
3998	8994	4239	9755	5567
4079	9079	4329	9849	5666
4160	9165	4418	9944	5766
4242	9250	4508	1.50038	5866
1.34323	1.39336	1.44598	1.50133	1.55966
4405	9421	4688	0228	6065
4487	9507	4778	0322	6165
4568	9593	4868	0417	6265
4650	9679	4958	0512	6366
4732	9764	5049	0607	6466
4814	9850	5139	0702	6566
4896	9936	5229	0797	6667
4978	1.40022	5320	0893	6767
5060	0109	5410	0988	6868
1.35142	1.40195	1.45501	1.51084	1.56969
5224	0281	5592	1179	7069
5307	0367	5682	1275	7170
5389	0454	5773	1370	7271
5472	0540	5864	1466	7372
5554	0627	5955	1562	7474
5637	0714	6046	1658	7575
5719	0800	6137	1754	7676
5802	0887	6229	1850	7778
5885	0974	6320	1946	7879
1.35968	1.41061	1.46411	1.52043	1.57981
6051	1148	6503	2139	8083
6134	1235	6595	2235	8184
6217	1322	6686	2332	8286
6300	1409	6778	2429	8388
6383	1497	6870	2525	8490
6466	1584	6962	2622	8593
6549	1672	7053	2719	8695
6633	1759	7146	2816	8797
6716	1847	7238	2913	8900
1.36800	1.41934	1.47330	1.53010	1.59002
6883	2022	7422	3107	9105
6967	2110	7514	3205	9208
7050	2198	7607	3302	9311
7134	2286	7699	3400	9414
7218	2374	7792	3497	9517
7302	2462	7885	3595	9620
7386	2550	7977	3693	9723
7470	2638	8070	3791	9826
7554	2726	8163	3888	9930
36°	35°	34°	33°	32°

Natural Co-tangents.

M.	60°	61°	62°	63°	64°	65°	66°
0	866025	874620	882948	891007	898794	906308	913545
1	6171	4761	3084	1139	8922	6431	3664
2	6316	4902	3221	1270	9049	6554	3782
3	6461	5042	3357	1402	9176	6676	3900
4	6607	5183	3493	1534	9304	6799	4018
5	6752	5324	3629	1666	9431	6922	4136
6	6897	5465	3766	1798	9558	7044	4254
7	7042	5605	3902	1929	9685	7166	4372
8	7187	5746	4038	2061	9812	7289	4490
9	7331	5886	4174	2192	9939	7411	4607
10	867476	876026	884309	892323	900065	907533	914725
11	7621	6167	4445	2455	0192	7655	4842
12	7765	6307	4581	2586	0319	7777	4960
13	7910	6447	4717	2717	0445	7899	5077
14	8054	6587	4852	2848	0572	8021	5194
15	8199	6727	4988	2979	0698	8143	5311
16	8343	6867	5123	3110	0825	8265	5429
17	8487	7006	5258	3241	0951	8387	5546
18	8632	7146	5394	3371	1077	8508	5663
19	8776	7286	5529	3502	1203	8630	5779
20	868920	877425	885664	893633	901329	908751	915896
21	9064	7565	5799	3763	1455	8872	6013
22	9207	7704	5934	3894	1581	8994	6130
23	9351	7844	6069	4024	1707	9115	6246
24	9495	7983	6204	4154	1833	9236	6363
25	9639	8122	6338	4284	1958	9357	6479
26	9782	8261	6473	4415	2084	9478	6595
27	9926	8400	6608	4545	2209	9599	6712
28	870069	8539	6742	4675	2335	9720	6828
29	0212	8678	6876	4805	2460	9841	6944
30	870356	878817	887011	894934	902585	909961	917060
31	0499	8956	7145	5064	2710	910082	7176
32	0642	9095	7279	5194	2836	0202	7292
33	0785	9233	7413	5323	2961	0323	7408
34	0928	9372	7548	5453	3086	0443	7523
35	1071	9510	7681	5582	3210	0563	7639
36	1214	9649	7815	5712	3335	0684	7755
37	1357	9787	7949	5841	3460	0804	7870
38	1499	9925	8083	5970	3585	0924	7986
39	1642	880063	8217	6099	3709	1044	8101
40	871784	880201	888350	896229	903834	911164	918216
41	1927	0339	8484	6358	3958	1284	8331
42	2069	0477	8617	6486	4083	1403	8446
43	2212	0615	8751	6615	4207	1523	8561
44	2354	0753	8884	6744	4331	1643	8676
45	2496	0891	9017	6873	4455	1762	8791
46	2638	1028	9150	7001	4579	1881	8906
47	2780	1166	9283	7130	4703	2001	9021
48	2922	1303	9416	7258	4827	2120	9135
49	3064	1441	9549	7387	4951	2239	9250
50	873206	881578	889682	897515	905075	912358	919364
51	3347	1716	9815	7643	5198	2477	9479
52	3489	1853	9948	7771	5322	2596	9593
53	3631	1990	890080	7900	5445	2715	9707
54	3772	2127	0213	8028	5569	2834	9821
55	3914	2264	0345	8156	5692	2953	9936
56	4055	2401	0478	8283	5815	3072	920050
57	4196	2538	0610	8411	5939	3190	0164
58	4338	2674	0742	8539	6062	3309	0277
59	4479	2811	0874	8666	6185	3427	0391
	29°	28°	27°	26°	25°	24°	23°

Natural Co-sines.

63°	64°	65°	66°	67°	68°	69°	
1.96261	2.05030	2.14451	2.24604	2.35585	2.47509	2.60509	60
6402	5182	4614	4780	5776	7716	0736	59
6544	5333	4777	4956	5967	7924	0963	58
6685	5485	4940	5132	6158	8132	1190	57
6827	5637	5104	5309	6349	8340	1418	56
6969	5790	5268	5486	6541	8549	1646	55
7111	5942	5432	5663	6733	8758	1874	54
7253	6094	5596	5840	6925	8967	2103	53
7395	6247	5760	6018	7118	9177	2332	52
7538	6400	5925	6196	7311	9386	2561	51
1.97681	2.06553	2.16090	2.26374	2.37504	2.49597	2.62791	50
7823	6706	6255	6552	7697	9807	3021	49
7966	6860	6420	6730	7891	2.50018	3252	48
8110	7014	6585	6909	8084	0229	3483	47
8253	7167	6751	7088	8279	0440	3714	46
8396	7321	6917	7267	8473	0652	3945	45
8540	7476	7083	7447	8668	0864	4177	44
8684	7630	7249	7626	8863	1076	4410	43
8828	7785	7416	7806	9058	1289	4642	42
8972	7939	7582	7987	9253	1502	4875	41
1.99116	2.08094	2.17749	2.28167	2.39449	2.51715	2.65109	40
9261	8250	7916	8348	9645	1929	5342	39
9406	8405	8084	8528	9841	2142	5576	38
9550	8560	8251	8710	2.40038	2357	5811	37
9695	8716	8419	8891	0235	2571	6046	36
9841	8872	8587	9073	0432	2786	6281	35
9986	9028	8755	9254	0629	3001	6516	34
2.00131	9184	8923	9437	0827	3217	6752	33
0277	9341	9092	9619	1025	3432	6989	32
0423	9498	9261	9801	1223	3648	7225	31
2.00569	2.09654	2.19430	2.29984	2.41421	2.53865	2.67462	30
0715	9811	9599	2.30167	1620	4082	7700	29
0862	9969	9769	0351	1819	4299	7937	28
1008	2.10126	9938	0534	2019	4516	8175	27
1155	0284	2.20108	0718	2218	4734	8414	26
1302	0442	0278	0902	2418	4952	8653	25
1449	0600	0449	1086	2618	5170	8892	24
1596	0758	0619	1271	2819	5389	9131	23
1743	0916	0790	1456	3019	5608	9371	22
1891	1075	0961	1641	3220	5827	9612	21
2.02039	2.11233	2.21132	2.31826	2.43422	2.56046	2.69853	20
2187	1392	1304	2012	3623	6266	2.70094	19
2335	1552	1475	2197	3825	6487	0335	18
2483	1711	1647	2383	4027	6707	0577	17
2631	1871	1819	2570	4230	6928	0819	16
2780	2030	1992	2756	4433	7150	1062	15
2929	2190	2164	2943	4636	7371	1305	14
3078	2350	2337	3130	4839	7593	1548	13
3227	2511	2510	3317	5043	7815	1792	12
3376	2671	2683	3505	5246	8038	2036	11
2.03526	2.12832	2.22857	2.33693	2.45451	2.58261	2.72281	10
3675	2993	3030	3881	5655	8484	2526	9
3825	3154	3204	4069	5860	8708	2771	8
3975	3316	3378	4258	6065	8932	3017	7
4125	3477	3553	4447	6270	9156	3263	6
4276	3639	3727	4636	6476	9381	3509	5
4426	3801	3902	4825	6682	9606	3756	4
4577	3963	4077	5015	6888	9831	4004	3
4728	4125	4252	5205	7095	2.60057	4251	2
4879	4288	4428	5395	7302	0283	4499	1
26°	25°	24°	23°	22°	21°	20°	

Natural Co-tangents.

Natural Sines.

73°	74°	75°	76°
956305	961262	965926	970296
6390	1342	6001	0366
6475	1422	6076	0436
6560	1502	6151	0506
6644	1582	6226	0577
6729	1662	6301	0647
6814	1741	6376	0716
6898	1821	6451	0786
6983	1901	6526	0856
7067	1980	6600	0926
957151	962059	966675	970995
7235	2139	6749	1065
7319	2218	6823	1134
7404	2297	6898	1204
7487	2376	6972	1273
7571	2455	7046	1342
7655	2534	7120	1411
7739	2613	7194	1480
7822	2692	7268	1549
7906	2770	7342	1618
957990	962849	967415	971687
8073	2928	7489	1755
8156	3006	7562	1824
8239	3084	7636	1893
8323	3163	7709	1961
8406	3241	7782	2029
8489	3319	7856	2098
8572	3397	7929	2166
8654	3475	8002	2234
8737	3553	8075	2302
958820	963630	968148	972370
8902	3708	8220	2438
8985	3786	8293	2506
9067	3863	8366	2573
9150	3941	8438	2641
9232	4018	8511	2708
9314	4095	8583	2776
9396	4173	8656	2843
9478	4250	8728	2911
9560	4327	8800	2978
959642	964404	968872	973045
9724	4481	8944	3112
9805	4557	9016	3179
9887	4634	9088	3246
9968	4711	9159	3313
960050	4787	9231	3379
0131	4864	9302	3446
0212	4940	9374	3512
0294	5016	9445	3579
0375	5093	9517	3645
960456	965169	969588	973712
0537	5245	9659	3778
0618	5321	9730	3844
0698	5397	9801	3910
0779	5473	9872	3976
0860	5548	9943	4042
0940	5624	970014	4108
1021	5700	0084	4173
1101	5775	0155	4239
1181	5850	0225	4305

Natural Tangents.

73°	74°	75°	76°
3.27085	3.48741	3.73205	4.01078
7426	9125	3640	1576
7767	9509	4075	2074
8109	9894	4512	2574
8452	3.50279	4950	3076
8795	0666	5388	3578
9139	1053	5828	4081
9483	1441	6268	4586
9829	1829	6709	5092
3.30174	2219	7152	5599
3.30521	3 52609	3.77595	4.06107
0868	.3001	8040	6616
1216	3393	8485	7127
1565	3785	8931	7639
1914	4179	9378	8152
2264	4573	9827	8666
2614	4968	3.80276	9182
2965	5364	0726	9699
3317	5761	1177	4.10216
3670	6159	1630	0736
3.34023	3.56557	3.82083	4.11256
4377	6957	2537	1778
4732	7357	2992	2301
5087	7758	3449	2825
5443	8160	3906	3350
5800	8562	4364	3877
6158	8966	4824	4405
6516	9370	5284	4934
6875	9775	5745	5465
7234	3.60181	6208	5997
3.37594	3.60588	3.86671	4.16530
7955	0996	7136	7064
8317	1405	7601	7600
8679	1814	8068	8137
9042	2224	8536	8675
9406	2636	9004	9215
9771	3048	9474	9756
3.40136	3461	9945	4.20298
0502	3874	3.90417	0842
0869	4289	0890	1387
3.41236	3.64705	3.91364	4.21933
1604	5121	1839	2481
1973	5538	2316	3030
2343	5957	2793	3580
2713	6376	3271	4132
3084	6796	3751	4685
3456	7217	4232	5239
3829	7638	4713	5795
4202	8061	5196	6352
4576	8485	5680	6911
3.44951	3.68909	3.96165	4.27471
5327	9335	6651	8032
5703	9761	7139	8595
6080	3.70188	7627	9159
6458	0616	8117	9724
6837	1046	8607	4.30291
7216	1476	9099	0860
7596	1907	9592	1430
7977	2338	4.00086	2001
8359	2771	0582	2573
16°	15°	14°	13°

Natural Co-tangents.

NATURAL SINES.

83°	84°	85°	86°	87°	88°	89°	
992546	994522	996195	997564	998630	999391	999848	60
2582	4552	6220	7584	8645	9401	9853	59
2617	4583	6245	7604	8660	9411	9858	58
2652	4613	6270	7625	8675	9421	9863	57
2687	4643	6295	7645	8690	9431	9867	56
2722	4673	6320	7664	8705	9441	9872	55
2757	4703	6345	7684	8719	9450	9877	54
2792	4733	6370	7704	8734	9460	9881	53
2827	4762	6395	7724	8749	9469	9886	52
2862	4792	6419	7743	8763	9479	9890	51
992896	994822	996444	997763	998778	999488	999894	50
2931	4851	6468	7782	8792	9497	9898	49
2966	4881	6493	7801	8806	9507	9903	48
3000	4910	6517	7821	8820	9516	9907	47
3034	4939	6541	7840	8834	9525	9910	46
3068	4969	6566	7859	8848	9534	9914	45
3103	4998	6590	7878	8862	9542	9918	44
3137	5027	6614	7897	8876	9551	9922	43
3171	5056	6637	7916	8890	9560	9925	42
3205	5084	6661	7934	8904	9568	9929	41
993238	995113	996685	997953	998917	999577	999932	40
3272	5142	6709	7972	8931	9585	9936	39
3306	5170	6732	7990	8944	9594	9939	38
3339	5199	6756	8008	8957	9602	9942	37
3373	5227	6779	8027	8971	9610	9945	36
3406	5256	6802	8045	8984	9618	9948	35
3439	5284	6825	8063	8997	9626	9951	34
3473	5312	6848	8081	9010	9634	9954	33
3506	5340	6872	8099	9023	9642	9957	32
3539	5368	6894	8117	9035	9650	9959	31
993572	995396	996917	998135	999048	999657	999962	30
3605	5424	6940	8153	9061	9665	9964	29
3638	5452	6963	8170	9073	9672	9967	28
3670	5479	6985	8188	9086	9680	9969	27
3703	5507	7008	8205	9098	9687	9971	26
3735	5535	7030	8223	9111	9694	9974	25
3768	5562	7053	8240	9123	9701	9976	24
3800	5589	7075	8257	9135	9709	9978	23
3833	5617	7097	8274	9147	9716	9980	22
3865	5644	7119	8291	9159	9722	9981	21
993897	995671	997141	998308	999171	999729	999983	20
3929	5698	7163	8325	9183	9736	9985	19
3961	5725	7185	8342	9194	9743	9986	18
3993	5752	7207	8359	9206	9749	9988	17
4025	5778	7229	8375	9218	9756	9989	16
4056	5805	7250	8392	9229	9762	9990	15
4088	5832	7272	8408	9240	9768	9992	14
4120	5858	7293	8425	9252	9775	9993	13
4151	5884	7314	8441	9263	9781	9994	12
4182	5911	7336	8457	9274	9787	9995	11
994214	995937	997357	998473	999285	999793	999996	10
4245	5963	7378	8489	9296	9799	9997	9
4276	5989	7399	8505	9307	9804	9997	8
4307	6015	7420	8521	9318	9810	9998	7
4338	6041	7441	8537	9328	9816	9998	6
4369	6067	7462	8552	9339	9821	9999	5
4400	6093	7482	8568	9350	9827	9999	4
4430	6118	7503	8583	9360	9832	1.00000	3
4461	6144	7523	8599	9370	9837	0000	2
4491	6169	7544	8614	9381	9843	0000	1

NATURAL TANGENTS.

83°	84°	85°	86°
8.14435	9.51436	11.4301	14.3007
6398	4106	4685	3607
8370	6791	5072	4212
8.20352	9490	5461	4823
2344	9.62205	5853	5438
4345	4935	6248	6059
6355	7680	6645	6685
8376	9.70441	7045	7317
8.30406	3217	7448	7954
2446	6009	7853	8596
8.34496	9.78817	11.8262	14.9244
6555	9.81641	8673	9898
8625	4482	9087	15.0557
8.40705	7338	9504	1222
2795	9.90211	9923	1893
4896	3101	12.0346	2571
7007	6007	0772	3254
9128	8931	1201	3943
8.51259	10.0187	1632	4638
3402	0483	2067	5340
8.55555	10.0780	12.2505	15.6048
7718	1080	2946	6762
9893	1381	3390	7483
8.62078	1683	3838	8211
4275	1988	4288	8945
6482	2294	4742	9687
8701	2602	5199	16.0435
8.70931	2913	5660	1190
3172	3224	6124	1952
5425	3538	6591	2722
8.77689	10.3854	12.7062	16.3499
9964	4172	7536	4283
8.82252	4491	8014	5075
4551	4813	8496	5874
6862	5136	8981	6681
9185	5462	9469	7496
8.91520	5789	9962	8319
3867	6118	13.0458	9150
6227	6450	0958	9990
8598	6783	1461	17.0837
9.00983	10.7119	13.1969	17.1693
3379	7457	2480	2558
5789	7797	2996	3432
8211	8139	3515	4314
9.10646	8483	4039	5205
3093	8829	4566	6106
5554	9178	5098	7015
8028	9529	5634	7934
9.20516	9882	6174	8863
3016	11.0237	6719	9802
9.25530	11.0594	13.7267	18.0750
8058	0954	7821	1708
9.30599	1316	8378	2677
3155	1681	8940	3655
5724	2048	9507	4645
8307	2417	14.0079	5645
9.40904	2789	0655	6656
3515	3163	1235	7678
6141	3540	1821	8711
8781	3919	2411	9755
6°	5°	4°	3°

Natural Co-tangents.

Deg.	0′	10′	20′	30′	40′	50′		P. Part to 1′.
0	1.000000	1.000004	1.000017	1.000038	1.000068	1.000106	89	2.5
1	000152	000207	000271	000343	000423	000512	88	7.6
2	000609	000715	000830	000953	001084	001224	87	12.7
3	001372	001529	001695	001869	002051	002242	86	17.8
4	002442	002650	002867	003092	003326	003569	85	22.9
5	003820	004080	004348	004625	004911	005205	84	28.1
6	005508	005820	006141	006470	006808	007154	83	33.3
7	007510	007874	008247	008629	009020	009419	82	38.6
8	009828	010245	010671	011106	011550	012003	81	43.9
9	012465	012936	013416	013905	014403	014910	80	49.3
10	1.015427	1.015952	1.016487	1.017030	1.017583	1.018145	79	54.8
11	018717	019297	019887	020487	021095	021713	78	60.4
12	022341	022977	023624	024280	024945	025620	77	66.0
13	026304	026998	027702	028415	029138	029871	76	71.8
14	030614	031366	032128	032900	033682	034474	75	77.7
15	035276	036088	036910	037742	038584	039437	74	83.7
16	040299	041172	042055	042949	043853	044767	73	89.8
17	045692	046627	047573	048529	049496	050474	72	96.1
18	051462	052461	053471	054492	055524	056567	71	102.6
19	057621	058686	059762	060849	061947	063057	70	109.2
20	1.064178	1.065310	1.066454	1.067609	1.068776	1.069955	69	116.1
21	071145	072347	073561	074786	076024	077273	68	123.2
22	078535	079808	081094	082392	083703	085025	67	130.4
23	086360	087708	089068	090441	091827	093225	66	137.9
24	094636	096060	097498	098948	100411	101888	65	145.6
25	103378	104881	106398	107929	109473	111030	64	153.7
26	112602	114187	115787	117400	119028	120670	63	162.0
27	122326	123997	125682	127382	129096	130826	62	170.7
28	132570	134329	136104	137893	139698	141518	61	179.7
29	143354	145205	147073	148956	150854	152769	60	189.1
30	1.154701	1.156648	1.158612	1.160592	1.162589	1.164603	59	198.9
31	166633	168681	170746	172828	174927	177044	58	209.1
32	179178	181331	183501	185689	187895	190120	57	219.7
33	192363	194625	196906	199205	201523	203861	56	230.9
34	206218	208594	210991	213406	215842	218298	55	242.6
35	220775	223271	225789	228327	230886	233466	54	254.8
36	236068	238691	241336	244003	246691	249402	53	267.7
37	252136	254892	257671	260472	263298	266146	52	281.3
38	269018	271914	274834	277779	280748	283741	51	295.6
39	286760	289803	292872	295967	299088	302234	50	310.7
40	1.305407	1.308607	1.311833	1.315087	1.318368	1.321677	49	326.7
41	325013	328378	331771	335192	338643	342123	48	343.6
42	345633	349172	352742	356342	359972	363634	47	361.5
43	367327	371052	374809	378598	382420	386275	46	380.5
44	390164	394086	398042	402032	406057	410118	45	400.7
45	414214	418345	422513	426718	430960	435239	44	422.3
46	439557	443912	448306	452740	457213	461726	43	445.3
47	466279	470874	475509	480187	484907	489670	42	469.8
48	494477	499327	504221	509160	514145	519176	41	496.2
49	524253	529377	534549	539769	545038	550356	40	524.4
50	1 555724	1.561142	1.566612	1.572134	1.577708	1.583335	39	554.7
51	589016	594751	600542	606388	612291	618251	38	587.4
52	624269	630346	636483	642680	648938	655258	37	622.7
53	661640	668086	674597	681173	687815	694524	36	660.9
54	701302	708148	715064	722051	729110	736241	35	702.2
55	743447	750727	758084	765517	773029	780620	34	747.2
56	788292	796045	803881	811801	819806	827899	33	796.2
57	836078	844348	852707	861159	869704	878344	32	849.8
58	887080	895914	904847	913881	923017	932258	31	908.5
59	941604	951058	960621	970294	980081	989982	30	973.0
	60′	50′	40′	30′	20′	10′	Deg.	

Deg.	0′	10′	20′	30′	40′	50′		P. Part to 1′.
60	2.000000	2.010136	2.020393	2.030772	2.041276	2.051906	29	1044
61	062665	073556	084579	095739	107036	118474	28	1123
62	130054	141781	153655	165681	177859	190195	27	1210
63	202689	215346	228168	241158	254320	267657	26	1308
64	281172	294869	308750	322820	337083	351542	25	1417
65	366202	381065	396137	411421	426922	442645	24	1539
66	458593	474773	491187	507843	524744	541896	23	1678
67	559305	576975	594914	613126	531618	650396	22	1835
68	669467	688387	708514	728504	748814	769453	21	2015
69	790428	811747	833419	855451	877853	900635	20	2222
70	2.923804	2.947372	2.971349	2.995744	3.020569	3.045835	19	2461
71	3.071553	3.097736	3.124396	3.151545	179198	207367	18	2740
72	236068	265315	295123	325510	356490	388082	17	3068
73	420304	453173	486711	520937	555871	591536	16	3458
74	627955	665152	703151	741978	781660	822225	15	3925
75	863703	906125	949522	993929	4.039380	4.085913	14	4492
76	4.133565	4.182378	4.232394	4.283658	336215	390116	13	5190
77	445411	502157	560408	620226	681675	744821	12	6062
78	809734	876491	945169	5.015852	5.088628	5.163592	11	7171
79	5.240843	5.320486	5.402633	487404	574926	665333	10	8612
80	5.758770	5.855392	5.955362	6.058858	6.166067	6.277193	9	
81	6.392453	6.512081	6.636329	6.765469	6.899794	7.039622	8	
82	7.185297	7.337191	7.495711	7.661298	7.834433	8.015645	7	
83	8.205509	8.404659	8.613790	8.833671	9.065151	9.309170	6	
84	9.566772	9.839123	10.12752	10.43343	10.75849	11.10455	5	
85	11.47371	11.86837	12.29125	12.74549	13.23472	13.76311	4	
86	14.33559	14.95788	15.63679	16.38041	17.19843	18.10262	3	
87	19.10732	20.23028	21.49368	22.92559	24.56212	26.45051	2	
88	28.65371	31.25758	34.38232	38.20155	42.97571	49.11406	1	
89	57.29869	68.75736	85.94561	114.5930	171.8883	343.7752	0	
	60′	50′	40′	30′	20′	10′	Deg.	

Natural Co-secants.

LENGTHS OF CIRCULAR ARCS.

Degrees.						Minutes.		Seconds.	
°		°		°		′		′	
1	.0174533	26	.4537856	51	.8901179	1	.0002909	1	.0000048
2	.0349066	27	.4712389	52	.9075712	2	.0005818	2	.0000097
3	.0523599	28	.4886922	53	.9250245	3	.0008727	3	.0000145
4	.0698132	29	.5061455	54	.9424778	4	.0011636	4	.0000194
5	.0872665	30	.5235988	55	.9599311	5	.0014544	5	.0000242
6	.1047198	31	.5410521	56	.9773844	6	.0017453	6	.0000291
7	.1221730	32	.5585054	57	.9948377	7	.0020362	7	.0000339
8	.1396263	33	.5759587	58	1.0122910	8	.0023271	8	.0000388
9	.1570796	34	.5934119	59	1.0297443	9	.0026180	9	.0000436
10	.1745329	35	.6108652	60	1.0471976	10	.0029089	10	.0000485
11	.1919862	36	.6283185	65	1.1344640	11	.0031998	11	.0000533
12	.2094395	37	.6457718	70	1.2217305	12	.0034907	12	.0000582
13	.2268928	38	.6632251	75	1.3089969	13	.0037815	13	.0000630
14	.2443461	39	.6806784	80	1.3962634	14	.0040724	14	.0000679
15	.2617994	40	.6981317	85	1.4835299	15	.0043633	15	.0000727
16	.2792527	41	.7155850	90	1.5707963	16	.0046542	16	.0000776
17	.2967060	42	.7330383	100	1.7453293	17	.0049451	17	.0000824
18	.3141593	43	.7504916	110	1.9198622	18	.0052360	18	.0000873
19	.3316126	44	.7679449	120	2.0943951	19	.0055269	19	.0000921
20	.3490659	45	.7853982	130	2.2689280	20	.0058178	20	.0000970
21	.3665191	46	.8028515	140	2.4434610	25	.0072722	25	.0001212
22	.3839724	47	.8203047	150	2.6179939	30	.0087266	30	.0001454
23	.4014257	48	.8377580	160	2.7925268	40	.0116355	40	.0001939
24	.4188790	49	.8552113	170	2.9670597	50	.0145444	50	0002424

Dist. 3.		Dist. 4.		Dist. 5.			
Lat.	Dep.	Lat.	Dep.	Lat.	Dep.	°	′
3.0000	0.0131	4.0000	0.0175	5.0000	0.0218	89	45
2.9999	0262	3.9998	0349	4.9998	0436		30
9997	0393	9997	0524	9996	0654		15
9995	0524	9994	0698	9992	0873	89	0
9993	0654	9990	0873	9988	1091		45
9990	0785	9986	1047	9983	1309		30
9986	0916	9981	1222	9977	1527		15
9982	1047	9976	1396	9970	1745	88	0
9977	1178	9969	1570	9961	1963		45
9971	1309	9962	1745	9952	2181		30
2.9965	0.1439	3.9954	0.1919	4.9942	0.2399		15
9959	1570	9945	2093	9931	2617	87	0
9952	1701	9936	2268	9920	2835		45
9944	1831	9925	2442	9907	3052		30
9936	1962	9914	2616	9893	3270		15
9927	2093	9903	2790	9878	3488	86	0
9918	2223	9890	2964	9863	3705		45
9908	2354	9877	3138	9846	3923		30
9897	2484	9863	3312	9828	4140		15
9886	2615	9848	3486	9810	4358	85	0
2.9874	0.2745	3.9832	0.3660	4.9790	0.4575		45
9862	2875	9816	3834	9770	4792		30
9849	3006	9799	4008	9748	5009		15
9836	3136	9781	4181	9726	5226	84	0
9822	3266	9762	4355	9703	5443		45
9807	3396	9743	4528	9679	5660		30
9792	3526	9723	4701	9653	5877		15
9776	3656	9702	4875	9627	6093	83	0
9760	3786	9680	5048	9600	6310		45
9743	3916	9658	5221	9572	6526		30
2.9726	0.4046	3.9635	0.5394	4.9543	0.6743		15
9708	4175	9611	5567	9513	6959	82	0
9690	4305	9586	5740	9483	7175		45
9670	4434	9561	5912	9451	7390		30
9651	4564	9534	6085	9418	7606		15
9631	4693	9508	6257	9384	7822	81	0
9610	4822	9480	6430	9350	8037		45
9589	4951	9451	6602	9314	8252		30
9567	5080	9422	6774	9278	8467		15
9544	5209	9392	6946	9240	8682	80	0
2.9521	0.5338	3.9362	0.7118	4.9202	0.8897		45
9498	5467	9330	7289	9163	9112		30
9474	5596	9298	7461	9123	9326		15
9449	5724	9265	7632	9081	9540	79	0
9424	5853	9231	7804	9039	9755		45
9398	5981	9197	7975	8996	9968		30
9371	6109	9162	8146	8952	1.0182		15
9344	6237	9126	8316	8907	0396	78	0
9317	6365	9089	8487	8862	0609		45
9289	6493	9052	8658	8815	0822		30
2.9260	0.6621	3.9014	0.8828	4.8767	1.1035		15
9231	6749	8975	8998	8719	1248	77	0
9201	6876	8935	9168	8669	1460		45
9171	7003	8895	9338	8618	1672		30
9140	7131	8854	9507	8567	1884		15
9109	7258	8812	9677	8515	2096	76	0
9077	7385	8769	9846	8462	2308		45
9044	7511	8726	1.0015	8407	2519		30
9011	7638	8682	0184	8352	2730		15
8978	7765	8637	0353	8296	2941	75	0
Dep.	Lat.	Dep.	Lat.	Dep.	Lat.	Course.	
Dist. 3.		Dist. 4.		Dist. 5.			

Traverse Table.

Dist. 7.		Dist. 8.		Dist. 9.		Dist. 10.	
Lat.	Dep.	Lat.	Dep.	Lat.	Dep.	Lat.	Dep.
.9999	0.0305	7.9999	0.0349	8.9999	0.0393	9.9999	0.0436
9997	0611	9997	0698	9997	0785	9996	0873
9994	0916	9993	1047	9992	1178	9991	1309
9989	1222	9988	1396	9986	1571	9985	1745
9983	1527	9981	1745	9979	1963	9976	2181
9976	1832	9973	2094	9969	2356	9966	2618
9967	2138	9963	2443	9958	2748	9953	3054
9957	2443	9951	2792	9945	3141	9939	3490
9946	2748	9938	3141	9931	3533	9923	3926
9933	3053	9924	3490	9914	3926	9905	4362
.9919	0.3358	7.9908	0.3838	8.9896	0.4318	9.9885	0.4798
9904	3664	9890	4187	9877	4710	9863	5234
9887	3968	9871	4535	9855	5102	9839	5669
9869	4273	9851	4884	9832	5494	9813	6105
9850	4578	9829	5232	9807	5886	9786	6540
9829	4883	9805	5581	9781	6278	9756	6976
9808	5188	9780	5929	9753	6670	9725	7411
9784	5492	9753	6277	9723	7061	9692	7846
9760	5797	9725	6625	9691	7453	9657	8281
9734	6101	9696	6972	9658	7844	9619	8716
.9706	0.6405	7.9664	0.7320	8.9622	0.8235	9.9580	0.9150
9678	6709	9632	7668	9586	8626	9540	9585
9648	7013	9597	8015	9547	9017	9497	1.0019
9617	7317	9562	8362	9507	9408	9452	0453
9584	7621	9525	8709	9465	9798	9406	0887
9550	7924	9486	9056	9421	1.0188	9357	1320
9515	8228	9445	9403	9376	0578	9307	1754
9478	8531	9404	9750	9329	0968	9255	2187
9440	8834	9360	1.0096	9280	1358	9200	2620
9401	9137	9316	0442	9230	1747	9144	3053
.9361	0.9440	7.9269	1.0788	8.9178	1.2137	9.9087	1 3485
9319	9742	9221	1134	9124	2526	9027	3917
9276	1.0044	9172	1479	9069	2914	8965	4349
9231	0347	9121	1825	9011	3303	8902	4781
9185	0649	9069	2170	8953	3691	8836	5212
9138	0950	9015	2515	8892	4079	8769	5643
9090	1252	8960	2859	8830	4467	8700	6074
9040	1553	8903	3204	8766	4854	8629	6505
8989	1854	8844	3548	8700	5241	8556	6935
8937	2155	8785	3892	8633	5628	8481	7365
.8883	1.2456	7.8723	1.4235	8.8564	1.6015	9.8404	1.7794
8828	2756	8660	4579	8493	6401	8325	8224
8772	3057	8596	4922	8421	6787	8245	8652
.8714	3357	8530	5265	8346	7173	8163	9081
8655	3656	8463	5607	8271	7558	8079	9509
8595	3956	8394	5949	8193	7943	7992	9937
8533	4255	8324	6291	8114	8328	7905	2.0364
8470	4554	8252	6633	8033	8712	7815	0791
8406	4852	8178	6974	7951	9096	7723	1218
8341	5151	8104	7315	7867	9480	7630	1644
.8274	1.5449	7.8027	1.7656	8.7781	1.9863	9.7534	2.2070
8206	5747	7950	7996	7693	2.0246	7437	2495
8137	6044	7870	8336	7604	0628	7338	2920
8066	6341	7790	8676	7513	1010	7237	3345
7994	6638	7707	9015	7421	1392	7134	3769
7921	6935	7624	9354	7327	1773	7030	4192
7846	7231	7538	9692	7231	2154	6923	4615

Course.		Dist. 1.		Dist. 2.		Dist. 3.		Dist. 4.		Dist. 5.			
		Lat.	Dep.	Lat.	Dep.	Lat.	Dep.	Lat.	Dep.	Lat.	Dep.		
°	′											°	′
15	15	0.9648	0.2630	1.9296	0.5261	2.8944	0.7891	3.8591	1.0521	4.8239	1.3152	74	45
	30	9636	2672	9273	5345	8909	8017	8545	0690	8182	3362		30
	45	9625	2714	9249	5429	8874	8143	8498	0858	8123	3572		15
16	0	9613	2756	9225	5513	8838	8269	8450	1025	8063	3782	74	0
	15	9600	2798	9201	5597	8801	8395	8402	1193	8002	3991		45
	30	9588	2840	9176	5680	8765	8520	8353	1361	7941	4201		30
	45	9576	2882	9151	5764	8727	8646	8303	1528	7879	4410		15
17	0	9563	2924	9126	5847	8689	8771	8252	1695	7815	4619	73	0
	15	9550	2965	9100	5931	8651	8896	8201	1862	7751	4827		45
	30	9537	3007	9074	6014	8612	9021	8149	2028	7686	5035		30
	45	0.9524	0.3049	1.9048	0.6097	2.8572	0.9146	3.8096	1.2195	4.7620	1.5243		15
18	0	9511	3090	9021	6180	8532	9271	8042	2361	7553	5451	72	0
	15	9497	3132	8994	6263	8491	9395	7988	2527	7485	5658		45
	30	9483	3173	8966	6346	8450	9519	7933	2692	7416	5865		30
	45	9469	3214	8939	6429	8408	9643	7877	2858	7347	6072		15
19	0	9455	3256	8910	6511	8366	9767	7821	3023	7276	6278	71	0
	15	9441	3297	8882	6594	8323	9891	7764	3188	7204	6485		45
	30	9426	3338	8853	6676	8279	1.0014	7706	3352	7132	6690		30
	45	9412	3379	8824	6758	8235	0138	7647	3517	7059	6896		15
20	0	9397	3420	8794	6840	8191	0261	7588	3681	6985	7101	70	0
	15	0.9382	0.3461	1.8764	0.6922	2.8146	1.0384	3.7528	1.3845	4.6910	1.7306		45
	30	9367	3502	8733	7004	8100	0506	7467	4008	6834	7510		30
	45	9351	3543	8703	7086	8054	0629	7405	4172	6757	7715		15
21	0	9336	3584	8672	7167	8007	0751	7343	4335	6679	7918	69	0
	15	9320	3624	8640	7249	7960	0873	7280	4498	6600	8122		45
	30	9304	3665	8608	7330	7913	0995	7217	4660	6521	8325		30
	45	9288	3706	8576	7411	7864	1117	7152	4822	6440	8528		15
22	0	9272	3746	8544	7492	7816	1238	7087	4984	6359	8730	68	0
	15	9255	3786	8511	7573	7766	1359	7022	5146	6277	8932		45
	30	9239	3827	8478	7654	7716	1481	6955	5307	6194	9134		30
	45	0.9222	0.3867	1.8444	0.7734	2.7666	1.1601	3.6888	1.5468	4.6110	1.9336		15
23	0	9205	3907	8410	7815	7615	1722	6820	5629	6025	9537	67	0
	15	9188	3947	8376	7895	7564	1842	6752	5790	5940	9737		45
	30	9171	3987	8341	7975	7512	1962	6682	5950	5853	9937		30
	45	9153	4027	8306	8055	7459	2082	6612	6110	5766	2.0137		15
24	0	9135	4067	8271	8135	7406	2202	6542	6269	5677	0337	66	0
	15	9118	4107	8235	8214	7353	2322	6470	6429	5588	0536		45
	30	9100	4147	8199	8294	7299	2441	6398	6588	5498	0735		30
	45	9081	4187	8163	8373	7244	2560	6326	6746	5407	0933		15
25	0	9063	4226	8126	8452	7189	2679	6252	6905	5315	1131	65	0
	15	0.9045	0.4266	1.8089	0.8531	2.7134	1.2797	3.6178	1.7063	4.5223	2.1328		45
	30	9026	4305	8052	8610	7078	2915	6103	7220	5129	1526		30
	45	9007	4344	8014	8689	7021	3033	6028	7378	5035	1722		15
26	0	8988	4384	7976	8767	6964	3151	5952	7535	4940	1919	64	0
	15	8969	4423	7937	8846	6906	3269	5875	7692	4844	2114		45
	30	8949	4462	7899	8924	6848	3386	5797	7848	4747	2310		30
	45	8930	4501	7860	9002	6789	3503	5719	8004	4649	2505		15
27	0	8910	4540	7820	9080	6730	3620	5640	8160	4550	2700	63	0
	15	8890	4579	7780	9157	6671	3736	5561	8315	4451	2894		45
	30	8870	4617	7740	9235	6610	3852	5480	8470	4351	3087		30
	45	0.8850	0.4656	1.7700	0.9312	2.6550	1.3968	3.5400	1.8625	4.4249	2.3281		15
28	0	8829	4695	7659	9389	6488	4084	5318	8779	4147	3474	62	0
	15	8809	4733	7618	9466	6427	4200	5236	8933	4045	3666		45
	30	8788	4772	7576	9543	6365	4315	5153	9086	3941	3858		30
	45	8767	4810	7535	9620	6302	4430	5069	9240	3836	4049		15
29	0	8746	4848	7492	9696	6239	4544	4985	9392	3731	4240	61	0
	15	8725	4886	7450	9772	6175	4659	4900	9545	3625	4431		45
	30	8704	4924	7407	9848	6111	4773	4814	9697	3518	4621		30
	45	8682	4962	7364	9924	6046	4886	4728	9849	3410	4811		15
30	0	8660	5000	7321	1.0000	5981	5000	4641	2.0000	3301	5000	60	0

Course.		Lat.	Dep.		
° 15	15	5.7[illegible]	1.0[illegible]	[illegible]	[illegible]
	30	7815	[illegible]	[illegible]	[illegible]
	45	77[illegible]	[illegible]	[illegible]	[illegible]
16	0	76[illegible]	[illegible]	[illegible]	[illegible]
	15	760[illegible]	[illegible]	[illegible]	[illegible]
	30	752[illegible]	[illegible]	[illegible]	[illegible]
	45	7[illegible]	[illegible]	[illegible]	[illegible]
17	0	[illegible]	[illegible]	[illegible]	[illegible]
	15	75[illegible]	[illegible]	[illegible]	[illegible]
	30	7223	[illegible]	[illegible]	[illegible]
	45	5.71[illegible]	1.[illegible]	5.[illegible]	[illegible]
18	0	706[illegible]	[illegible]	[illegible]	[illegible]
	15	6[illegible]	[illegible]	[illegible]	[illegible]
	30	6[illegible]	[illegible]	[illegible]	[illegible]
	45	6[illegible]	[illegible]	[illegible]	[illegible]
19	0	6[illegible]	[illegible]	6[illegible]	2701
	15	66[illegible]	[illegible]	60[illegible]	[illegible]
	30	655[illegible]	2.[illegible]	5055	[illegible]
	45	6[illegible]	[illegible]	5[illegible]	[illegible]
20	0	6[illegible]	[illegible]	5[illegible]	[illegible]
	15	5.6[illegible]	2.076[illegible]	6.56[illegible]	[illegible]
	30	6200	1012	55[illegible]	[illegible]
	45	6105	1257	5[illegible]	[illegible]
21	0	6015	1502	5[illegible]	[illegible]
	15	5920	1746	[illegible]	[illegible]
	30	5825	1990	[illegible]	[illegible]
	45	5729	2233	[illegible]	[illegible]
22	0	5631	2476	[illegible]	[illegible]
	15	5532	[illegible]	[illegible]	[illegible]
	30	5433	29[illegible]	[illegible]	[illegible]
	45	5.5332	2.[illegible]3	[illegible]	[illegible]
23	0	5230	[illegible]	[illegible]	[illegible]
	15	51[illegible]	3[illegible]	[illegible]	[illegible]
	30	502[illegible]	3[illegible]	[illegible]	[illegible]
	45	49[illegible]	[illegible]	[illegible]	[illegible]
24	0	4813	[illegible]	[illegible]	[illegible]
	15	4706	[illegible]	[illegible]	[illegible]
	30	4[illegible]	[illegible]	[illegible]	[illegible]
	45	4[illegible]	[illegible]	[illegible]	[illegible]
25	0	4[illegible]	[illegible]	[illegible]	[illegible]
	15	5.4[illegible]	[illegible]	[illegible]	[illegible]
	30	4[illegible]	[illegible]	[illegible]	[illegible]
	45	4[illegible]	[illegible]	[illegible]	[illegible]
26	0	39[illegible]	[illegible]	[illegible]	[illegible]
	15	3812	[illegible]	[illegible]	[illegible]
	30	3696	[illegible]	[illegible]	[illegible]
	45	3579	[illegible]	[illegible]	[illegible]
27	0	3460	[illegible]	[illegible]	[illegible]
	15	3341	[illegible]	[illegible]	[illegible]
	30	3221	[illegible]	[illegible]	[illegible]
	45	[illegible]	[illegible]	[illegible]	[illegible]
28	0	[illegible]	[illegible]	[illegible]	[illegible]
	15	[illegible]	[illegible]	[illegible]	[illegible]
	30	[illegible]	[illegible]	[illegible]	[illegible]
	45	2004	[illegible]	[illegible]	[illegible]
29	0	247[illegible]	[illegible]	[illegible]	[illegible]
	15	2350	[illegible]	[illegible]	[illegible]
	30	2221	[illegible]	[illegible]	[illegible]
	45	2092	[illegible]	[illegible]	[illegible]
30	0	1962	[illegible]	[illegible]	[illegible]
		Dep.	Lat.		

Dist. 3.		Dist. 4.		Dist. 5.			
Lat.	Dep.	Lat.	Dep.	Lat.	Dep.	°	′
2.5915	1.5113	3.4553	2.0151	4.3192	2.5189	59	45
5849	5226	4465	0302	3081	5377		30
5782	5339	4376	0452	2970	5565		15
5715	5451	4287	0602	2858	5752	59	0
5647	5563	4196	0751	2746	5939		45
5579	5675	4106	0900	2632	6125		30
5511	5786	4014	1049	2518	6311		15
5441	5898	3922	1197	2402	6496	58	0
5372	6008	3829	1345	2286	6681		45
5302	6119	3736	1492	2170	6865		30
2.5231	1.6229	3.3642	2.1639	4.2052	2.7049		15
5160	6339	3547	1786	1934	7232	57	0
5089	6449	3451	1932	1814	7415		45
5017	6558	3355	2077	1694	7597		30
4944	6667	3259	2223	1573	7779		15
4871	6776	3162	2368	1452	7960	56	0
4798	6884	3064	2512	1329	8140		45
4724	6992	2965	2656	1206	8320		30
4649	7100	2866	2800	1082	8500		15
4575	7207	2766	2943	0958	8679	55	0
2.4499	1.7314	3.2666	2.3086	4.0832	2.8857		45
4423	7421	2565	3228	0706	9035		30
4347	7527	2463	3370	0579	9212		15
4271	7634	2361	3511	0451	9389	54	0
4193	7739	2258	3652	0322	9565		45
4116	7845	2154	3793	0193	9741		30
4038	7950	2050	3933	0063	9916		15
3959	8054	1945	4073	3.9932	3.0091	53	0
3880	8159	1840	4212	9800	0265		45
3801	8263	1734	4350	9668	0438		30
2.3721	1.8367	3.1628	2.4489	3.9534	3.0611		15
3640	8470	1520	4626	9401	0783	52	0
3560	8573	1413	4764	9266	0955		45
3478	8675	1304	4901	9130	1126		30
3397	8778	1195	5037	8994	1296		15
3314	8880	1086	5173	8857	1466	51	0
3232	8981	0976	5308	8720	1635		45
3149	9082	0865	5443	8581	1804		30
3065	9183	0754	5578	8442	1972		15
2981	9284	0642	5712	8302	2139	50	0
2.2897	1.9384	3.0529	2 5845	3.8162	3.2306		45
2812	9483	0416	5978	8020	2472		30
2727	9583	0303	6110	7878	2638		15
2641	9682	0188	6242	7735	2803	49	0
2555	9780	0074	6374	7592	2967		45
2469	9879	2.9958	6505	7448	3131		30
2382	9976	9842	6635	7303	3294		15
2294	2.0074	9726	6765	7157	3457	48	0
2207	0171	9609	6895	7011	3618		45
2118	0268	9491	7024	6864	3780		30

Dist. 8.		Dist. 9.		Dist. 10.			
Lat.	Dep.	Lat.	Dep.	Lat.	Dep.	°	′
6.9107	4.0302	7.7745	4.5340	8.6384	5.0377	59	45
8930	0603	7547	5678	6163	0754		30
8753	0903	7347	6016	5941	1129		15
8573	1203	7145	6353	5717	1504	59	0
8393	1502	6942	6690	5491	1877		45
8211	1800	6738	7025	5264	2250		30
8028	2097	6532	7359	5035	2621		15
7844	2394	6324	7693	4805	2992	58	0
7658	2689	6116	8025	4573	3361		45
7471	2984	5905	8357	4339	3730		30
6.7283	4.3278	7.5694	4.8688	8.4104	5.4097		15
7094	3571	5480	9018	3867	4464	57	0
6903	3863	5266	9346	3629	4829		45
6711	4155	5050	9674	3389	5194		30
6518	4446	4832	5.0001	3147	5557		15
6323	4735	4613	0327	2904	5919	56	0
6127	5024	4393	0652	2659	6280		45
5930	5312	4171	0977	2413	6641		30
5732	5600	3948	1300	2165	7000		15
5532	5886	3724	1622	1915	7358	55	0
6.5331	4.6172	7.3498	5.1943	8.1664	5.7715		45
5129	6456	3270	2263	1412	8070		30
4926	6740	3042	2582	1157	8425		15
4721	7023	2812	2901	0902	8779	54	0
4516	7305	2580	3218	0644	9131		45
4309	7586	2347	3534	0386	9482		30
4100	7866	2113	3849	0125	9832		15
3891	8145	1877	4163	7.9864	6.0182	53	0
3680	8424	1640	4476	9600	0529		45
3468	8701	1402	4789	9335	0876		30
6.3255	4.8977	7.1162	5.5100	7.9069	6.1222		15
3041	9253	0921	5410	8801	1566	52	0
2825	9528	0679	5718	8532	1909		45
2609	9801	0435	6026	8261	2251		30
2391	5.0074	0190	6333	7988	2592		15
2172	0346	6.9943	6639	7715	2932	51	0
1951	0616	9695	6943	7439	3271		45
1730	0886	9446	7247	7162	3608		30
1507	1155	9196	7550	6884	3944		15
1284	1423	8944	7851	6604	4279	50	0
6.1059	5.1690	6.8691	5.8151	7.6323	6.4612		45
0832	1956	8437	8450	6041	4945		30
0605	2221	8181	8748	5756	5276		15
0377	2485	7924	9045	5471	5606	49	0
0147	2748	7666	9341	5184	5935		45
5.9916	3010	7406	9636	4896	6262		30
9685	3271	7145	9929	4606	6588		15
9452	3530	6883	6.0222	4314	6913	48	0
9217	3789	6620	0513	4022	7237		45
8982	4047	6355	0803	3728	7559		30
5.8746	5.4304	6.6089	6.1092	7.3432	6.7880		15
8508	4560	5822	1380	3135	8200	47	0
8270	4815	5553	1666	2837	8518		45
8030	5068	5284	1952	2537	8835		30
7789	5321	5013	2236	2236	9151		15
7547	5573	4741	2519	1934	9466	46	0
7304	5823	4467	2801	1630	9779		45
7060	6073	4193	3082	1325	7.0091		30
6815	6321	3917	3361	1019	0401		15
6569	6569	3640	3640	0711	0711	45	0
Dep.	Lat.	Dep.	Lat.	Dep.	Lat.		

LATITUDE.														
Min.	0°	1°	2°	3°	4°	5°	6°	7°	8°	9°	10°	11°	12°	Min.
0	0.0	60.0	120.0	180.1	240.2	300.4	360.7	421.1	481.6	542.2	603.1	664.1	725.3	0
1	1.0	61.0	21.0	81.1	41.2	01.4	61.7	22.1	82.6	43.3	04.1	65.1	26.3	1
2	2.0	62.0	22.0	82.1	42.2	02.4	62.7	23.1	83.6	44.3	05.1	66.1	27.4	2
3	3.0	63.0	23.0	83.1	43.2	03.4	63.7	24.1	84.6	45.3	06.1	67.1	28.4	3
4	4.0	64.0	24.0	84.1	44.2	04.4	64.7	25.1	85.6	46.3	07.1	68.2	29.4	4
5	5.0	65.0	25.0	85.1	45.2	05.4	65.7	26.1	86.6	47.3	08.2	69.2	30.5	5
6	6.0	66.0	26.0	86.1	46.2	06.4	66.7	27.1	87.6	48.3	09.2	70.2	31.5	6
7	7.0	67.0	27.0	87.1	47.2	07.4	67.7	28.1	88.6	49.3	10.2	71.2	32.5	7
8	8.0	68.0	28.0	88.1	48.2	08.4	68.7	29.1	89.6	50.3	11.2	72.2	33.5	8
9	9.0	69.0	29.0	89.1	49.2	09.4	69.7	30.1	90.7	51.4	12.2	73.3	34.5	9
10	10.0	70.0	130.0	190.1	250.2	310.4	370.7	431.1	491.7	552.4	613.2	674.3	735.6	10
11	11.0	71.0	31.0	91.1	51.2	11.4	71.7	32.1	92.7	53.4	14.2	75.3	36.6	11
12	12.0	72.0	32.0	92.1	52.2	12.4	72.7	33.1	93.7	54.4	15.3	76.3	37.6	12
13	13.0	73.0	33.0	93.1	53.2	13.4	73.7	34.2	94.7	55.4	16.3	77.3	38.6	13
14	14.0	74.0	34.0	94.1	54.2	14.4	74.7	35.2	95.7	56.4	17.3	78.4	39.6	14
15	15.0	75.0	35.0	95.1	55.2	15.4	75.7	36.2	96.7	57.4	18.3	79.4	40.7	15
16	16.0	76.0	36.0	96.1	56.2	16.4	76.8	37.2	97.7	58.4	19.3	80.4	41.7	16
17	17.0	77.0	37.0	97.1	57.2	17.5	77.8	38.2	98.7	59.4	20.3	81.4	42.7	17
18	18.0	78.0	38.0	98.1	58.2	18.5	78.8	39.2	99.8	60.5	21.3	82.4	43.7	18
19	19.0	79.0	39.0	99.1	59.2	19.5	79.8	40.2	500.8	61.5	22.4	83.5	44.8	19
20	20.0	80.0	140.0	200.1	260.2	320.5	380.8	441.2	501.8	562.5	623.4	684.5	745.8	20
21	21.0	81.0	41.0	01.1	61.3	21.5	81.8	42.2	02.8	63.5	24.4	85.5	46.8	21
22	22.0	82.0	42.0	02.1	62.3	22.5	82.8	43.2	03.8	64.5	25.4	86.5	47.8	22
23	23.0	83.0	43.0	03.1	63.3	23.5	83.8	44.2	04.8	65.5	26.4	87.5	48.9	23
24	24.0	84.0	44.0	04.1	64.3	24.5	84.8	45.2	05.8	66.6	27.4	88.6	49.9	24
25	25.0	85.0	45.0	05.1	65.3	25.5	85.8	46.3	06.8	67.6	28.5	89.6	50.9	25
26	26.0	86.0	46.0	06.1	66.3	26.5	86.8	47.3	07.8	68.6	29.5	90.6	51.9	26
27	27.0	87.0	47.0	07.1	67.3	27.5	87.8	48.3	08.9	69.6	30.5	91.6	53.0	27
28	28.0	88.0	48.0	08.1	68.3	28.5	88.8	49.3	09.9	70.6	31.5	92.6	54.0	28
29	29.0	89.0	49.0	09.1	69.3	29.5	89.8	50.3	10.9	71.6	32.5	93.6	55.0	29
30	30.0	90.0	150.0	210.1	270.3	330.5	390.8	451.3	511.9	572.6	633.5	694.7	756.0	30
31	31.0	91.0	51.0	11.1	71.3	31.5	91.8	52.3	12.9	73.7	34.6	95.7	57.1	31
32	32.0	92.0	52.0	12.1	72.3	32.5	92.9	53.3	13.9	74.7	35.6	96.7	58.1	32
33	33.0	93.0	53.1	13.1	73.3	33.5	93.9	54.3	14.9	75.7	36.6	97.7	59.1	33
34	34.0	94.0	54.1	14.1	74.3	34.5	94.9	55.3	15.9	76.7	37.6	98.7	60.1	34
35	35.0	95.0	55.1	15.1	75.3	35.5	95.9	56.3	16.9	77.7	38.6	99.8	61.1	35
36	36.0	96.0	56.1	16.1	76.3	36.5	96.9	57.3	18.0	78.7	39.6	700.8	62.2	36
37	37.0	97.0	57.1	17.1	77.3	37.5	97.9	58.4	19.0	79.7	40.7	01.8	63.2	37
38	38.0	98.0	58.1	18.1	78.3	38.5	98.9	59.4	20.0	80.8	41.7	02.8	64.2	38
39	39.0	99.0	59.1	19.1	79.3	39.6	99.9	60.4	21.0	81.8	42.7	03.8	65.2	39
40	40.0	100.0	160.1	220.2	280.3	340.6	400.9	461.4	522.0	582.8	643.7	704.9	766.3	40
41	41.0	01.0	61.1	21.2	81.3	41.6	01.9	62.4	23.0	83.8	44.7	05.9	67.3	41
42	42.0	02.0	62.1	22.2	82.3	42.6	02.9	63.4	24.0	84.8	45.8	06.9	68.3	42
43	43.0	03.0	63.1	23.2	83.3	43.6	03.9	64.4	25.0	85.8	46.8	07.9	69.3	43
44	44.0	04.0	64.1	24.2	84.3	44.6	04.9	65.4	26.0	86.8	47.8	09.0	70.4	44
45	45.0	05.0	65.1	25.2	85.3	45.6	05.9	66.4	27.1	87.9	48.8	10.0	71.4	45
46	46.0	06.0	66.1	26.2	86.3	46.6	07.0	67.4	28.1	88.9	49.8	11.0	72.4	46
47	47.0	07.0	67.1	27.2	87.3	47.6	08.0	68.4	29.1	89.9	50.8	12.0	73.4	47

LATITUDE.

13°	14°	15°	16°	17°	18°	19°	20°	21°
86.8	848.5	910.5	972.7	1035.3	1098.2	1161.5	1225.1	1289.
87.8	49.5	11.5	73.8	36.3	99.3	62.5	26.2	90.
88.8	50.5	12.6	74.8	37.4	1100.3	63.6	27.3	91.
89.9	51.6	13.6	75.9	38.4	01.4	64.7	28.3	92.
90.9	52.6	14.6	76.9	39.5	02.4	65.7	29.4	93.
91.9	53.6	15.7	78.0	40.5	03.5	66.8	30.4	94.
92.9	54.7	16.7	79.0	41.6	04.5	67.8	31.5	95.
94.0	55.7	17.7	80.0	42.6	05.6	68.9	32.6	96.
95.0	56.7	18.8	81.1	43.7	06.6	70.0	33.6	97.
96.0	57.8	19.8	82.1	44.7	07.7	71.0	34.7	98.
97.0	858.8	920.8	983.2	1045.8	1108.7	1172.1	1235.8	1299.
98.1	59.8	21.9	84.2	46.8	09.8	73.1	36.8	1301.
99.1	60.9	22.9	85.2	47.9	10.8	74.2	37.9	02.
00.1	61.9	23.9	86.3	48.9	11.9	75.2	39.0	03.
01.2	62.9	25.0	87.3	49.9	12.9	76.3	40.0	04.
02.2	64.0	26.0	88.4	51.0	14.0	77.4	41.1	05.
03.2	65.0	27.0	89.4	52.0	15.0	78.4	42.2	06.
04.2	66.0	28.1	90.4	53.1	16.1	79.5	43.2	07.
05.3	67.1	29.1	91.5	54.1	17.1	80.5	44.3	08.
06.3	68.1	30.1	92.5	55.2	18.2	81.6	45.4	09.
07.3	869.1	931.2	993.6	1056.2	1119.2	1182.7	1246.4	1310.
08.4	70.1	32.2	94.6	57.3	20.3	83.7	47.5	11.
09.4	71.2	33.3	95.6	58.3	21.3	84.8	48.6	12.
10.4	72.2	34.3	96.7	59.4	22.4	85.8	49.6	13.
11.4	73.2	35.3	97.7	60.4	23.4	86.9	50.7	14.
12.5	74.3	36.3	98.8	61.4	24.5	88.0	51.8	16.
13.5	75.3	37.4	99.8	62.5	25.5	89.0	52.8	17.
14.5	76.3	38.4	1000.8	63.5	26.6	90.1	53.9	18.
15.5	77.4	39.5	01.9	64.6	27.6	91.1	55.0	19.
16.6	78.4	40.5	02.9	65.6	28.7	92.2	56.0	20.
17.6	879.4	941.6	1004.0	1066.7	1129.7	1193.2	1257.1	1321.
18.6	80.5	42.6	05.0	67.7	30.8	94.3	58.2	22.
19.6	81.5	43.6	06.1	68.8	31.8	95.4	59.2	23.
20.7	82.5	44.7	07.1	69.8	32.9	96.4	60.3	24.
21.7	83.6	45.7	08.1	70.9	34.0	97.5	61.4	25.
22.7	84.6	46.7	09.2	72.0	35.1	98.5	62.4	26.
23.8	85.6	47.8	10.2	73.0	36.1	99.6	63.5	27.
24.8	86.7	48.8	11.3	74.1	37.2	1200.7	64.6	28.
25.8	87.7	49.9	12.3	75.1	38.2	01.7	65.6	30.
26.9	88.7	50.9	13.4	76.2	39.3	02.8	66.7	31.
27.9	889.8	951.9	1014.4	1077.2	1140.3	1203.9	1267.8	1332.
28.9	90.8	53.0	15.4	78.3	41.4	04.9	68.8	33.
29.9	91.8	54.0	16.5	79.3	42.4	06.0	69.9	34.
31.0	92.9	55.1	17.5	80.4	43.5	07.1	71.0	35.
32.0	93.9	56.1	18.6	81.4	44.6	08.1	72.1	36.
33.0	94.9	57.1	19.6	82.5	45.6	09.2	73.1	37.
34.1	96.0	58.2	20.6	83.5	46.7	10.2	74.2	38.
35.1	97.0	59.2	21.7	84.6	47.7	11.3	75.3	39.
36.1	98.0	60.2	22.7	85.6	48.8	12.4	76.3	40.
37.2	99.1	61.3	23.8	86.7	49.9	13.4	77.4	41.
38.2	900.1	962.3	1024.8	1087.7	1150.9	1214.5	1278.5	1342.
39.2	01.1	63.4	25.9	88.8	52.0	15.5	79.5	44.
40.2	02.2	64.4	26.9	89.8	53.0	16.6	80.6	45.
41.3	03.2	65.5	28.0	90.9	54.1	17.7	81.7	46.
42.3	04.3	66.5	29.0	91.9	55.1	18.7	82.8	47.
43.3	05.3	67.5	30.1	93.0	56.2	19.8	83.8	48.
44.4	06.3	68.6	31.1	94.0	57.2	20.9	84.9	49.
45.4	07.4	69.6	32.2	95.1	58.3	21.9	86.0	50.

LATITUDE.

28°	29°	30°	31°
1751.2	1819.4	1888.4	1958.0
52.3	20.6	89.5	59.2
53.4	21.7	90.7	60.4
54.6	22.9	91.9	61.6
55.7	24.0	93.0	62.7
56.8	25.2	94.1	63.9
58.0	26.3	95.3	65.0
59.1	27.5	96.5	66.2
60.2	28.6	97.6	67.4
61.4	29.7	98.8	68.5
1762.5	1830.9	1899.9	1969.7
63.6	32.0	1901.1	70.9
64.8	33.2	02.3	72.0
65.9	34.3	03.4	73.2
67.0	35.5	04.6	74.4
68.2	36.6	05.7	75.6
69.3	37.8	06.9	76.8
70.5	38.9	08.1	77.9
71.6	40.1	09.2	79.1
72.7	41.2	10.4	80.2
1773.9	1842.4	1911.5	1981.4
75.0	43.5	12.7	82.6
76.1	44.6	13.8	83.7
77.2	45.8	15.0	84.9
78.4	46.9	16.2	86.1
79.5	48.1	17.3	87.3
80.6	49.2	18.5	88.4
81.8	50.4	19.6	89.6
83.0	51.5	20.8	90.8
84.1	52.7	21.9	92.0

LATITUDE.

Min.	36°	37°	38°	39°	40°	41°	42°	43°	44°	45°	46°	Min.
0	2318.0	2392.6	2468.3	2545.0	2622.7	2701.6	2781.7	2863.1	2945.8	3030.0	3115.6	0
1	19.2	93.9	69.5	46.2	24.0	02.9	83.1	64.5	47.2	31.4	17.0	1
2	20.5	95.1	70.8	47.5	25.3	04.3	84.4	65.8	48.6	32.8	18.5	2
3	21.7	96.4	72.1	48.8	26.6	05.6	85.8	67.2	50.0	34.2	19.9	3
4	23.0	97.7	73.4	50.1	27.9	06.9	87.1	68.5	51.4	35.6	21.4	4
5	24.2	98.9	74.6	51.4	29.2	08.3	88.5	70.0	52.8	37.0	22.8	5
6	25.4	2400.2	75.9	52.7	30.5	09.6	89.8	71.3	54.2	38.4	24.2	6
7	26.7	01.4	77.1	54.0	31.9	10.9	91.2	72.7	55.6	39.8	25.7	7
8	27.9	02.7	78.5	55.3	33.2	12.2	92.5	74.1	57.0	41.3	27.1	8
9	29.1	03.9	79.7	56.6	34.5	13.5	93.8	75.4	58.3	42.7	28.5	9
10	2330.4	2405.2	2481.0	2557.8	2635.8	2714.9	2795.1	2876.8	2959.8	3044.1	3130.0	10
11	31.6	06.4	82.2	59.1	37.1	16.2	96.5	78.2	61.1	45.5	31.5	11
12	32.9	07.7	83.5	60.4	38.4	17.5	97.9	79.5	62.5	47.0	32.9	12
13	34.1	09.0	84.8	61.7	39.7	18.9	99.3	80.9	63.9	48.4	34.3	13
14	35.3	10.2	86.1	63.0	41.0	20.2	2800.6	82.3	65.3	49.8	35.8	14
15	36.6	11.5	87.4	64.3	42.3	21.5	02.0	83.7	66.7	51.2	37.2	15
16	37.8	12.7	88.6	65.6	43.6	22.9	03.3	85.0	68.1	52.6	38.7	16
17	39.0	14.0	89.9	66.9	44.9	24.2	04.7	86.4	69.5	54.1	40.1	17
18	40.3	15.2	91.2	68.2	46.3	25.5	06.0	87.8	70.9	55.5	41.6	18
19	41.5	16.5	92.4	69.5	47.6	26.8	07.3	89.1	72.3	56.9	43.0	19
20	2342.8	2417.8	2493.7	2570.7	2648.9	2728.2	2808.8	2890.5	2973.7	3058.3	3144.5	20
21	44.0	19.0	95.0	72.0	50.2	29.5	10.1	91.9	75.1	59.7	45.9	21
22	45.3	20.3	96.3	73.3	51.5	30.8	11.4	93.3	76.5	61.2	47.4	22
23	46.5	21.5	97.6	74.6	52.8	32.2	12.8	94.7	77.9	62.6	48.8	23
24	47.7	22.8	98.8	75.9	54.1	33.5	14.1	96.0	79.3	64.0	50.3	24
25	49.0	24.0	2500.1	77.2	55.5	34.8	15.5	97.4	80.7	65.4	51.7	25
26	50.2	25.3	01.4	78.5	56.8	36.2	16.8	98.8	82.1	66.9	53.2	26
27	51.5	26.5	02.7	79.8	58.1	37.5	18.2	2900.2	83.5	68.3	54.6	27
28	52.7	27.8	03.9	81.1	59.4	38.8	19.5	01.5	84.9	69.7	56.1	28
29	54.0	29.1	05.2	82.4	60.7	40.2	20.9	02.9	86.3	71.1	57.5	29
30	2355.2	2430.3	2506.5	2583.7	2662.0	2741.5	2822.3	2904.3	2987.7	3072.6	3159.0	30
31	56.5	31.6	07.8	85.0	63.3	42.9	23.6	05.7	89.1	74.0	60.4	31
32	57.7	32.9	09.0	86.3	64.6	44.2	25.0	07.1	90.5	75.4	61.9	32
33	58.9	34.1	10.3	87.6	66.0	45.5	26.3	08.4	91.9	76.9	63.3	33
34	60.2	35.4	11.6	88.9	67.3	46.9	27.7	09.7	93.3	78.3	64.8	34
35	61.4	36.7	12.9	90.2	68.6	48.2	29.0	11.2	94.7	79.7	66.2	35
36	62.7	37.9	14.2	91.5	69.9	49.5	30.4	12.6	96.1	81.1	67.7	36
37	63.9	39.2	15.4	92.8	71.2	50.9	31.7	14.0	97.5	82.6	69.1	37
38	65.2	40.4	16.7	94.1	72.5	52.2	33.1	15.3	98.9	84.0	70.6	38
39	66.4	41.7	18.0	95.4	73.9	53.5	34.5	16.7	3000.3	85.4	72.0	39
40	2367.6	2443.0	2519.3	2596.7	2675.2	2754.9	2835.8	2918.1	3001.8	3086.9	3173.5	40
41	68.9	44.2	20.5	98.0	76.5	56.2	37.2	19.5	03.2	88.3	75.0	41
42	70.2	45.5	21.8	99.3	77.8	57.6	38.6	20.9	04.6	89.7	76.4	42
43	71.4	46.8	23.1	2600.5	79.1	58.9	39.9	22.3	06.0	91.2	77.9	43
44	72.6	48.0	24.4	01.9	80.5	60.2	41.3	23.6	07.4	92.6	79.3	44
45	73.9	49.3	25.7	03.2	81.8	61.5	42.6	25.0	08.8	94.0	80.8	45
46	75.1	50.6	27.0	04.5	83.1	62.9	44.0	26.4	10.2	95.5	82.3	46
47	76.4	51.8	28.3	05.8	84.4	64.3	45.4	27.8	11.6	96.9	83.7	47
48	77.6	53.1	29.5	07.1	85.7	65.6	46.7	29.2	13.0	98.3	85.2	48
49	78.9	54.3	30.8	08.4	87.1	66.9	48.1	30.6	14.4	99.7	86.6	49
50	2380.1	2455.6	2532.1	2609.7	2688.4	2768.3	2849.5	2932.0	3015.8	3101.2	3188.1	50
51	81.4	56.9	33.4	11.0	89.7	69.6	50.8	33.3	17.2	02.6	89.6	51
52	82.6	58.1	34.7	12.3	91.0	71.0	52.2	34.7	18.7	04.1	91.0	52
53	83.9	59.4	36.0	13.6	92.3	72.3	53.5	36.1	20.1	05.6	92.5	53
54	85.1	60.7	37.2	14.9	93.7	73.7	54.9	37.5	21.5	07.0	94.0	54
55	86.4	61.9	38.5	16.2	95.0	75.0	56.3	38.9	22.9	08.4	95.4	55
56	87.6	63.2	39.8	17.5	96.3	76.3	57.7	40.3	24.3	09.8	96.9	56
57	88.9	64.5	41.1	18.8	97.6	77.7	59.0	41.7	25.7	11.2	98.4	57
58	90.2	65.8	42.4	20.1	99.0	79.0	60.5	43.1	27.1	12.7	99.8	58

LATITUDE.

Min.	47°	48°	49°	50°	51°	52°	53°	54°	55°	56°	57°	Min.
0	3202.7	3291.5	3382.1	3474.5	3568.8	3665.2	3763.8	3864.6	3968.0	4073.9	4182.6	0
1	04.2	93.0	83.6	76.0	70.4	66.8	65.4	66.3	69.7	75.7	84.5	1
2	05.7	94.5	85.1	77.6	72.0	68.4	67.1	68.0	71.5	77.5	86.3	2
3	07.1	96.0	86.7	79.1	73.6	70.1	68.8	69.7	73.2	79.3	88.1	3
4	08.6	97.5	88.2	80.7	75.2	71.7	70.4	71.5	75.0	81.1	90.0	4
5	10.1	99.0	89.7	82.3	76.8	73.3	72.1	73.2	76.7	82.9	91.8	5
6	11.5	3300.5	91.3	83.8	78.4	75.0	73.7	74.9	78.4	84.7	93.7	6
7	13.0	02.0	92.8	85.4	79.9	76.6	75.4	76.6	80.2	86.4	95.5	7
8	14.5	03.5	94.3	87.0	81.5	78.2	77.1	78.3	82.0	88.2	97.3	8
9	15.9	05.0	95.8	88.5	83.1	79.8	78.7	80.0	83.7	90.0	99.2	9
10	3217.4	3306.5	3397.4	3490.1	3584.7	3681.5	3780.4	3881.7	3985.4	4091.8	4201.0	10
11	18.9	08.0	98.9	91.6	86.3	83.1	82.1	83.4	87.2	93.6	02.9	11
12	20.3	09.5	3400.4	93.2	87.9	84.7	83.7	85.1	88.9	95.4	04.7	12
13	21.8	11.0	02.0	94.7	89.5	86.4	85.4	86.8	90.7	97.2	06.6	13
14	23.3	12.5	03.5	96.3	91.1	88.0	87.1	88.5	92.5	99.0	08.4	14
15	24.8	14.0	05.0	97.9	92.7	89.6	88.8	90.2	94.2	4100.8	10.3	15
16	26.2	15.5	06.6	99.4	94.3	91.3	90.4	92.0	96.0	02.6	12.1	16
17	27.7	17.0	08.1	3501.0	95.9	92.9	92.1	93.7	97.7	04.4	14.0	17
18	29.2	18.5	09.6	02.6	97.5	94.5	93.8	95.4	99.5	06.2	15.8	18
19	30.7	20.0	11.1	04.1	99.1	96.2	95.5	97.1	4001.2	08.0	17.7	19
20	3232.1	3321.5	3412.7	3505.7	3600.7	3697.8	3797.1	3898.8	4003.0	4109.8	4219.5	20
21	33.6	23.0	14.2	07.3	02.3	99.4	98.8	3900.5	04.7	11.6	21.4	21
22	35.1	24.5	15.7	08.8	03.9	3701.1	3800.5	02.2	06.5	13.4	23.2	22
23	36.6	26.0	17.3	10.4	05.5	02.7	02.2	04.0	08.3	15.2	25.1	23
24	38.0	27.5	18.8	12.0	07.1	04.4	03.8	05.7	10.0	17.1	27.0	24
25	39.5	29.0	20.4	13.5	08.7	06.0	05.5	07.4	11.8	18.9	28.8	25
26	41.0	30.6	21.9	15.1	10.3	07.6	07.2	09.1	13.5	20.7	30.6	26
27	42.5	32.1	23.5	16.7	11.9	09.3	08.9	10.8	15.3	22.5	32.5	27
28	44.0	33.6	25.0	18.3	13.6	10.9	10.5	12.5	17.1	24.3	34.4	28
29	45.4	35.1	26.5	19.8	15.1	12.6	12.2	14.3	18.8	26.1	36.2	29
30	3246.9	3336.6	3428.0	3521.4	3616.7	3714.2	3813.9	3916.0	4020.6	4127.9	4238.1	30
31	48.4	38.1	29.6	23.0	18.4	15.8	15.6	17.7	22.4	29.7	40.0	31
32	49.9	39.6	31.1	24.6	20.0	17.5	17.3	19.4	24.1	31.5	41.8	32
33	51.4	41.1	32.7	26.1	21.6	19.1	18.9	21.2	25.9	33.3	43.7	33
34	52.8	42.6	34.2	27.7	23.2	20.8	20.6	22.9	27.7	35.2	45.5	34
35	54.3	44.1	35.8	29.3	24.8	22.4	22.3	24.6	29.4	37.0	47.4	35
36	55.8	45.7	37.3	30.8	26.4	24.1	24.0	26.3	31.2	38.8	49.3	36
37	57.3	47.2	38.8	32.4	28.0	25.7	25.7	28.1	33.0	40.6	51.1	37
38	58.8	48.7	40.4	34.0	29.6	27.4	27.4	29.8	34.8	42.4	53.0	38
39	60.3	50.2	41.9	35.6	31.3	29.0	29.1	31.5	36.5	44.2	54.9	39
40	3261.7	3351.7	3443.5	3537.1	3632.8	3730.7	3830.8	3933.2	4038.3	4146.1	4256.7	40
41	63.2	53.2	45.0	38.7	34.4	32.3	32.4	35.0	40.1	47.9	58.6	41
42	64.7	54.7	46.6	40.3	36.1	34.0	34.1	36.7	41.8	49.7	60.5	42
43	66.2	56.2	48.1	41.9	37.7	35.6	35.8	38.4	43.6	51.5	62.3	43
44	67.7	57.8	49.7	43.5	39.3	37.3	37.5	40.2	45.4	53.4	64.2	44
45	69.2	59.3	51.2	45.0	40.9	38.9	39.2	41.9	47.2	55.2	66.1	45
46	70.7	60.8	52.8	46.6	42.5	40.6	40.9	43.6	49.0	57.0	68.0	46
47	72.1	62.3	54.3	48.2	44.1	42.2	42.6	45.4	50.7	58.8	69.8	47
48	73.6	63.8	55.8	49.8	45.8	43.9	44.3	47.1	52.5	60.7	71.7	48
49	75.1	65.4	57.4	51.4	47.4	45.5	46.0	48.8	54.3	62.5	73.6	49
50	3276.6	3366.9	3458.9	3553.0	3649.0	3747.2	3847.7	3950.6	4056.1	4164.3	4275.5	50
51	78.1	68.4	60.5	54.6	50.6	48.8	49.4	52.3	57.8	66.1	77.4	51
52	79.6	69.9	62.0	56.1	52.2	50.5	51.1	54.0	59.6	68.0	79.2	52
53	81.1	71.4	63.6	57.7	53.8	52.1	52.8	55.8	61.4	69.8	81.1	53
54	82.6	73.0	65.2	59.3	55.5	53.8	54.4	57.5	63.2	71.6	83.0	54
55	84.1	74.5	66.7	60.9	57.1	55.5	56.1	59.3	65.0	73.5	84.9	55
56	85.6	76.0	68.3	62.5	58.7	57.1	57.8	61.0	66.8	75.3	86.8	56
57	87.1	77.5	69.8	64.0	60.3	58.8	59.5	62.7	68.5	77.1	88.6	57
58	88.5	79.0	71.4	65.6	61.9	60.4	61.2	64.5	70.3	79.0	90.5	58

LATITUDE.

61°	62°	63°	64°	65°
649.2	4775.0	4904.9	5039.4	5178.8
51.3	77.1	07.1	41.7	81.2
53.4	79.3	09.4	44.0	83.5
55.4	81.4	11.6	46.3	85.9
57.5	83.5	13.8	48.6	88.3
59.6	85.6	16.0	50.9	90.7
61.6	87.8	18.2	53.1	93.1
63.7	89.9	20.4	55.4	95.4
65.8	92.1	22.6	57.7	97.8
67.8	94.2	24.8	60.0	5200.2
669.9	4796.3	4927.0	5062.3	5202.6
72.0	98.5	29.2	64.6	04.9
74.1	4800.6	31.5	66.9	07.3
76.1	02.8	33.7	69.2	09.7
78.2	04.9	35.9	71.5	12.1
80.3	07.1	38.1	73.8	14.5
82.4	09.2	40.3	76.1	16.9
84.5	11.4	42.6	78.4	19.3
86.5	13.5	44.8	80.7	21.6
88.6	15.7	47.0	83.0	24.0

LATITUDE.

Min.	69°	70°	71°	72°	73°	74°	75°	76°	77°	78°	79°	Min.
0	5794.6	5965.9	6145.7	6334.8	6534.4	6745.7	6970.3	7210.1	7467.2	7744.6	8045.7	0
1	97.4	68.8	48.8	38.1	37.8	49.4	74.2	14.2	71.7	49.4	51.0	1
2	5800.1	71.8	51.9	41.3	41.3	53.0	78.1	18.3	76.1	54.2	56.2	2
3	02.9	74.7	54.9	44.6	44.7	56.6	81.9	22.5	80.6	59.0	61.5	3
4	05.7	77.6	58.0	47.8	48.1	60.3	85.8	26.6	85.0	63.9	66.7	4
5	08.5	80.6	61.1	51.1	51.6	63.9	89.7	30.8	89.5	68.7	72.0	5
6	11.3	83.5	64.2	54.3	55.0	67.6	93.6	35.0	94.0	73.5	77.3	6
7	14.2	86.4	67.3	57.6	58.5	71.2	97.5	39.1	98.5	78.4	82.6	7
8	17.0	89.4	70.4	60.8	61.9	74.9	7001.4	43.3	7503.0	83.3	87.9	8
9	19.8	92.3	73.5	64.1	65.3	78.5	05.3	47.5	07.4	88.1	93.2	9
10	5822.6	5995.3	6176.6	6367.4	6568.8	6782.2	7009.2	7251.7	7511.9	7793.0	8098.5	10
11	25.4	98.2	79.7	70.6	72.3	85.9	13.1	55.8	16.5	97.9	8103.8	11
12	28.2	6001.2	82.8	73.9	75.7	89.6	17.0	60.0	21.0	7802.8	09.2	12
13	31.0	04.1	85.9	77.2	79.2	93.2	20.9	64.2	25.5	07.7	14.5	13
14	33.8	07.1	89.0	80.4	82.6	96.9	24.8	68.4	30.0	12.6	19.9	14
15	36.7	10.0	92.1	83.7	86.1	6800.6	28.8	72.6	34.5	17.5	25.2	15
16	39.5	13.0	95.2	87.0	89.6	04.3	32.7	76.8	39.1	22.4	30.6	16
17	42.3	16.0	98.3	90.3	93.0	08.0	36.6	81.1	43.6	27.3	36.0	17
18	45.1	18.9	6201.4	93.6	96.5	11.6	40.6	85.3	48.1	32.2	41.3	18
19	48.0	21.9	04.5	96.9	6600.0	15.4	44.5	89.5	52.7	37.2	46.7	19
20	5850.8	6024.9	6207.7	6400.2	6603.5	6819.1	7048.5	7293.7	7557.3	7842.1	8152.1	20
21	53.6	27.8	10.8	03.4	07.0	22.8	52.4	98.0	61.8	47.1	57.5	21
22	56.5	30.8	13.9	06.7	10.5	26.5	56.4	7302.2	66.4	52.0	62.9	22
23	59.3	33.8	17.0	10.1	14.0	30.2	60.3	06.4	71.0	57.0	68.4	23
24	62.2	36.8	20.2	13.4	17.5	33.9	64.3	10.7	75.5	61.9	73.8	24
25	65.0	39.8	23.3	16.7	21.0	37.6	68.3	15.0	80.1	66.9	79.2	25
26	67.8	42.7	26.5	20.0	24.5	41.3	72.2	19.2	84.7	71.9	84.7	26
27	70.7	45.7	29.6	23.3	28.0	45.1	76.2	23.5	89.3	76.9	90.1	27
28	73.5	48.7	32.7	26.6	31.5	48.8	80.2	27.7	93.9	81.9	95.6	28
29	76.4	51.7	35.9	29.9	35.0	52.5	84.2	32.0	98.5	86.9	8201.1	29

Mid. Lat.	DIFFERENCE OF LATITUDE.																				Mid. Lat.
	1°	2°	3°	4°	5°	6°	7°	8°	9°	10°	11°	12°	13°	14°	15°	16°	17°	18°	19°	20°	
°	′	′	′	′	′	′	′	′	′	′	′	′	′	′	′	′	′	′	′	′	°
15	0	1	2	3	5	7	9	12	15	18	22	26	31	36	41	47	52	59	65	72	15
16	0	1	2	3	4	6	9	11	14	18	21	25	30	34	39	44	50	56	62	69	16
17	0	1	2	3	4	6	8	11	14	17	20	24	28	33	38	43	48	54	60	66	17
18	0	1	1	3	4	6	8	10	13	16	20	23	27	32	36	41	46	52	58	64	18
19	0	1	1	3	4	6	8	10	13	16	19	22	26	30	35	40	45	50	56	61	19
20	0	1	1	2	4	5	7	10	12	15	18	22	25	29	34	38	43	48	54	60	20
21	0	1	1	2	4	5	7	9	12	15	18	21	25	29	33	37	42	47	52	58	21
22	0	1	1	2	4	5	7	9	12	14	17	21	24	28	32	36	41	46	51	56	22
23	0	1	1	2	3	5	7	9	11	14	17	20	23	27	31	35	40	45	50	55	23
24	0	1	1	2	3	5	7	9	11	14	16	20	23	27	31	35	39	44	49	54	24
25	0	1	1	2	3	5	7	9	11	13	16	19	23	26	30	34	39	43	48	53	25
26	0	1	1	2	3	5	6	8	11	13	16	19	22	26	30	34	38	42	47	52	26
27	0	1	1	2	3	5	6	8	11	13	16	19	22	25	29	33	37	42	47	52	27
28	0	1	1	2	3	5	6	8	10	13	16	18	22	25	29	33	37	41	46	51	28
29	0	1	1	2	3	5	6	8	10	13	15	18	21	25	28	32	37	41	46	51	29
30	0	1	1	2	3	5	6	8	10	13	15	18	21	25	28	32	36	41	45	50	30
31	0	1	1	2	3	5	6	8	10	12	15	18	21	24	28	32	36	40	45	50	31
32	0	0	1	2	3	4	6	8	10	12	15	18	21	24	28	32	36	40	45	50	32
33	0	0	1	2	3	4	6	8	10	12	15	18	21	24	28	32	36	40	45	49	33
34	0	0	1	2	3	4	6	8	10	12	15	18	21	24	28	32	36	40	45	49	34
35	0	0	1	2	3	4	6	8	10	12	15	18	21	24	28	32	36	40	45	49	35
36	0	1	1	2	3	4	6	8	10	12	15	18	21	24	28	32	36	40	45	49	36
37	0	1	1	2	3	4	6	8	10	12	15	18	21	24	28	32	36	40	45	49	37
38	0	1	1	2	3	4	6	8	10	12	15	18	21	24	28	32	36	40	45	50	38
39	0	1	1	2	3	4	6	8	10	12	15	18	21	24	28	32	36	40	45	50	39
40	0	1	1	2	3	5	6	8	10	13	15	18	21	25	28	32	36	41	45	50	40
41	0	1	1	2	3	5	6	8	10	13	15	18	21	25	28	32	37	41	46	51	41
42	0	1	1	2	3	5	6	8	10	13	15	18	22	25	29	33	37	41	46	51	42
43	0	1	1	2	3	5	6	8	10	13	16	18	22	25	29	33	37	42	46	52	43
44	0	1	1	2	3	5	6	8	10	13	16	19	22	25	29	33	38	42	47	52	44
45	0	1	1	2	3	5	6	8	11	13	16	19	22	26	30	34	38	43	48	53	45
46	0	1	1	2	3	5	6	8	11	13	16	19	22	26	30	34	38	43	48	53	46
47	0	1	1	2	3	5	7	9	11	13	16	19	23	26	30	35	39	44	49	54	47
48	0	1	1	2	3	5	7	9	11	14	17	20	23	27	31	35	40	44	50	55	48
49	0	1	1	2	3	5	7	9	11	14	17	20	23	27	31	36	40	45	50	56	49
50	0	1	1	2	4	5	7	9	11	14	17	20	24	28	32	36	41	46	51	57	50
51	0	1	1	2	4	5	7	9	12	14	17	21	24	28	32	37	42	47	52	58	51
52	0	1	1	2	4	5	7	9	12	15	18	21	25	29	33	38	43	48	53	59	52
53	0	1	1	2	4	5	7	10	12	15	18	21	25	29	34	38	43	49	54	[illegible]	53
54	0	1	1	2	4	5	7	10	12	15	18	22	26	30	34	39	44	50	56	[illegible]	54
55	0	1	1	2	4	6	8	10	13	16	19	22	26	31	35	40	45	51	57	63	55
56	0	1	1	3	4	6	8	10	13	16	19	23	27	31	36	41	46	52	58	65	56
57	0	1	1	3	4	6	8	10	13	16	20	24	28	32	37	42	48	54	60	66	57
58	0	1	2	3	4	6	8	11	14	17	20	24	28	33	38	43	49	55	61	68	58
59	0	1	2	3	4	6	8	11	14	17	21	25	29	34	39	45	50	57	63	70	59
60	0	1	2	3	4	6	9	11	14	18	22	26	30	35	40	46	52	58	65	72	60
61	0	1	2	3	5	7	9	12	15	18	22	26	31	36	42	47	53	60	67	75	61
62	0	1	2	3	5	7	9	12	15	19	23	27	32	37	43	49	55	62	70	77	62
63	0	1	2	3	5	7	10	12	16	20	24	28	33	39	44	51	57	64	72	80	63
64	0	1	2	3	5	7	10	13	16	20	24	29	34	40	46	52	59	67	75	83	64
65	0	1	2	3	5	7	10	13	17	21	25	30	36	41	48	54	62	69	78	86	65
66	0	1	2	3	5	8	11	14	18	22	26	32	37	43	50	57	64	72	81	90	66
67	0	1	2	4	6	8	11	14	18	23	28	33	39	45	52	59	67	76	85	94	67
68	0	1	2	4	6	8	12	15	19	24	29	34	40	47	54	62	70	79	89	99	68
69	0	1	2	4	6	9	12	16	20	25	30	36	42	49	57	65	74	83	93	104	69
70	0	1	2	4	6	9	13	16	21	26	32	38	44	52	60	68	78	88	98	110	70
71	0	1	2	4	7	10	13	17	22	27	33	40	47	55	63	72	82	93	104	116	71

In computing compound interest for long periods of time, it is necessary to have the following logarithms to more than six places.

Number.	Logarithm.	Number.	Logarithm.
1.0025	.00108 43813	1.0425	.01807 60636
1.0050	.00216 60618	1.0450	.01911 62904
1.0075	.00324 50548	1.0475	.02015 40316
1.0100	.00432 13738	1.0500	.02118 92991
1.0125	.00539 50319	1.0525	.02222 21045
1.0150	.00646 60422	1.0550	.02325 24596
1.0175	.00753 44179	1.0575	.02428 03760
1.0200	.00860 01718	1.0600	.02530 58653
1.0225	.00966 33167	1.0625	.02632 89387
1.0250	.01072 38654	1.0650	.02734 96078
1.0275	.01178 18305	1.0675	.02836 78837
1.0300	.01283 72247	1.0700	.02938 37777
1.0325	.01389 00603	1.0725	.03039 73009
1.0350	.01494 03498	1.0750	.03140 84643
1.0375	.01598 81054	1.0775	.03241 72788
1.0400	.01703 33393	1.0800	.03342 37555

NUMBERS OFTEN USED IN CALCULATIONS.

		Logarithms.
Circumference of a circle to diameter 1 Surface of a sphere to diameter 1 Area of a circle to radius 1	= 3.1415926	0.497150
Area of a circle to diameter 1	= .7853982	9.895090
Capacity of a sphere to diameter 1	= .5235988	9.718999
Capacity of a sphere to radius 1	= 4.1887902	0.622089
1 ÷ 3.1415926	= 0.3183099	9.502850
Arc equal to radius expressed in degrees	= 57°.2957795	1.758123
Arc equal to radius expressed in seconds	= 206264".8	5.314425
Length of 1 degree in parts of radius	= .0174533	8.241877
Length of 1 minute in parts of radius	= .0002909	6.463726
Sine of 1 second	= .00000485	4.685575
Sine of 2 seconds	= .00000970	4.986605
Sine of 3 seconds	= .00001454	5.162696
Sine of 4 seconds	= .00001939	5.287635
Sine of 5 seconds	= .00002424	5.384545
Sine of 6 seconds	= .00002909	5.463726
Sine of 7 seconds	= .00003394	5.530673
Sine of 8 seconds	= .00003879	5.588665
Sine of 9 seconds	= .00004363	5.639817
Base of Napier's system of logarithms	= 2.7182818	0.434294
Modulus of the common logarithms	= .4342945	9.637784
360 degrees expressed in seconds	= 1296000	6.112605
24 hours expressed in seconds	= 86400	4.936514
Number of feet in one mile	= 5280	3.722634

THE END.

www.ingramcontent.com/pod-product-compliance
Lightning Source LLC
Chambersburg PA
CBHW021503210326
41599CB00012B/1114
* 9 7 8 3 7 4 1 1 9 4 9 1 7 *